RAILWAY
ENGINEERING

SATISH CHANDRA
Professor, Department of Civil Engineering,
Indian Institute of Technology Roorkee

M.M. AGARWAL
Retired Chief Engineer,
Northern Railways

OXFORD
UNIVERSITY PRESS

Preface to the Second Edition

Railway engineering is a versatile subject that acquaints students systematically to the whole gamut of activities that take place in railways, starting from planning, design, and construction to operation and maintenance. Therefore, it is quite natural that the subject encompasses a wide range of engineering disciplines including civil, computer science, electrical, mechanical, and production engineering.

Indian Railways has long been regarded as instrumental in India's economic development. However, it is currently passing through a difficult phase where the challenges are many. In general, despite being the most energy-efficient as well as eco-friendly mode of transport, the share of the Railways in total traffic carried has gone down drastically. In addition, there is a serious resource crunch that is pulling Indian Railways down. To face this challenge head on, the Railways has plans to boost its traffic output by expanding the existing network. Further, modernizing the existing infrastructure, improving the passenger facilities, and most important of all, making rail travel safe and comfortable are seen as the long-term solutions to attract more passengers.

Indian Railways, over the past few years, has embraced many technological changes in order to meet the challenges of modern rail traffic. In fact, it has never shied away from modernization— always eager to evolve in new ways. The Indian Railways Vision 2020 acts as a blueprint for inclusive geographical and social development with environment sustainability. As part of Vision 2020, the Ministry of Railways seeks to raise railway revenues to a gross domestic product (GDP) ratio of 3 per cent, which has remained at around 1.3 per cent in the past decade. Such growth will naturally require the expansion and modernization of Indian Railways at a rapid pace.

Incorporating the latest developments that have taken place in Indian Railways during the past decade, this revised edition discusses the measures taken by the Railways, along with their technical details, to meet the challenges described.

NEW TO THE SECOND EDITION

The text has been extensively revised with suitable changes in most chapters. A balanced amalgamation of fundamental concepts and modern technological developments, with special focus on Indian Railways, the revised edition presents the following:

- **New chapter** on the Dedicated Freight Corridor project of Indian Railways
- **Updated statistical data** on Indian Railways and a detailed discussion on **Vision 2020**
- Detailed discussion on **modern methods of track maintenance**
- A complete procedure on **how to calculate stresses** in different components of a railway track

PEDAGOGY

The second edition comes with a rich pedagogy that includes the following:
- Colour photographs for better illustration of the latest machinery and modern methods of track maintenance
- Chapter-end exercises now include MCQs in addition to theoretical and numerical questions
- More than 250 illustrations and as many as 300 exercise questions for better comprehension and practice

EXTENDED CHAPTER MATERIAL

Chapter 1 is completely revised and includes the latest statistical data on various aspects of Indian Railways. The chapter gives details of the *Railway Vision 2020* which lays down the 'goals' of Indian Railways proposed to be achieved by 2020.

Chapter 5 has now additional material on stresses in a railway track. A solved example is included to help the students to calculate bending stress in the track.

Chapter 6 provides updated rail specifications as per the latest IRS code. High strength 110 UTS rail sections recently introduced on some important lines on Indian Railways are also explained.

Chapter 14 includes details of high-speed turnouts developed on Indian Railways for passenger yards. It also includes details of modern turnouts developed by DMRC.

Chapter 17 includes the procedure followed by Indian Railways for distressing of long-welded rails, with and without rail tensors.

Chapter 18 now covers the latest developments in track maintenance such as modification in the formula for calculating the gang strength. It also includes the work load of PWI and other field officers.

Chapter 20 provides the details of the latest track machines for mechanized track maintenance.

Chapter 22 presents the new classification of accidents. The remedial measures to be taken to avoid accidents are also highlighted.

Chapter 24 describes the new technology adopted by Indian Railways for design of rolling stock.

Chapter 26 explains the new concept of model and world-class stations with state-of-the-art facilities.

Chapter 32 explains the concepts of tilting trains and Maglev trains, which are the latest trends in high-speed railways.

CONTENT AND STRUCTURE

Chapter 1 presents a historical account of railways around the world with special focus on the important features of Indian Railways. It includes a discussion on various undertakings of the Ministry of Railways and highlights the important features of Vision 2020 document.

Chapter 2 discusses the various gauges adopted by Indian Railways and the problems associated with multi-gauge systems.

Chapter 3 explains the factors affecting track alignment. Mountain rails and rack railways are also discussed in this chapter.

Chapter 4 describes the different types of engineering surveys required to be undertaken before launching a new railway project as well as the construction of new lines including doubling and gauge conversions. It also includes modern survey techniques for alignment survey for new lines.

Chapter 5 presents details of track specifications and structures for different gauges on Indian Railways. It also discusses the forces acting on a track and the stresses generated in track components. The method of evaluating stresses and deflection in the track is explained with the help of illustrative examples.

Chapters 6 to 10 describe the various components of a railway track such as rails, sleepers, ballast, formation, and fittings and fastenings. The recent provisions adopted by Indian Railways with respect to these components are also discussed in these chapters.

Chapter 11 explains the causes and remedial measures for creep in rails. The method adopted on Indian Railways for field measurement of creep is also described in the chapter. Chapter 12 presents the basic aspects of the geometric design of railway tracks.

Chapter 13 discusses the detailed design of horizontal and vertical curves on a railway line. The method adopted by Indian Railways for calculating design speed for a curve and length of horizontal curve in restricted areas is described in detail.

Chapters 14 and 15 elaborate on the various types of track junctions and their designs, layouts of turnouts, and the factors affecting speed on turnouts.

Chapter 16 discusses rail joints and the methods used for welding rails in India. Recent developments in welding techniques are also included.

Chapter 17 describes the theory of long-welded rails and the track specifications for long and short-welded rails. It also includes the procedure followed by Indian Railways for distressing of long-welded rails, with and without rail tensors.

Chapter 18 describes the essentials of track maintenance and calendar system of track maintenance being followed by Indian Railways. It provides a detailed discussion on maintenance of rails, sleepers, and ballast. It has conventional type of track maintenance operations as well as the latest developments in track maintenance on Indian Railways. The work load of PWI and other field officers is also discussed.

Chapter 19 details the track drainage system, along with the drainage of station platforms, yards, and subsurface drains.

Chapter 20 describes the modern methods of track maintenance followed by Indian Railways employing track machines. It presents the details of measured shovel packing (MSP) equipment on Indian Railways. The chapter also describes Directed Track Maintenance (DTM), an annual programme conducted by Indian Railways and its future scope.

Chapter 21 elaborates on the methods of track rehabilitation and the renewal of track components. It also discusses some of the new machines used by Indian Railways for track renewal. Estimation of track material for a new track is also included here.

Chapter 22 discusses accident and disaster management, which assumes significance because of greater emphasis being placed on safety by Indian Railways. It gives a detailed classification of accidents and the remedial measures to be taken to avoid the accidents are also highlighted.

Chapter 23 describes level crossings and the measures taken to prevent accidents at such crossings. **Chapter 24** includes the design and maintenance features of locomotives and other rolling stock. It elaborates upon the new technology adopted by Indian Railways for design of rolling stock. **Chapter 25** provides details of train resistances and their evaluation. It also defines tractive effort and hauling capacity of locomotives.

Chapter 26 includes details of railway stations and yards. Facilities to be provided at station building and platforms are discussed here. Details of special facilities to be provided for model stations and world class stations have also been given.

Chapter 27 discusses various types of equipment used in railway stations. It includes foot overbridge, turntables, carriage washing platforms, buffer stops, and other such facilities. **Chapter 28** discusses the construction of new railway lines and doubling of railway lines. It also includes civil engineering works of gauge conversion projects.

Chapter 29 exclusively discusses suburban railways in metropolitan cities, which is the latest trend in the Indian context, so that this subject is fully appreciated by the readers. It gives details of Delhi Metro and Calcutta Metro Railways.

Chapter 30 is devoted to the specialized subject of railway tunnelling, describing the various techniques of construction and maintenance of tunnels in different types of soils.

Chapter 31 on signalling and interlocking gives the details of various types of signals, interlocking techniques, and modern developments concerning train control. The chapter also gives details of modern signalling techniques to control railway accidents such as train protection and warning system and anti-collision devices.

Chapter 32 discusses the various modernization plans of Indian Railways, covering tracks, locomotives, and rolling stock for high-speed trains, with the aim of widening the readers' understanding of the scope of railway engineering. The chapter also discusses specification for high-speed railways and some of the new concepts such as tilting trains and Maglev trains.

Chapter 33 is exclusively written on dedicated freight corridor, a flagship programme of Indian Railways. It provides a general and technical description of various freight corridor and link projects in India. The new developments on Indian Railways are also discussed in this chapter.

ACKNOWLEDGEMENTS

While revising this book, references have again been made to several Indian Railways codes and manuals and these are gratefully acknowledged. The suggestions and feedback received from different sources, which have helped the authors to a great extent during the revision of the book, are also thankfully acknowledged. We are grateful to our family members for their moral support and cooperation during the revision.

The authors are extremely grateful to all members of the team at the Oxford University Press, India, who have worked very hard to publish this second edition of the book in its current format.

The authors are also thankful to the reviewers of the first edition of the book for their valuable comments and suggestions. Many of these suggestions have been incorporated in the revised edition. Suggestions for further improvement of the content are welcome and will be gratefully acknowledged. These can be sent to us at satisfce@iitr.ernet.in and agarwalmm@sify.com.

Satish Chandra
M.M. Agarwal

Preface to the First Edition

There have been major technological developments in railways around the world in the recent past to meet the challenges of heavier traffic and higher speeds. In Indian Railways especially, the track structure has been modernized in a big way in the last three decades. Long welded rails, concrete sleepers, and elastic fastenings have been used on high-speed routes to provide stable and resilient structures. Metro railways are also being introduced in metropolitan cities to ease the problem of congestion on roads. Diesel and electric locomotives, which have superior performance capabilities, have replaced steam locomotives. In addition, modern signalling, automatic warning, and centralized traffic control systems are being adopted to ensure safety and maximum utilization of track capacity.

It is very important for engineering students and new entrants into the field of railways to be aware of not only the fundamentals of railway engineering but also latest developments with regard to railway tracks, locomotives and rolling stock, signalling and interlocking, etc.

ABOUT THE BOOK

This book deals with all aspects of railway engineering, from fundamental concepts to modern technological developments, with special focus on Indian Railways. It is an amalgamation of the vast experiences of the authors—of teaching the subject as well as of serving in Indian Railways. The text presents the theories and field practices as well as the modern techniques in detail.

CONTENT AND COVERAGE

The book treats the theoretical and practical aspects of the subject exhaustively and incorporates the latest provisions adopted by Indian Railways (IR).

Chapter 1 presents a historical account of railways around the world with special focus on the important features of IR. Chapter 2 discusses the various gauges adopted by IR and the problems associated with multigauge systems. Chapter 3 explains the factors affecting track alignment. Chapter 4 describes the different types of engineering surveys required to be undertaken before launching a new railway project as well as the construction of new lines including doubling and gauge conversions. Chapter 5 presents the details of track specifications and structures for different gauges on IR. It also discusses the forces acting on a track and the stresses generated in track components.

Chapters 6 to 10 describe the various components of a railway track—rails, sleepers, ballast, formation, and fittings and fastenings. It also discusses the recent provisions adopted by IR with respect to these components.

Chapter 11 explains the causes and remedial measures for creep in rails. Chapter 12 presents the basic aspects of the geometric design of railway tracks. Chapter 13 discusses the detailed design of the horizontal and vertical curves on a railway line. Chapter 14 and 15 elaborate on the various types of track junctions and their designs, layouts of turnouts, and the factors affecting speed on turnouts.

Chapter 16 discusses rail joints and the methods used for welding rail joints. Chapter 17 describes the developments in welded railway tracks including long welded rails (LWRs) and continuous welded rails (CWRs). It also explains the detailed procedure for laying LWRs and the track specifications for long and short welded rails. Chapter 18 describes the calendar system of track maintenance being followed by Indian Railways, which includes the conventional track maintenance operations of through packing, systematic overhauling, and picking up slacks. Chapter 19 details the track drainage system, along with the drainage of station platforms, yards, and subsurface drains. Chapter 20 describes the modern methods of track maintenance followed by Indian Railways employing track machines. It presents the details of measured shovel packing (MSP) equipment on IR. The chapter also describes Directed Track Maintenance (DTM)—IR's annual programme—and its future scope.

Chapter 21 elaborates on the methods of track rehabilitation and the renewal of track components. It also discusses some new machines used by IR for track renewal. Chapter 22 discusses accident and disaster management, which assumes significance because of greater emphasis being placed on safety by Indian Railways. Chapter 23 describes level crossings and the measures taken to prevent accidents at such crossings. Chapter 24 includes the design and maintenance features of locomotives and other rolling stock. Chapter 25 provides the details of train resistances and their evaluation. It also defines tractive effort and hauling capacity of locomotives.

Chapters 26 and 27 include details of station yards and the various equipment used in railway stations, respectively. Chapter 28 discusses the construction of new railway lines, track material required for BG tracks, doubling of railway lines, and gauge conversion. Chapter 29 exclusively discusses suburban railways in metropolitan cities, which is the latest trend in the Indian context, so that this subject is fully appreciated by the readers. Chapter 30 is devoted to the specialized subject of railway tunnelling, describing the various techniques of construction and maintenance of tunnels in different types of soils.

Chapter 31 on signalling and interlocking gives the details of various signals, interlocking techniques, and modern developments concerning train control. Chapter 32 discusses the various modernization plans of IR, covering tracks, locomotives, and rolling stock for high-speed trains, with the aim of widening the readers' understanding of the scope of railway engineering.

ACKNOWLEDGEMENTS

While writing this book, references have been made to several Indian Railways codes and manuals and other documents published by RDSO (Research Design and Standards Organisation, India). We have liberally used these documents and gratefully acknowledge RDSO. We are grateful to our family members for their moral support and cooperation while the book was being written.

Though every care has been taken to produce an error-free text, some errors may have gone unnoticed. We will be grateful to the users of the book for bringing any such errors to our notice, so that these can be rectified in subsequent editions. Constructive suggestions and comments for further improvement of the content are welcome.

Satish Chandra
M.M. Agarwal

Brief Contents

Detailed Contents

History and General Features of Indian Railways

INTRODUCTION

The history of railways is closely linked with the development of civilization. As the necessity arose, human beings developed various methods of transporting goods from one place to another. In the primitive days goods were carried as head loads or in carts drawn by men or animals. Then efforts were made to replace animal power with mechanical power. In 1769, Nicholes Carnot, a Frenchman, carried out the pioneering work of developing steam energy. This work had very limited success and it was only in 1804 that Richard Trevithick designed and constructed a steam locomotive. This locomotive, however, could be used for traction on roads only. The credit of perfecting the design goes to George (Stephenson) who in 1814 developed the first steam locomotive used for traction on railways.

The first public railway in the world was opened to traffic on 27 September 1825 between Stockton and Darlington in the UK. Simultaneously, other countries in Europe also developed such railway systems; most introduced trains for carriage of passenger traffic during that time. The first railway line in Germany was operated from Nurenberg to Furth in 1835. The US operated its first railway line between Mohawk and Hudson in 1833.

The first railway line in India became functional in 1853. The first train, consisting of one steam engine and four coaches, made its maiden trip on 16 April 1853, when it traversed a 34 km stretch between Bombay (now Mumbai) and Thane in 1.25 hours.

Starting from this humble beginning, the Indian Railways has grown today into a giant network consisting of over 64,460 km route criss-crossing the vast Indian terrain from the foothills of the Himalayas in the north to Kanyakumari in the south and Dibrugarh in the east to Dwarka in the west.

Indian Railways operates about 18,820 trains including 11,765 passenger trains every day, serving 7133 railway stations and carry about 7651 million passengers and about 926 million tonnes of goods traffic in a year. For moving this traffic,

Indian Railways deploys about 1.326 million regular employees and maintains 9213 locomotives consisting of 43 steam locomotives, 5137 diesel locomotives and 4033 electric locomotives. It also operates 2.23 lakh goods wagons, 45,123 conventional coaches and 7334 EMU and 763 DMU/DMHU coaches apart from 6505 other coaches. Today, Indian Railways is the largest railway network in the world under a single management.

Indian Railways, besides being the largest mode of transportation, also has a deep social obligation to serve efficiently and fulfil the increasing needs of the country and to provide the necessary infrastructure for the rapid industrialization and economic development of the country.

1.1 DEVELOPMENTS ON INDIAN RAILWAYS

Important developments on Indian Railways from 1831 to 2012 have been chronologically listed in Table 1.1.

Table 1.1 Important events in the history of Indian Railways

Year	Important events
1831–33	The first idea of a railway line from Madras (now Chennai) to Bangalore conceived to improve the transport system of southern India.
1844	RM Stephenson forms the East Indian Railway Company for construction of railway lines.
1845–46	Trial survey conducted for a new line from Calcutta (now Kolkata) to Delhi.
1848–49	Construction of a railway line from Howrah to Raniganj sanctioned.
1850	Construction of a railway line from Bombay to Thane started by the Great Indian Peninsula Railway Company.
1853	First railway line from Bombay to Thane opened for passenger traffic for a distance of 34 km on 16 April.
1854	Railway line between Howrah and Hoogly (39 km) opened for passenger traffic on 15 August.
1856	Railway line between Veyasarpady and Waljah road (100 km) opened for traffic under the banner of Madras Railway Company. In fact, this was the first proposal initiated in 1831 but could be completed only in 1854.
1856	First train in South India from Royapuram to Waljah road (Arcot).
1866	Calcutta linked with Delhi, Amritsar, and Bombay.
1850–68	First stage of development of Indian Railways classified as the Early Guarantee System. The government guaranteed a minimum percentage of return to shareholders in order to attract private enterprises to construct railways, but retained the right to purchase these railways at the end of 25 or 50 years. A number of railway companies were formed for the construction of railways, namely East India Railway (EIR), Great Indian Peninsula Railway (GIP), Bombay, Baroda, and Central India Railway (BB & CIR) and Madras State Railway (MSR), etc.
1869–81	In 1869, it was decided by the British Government that future railway projects should be either under the new guarantee system or under state-owned railways. Few states started construction of railway lines separately. The government, however, exercised a considerable measure of control to commercialize them. At this time there were company-managed railways under the new guarantee system as well as state-managed railways. After 1870 the railways developed very fast.
1871	Metre gauge in India was introduced on account of its being cheap and economic.

(Contd.)

Table 1.1 *(Contd.)*

Year	Important events
1873	First metre gauge line opened from Delhi to Farukanagar.
1881	First hill railway (Darjeeling Himalayan Railway; narrow gauge, 2' 0″ gauge) inaugurated.
1887	Victoria Terminus railway station constructed in Bombay.
1891	Toilets introduced in third class coaches.
1903	96-km-long Kalka–Shimla narrow gauge line opened to traffic on 9 November.
1905	Railway Board assumes office; established with one president and two members.
1922	Railway Board reconstructed and given a freehand with extra powers.
1924	Railway finances separated from general finances.
1924	Railway Board reconstituted with the chief commissioner as the president, an ex-officio secretary to the Government of India, and two members.
1925	12' wide, 1500-V dc stock started plying on the Mumbai–Kurla electrified section on 3 February.
1925	As a general policy to assume control over company railways, the Government took over the management of East India Railway and Great India Peninsula Railway.
1925	First railway line electrified, consisting of the harbour branch line of GIP.
1930	Central Standards Office (CSO) under Chief Controller of Standardization set up to standardize all equipment commonly in use by the Railways.
1930	Indian Railways stretched over 66,300 route km.
1931	Electrification of double track from Madras Beach to Tambaram and of sidings at Madras Beach, Madras Egmore, and Tambaram stations completed on 2 April; first electric train started on 11 May.
1936	Air conditioning introduced in passenger coaches.
1937	Burma separated from India and about 3200 km of railway lines taken out of Indian Railways.
1939–42	During World War II, the Indian Railways was called upon to release track material, locomotives, and wagons for construction of lines in the Middle East. This resulted in the closing down of 26 branch lines. Railway workshops were used for manufacture of defence material. At the end of the War, there were heavy arrears for the renewal and replacement of various assets.
1947	India became independent; due to partition of the country, railway lines and assets divided between India and Pakistan.
1947	Immediately after Independence, the Railway Board consisted of five members including a chief commissioner and financial commissioner.
1947–51	At the time of Independence, there were 42 railway systems consisting of 13 class I railways, 10 class II railways, and 19 class III railways. These included 32 lines owned by ex-Indian states. The Government of India decided to rationalize these railways to improve efficiency and facilitate better management.
	Regrouping of railways was completed and six zones were formed, namely Central Railway, Eastern Railway, Northern Railway, North Eastern Railway, Southern Railway, and Western Railway. The main objective of the planning of Indian Railways after Independence has been to develop rail transport to provide appropriate support for the planned growth of national resources as a whole. While doing so, the emphasis has been suitably readjusted during each 5-year period to take note of certain special features.

(Contd.)

Table 1.1 (*Contd.*)

Year	Important events
1950	Production of steam locomotives started in Chittaranjan Locomotive Works.
1952	Railway Staff College set up in Vadodara.
1952	Railway Testing and Research Centre (RTRC) was set up.
1952	Integral Coach Factory (ICF), Madras, set up as a production unit for all welded steel, lightweight integral coaches.
1954	Position of chief commissioner for railways renamed as the chairman of the Railway Board.
1955	Indian Railway Institute for Civil Engineering, Pune, set up.
1951–56	During the first Five Year Plan, there was special emphasis on rehabilitation and replacement of the assets overstrained and totally neglected during World War II. A sum of ₹ 2570 million was allotted to Indian Railways out of a total plan expenditure of ₹ 23,780 million.
1956–61	During the second Five Year Plan, the focus was on the development of rail transport capacity to meet the requirement of movement of raw materials and goods. A sum of ₹ 8960 million (18.7%) was allotted to Indian Railways (IR) out of a total plan expenditure of ₹ 48,000 million.
1957	Indian Railway Institute for Signal and Telecommunication, Secunderabad, set up.
1957	Research Design and Standards Organisation (RDSO), Lucknow, set up after the merger of various standards committees and RTRC.
1957	Indian Railways decides to adopt a 25-kV, 50-cycle, single-phase, ac system for electrification.
1957	Railway Protection Force (RPF) constituted.
1961	Diesel Locomotive Works (DLW) set up at Varanasi; Chittaranjan Locomotive Works (CLW) started manufacture of electric locomotives.
1961–66	The strategy adopted in the third Five Year Plan was to build up an adequate rail transport capacity to meet the traffic demands. It was proposed that this should be done through modernization of traction, i.e., by switching from steam traction to diesel or electric traction in a progressive manner. Track technology and signalling were also improved to match the new traction system. During the third Five Year Plan, a sum of ₹ 8900 million (11.9%) was allotted to Indian Railways out of a total grant of ₹ 75,000 million.
1969	Divisional system uniformly adopted on Indian Railways.
1969	New Delhi–Howrah Rajdhani Express running at a speed of 120 km/h introduced.
1969–74	The fourth Five Year Plan was drawn with a renewed emphasis on the twin objectives of modernization of the Railways and improving the operational efficiency of the system by more intensive utilization of the existing assets of the Railways. A sum of ₹ 10,500 million (6.6%) was allotted for the development of the Railways out of a total of ₹ 159,000 million.
1974	Rail India Technical and Economic Services (RITES) formed.
1976	Indian Railway Construction Company (IRCON) formed.
1974–78	The main emphasis of the fifth Five Year Plan was on the development of a rapid transport system in metropolitan cities, improvement in financial viability through cost reduction techniques, resource mobilization, optimum utilization of assets, and achievement of national self-sufficiency in railway equipment. A sum of ₹ 22,000 million (5.6%) was allotted to the Railways out of a total of ₹ 393,000 million.
1984	First Metro rail introduced in Kolkata.

(*Contd.*)

Table 1.1 *(Contd.)*

Year	Important events
1980–85	The sixth Five Year Plan was drawn up in the face of anticipated resource constraints and a heavy backlog of arrears of renewal of assets such as wagon and tracks. The main plan was that the limited resources of the Railways would be used for the rehabilitation of assets. A sum of ₹ 51,000 million was allotted to the Railways out of a total of ₹ 975,000 million for all the public sector undertakings for the entire plan period.
1985–90	The seventh Five Year Plan provided an outlay of ₹ 123,340 million. Freight traffic in the terminal year of the plan, namely 1989–90, was estimated to reach a level of 340 million tonnes.
1985	Computerized Passenger Reservation System introduced.
1987	Rail Coach Factory (RCF) established at Kapurthala.
1988	First Shatabdi train introduced between New Delhi and Jhansi.
1988	Container Corporation of India (CONCOR) established.
1992–97	The eighth Five Year Plan provided an outlay of ₹ 272,020 million (6.3 per cent) for Indian Railways out of a total outlay of ₹ 4,341,000 million for the full plan. Some of the main objectives of the eighth Plan were to generate adequate transport capacity, complete the process of rehabilitation of overaged assets, modernize the system to reduce cost and improve reliability, complete uni-gauge conversion of 6000 km of metre gauge (MG) and narrow gauge (NG) to broad gauge (BG), phase out steam locomotives, electrify 2700 route km, expand and upgrade intermodal operation, and improve manpower productivity.
1998	Konkan Railway system becomes fully operational on 26 January.
1998	Guiness Certificate for Fairy Queen—the oldest working steam locomotive in the world.
1999	Darjeeling Himalayan Railway declared World Heritage Site by UNESCO.
1999	Centenary celebrations of the Nilgiri Mountain Railway.
1999	Guiness certificate for Delhi main station equipped with the world's largest route relay interlocking system.
1997–2002	The ninth Plan envisaged, an outlay of ₹ 454,130 million, which is 14.1 per cent of the total outlay of ₹ 8,592,000 million for the full plan. Some of the main objectives were generation of adequate transport capacity for handling additional traffic, modernization and upgrading of the rail transport system, completion of the process of rehabilitation, replacement and renewal of overaged assets, and continuation of the policy of unit gauge.
2002	Jan Shatabdi trains introduced.
2002	East Cental Railway (HQ in Hazipur) and North Western Railway (HQ in Jaipur) become operational with effect from 1 October.
2003	Indian Railways has 16 (earlier 9) zones and 67 (earlier 59) divisions with effect from 1 April.
2002–07	The X plan envisaged an outlay of ₹ 840,030 million, for Indian Railways which is 5.5 per cent of total outlay of ₹ 15,256,390 million for the full plan.
2007–2012	The XI plan envisages an outlay of ₹ 2,332,890 million, for Indian Railways which is 5.6 per cent of total outlay of ₹ 41,185,310 million The main objectives of the plan are strengthening of high density network, technological upgradation of assets; utilizing information technology for better customer interface; improving rail safety by replacement of overaged assets and to increase the share of freight and passenger traffic during the plan.

(Contd.)

Table 1.1 *(Contd.)*

Year	Important events
2008	Garib Rath express between Secunderabad and Visakhapatnam inaugurated on 24 October.
2008	Kalka–Shimla Rail section declared as World Heritage site on 9 November.
2010	First regular passenger service in Kashmir Valley between Qazigund to Baramula opened for traffic.
2010	Metro Railway declared as new Zonal Railway with effect from 29 December.
2011–2020	The document **'Railway Vision 2020'** prepared and the document submitted to Parliament in December–2009. Brief details of this document are given separately.

1.2 DIFFERENT MODES OF TRANSPORT

Our environment consists of land, air, and water. These media have provided scope for three modes of transport—land transport, air transport, and water transport. Rail transport and road transport are the two modes of land transport. Each mode of transport, depending upon its various characteristics, has intrinsic strengths and weaknesses and can be best used for a particular type of traffic as given below.

Rail transport   Owing to the heavy expenditure on the basic infrastructure required, rail transport is best suited for carrying bulk commodities and a large number of passengers over long distances.

Road transport   Owing to flexibility of operation and the ability to provide door-to-door service, road transport is ideally suited for carrying light commodities and a small number of passengers over short distances.

Air transport   Owing to the heavy expenditure on the sophisticated equipment required and the high fuel costs, air transport is better suited for carrying passengers or goods that have to reach their destinations in a very short period of time.

Water transport   Owing to low cost of infrastructure and relatively slow speeds, water transport is best suited for carrying heavy and bulky goods over long distances, provided there is no consideration of the time factor.

1.2.1 Railway as a Mode of Land Transport

There are two modes of land transport, railways and roads, and each has its relative advantages and disadvantages. These have been summarized in Table 1.2.

1.2.2 Role of Indian Railways

Since its inception, Indian Railways has successfully played the role of the prime carrier of goods and passengers in the Indian subcontinent. As the principal constituent of the nation's transport infrastructure, the Railways has many roles to play as given below.

(a) It helps integrate fragmented markets and thereby stimulates the emergence of a modern market economy.

Table 1.2 Rail transport versus road transport

Feature	Rail transport	Road transport
Tractive resistance	The movement of steel wheels on steel rails has the basic advantage of low rolling resistance. This reduces haulage costs because of low tractive resistance.	The tractive resistance of a pneumatic tyre on metalled roads is almost five times compared to that of wheels on rails.
Right of way	A railway track is defined on two rails and is within protected limits. Trains work as per a prescribed schedule and no other vehicle has the right of way except at specified level crossings.	Roads, though having well-defined limits, can be used by any vehicular traffic and even by pedestrians; they are open to all.
Cost analysis	Owing to the heavy infrastructure, the initial as well as maintenance cost of a railway line is high.	The cost of construction and maintenance of roads is comparatively low.
Gradients and curves	The gradients of railways tracks are flatter (normally not more than 1 in 100) and curves are limited to 10° on broad gauge.	Roads are constructed normally with steeper gradients of up to 1 in 30 and relatively much sharper curves.
Flexibility of movement	Due to the defined routes and facilities required for the reception and dispatch of trains, railways can be used only between fixed points.	Road transports have much more flexibility in movement and can provide door-to-door services.
Environment pollution	Railways have minimum adverse effects on the environment.	Road transport creates comparatively greater pollution than the railways.
Organization and control	Railways are government undertakings, with their own organization.	Barring member state government transport, road transport is managed by the private sector.
Suitability	Railways are best suited for carrying heavy goods and large numbers of passengers over long distances.	Road transport is best suited for carrying lighter goods and smaller numbers of passengers over shorter distances.

(b) It connects industrial production centres with markets as well as sources of raw material and thereby facilitates industrial development.

(c) It links agricultural production centres with distant markets as well as sources of essential inputs, thereby promoting rapid agricultural growth.

(d) It provides rapid, reliable, and cost-effective bulk transportation to the energy sector; for example, to move coal from the coalfield to power plants and petroleum products from refineries to consumption centres.

(e) It links people with places, enabling large-scale, rapid, and low-cost movement of people across the length and breadth of the country.

(f) In the process, Indian Railways has become a symbol of national integration and a strategic instrument for enhancing our defence preparedness.

1.3 ORGANIZATION OF INDIAN RAILWAYS

Indian Railways is currently the largest public undertaking of the Government of India, having a capital-at-charge of about ₹ 1,230,000 million. The enactments regulating the construction and operation of railways in India are the Indian Tramway Act, 1816 and the Indian Railway Act, 1890 as amended from time to time. The executive authority in connection with the administration of the railways vests with the Central Government and the same has been delegated to the Railway Board as per the Indian Railway Act.

1.3.1 Railway Board

The responsibility of the administration and management of Indian Railways rests with the Railway Board under the overall supervision of the Minister for Railways. The Railway Board exercises all the powers of the Central Government in respect of the regulation, construction, maintenance, and operation of the Railways.

The Railway Board consists of a chairman, a financial commissioner for railways, and five other functional members. The chairman is the ex-officio principal secretary to the Government of India in the Ministry of Railways. He reports to the Minister for Railways and is responsible for making decisions on technical and administrative matters and advising the Government on matters of railway policy. All policies and other important matters are put up to the Minister through the chairman or other board members.

The Financial Commissioner, Railways is vested with the full powers by the Government of India to sanction railway expenditure and is the ex-officio secretary to the Government in financial matters. The members of the Railway Board are separately in charge of matters relating to staff, civil engineering, traffic, mechanical engineering, and electrical engineering. They function as ex-officio secretaries to the Government in their respective spheres.

To be able to effectively tackle the additional duties and responsibilities arising from increased tempo of work, the Railway Board is assisted by a number of technical officers designated additional members and executive directors, who are in charge of different directorates, such as civil engineering, mechanical, electrical, stores, traffic and transportation, commercial, and planning and are responsible for carrying out technical functions. These officers, however, do not make major policy decisions.

1.3.2 Research Design and Standards Organisation

Research Designs and Standards Organisation (RDSO) is headquartered at Lucknow. The Director general, RDSO heads a team of specialists from different fields of railways. RDSO functions as a technical adviser and consultant to the Railway Board, the zonal railways, and production units as well as to public and private sector undertakings with respect to the designs and standardization of railway equipment. RDSO has also been approved for its quality management system ISO 9001:2000.

1.3.3 Zonal Railways

The entire railway system was earlier divided into nine zonal railways. To increase efficiency, the Railway Ministry decided to set up seven new railway zones, namely

North Western Railway at Jaipur, East Central Railway at Hajipur, East Coast Railway at Bhubaneswar, North Central Railway at Allahabad, South Western Railway at Bangalore, West Central Railway at Jabalpur, and South East Central Railway at Bilaspur. All the new railway zones have been fully functional from 1 April 2003.

Currently, Indian Railways is divided into 16 zones, each having different territorial jurisdictions which vary from 2300 to 7000 route km. The route kilometres of various zonal railways (zone wise and gauge wise) are given in Table 1.3. Route kilometre indicates the length of a route from one point to another. The length of running track on a route is called the running track kilometre. On a double-line section, the running track kilometre will be almost twice the route kilometre. When the length of turnouts are also added to the running track kilometre it becomes total track kilometre. The three lengths on different gauges are given in Table 1.4.

Table 1.3 Railway-wise route kilometrage as on 31 March 2011

Railways	*Head quarters*	*Broad gauge (BG)*	*Metre gauge (MG)*	*Narrow gauge (NG)*	*Total*
Central (CR)	Mumbai	3606.86	0.00	298.61	3905.47
Eastern (ER)	Kolkata	2302.68	0.00	132.53	2435.21
East Central (ECR)	Hajipur	3217.19	438.31	0.00	3655.50
East Cost (ECOR)	Bhubaneswar	2646.40	0.00	0.00	2646.40
Northern (NR)	Delhi	6696.28	11.27	260.85	6968.40
North Central (NCR)	Allahabaad	2850.78	11.48	288.59	3150.85
North Eastern (NER)	Gorakhpur	2349.60	1371.14	0.00	3720.74
Northeast Frontier (NFR)	Maligaon	2496.90	1323.64	87.48	3908.02
North Western (NWR)	Jaipur	4297.28	1166.50	0.00	5463.78
Southern (SR)	Chennai	4203.86	898.26	0.00	5102.12
South Central (SCR)	Secunderabad	5634.06	175.93	0.00	5809.99
South Eastern (SER)	Kolkata	2631.61	0.00	0.00	2631.61
South East Central (SECR)	Bilaspur	1743.97	0.00	711.01	2454.98
South Western (SWR)	Hubli	3176.50	0.00	0.00	3176.50
Western (WR)	Mumbai	4343.90	1412.39	684.06	6440.35
West Central (WCR)	Jabalpur	2964.84	0.00	0.00	2964.84
Grand Total		55162.71	6808.92	2463.13	64434.76

Table 1.4 Track lengths on different gauges as on March 2011

Gauge	*Route km*	*Running track km*	*Total track km*
BG (1676 mm)	55,188	77,347	102,680
MG (1000 mm)	6,809	7,219	8,561
NG (762 mm and 610 mm)	2,463	2,474	2,753
Total	64,460	87,040	113,994

The zonal railways take care of the railway business in their respective areas and are responsible for management and planning of all work. Each zonal railway is administered by a general manager assisted by additional general managers and heads of departments of different disciplines. These disciplines include civil engineering, mechanical, operating, commercial, accounts, security, signals and telecommunications, electrical, personnel, and medical. The typical organization of a zonal railway is given in Fig. 1.1. The duties of the various heads of departments are given in Table 1.5.

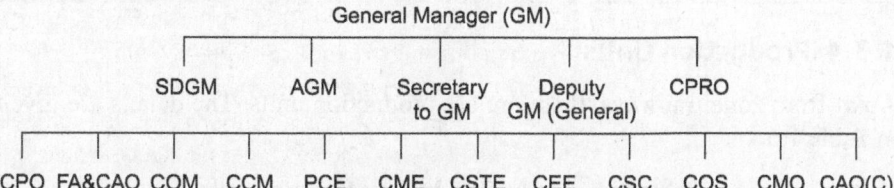

Fig. 1.1 Typical organization chart of a zonal railway

Table 1.5 Duties of the various principal heads of the department of zonal railways

Designation	Abbreviation	Brief duties
General manager	GM	Overall in-charge of a zonal railway of about 2500 to 7000 routes
Additional general manager	AGM	Second in position in-charge of general administration
Senior deputy general Manager	SDGM	Chief vigilance officer and in charge of allotment of houses, etc.
Chief public relations officer	CPRO	Public relations work
Chief personal officer	CPO	Establishment matters and labour relations
Financial adviser and chief accounts officer	FA&CAO	Accounting and all financial matters
Chief operations manager	COM	Running of trains and transport of passengers and goods
Chief commercial manager	CCM	Sales and marketing of passenger and goods services
Principal chief engineer	PCE	Maintenance and management of track, bridges and civil engineering assets
Chief mechanical engineer	CME	Maintenance and repair of locomotives, coaches, and wagons
Chief signal and telecommunication engineer	CSTE	Maintenance and construction of signalling and telecommunication facilities
Chief electrical engineer	CEE	Maintenance and repair of electric locomotives, electric multiple units (EMUs) stock, and all electric installations
Chief security commissioner	CSC	Security of railway installations

(Contd.)

Table 1.5 *(Contd.)*

Designation	Abbreviation	Brief duties
Controller of stores	COS	Procurement and supply of all stores items
Chief medical officer	CMO	Medical attention and healthcare of all railway employees
Chief administrative officer (construction)	CAO (c)	Construction of all major engineering projects

1.3.4 Production Units

Apart from zonal railways, there are six production units. The details are given in Table 1.6.

Table 1.6 Production or manufacturing units

Unit	Headquarters	Production
Chittaranjan Locomotive Works	Chittaranjan	Electric locomotives
Diesel Locomotive Works	Varanasi	Diesel locomotives
Integral Coach Factory	Chennai	Coaches
Diesel Components Works	Patiala	Diesel components
Rail Coach Factory	Kapurthala	Coaches
Wheel and Axle Plant	Bengaluru	Wheels and axles

1.3.5 COFMOW

The central organization for modernization of workshops (COFMOW) was set up in 1979 as a specialized agency to implement the various workshop modernization programmes of Indian Railways. Most of the workshops of Indian Railways are over 100 years old and COFMOW is in a planned way carrying out the modernization of these workshops with the assistance of World Bank. COFMOW is now the designated organization of Indian Railways for the selection, procurement, and induction of modern workshop technologies and new machinery and plant (M&P).

The various activities of COFMOW are summarized as follows:

(i) Guide railway customers in the selection of manufacturing technologies and M&P.
(ii) Prepare detailed technical specifications for procurement of M&P.
(iii) Procurement of M&P and commissioning at site.
(iv) Coordinate turnkey works associated with M&P.
(v) Provide training to workshop personnel on the operation and maintenance of machines.
(vi) Coordinate warranty services with vendors.
(vii) Undertake industrial engineering and layout studies.
(viii) Convert/recondition machines and manufacturing lines.
(ix) Design standard processes, layouts, and facilities for overhaul/maintenance of critical sub-assemblies.

(x) Promote indigenization of M&P items.
(xi) Support rolling stock of projects such as LHB coach, GM diesel loco, and ABB loco by purchasing special-purpose M&P.

1.3.6 Divisions

Zonal railways work on the divisional system. Each railway is divided into three to six divisions, each having approximately 700 to 1000 route km in its territory. There are about 67 divisions on Indian Railways. Each division works under the overall control of a divisional railway manager, who is assisted by one or two additional divisional railway managers. There are divisional officers in charge of each discipline either in the junior administrative grade or the senior scale, namely Divisional Superintending Engineer (DSE) or senior divisional executive, divisional engineer for civil engineering, senior divisional mechanical engineer or divisional mechanical engineer for mechanical engineering, senior divisional commercial manager or divisional commercial manager for commercial work, etc.

The engineering branch of each unit is headed by DSE or senior divisional engineer. Under each DSE, there are two or three divisional engineers (DENs), each in charge of approximately a distance of 800 to 1000 integrated track kilometres and assisted by two or three assistant engineers (AENs) in the maintenance of track and works. An AEN has about 400 integrated track kilometres under his charge. The total number of DENs and AENs for maintenance work on Indian Railways is approximately 300 and 600, respectively. AENs are assisted by permanent way inspectors (PWI) for maintenance of the track structure. Each PWI has a jurisdiction of 50–70 route km of the track. The total number of PWIs for normal maintenance work on Indian Railways is roughly 3000.

1.4 INDIAN RAILWAY FINANCES AND THEIR CONTROL

The finances of Indian Railways were separated from the general finances of the country in 1924 by a resolution of the Central Legislature. Under the separation convention, the Railways is required to pay dividends at a fixed rate on their capital, which has been advanced by the Central Government, subject to the obligation to pay dividends at the prescribed rates to the General Exchequer and observance of the national economic policies. The Railways is free to pursue its own financial policies to its best advantage. The dividend rates as well as other financial arrangements between the Railways and General Finance are determined periodically by the convention committee, which is an Indian Parliament committee. Indian Railways was required to pay a dividend at 4.5 per cent on capital invested up to March 1964, 5.5 per cent for fresh investments up to the period 1980–85, and 6.5 per cent thereafter for the seventh plan period, i.e., for the period 1985–90. Railway finances and policies are controlled by the Parliament through discussions and debates on the Annual Railway Budget, interpellations during the question hour whenever the Parliament is in session, and parliamentary committees, such as the Railway Convention Committee, the Estimates Committee, and the Public Accounts Committee.

The railway administration also gets a feel of public opinion and secures people's cooperation through Railway Users' Consultative Committees at various levels and also through advisory committees for specific purposes, such as the Passenger Amenities Committee and Time Table Advisory Committee.

During the financial year 2010–11 the revenue of Railways from freight accounted for 66.5 per cent of gross earnings; 87.8 per cent of goods earnings was realized from goods hauled in bulk, 8.49 per cent from general merchandise, and 2.71 per cent from miscellaneous charges, passenger earnings constituted 27.19 per cent of the gross earnings, of which 6.95 per cent was from suburban services, 75.61 per cent from express long distance, and 17.44 per cent from ordinary short-distance traffic.

1.5 COMMISSION OF RAILWAY SAFETY

Initially, to exercise effective control over the construction and operation of the railways in India, which were entrusted to private companies, consulting engineers were appointed under the Government of India. Later, when the Government undertook the construction of railway lines, the consulting engineers were designated Government inspectors. In 1883, their position was statutorily recognized. Later, the Railway Inspectorate was placed under the Railway Board, which was established in 1905.

The Railway Inspectorate was subsequently placed under the Ministry of Transport and Communication in 1940. In an attempt to give it a better status, the Railway Inspectorate was re-designated as the Commission of Railway Safety in 1961.

Today, the Commission of Railway Safety is headed by a chief commissioner; the commission headquartered at Lucknow under the Ministry of Civil Aviation. Under the administrative control of the chief commissioner of Railway Safety, who looks after the safety aspects and other statutory functions of the various zonal railways, nine commissioners of railway safety are stationed at various locations such as Lucknow, Mumbai, Delhi, Kolkata, Chennai, Bengaluru, and Secunderabad.

The responsibility for safety in the working and operation of the Railways rests solely with the Railway Board and the zonal railway administrations. The main task of the Commission of Railway Safety, however, is to direct, advise, and caution railway executives with a view to ensure that all reasonable precautions are taken with regard to the soundness of rail construction and safety of train operation.

1.5.1 Functions of Commission of Railway Safety

The principal functions of the Commision of Railway Safety are as follows:
 (i) Inspection of new railway lines prior to authorization for passenger traffic
 (ii) Periodical inspection of open lines
(iii) Approval of new works and renewals affecting passenger carrying lines
(iv) Investigations into accidents, including inquiries into such accidents of passenger trains as are considered to be of a serious nature
 (v) General advice on matters concerning safety of train operation
(vi) Statutory powers under Sections 4, 5, and 6 of Indian Railway Act to inspect the railway systems, conduct inquiries, causes of accidents, and sanction execution of all works affecting the safety of the running line

1.6 LONG-TERM CORPORATE PLAN OF INDIAN RAILWAYS

The Railways has a crucial role to play in the planned economic development of the country. In an environment of rapid technological change, faster economic growth, and enhanced expectations of the people, a long-term perspective of the railway development is vital. The corporate plan provides this perspective, spells out the objectives to be achieved, and strategies to be followed to the end of the century.

The first corporate plan of Indian Railways, covering a period of 15 years upto 1988–89, was published in 1976. Subsequently, a fresh corporate plan for the period 1985–2000 was prepared, which had envisaged an investment of about ₹ 461,700 million in the railways upto 2000.

A new corporate plan for the period 2001–2012 has been prepared recently by Indian Railways. Some of the salient features of the new corporate plan are as follows:

(i) *Core value–customer above everything else–* Creation of value of the customer and provide them safety, security, punctuality and reliability.

(ii) *Core purpose–*to provide cost-effective rail and integrated inter/multi-modal transportation/logistics service with state-of-the art, eco-friendly technology (including information technology) while maintaining the financial viability of the system.

(iii) *The toal requirement of funds* for the 12-year period for 2001–2012 at 2000–01 prices is approximately ₹ 1,750,000 million at normal growth.

1.6.1 Strengths of Indian Railways

The following are the strengths of the Indian Railways.

(a) For a vast country with long distances and a large population, the Railways has an inherent advantage over other modes of transport in its suitability for movement of large volumes of passenger and goods traffic over long distances.

(b) The movement of steel wheels on steel rails in the railway system has the basic advantage of low rolling resistance, which reduces energy requirements and haulage costs.

(c) Rail transport is more efficient than road transport in terms of land use.

(d) Railways is an energy-efficient mode of transport, particularly for freight traffic, and can use different forms of energy. It also causes relatively less environmental pollution than road transport.

(e) In densely populated urban centres, a rapid transit rail-based system is the most appropriate mode of transport for suburban intra-urban travel, as part of a city's integrated transport system.

(f) Indian Railways is a well-established organization with a large pool of skilled and trained personnel.

(g) Being part of the Central Government, Indian Railways has the government's financial backing. At the same time, it has considerable financial autonomy.

(h) Indian Railways is a self-reliant system with respect to its major equipment needs.

1.6.2 Weaknesses of Indian Railways

The following are the weaknesses of the Indian Railways system.

(a) A large portion of the Railways infrastructure is overaged, and in urgent need of replacement or rehabilitation. This includes track, motive power and rolling stock, signalling, operational, and maintenance equipment.

(b) In certain parts of the infrastructure, the technology lags by 20–25 years behind some of the developed railway systems. Consequently the productivity levels are comparatively low.

(c) Indian Railways has a large force of unskilled manpower. The training facilities need augmentation and modernization.

(d) A persistent resource constraint in the past has adversely affected the Railways' development.

(e) Indian Railways carries a substantial 'social burden' in the form of continued operation of un-remunerative branch lines, subsidies on passenger and suburban travel, and even freight subsidy on certain commodities.

(f) In certain areas, pilferage and vandalism seriously affect operational efficiency.

(g) The Railways is not suited for carriage of small quantities of freight particularly over short distances.

(h) Heavy investments are needed to build up railway transport capacity and the gestation periods are long.

(i) Transport capacity is volatile and cannot be recouped if not utilized continuously.

1.6.3 Planning Strategy

The development plans of Indian Railways have been drawn up within the framework of the National Five Year plans of the country. The plan outlay of the Indian Railways as well as those for the transport sector as a whole are given in Table 1.7.

Table 1.7 Plan outlay of funds (₹ in crores)*

Item	Upto Fifth Plan 1950–80	Sixth Plan 1980–85	Seventh Plan 1985–90	Eighth Plan 1992–97	Ninth Plan 1997–02	Tenth Plan 2002–02	Eleventh Plan 2007–12
Railways	4723	6,585	16,549	32,306	45,725	84,003	233,289
Transport sector	10.117	13,962	29,548	65,173	117,563	225,977	448.987
Total Plan outlay	50.979	109,292	218,729	485,457	81,398	1525,639	4118,531
Transport as percentage of total Plan	16.9	12.8	13.5	13.4	14.4	17.0	10.9
Railways as percentage of total Plan	7.9	6.0	7.6	6.7	5.6	5.5	5.6

* One crore is equal to 10 million (100,00,000).

1.7 INDIAN RAILWAYS VISION 2020

The document 'Indian Railway Vision 2020' is intended to lay down the vision, which would enable Indian Railways to meet the expectations of the nation and play its rightful role as the catalyst of economic development of the country in times to come.

Vision statement

Indian Railways shall provide efficient, affordable, customer-focused, and environmentally sustainable integrated transportation solutions. It shall be a vehicle of growth, connecting regions, communities, ports and centres of industry, commerce, tourism and pilgrimage across the country. The reach and access of its services will be continuously expanded and improved by its integrated team of committee, empowered and satisfied employees and by use of cutting-edge technology.

National goal

Vision 2020 will address the following four strategic national goals:
  (i) Inclusive development, both geographically and socially
 (ii) Strengthening national integration
(iii) Large-scale generation of productive employment
(iv) Environmental sustainability

Goals of Railway Vision 2020

By 2020 Indian Railways would strive to achieve the following goals.

Quality of service   Establish quality of service benchmarked to the best of the railway systems in the world.

Expansion of network   Expand its route network at the rate of 2500 km, per annum. By 2020, 25000 km of new line will be added and almost the entire network (barring the hill and heritage railways) would be broad gauge. This would include completion of the pending shelf of new line projects of 11985 km. More than 30,000 km of route would be of double/multiple lines. Electrification of 14,000 km of routes would take the total length of electrified route to 33,000 km. This would include all inter-metro links and the other busy corridors.

Multiplication of lines   Have more than 6000 km, for quadrupled lines with segregation of passenger and freight services into separate double-line corridors. This shall include Delhi–Kolkata, Delhi–Mumbai, Kolkata–Mumbai and Delhi–Chennai routes. All these routes would have separate dedicated freight corridors and high-speed passenger corridors.

Increasing speeds   Raise speeds of passenger trains from 110–130 km per hour to 160/200 km per hour on segregated routes and speed of freight trains from 60–70 km per hour to 100 km per hour.

Availability on demand   Virtually attain state of 'availability on demand' in freight, passenger, and parcel services.

Design and deliver targeted services   For transport of perishables, agri-produce, and products of small and medium enterprises (SMEs), such as auto-hubs and other similar clusters.

Zero accidents   Target to achieve zero accidents.

Zero failure   Target to achieve zero failure in equipment.

Environmental development   Utilize at least 10 per cent of its energy requirement from renewable sources and institute a foolproof eco-friendly waste management system.

High-speed corridors   Complete four high-speed corridors of 2000 km and plan development of eight others.

Summary of broad goals

A summary of the broad goals of Railway Vision 2020 is given in Table 1.8.

Table 1.8   Summary of broad goals of Railway Vision 2020

Broad category	Short-term target (2010–11 to 2011–12)	Long-term target (2012–13 to 2019–20)	Total target
Doubling (including DFC)	1000 km	11000 km	12000 km
Gauge	2500 km	9500 km	12000 km
New line	1000 km	24000 km	25000 km
Electrification	2000 km	12000 km	14000 km
Procurement of wagons	33909	2,55,227	2,89,136
Procurement of diesel locomotives	690	4,644	5,334
Procurement of electric locomotives	555	3,726	4,281
Procurement of passenger coaches	6,912	43,968	50,880
World-class stations (bid-out/cession)	12 stations	38 stations	50 stations

1.8 CLASSIFICATION OF RAILWAY LINES IN INDIA

The Railway Board has classified the railway lines in India based on the importance of the route, the traffic carried, and the maximum permissible speed on the route. The complete classification is given below.

1.8.1 Broad Gauge Routes

All the broad gauge (BG) routes on Indian Railways have been classified into five different groups based on speed criteria as given below.

Group A lines

Group A lines are meant for a sanctioned speed of 160 km per hour.
- New Delhi – Howrah by Rajdhani route
- New Delhi – Mumbai Central by Frontier Mail/Rajdhani route
- New Delhi – Chennai Central by Grand Trunk route
- Howrah – Mumbai VT via Nagpur

Group B lines

Group B lines are meant for a sanctioned speed of 130 km per hour.
- Allahabad–Itarsi–Bhusaval

- Kalyan–Wadi Raichur–Madras
- Kharagpur–Waltair–Vijayawada
- Wadi–Secunderabad–Kazipet
- Howrah–Bandel–Burdwan–Barharwa over Farakka–Malda town
- Barsoi–New Jalpaiguri
- Sitarampur–Kiul–Patna–Mughalsarai
- Kiul–Sahibganj–Barharwa
- Delhi–Ambala Cantt–Kalka
- Ambala Cantt–Ludhiana–Pathankot
- Ambala Cantt–Moradabad–Lucknow–Paratapgarh–Mughalsarai
- Arakonam–Erode–Coimbatore
- Vadodara–Ahemadabad
- Jalapet–Bangalore

Group C lines

Group C lines are meant for suburban sections of Mumbai, Kolkata, and Delhi.

Group D and D Spl lines

Group D lines are meant for sections where the maximum sanctioned speed is 100 km per hour.

Group E and E Spl lines

These lines are meant for other sections and branch lines.

D Spl and E Spl routes Based on the importance of routes, it has been decided that few selected routes currently falling under D and E routes will be classified as D special and E special routes. This has been done for the purpose of track renewal and priority allotment of funds. The track standards for these routes will be 60 kg 90 ultimate tensile strength (UTS) rails and prestressed concrete (PSC) sleepers with sleeper density of 1660 per km.

1.8.2 Metre Gauge Routes

Depending upon the importance of routes, traffic carried, and maximum permissible speed, the metre gauge (MG) tracks on Indian Railways were earlier classified into three main categories, namely trunk routes, main lines, and branch lines. These track standards have since been revised and now MG routes have been classified as Q, R1, R2, R3, and S routes as discussed below.

Review of track standard for MG routes

A committee of directors, chief engineers, and additional commissioner of railway safety (ACRS) was formed in 1977 to review the track standards for MG routes. The committee submitted its report in December 1981, in which it recommended that MG routes be classified into three categories, namely Q, R, and S routes, based on speed criteria. The committee's recommendations were accepted by the Railway Board after certain modifications. The final categories are as follows.

Q routes Routes with a maximum permissible speed of more than 75 km per hour. The traffic density is generally more than 2.5 GMT [gross million tonne(s) per km/annum].

R routes   Routes with a speed potential of 75 km per hour and a traffic density of more than 1.5 GMT. R routes have further been classified into three categories depending upon the volume of traffic:
   (i) R1—traffic density more than 5 GMT
   (ii) R2—traffic density between 2.5 and 5 GMT
   (iii) R3—traffic density between 1.5 and 2.5 GMT

S routes   Routes with a speed potential of less than 75 km per hour and a traffic density of less than 1.5 GMT. These consist of routes that are not covered in Q, R1, R2, and R3 routes. S routes have been further subclassified into three routes, namely S1, S2, and S3. S1 routes are used for the through movement of freight traffic, S3 routes are uneconomical branch lines, and S2 routes are those which are neither S1 nor S3 routes.

1.9 GENERAL FEATURES OF INDIAN RAILWAYS

Indian Railways is the second largest state-owned railway system in the world (after Russian Railways) under unitary management. The important features of Indian Railways are described here.

1.9.1 Track

Track or permanent way is the single costliest asset of Indian Railways. It consists of rails, sleepers, fittings and fastenings, ballast, and formation. Complete details of the track are given in Chapter 5.

1.9.2 Locomotives

In 2010–11 Indian Railways owned a total fleet of 9,213 locomotives including 43 steam locomotives, 5137 diesel locomotives, and 4033 electric locomotives. The steam locomotives ownership reached a peak in 1963–64 with 10,810 units. It then declined gradually as production of steam locomotives was stopped in 1971. Diesel and electric locomotives which are more than twice as powerful as steam locos are progressively replacing steam locos. Due to the heavy investments involved in replacing all the existing steam locomotives by diesel and electric locomotives, steam locos are being gradually phased out and it was planned that these should be retained in service till the expiry of their codal life or 2000 whichever is earlier. Accordingly, most of the steam locos have been phased out of service on Indian Railways.

Apart from replacing steam locomotives by diesel and electric locomotives in areas of heavy traffic density, a large number of diesel shunting engines are also being introduced to replace steam shunting locomotives. This has enabled the Indian Railways to improve operational efficiency in both passenger and freight operations.

1.9.3 Traction

The traction mix has significantly changed in the past three decades and the railways have been progressively switching over to diesel and electric traction. Steam locomotion, though it involves the least initial costs, is technologically inferior to diesel and electric traction in many respects. On the other hand, diesel and electric locomotives have much superior performance capabilities, electric locomotives being more powerful of the two. Electric traction is also the most capital intensive and therefore requires a certain minimum level of traffic density for its economic use. In broad terms, the traction policy in Indian Railways envisages the use of steam locomotives till the end of their codal life but maximum upto 2000, extension of electrification of high-density routes as dictated by economic and resource considerations and dieselization of the remaining services.

1.9.4 Electrification and Electric Traction

Electric traction on 1500 volts dc was first introduced in 1925 on a small section of the Bombay area and till 1957 it was confined to less than 466 kilometres, comprising mainly the suburban sections of Bombay and Madras. Electrification on the main line sections was, however, taken up towards the end of the Second Five Year Plan on 25 KV single-phase AC system.

The electrification of Howrah–Burdwan suburban section of Calcutta on Eastern Railway was taken up during the First Five Year Plan (1951–56) and was completed in 1958. Thereafter, electrification on Indian Railways has continued in a planned manner on the trunk routes connecting the four metropolitan cities of Calcutta, Delhi, Mumbai, and Chennai and other high-density routes.

The number of electrified route kilometres on Indian Railways is currently 18,927 km out of a total route km of 64,465 and constitute about 29 per cent of the total route kilometres. During 2009–10, 51.4 per cent of passenger train km and 63.1 per cent of BG freight gross tonne km were operated on electric traction.

Broad gauge, though forming 85.6 per cent of the route, generated 99.9 per cent of freight output (net–tonne km.) and 97.9 per cent of the passenger output (passenger km).

The route length as on 31 March 2011 in each gauge indicating double/multiple line, single line and electrified route, is given in Table 1.9.

Table 1.9 Electrified and non-electrified route km on Indian Railways

Gauge	Single line		Total km	Double/multiple line		Total km	Grand total
	Elec-trified	Non-electrified		Electri-fied	Non-electrified		
BG (1676 mm)	5242	30,723	33,965	14,365	4858	19,223	55,188
MG (1000 mm)	—	6809	6809	—	—	—	6,809
NG (762/610 mm)	—	2463	2463	—	—	—	2,463
Total	5242	39,995	45,237	14,365	4858	19,223	64,465

Almost all double/multiple track sections and electrified routes are with broad gauge. However, metre and narrow gauges are mostly single line and non-electrified.

Indian Railways has successfully absorbed the technology for production of high horse power and state-of-the art technology based on three-phase electric locomotive serial production of four-axle passenger locos type WAP-5, six-axle passenger locos type WAP-7, and freight locos type WAG-9 is going on at Chittaranjan Locomotives Works. 125 such locomotives are targeted to be manufactured during the Xth Plan.

Indian Railways has also embarked upon adoption of next generation Insulated Gate Bipolar Junction Transistor (IGBT) technology for traction propulsion system for three-phase electric locomotives. This technology would further improve the efficiency of propulsion system and also address the problem of future obsolescence of Gate Turnoff Thyristor (GTO) devices.

1.9.5 Dieselization and Diesel Traction

Diesel and electric locomotives are comparatively more efficient than steam locomotives. They provide greater hauling capacity, have better acceleration and deceleration, and are capable of higher speeds. They have less servicing needs, and, therefore, their availability for traffic is comparatively more. Thus, electrification and dieselization lead to considerable savings as well as improvement of line capacity.

Diesel traction started off on the ex North Western Railway prior to World War II in a very small way with the introduction of diesel shunting engines. Diesel traction got off to a real start with the use of diesel locomotives on the newly laid MG line to Gandhidham in 1955–56. Diesel locomotion then progressed rapidly after the introduction of BG main line locomotives on the heavy-density sections of the Eastern region in 1958–59 to ensure speedy and adequate transportation of raw materials to steel plants and finished products. Diesel traction has subsequently been extended to other high-density routes and routes situated away from coalfields. Today, diesel electric traction is significant in the motive power scene on Indian Railways.

New generation diesel locos

Indian Railways has successfully absorbed the technology of new generation 4,000 HP, three-phase, AC-AC diesel locomotives of General Motors. The serial production of freight locos type WDG-4 and passenger locos type WDP-4 is going on at Diesel Locomotive Works, Varanasi. These locomotives have a number of state-of-the-art features, such as computer-controlled brakes, AC-AC transmission, microprocessor-based control with self-diagnostic feature, 100 per cent dynamic braking, creep control for better adhesion, and so on. Passenger locomotives have a speed potential, of 160 kmph. Freight locomotives are capable of hauling 58 BOXN load on 1:150 gradient and short stretches of 1:100.

Traction Mix　Indian Railways uses a mix of electric and diesel traction. The share of traffic for passenger train km and gross tonnes km of freight traffic is given in Table 1.10.

Table 1.10　Share of traffic for passenger train and gross freight

Type of traffic	Diesel traction (%)	Electric traction (%)
Passenger train km by types of traction	50	50*
Gross freight tonnes km by types of traction	36	64

* *This consists of 38% by loco & 12% by EMU*

1.9.6 Rolling Stock

The fleet of Indian Railways rolling stock is presented in Table 1.11. Indian Railways is modernizing its rolling stock by inducting newer designs of fuel

Table 1.11 Details of Indian Railways rolling stock

Type of rolling stock	Number on 31 March 2011	Details of rolling stock
Locomotives	9,213	43 steam, 5137 diesel, and 4033 electric locos.
Total coaching stock	59,713	7334 EMU, 45,123 conventional coaches. 763 DMU and 6493 other coaching vehicles.
Total goods stock	229,381	26.5 % covered 52.9 % open 20.6 % other type brake vans departmental wagons.

efficient locomotives, high-speed coaches, and modern high-speed bogies for freight traffic. The Railways is also gradually replacing four-wheeler stock by bogie wagons having a higher pay load and speed potential for optimum utilization of line capacity. These include BCN, BTPN, and BOXN wagons.

1.9.7 Railway Safety

Indian Railways is striving to improve the standard of railway safety in an attempt to provide an accident-free travel. During 2010–11, there were 139 consequential train accidents causing death of 235 passengers, injury to 358 passengers and a sum of ₹ 58.6 million was paid as compensation.

Train accidents per million train km, an important index of safety on Indian Railways has been dropping from year to year. It has dropped from 5.50 in 1960–61 to 0.14 in 2010–11.

Out of 139 accidents which occurred on Indian Railways during 2010–11, 116 (83.5 per cent) were due to human failure, including 58(41.7 per cent) due to the failure of railway staff and 58(41.8 per cent) due to the failure of people other than the railway staff. Other accidents were due to equipment failure and other causes. Indian Railways have taken the following steps to improve safety:

Railway safety fund   A non-lapsable Special Railway Safety Fund (SRSF) of ₹ 170,000 million has been set up to wipe out the arrears of replacement of averaged assets like track, bridges, rolling stock and signalling equipment within a fixed timeframe of six years. The fund is operational since October 2001 and it has been positive to wipe off almost all the arrears of overaged assets.

New technological innovations   A number of new technological innovations has been made to prevent accidents even if there is human failure. Some of the devices are Train Protection and Warning System (TPWS), Anti-Collision Devices (ACD), Continuous Track Circuiting, Auxiliary Warning System (AWS), last vehicle check by axle counters and such other means.

Modernization of infrastructure   To modernize track, rolling stock, signalling equipment, etc., so as to improve safety standards.

Preventive maintenance   Steps have been adopted towards preventive mainte-nance so as to improve the reliability of assests.

1.10 IMPORTANT STATISTICS OF INDIAN RAILWAYS

Certain statistical data of Indian Railways regarding passenger and freight traffic, operating efficiency, railway employees and their training, social costs of Indian

Railways as well as engineering data is given in subsequent sections of this chapter. Technical and statistical details about various track items are given in various tables at the end of the chapter.

1.10.1 Passenger Traffic

Indian Railways carried 7,651 million passengers during 2010–11, some of the salient features of passenger traffic are given in Table 1.12.

Table 1.12 Details of passenger traffic on Indian Railways

Item	Unit	Non-suburban traffic	Suburban traffic	Total
Number of originating passengers	Million	3,590	4,061	7,651
Total passengers	Million	779,178	131,127	910,305
Revenue earned (₹)	Million	239,190	17,860	257,050

The analysis of passenger figures for the past 50 years leads to the following conclusions:
  (i) Over the years there has been a steady increase in passenger traffic output in terms of passengers kilometres.
 (ii) Since 1950–51 originating passengers have increased by 564 per cent and passenger kilometres by 1358 per cent.
(iii) Suburban as well as mail/express traffic have shown a higher rate of growth since 1950–51 than the overall average.

The special features of new passenger amenity services are as follows:

Onboard housekeeping service   Onboard house-keeping service scheme has been launched by the Railways to carry out frequent onboard cleaning of mail/express coaches through professional agencies.

Duronto trains   These have recently been introduced on Indian Railways. These superfast trains, with fully reserved accommodation, run as end-to-end non-stop service. A total of 14 such trains were introduced during 2009–10.

Air-conditioned double-decker coaches   Earnest efforts were put in and a prototype air-conditioned double-decker coach was conceived, designed, and manufactured during 2009–10.

Adarsh stations   To improve passenger amenity. 'Adarsh Stations' are being developed with basic facilities, such as drinking water, adequate toilets, catering services, waiting rooms and dormitories especially for female passengers, better signage and other basic facilities including escalators at selected stations. Of the 378 stations identified as Adarsh Stations, 333 stations have been developed. It is proposed to develop 94 more stations during 2010–11.

1.10.2 Freight Traffic

With the pace of economic growth Indian Railways is being called upon to provide much higher and stronger infrastructure support to the economy.

Some of the important statistics about revenue earning freight traffic for 1950–51 as well as those for 2010–11 are listed in Table 1.13.

Table 1.13 Details of revenue earning traffic on Indian Railways

Item	Unit	1950–51	2006–07	2008–09	2010–11
Revenue traffic	Million tonnes	73.2	727.7	833.4	921.8
Average lead	Kilometres	513	661.0	662.0	679.0
Net tonnes kilometres	Million	37,565	4,80,993	5,51,448	6,25,723
Freight train kilometres	Million	112	305.0	340.0	368.0

Note: There has been a substantial increase in revenue earning traffic in the recent past because of certain innovative steps like increase of loading in wagons (CC + 8 + 2). Details of these steps, which have resulted in a turnround of Indian Railways are given in Chapter 27.

1.10.3 Operating Efficiency Indices of Indian Railways* vis-a-vis those of Advanced Railways**

The operating efficiency indices of Indian Railways compare favourably even with some advanced railway systems of the world as presented in Table 1.14.

Table 1.14 Operating efficiency of world railways

Railway (2003–04)	Wagon km/ wagon/day	Net tonne km/ wagon/day	Net tonne km of wagon capacity/annum
French National Railways	73.4	1600	11,681
German Federal Railways	70.6	1115	9,139
Italian State Railways	53.4	962	8,006
Japanese National Railways	258.4	3481	36,713
Indian Railways (BG) (2003–04)	187.3	2570	42,237
(2010–11)	262.1	9247	57,953

1.10.4 Railway Employees

Indian Railways has a workforce of 13.68 million employees with a wage bill of ₹ 537,069 million during 2010–11 as per the break-up given in Table 1.15.

Table 1.15 Staff strength as on 31 March 2011 (in thousands)

Year	Groups A & B	Group C	Group D	Total	Expenditure on staff (₹ in million)
1950–51	2.3	223.5	687.8	913.6	1,138
1970–71	8.1	583.2	782.9	1,374.2	4,599
1990–91	14.3	891.4	746.1	1,651.8	51,663
2000–01	14.8	900.3	630.2	1,545.3	188,414
2010–11	16.8	1076.9	234.8	1328.5	537,069

* *Indian Railways Year Book* 2010–11.
** *International Railway Statistical Book.*

1.10.5 Training of Railway Employees

Indian Railways has developed their own facilities for conducting extensive training programmes for their officers and staff to enable them to improve upon their skill/abilities and equip them with the latest technological developments. Training for officers is organized in the following six centralized training institutes (CTIs).

   (i) The Railway Staff College, Vadodra.

   (ii) Indian Railways Institute of Civil Engineering, Pune.

   (iii) Indian Railways Institute of Signal and Telecommunication Engineering, Secunderabad.

   (iv) Indian Railways Institute of Mechanical Engineering, Jamalpur.

   (v) Indian Railways Institute of Electrical Engineering, Nasik.

   (vi) Indian Railways Institute of Transport Management, Lucknow.

Apart from probationary training, these institutes also cater to the various specialized training needs of serving officers. Railway Staff College, provides inputs in general management, strategic management and function-related areas for serving Railway officers. Other institutes conduct specialized technical training courses in respective functional areas.

Training needs of non-gazetted staff are being taken care of by over 200 training centres located on Indian Railways. Training has been made mandatory at different stages to make it more effiective for staff belonging to some of the safety categories.

1.10.6 Social Costs of Indian Railways

Indian Railways is a public utility undertaking of the Government of India. It does not have the freedom to adjust the freight and fare rates corresponding to increase in the prices of various inputs used. It also carries certain essential commodities, as well as passenger traffic, at rates which do not even cover the cost of movement. In addition, the traffic bound for flood effected and drought hit areas is carried at concessional rates. Certain unremunerative branch lines are also being operated purely in public interest. Such social obligations, which the Indian Railways has carried all along, are not usually borne by purely commercial undertakings. Losses incurred on this account are termed as 'Social Service Obligation'.

Net Social Service Obligation borne by Indian Railways in 2010–11 is assessed at about ₹ 1,51,730 million excluding staff welfare cost (₹ 35,550 million) and law and order cost (₹ 21,150 million).

The main items involved in this social service obligation are as follows:

1. Essential commodities carried at concessional rates below actual costs, such as food grains, salt, fodder, sugarcane, edible oils, and fruits and vegetables.
2. Subsidized suburban and some other passenger services.
3. Uneconomic branch lines totalling about 100 in number.
4. Freight concessions on relief materials and other miscellaneous social costs including expenditure on RPF and GRP.
5. Staff welfare measures, such as health, education, and subsidized housings.
6. Law and order costs.

Table 1.16   Important engineering data of Indian Railways as on 31 March 2011

Item	Quantity	Details
Total route kilometres	64,460	BG 55,188, MG 6,809, NG 2463
Running track kilometres	87,040	BG 77,347, MG 7219, NG 2474
Total track kilometres	113,994	BG 133,160, MG 8561, NG 2763
Number of bridges	133,160	Important–720, major–10,828 and minor–121,612
Number of level crossings	32,735	Manned–17560 and unmanned 14,896
Total land owned by Indian Railways	4,31,000 hectares	Track & structure 3,27,000, afforestation 45,000, vacant 46,000, others 13,000
Total engineering employees	346,119	24.61% out of total employees on Indian Railways
Length of welded rails on BG	—	63385 km of LWR 13655 km of SWR
Track upgradation		87.4 % of track as LWR; 97.0 % with PSC sleepers and with 52 kg/60 kg rails

1.11 VARIOUS UNDERTAKINGS UNDER MINISTRY OF RAILWAYS

This section elaborates on the various undertakings under the Ministry of Railways.

1.11.1 Rail India Technical and Economic Services Ltd

Rail India Techincal and Economic Services Ltd. (RITES), which is a Government of India Undertaking, provides consultancy services on all aspects of railways from concept to completion. RITES is closely linked with Indian Railways and is in a privileged position to draw freely upon the huge pool of experience, expertise, and technical know-how acquired over a century of operations on Indian Railways.

RITES is a multi-disciplinary, ISO 9001–2000 certified, consultancy organization in the field of transportation infrastructure and related technologies. It is a '*Mini Ratna*' company and provides consultancy services from concept to commissioning in the fields of railways, urban transport, urban development and urban engineering, roads and highways, airports, ropeways, inland waterways, ports and harbour, information technology and export packages of rolling stock and railway-related equipment. Its diversfied device packages, among others, include feasibility, design and detailed engineering, multi-modal transport studies, project management and construction supervision, quality assurance and management, materials management, workshop management, operation and maintenance, system engineering, economic and financial evolution, financing plan and privatization, property development, railway electrification, signalling and telecommunication, environment impact assessment, and training and human resources development.

Some of the assignments undertaken by RITES in the recent past include providing consultancy services, supply of locomotives, executions of construction works, management and operation of railways and similar services of different countries like Afghanistan, Botswana, Ethiopia, Indonesia, Mozambique, Nepal, Saudi Arabia, Senegal, Sri Lanka, and such other countries.

RITES, with recognition from multi-lateral funding agencies has experience in over 62 countries in Africa, Middle East, Latin America, South East Asia, UK, USA and Europe and provides a comprehensive array of services under a single roof.

1.11.2 Indian Railways Construction Company Ltd (IRCON)

Indian Railways Construction Company Ltd (IRCON), which is a public sector undertaking under the Ministry of Railways was incorporated in 1976 as a specialized agency to undertake construction of major railway projects both in India and abroad. It was set up with a view to channelize the export of construction services, technological know-how, and special skills gained by the Indian Railways in over 150 years. IRCON is in an ideal position to undertake the entire spectrum of construction activities concerning various aspects of railway disciplines, such as civil, mechanical, electrical, and signalling and telecommunications.

From being an exclusively railway construction company, IRCON diversified its activities to other sectors such as roads, highways, expressways, road bridges, flyovers, cable stayed bridges, mass rapid transit system, buildings, industrial and residential complexes, airports, and hangars. In 1993, IRCON included BOT, BOOT, BOLT projects, business relating to leasing, real estate, etc. In 1997, business relating to commercial operations of air transport was also included and extended. In 1999, it entered the field of telecommunication also and now provides a full range of telecom and IT services in India and many other countries.

Currently, IRCON is listed 128th amongst the top 225 international construction contractors, and is also amongst the top ten in mass transit and rail in the McGraw Hill publication Engineering News Record rankings.

IRCON, which is an ISO 9002 certified construction company has completed projects in Algeria, Angola, Bangladesh, Indonesia, Iraq, Jordan, Italy, Lebanon, Malaysia, Nepal, Nigeria, Saudi Arabia, Syria, Tanzania, Turkey, United Kingdom, and Zambia.

During 2010–11 IRCON declared a profit of ₹ 4,010 million with a net worth of ₹ 13,820 million.

1.11.3 Dedicated Freight Corridor Corporation of India Ltd

Dedicated Freight Corridor Corporation of India Ltd (DFCCIL), a special purpose organization was incorporated for planning and development, mobilization of financial resources, and construction, maintenance and operation of Dedicated Freight Corridor. DFCCIL was registered as a company under the Company's Act, 1956 on 30 October 2006.

Dedicated Freight Corridor, primarily a double-line corridor was established exclusively for running freight trains at a maximum permissible speed of 100 kmph to cover approximately 3,300 route kilometres, which includes the Eastern Corridor (about 1,800 route km) from Ludhiana to Sonnagar via Dadri proposed to be extended up to Dankuni and the Western Corridor (about 1,490 route km) from Jawaharlal Nehru port, Mumbai to Tughlakabad and Dadri alongwith interlinking of the two corridors at Dadri.

DFCCL has set up field units at Mumbai, Surat, Vadodra, Ahmedabad, Ajmer, Jaipur, Kanpur, Allahabad, and Ludhiana on western and eastern corridors under the control of chief project managers.

Construction started on both eastern and western corridors during 2008–09. Civil engineering work consisting of earthwork, bridges, etc., on approximately 105 km of Sonnagar–Mughalsarai section of the eastern Corridor and the construction work of 54 major bridges on Virar–Surat section on the western corridor have started. As per latest estimates, the western corridor will cost approximately ₹ 2,61,250 million which will be funded by JICA (Government of Japan). The Khurja–Mughalsarai section of the eastern corridor will cost about ₹ 1,29,400 million which will be funded by the World Bank.

1.11.4 Indian Railway Finance Corporation Limited

Indian Railways Finance Corporation Limited (IRFC) was set up as a public limited company in December 1986, with the sole objective of raising money from the market for meeting the development needs of Indian Railways. Funds are raised through issue of bonds, term loans from banks/financial institutions and through external commercial borrowings/export credit. The Department of Public Enterprises has rated it as 'Excellent' for the ninth year in succession.

IRFC has had a consistent profit-earning track record. It has so far paid the year ₹ 14,680 million as dividend to the government on the paid-up capital of ₹ 2,320 million

IRFC has earned a net profit of ₹ 4,850 million during the year 2010–11. The overhead-to-turnover ratio of the company is 0.12 per cent, which sets an industry benchmark.

1.11.5 Container Corporation of India Ltd

Container Corporation of India Ltd (CONCOR), a Mini Ratna (category-I) company, was set up in 1989 with the prime objectives of developing multimodal transport and logistics infrastructure to support the country's growing international trade as well as for transportation of domestic traffic in ISO containers by adopting latest technology and practice. It provides responsive, cost effective, efficient and reliable logistics, latest technology and practice. It provides responsive, cost effective, efficient and reliable logistics solutions to its customers through a synergy with its community partners, ensuring profitability and growth. CONCOR has been 'Excellent' in its MOU since 1992–93.

The mainstay of CONCOR's business is setting up and managing a network of rail-linked and road-based Inland container depots/container freight stations in the country for handling export/import domestic cargo and transportation of export-import-domestic containers by rail and road.

Today, CONCOR has a nation-wide network of 56 container terminals. It is today the most dominant provider of logistics services and support for both international and domestic cargo industry in India. The company operates CONRAJ and CONTRACK train services, to meet specific customer requirements, particularly of large companies between predetermined pairs of points with assured transit time for timely delivery of consignments at ports/destinations and for regaining market for piecemeal general goods cargo for rail.

Recently, CONCOR has successfully run double-stack container trains to Pipavan as well as coastal shipping services. It has also diversified its business by foraying into Cold Chain logistics.

CONCOR has paid a total dividend of ₹ 2,015 million for 2010–11 to the government on its paid-up capital. CONCOR had a turnover ₹ 38,850 million and Net Profit was ₹ 22.88 per cent of turnover during the year.

1.11.6 Konkan Railway Corporation Ltd (KRCL)

The Konkan Railway is the first railway project in the country to be executed on the BOT (Build, Operate and Transfer) principle. The company was formed with the participation of four states viz., Maharashtra, Goa, Karnataka, and Kerala, along with the Ministry of Railways. The construction work began in 1990 and the entire 760 km line was completed in 1998 and dedicated to the nation. on 1 May1998. Konkan Railway has considerably reduced the distance and travel time to southern India.

New technologies on Konkan Railways

The project of Anti-Collision Device Network (ACD) ACD, also known as the 'Raksha kavach' was launched on 20 Janaury 2004. ACD was installed for the first time, on the North East Frontier Railway on its entire BG route to prevent collisions well in time.

Sky Bus Metro It is an economic and eco-friendly mass rapid transportation soluton. The Sky Bus Test Track covering a length of 1.6 km has been constructed at Madgaon (Goa).

The Corporation has completed several projects, such as the tunnlling work on the Mumbai Pune Expressway, ventilation and lighting of Jawahar tunnel in Jammu and Kashmir and Owk tunnel in Andhra Pradesh. The Katra–Laole section of USBRL project in Jammu and Kashmir and 14 ROBs in Jharkhand are under construction.

1.11.7 Indian Railways Catering and Tourism Corporation Ltd

Indian Railways Catering and Tourism Corporation Ltd (IRCTC) is a new corporation under the Ministry of Railways. Some of the important projects undertaken by corporation are as follows:

Catering services The catering service managed by the Indian Railways: The Catering and Tourism Corporation Ltd (IRCTC) is operational on 14 Rajdhani, 13 Shatabdi, 16 Sampark Kranti, 16 Jan Shatabdi, and 196 Mail/Express trains as on 31 March 2009. Over the years, the corporation has awarded 349 contracts of AVM kiosks for sale stations.

Rail tourism IRTC's tourism portal, www.railtourismindia.com, is fast growing into a one-stop travel shop catering to all the travel and tourism needs of a customer. All the three Bharat Darshan rakes are now being operated and marketed by IRCTC.

Internet ticketing system During the year 2008–09, the number of railway tickets booked through the IRCTC website www.irct.co.in went up to 44.1 million. The value of the tickets booked has also gone up to ₹ 38,890 million during the year. In recognition of the excellent performance the website was awarded the 'Indian Express Indian Innovation Award 2008–09, and the 'CNBC Awaz-Special Commendation for redefining Indian Railways Award 2009.'

Packaged drinking water project (Rail Neer) The first Rail Neer plant of the corporation was inaugurated in May 2003 at Nangloi and another at Khagul, Danapur (Bihar) in February 2004. Good quality drinking water is supplied to all passengers and other rail users from these plants.

1.11.8 RailTel Corporation of India Ltd

RailTel Corporation of India Ltd, (RailTel) is one of the PSUs under the administrative control of the Ministry of Railways. RailTel was incorporated in September, 2000 with an authorized capital of ₹ 10,000 million.

The main objects of the Company, include building a nationwide telecom multimedia network for laying of Optical Fibre Cable (OFC) with a view to modernize Indian Railway's communication systems for safe and efficient train operation and to generate revenue through commercial exploitation of the system. RailTel is posed to emerge as a national level operator in the sector by using the railway right of way on about 63,000 route km by creating an Optic Fibre Cable (OFC) network.

RailTel has modernized train control and emergency communication system of Indian Railways by providing OFC network along the railway track and high band width Point of Presence (PoP) at more than 3,774 stations. It has also connected railways electronic telephone exchanges at E1 level to provide seamless railway STD service. The long-haul network coverage for STM 16 is 24,954 route kms with multiple things on common section of 17,533 route km.

RailTel has also set up cyber cafes/Internet kiosks to provide the facility of high-speed Internet browsing, e-mail, audio chatting, videoconferencing, IP telephony, etc.

1.11.9 Rail Vikas Nigam Ltd

Rail Vikas Nigam Limited (RVNL) is a special-purpose vehicle to execute projects with a view to strengthen the golden quadrilateral and its diagonals and to augment port connectivity under National Rail Vikas Yojna launched by the Government of India. Its main objectives are to undertake project development, financial resource mobilization, and execution of projects on a commercial format using largely non-budgetary funds.

Till 31 March 2011 RVNL had completed 22 projects covering 185 km of new line, 1571 km gauge conversion projects, 422 km doubling of railways lines, and 1152 km of railway electrification.

1.11.10 Pipavan Railway Corporation Ltd

Pipavan Railway Corporation Limited (PRCL), a joint venture company of the Ministry of Railways and Gujarat Pipavan Port Limited (GPPL) with an equal equity participation has been formed to execute the Surendranagar-Rajula-Pipavan Port Gauge Conversion/New line project. This is the first railway infrastructure executed through a private sector paticipation. PRCL has concessionary rights to construct, operate and maintain this project line for 33 years. PRCL is entitled to the rights, obligations and duties of Railway Administration enumerated in the Railways Act, 1989.

During 2010–11, PRCL, handled 2483 trains including 1302 container trains and transported 2.94 million tonnes of cargo yielding an apportioned earnings of ₹ 65,290 million. The first double stacked container train was run from Jaipur to Pipavan on 23 March 2007. Twelve pairs of passenger trains are also running on different sections of the Pipavan Railway.

PRCL has also been granted permission to run container trains on rail corridors serving the ports of Pipavan, Mundra, Chennai, Ennore, Vizag and their hinterlands.

1.11.11 Centre for Railway Information System

Centre for Railway Information System (CRIS), which is an autonomous body under the patronage of the Ministry of Railways, has been established as a non-profit-making organization. It has been entrusted with the design, development, ,and information of all major computer services on the railways. During the last few years CRIS has made substantial progress in Freight Operation Information System (FOIS), Passenger Reservation System (PRS), National Train Enquiry System (NTES), micro-processor based self printing ticketing Machines and Track Management System.

Unreserved Ticketing Systems (UTS)　The Unreserved Ticketing System (UTS) has also been computerized by CRIS and expanded to cover 3614 stations on Indian Railways to enable passengers to puchase unreserved tickets in advance from the date of journey from any booking station to any station during the year. UTS has won the Prime Minister's award for Excellence in Public Administration in April 2008.

Training　CRIS has launched a scheme to provide high quality training to Indian Railway personnel at their door steps on demand and through regular courses at four major metropolises, viz., Delhi, Mumbai, Kolkata and Chennai to build an IT enabled workforce to carry forward IT initiatives.

CRIS has in recent past taken many important projects at few important locations, such as Track Management system (TMS) introduced in 2008–09 and implemented in many divisions on Indian Railways, Parcel Management System (PMS), Rake Management System, Integrated Coaching Management System (ICMS), Coaching Informations Systems (COIS), and Crew Management Systems (CMS).

1.11.12 Rail Land Development Authority

Rail Land Development Authority (RLDA) was set up in 2005 through an amendment of the Indian Railways Act 1989 for development of vacant railway land for commercial use for the purpose of generating revenue by non-tariff measures for Indian Railways. Railway land sites, not required for operational purposes for future expansion, are identified by Zonal Railways and entrusted to

RLDA for commercial development. The Authority initially engaged a consultant to ascertain the suitability and potential for commercial development of the site and thereafter based on the feasibility report, identify a suitable development model for its commercial development through an open and transparent bidding process to generate maximum revenue for the railways.

1.11.13 Indian Railway Welfare Organisation

An autonomous body under the patronage of Ministry of Railways called 'Indian Railway Welfare Organisation (IRWO) was registered on 25 September 1989 under the 'Societies Registration Act' for meeting specific needs of housing for serving retired railway employees with its headquarters at New Delhi. IRWO has been constructing houses in different cities and towns all over the country on a self-financing basis for serving retired railway employees purely as a welfare activity on 'no profit no loss basis'.

In the last few years, IRWO has already acquired land at about 20 places and has announced group housing schemes at Noida, Gurgaon, Gorakhpur, Kolkata, Madras, Indrapuram (Ghaziabad), Hyderabad, Chandigarh and few other places. They have already completed the housing projects, at Noida, Gurgaon, Gorakhpur, Hyderabad, Kolkata, Chandigarh, Chennai, and Mumbai in record time and handed over the flats to owners.

Housing schemes have also been progressing or have commenced at Allahabad, Asansol, Podanur (Coimbatore), Gurgaon Phase IV, kalkata Phase III, Lucknow Phase II, Visakhapatnam Phase III, Vasundra (Ghaziabad) Phase II, Chennai Phase II, Lucknow Phase III, Sonepat/Kundli (Haryana), Mumbai Phase II, Meerut, Hyderabad, Ajmer, Moradabad, Zerakpur (Chandigarh).

1.12 RECORD OF NOTABLE FACTS ON INDIAN RAILWAYS

First railway trains in Asia

Broad gauge 34 km track between Boribunder and Thane was opened to traffic on 16 April 1853.

Metre gauge The 84 km track between Delhi and Rewari and 12.3 km track (Faruknagar salt branch) of the Rajputana–Malwa Railway were opened for traffic on 14 February 1873.

Narrow gauge First line opened in 1862 near Baroda by Gaekwad Baroda State Railway.

Only rack-railway The Nilgiri Mountain Railway uses the rack-railway system between Mettupalayam and Udhagamandalam covering a distance of 46 km.

First ladies special train The first ladies special train between Churchgate and Borivali station.

First woman station master Rinku Sinha Roy joined the Eastern Railway in August 1994 in the Dum Dum Cantt station of the Sealdah Bangaon section.

First woman engine driver Surekha Bhonsle of Mumbai joined the Indian Railways as trainee assistant driver in 1990 when she was 24.

Fastest train   The Shatabdi Express touches 150 km per hour on New Delhi–Bhopal route. The tracks in this route can withstand speed up to 160 km per hour.

Slowest train   The Mettupalayam–Udhagamandalam train of the Nilgiri Railway on NG is the slowest train, covering 76 km in 7 hours and 25 minutes, at an average speed of 10.24 km per hour.

Oldest working locomotive   The Fairy Queen built in 1855 by Kitson, Thompson and Hewitson of UK, is the oldest working locomotive on Indian Railways.

Railway station at the highest attitude   Ghoom station on the Darjeeling Himalayan Railway (gauge 610 mm) is 2,258 m high.

Railway section with maximum tunnels and bridges   The Kalka–Shimla route has 102 tunnels on one section, which is less than 100 km. It has 809 bridges and 919 curves which cover about 30 per cent of the length of the track.

Longest platform   The platform at Kharagpur in West Bengal is 833 m (2.733 ft) long.

Longest tunnel   Kharbude tunnel on Konkan Railway connecting Mumbai with Goa is 645 m long.

Longest bridge   The Dehri-on-Sone railway bridge over the river Sone near Sasaram, Bihar, on the Calcutta–Delhi mainline is 3.06 km (10.052 ft.) long and has 93 spans of 105 ft each. It was opened for traffic on 27 February 1900.

Tallest viaduct in Asia   The Panvel Nadi Viaduct on Konkan Railway is 64 m high.

Deepest and longest span railway bridge   Chenab bridge on Jammu–Kashmir Rail link has a height of 359 m (bed level to formation level). Its total length is 1315 m with the main central arch of 467 m long.

Longest passenger train   The New Delhi–Allahabad Prayag Raj Express with 24/26 coaches is the longest passenger train.

Ticket to the millennium   Fakhruddin Takulla of Mumbai purchased on 15 July 1973 a Rajdhani ticket (No. 35582) for 19 January 2000 to witness the golden jubilee celebrations of the Republic Day in 2000.

Largest railway yard   Mughalsarai is the largest Railway Yard on Indian Railways.

Oldest working steam locomotive   Fairy Queen service between Delhi Cantt and Alwar (Guinness certified as the oldest 152 years) working steam locomotive.

UNESCO World Heritage Site   The Darjeeling Himalayan Railway (DHR) from New Jalpaiguri to Darjeeling, now in its 128th year, and a UNESCO World Heritage Site.

Shimla–Kanda Ghat on Kalka–Shimla Railway (KSR)   The Shimla–Kalka section is now a UNESCO World Heritage Site.

Neral–Matheran on Matheran Light Railway (MLR)   It is proposed to enscript this railway as a World Heritage Site.

SUMMARY

Indian Railways has played an integral role in the social and economic development of the country. The Railways has been performing the dual role of functioning as a commercial undertaking and a provider of public utility service. The key challenge for the Indian Railways in the passenger traffic segment is to maintain its social obligation on low-price service lines while at the same time increasing both capacity and utilization in the upper class services, through a strategy for higher growth in traffic as well as appropriate tariff balancing. The administrative costs of the Railways have been increasing rapidly. With about 1.41 million employees, Indian Railways is the largest employer among public sector undertakings in the country.

REVIEW QUESTIONS

1. What are the different modes of transport? Compare rail transport with road transport, listing advantages and disadvantages of both.
2. What are the objectives of the long-term corporate plans of Indian Railways? How do they compare with the existing level of achievements?
3. Give the complete classification of a railway line as adopted by Indian Railways.
4. Discuss the organizational structure of Indian Railways. How are the duties distributed in a typical zonal railway headquarters?
5. Discuss the merits and demerits of diesel and electric traction.
6. What are the social obligations of Indian Railways? Discuss the main components of Social Service Obligation.
7. What is the object of 'Indian Railways Vision 2020'. Discuss the important goals as listed in the document.
8. 'Safety on Indian Railways has improved in the past four decades.' Comment.
9. What do you understand by the 'social cost of Indian Railways'? Explain.

Choose the correct answer from the choices given. *

10. The first railway line in India was opened for traffic between:
    (a) Madras and Bombay   (b) Agra and Bombay
    (c) Bombay and Thane   (d) Khandwa and Indore
11. The total number of administrative zones on Indian railways is:
    (a) 5   (b) 16
    (c) 11   (d) 9
12. Metre gauge was introduced on Indian Railways in:
    (a) 1853   (b) 1865
    (c) 1871   (d) 1889
13. Toilets in third-class railway coaches on Indian Railways were introduced in:
    (a) 1870   (b) 1882
    (c) 1891   (d) 1900

* Answers to multiple choice questions for all chapters are provided in Appendix A.

14. The first Rajdhani train at 120 kmph was introduced in:
    (a) 1965   (b) 1969
    (c) 1975   (d) 1980
15. The longest railway platform on Indian Railways is at:
    (a) Sonepur   (b) Kharagpur
    (c) Mumbai VT   (d) Itarsi
16. The longest railway bridge on Indian Railways is:
    (a) Sone bridge at Dehri-on-Sone
    (b) Yamuna bridge at Kalpi
    (c) Ganga bridge near Patna
17. Railways originated in:
    (a) USA   (b) USSR
    (c) Germany   (d) England
18. The first rail on Indian Railways was opened in:
    (a) 1825   (b) 1853
    (c) 1854   (d) none of these
19. The headquarters of South Central Railway is at:
    (a) Maligaon   (b) Secunderabad
    (c) Chennai   (d) Kolkata
20. The headquarters of East Central Railway is at:
    (a) Hazipur   (b) Patna
    (c) Bhubaneshwar   (d) none of these

Railway Track Gauge

INTRODUCTION

Gauge is defined as the minimum distance between two rails. Indian Railways follows this standard practice and the gauge is measured as the clear minimum distance between the running faces of two rails as shown in Fig. 2.1.

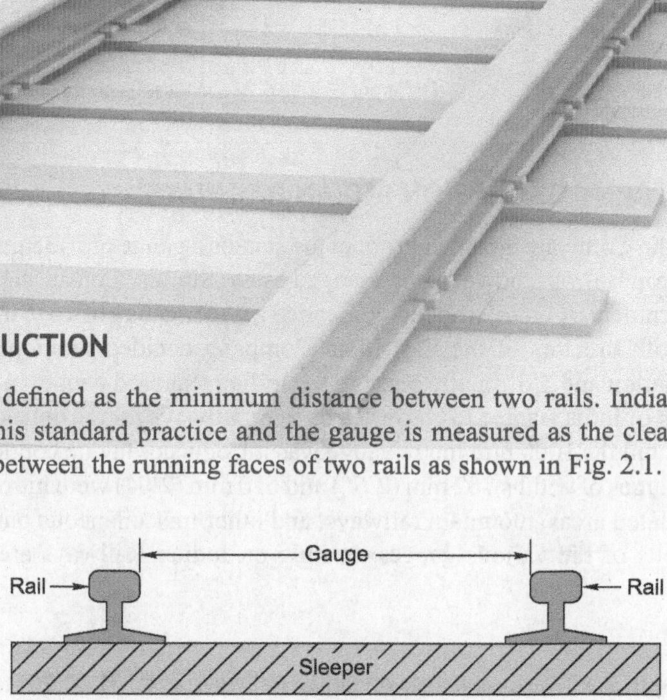

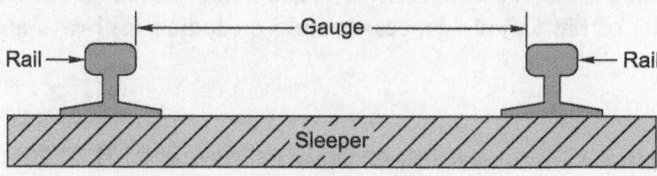

Fig. 2.1 Gauge

In European countries, the gauge is measured between the inner faces of two rails at a point 14 mm below the top of the rail. This chapter describes the different gauge widths prevalent in India and other countries. It also discusses the problems and implications of a multiple-gauge system as adopted in India.

2.1 GAUGES ON WORLD RAILWAYS

Various gauges have been adopted by different railways in the world due to historical and other considerations. Initially British Railways had adopted a gauge of 1525 mm (5 feet), but the wheel flanges at that time were on the outside of the rails. Subsequently, in order to guide the wheels better, the flanges were made inside the rails. The gauge then became 1435 mm (4′8.5″), as at that time the width of the rail at the top was 45 mm (1.75″). The 1435 mm gauge became the standard on most European Railways. The approximate proportions of various gauges on world railways are given in Table 2.1.

Table 2.1 Various gauges on world railways

Type of gauge	Gauge (mm)	Gauge (feet)	% of total length	Countries
Standard gauge	1435	4′8.5″	62	England, USA, Canada, Turkey, Persia, and China
Broad gauge	1676	5′6″	6	India, Pakistan, Sri Lanka, Brazil, Argentina
Broad gauge	1524	5′0″	9	Russia, Finland
Cape gauge	1067	3′6″	8	Africa, Japan, Java, Australia, and New Zealand
Metre gauge	1000	3′3.5″	9	India, France, Switzerland, and Argentina
23 various other gauges	Different gauges	Different gauges	6	Various countries

2.2 DIFFERENT GAUGES ON INDIAN RAILWAYS

The East India Company intended to adopt the standard gauge of 1435 mm in India also. This proposal was, however, challenged by W. Simms, Consulting Engineer to the Government of India, who recommended a wider gauge of 1676 mm (5′6″). The Court of Directors of the East India Company decided to adopt Simms's recommendation and 5′6″ finally became the Indian standard gauge. In 1871, the Government of India wanted to construct cheaper railways for the development of the country and the 1000 mm metre gauge was introduced. In due course of time, two more gauges of widths 762 mm (2′6″) and 610 mm (2′0″) were introduced for thinly populated areas, mountain railways, and other miscellaneous purposes.

The details of the various gauges existing on Indian Railways are given in Table 2.2.

Table 2.2 Various gauges on Indian Railways as on 31.03.2011

Name of gauge	Width (mm)	Width (feet)	Route (km)	% of route (km)
Broad gauge (BG)	1676	5′6″	55,188	85.6
Metre gauge (MG)	1000	3′3.37″	6809	10.6
Narrow gauge (NG)	762	2′6″	2463	3.8
	610	2′0″		
Total all gauges	Different gauges		64,460	100

2.3 CHOICE OF GAUGE

The choice of gauge is very limited, as each country has a fixed gauge and all new railway lines are constructed to adhere to the standard gauge. However, the following factors theoretically influence the choice of the gauge:

Cost considerations

There is only a marginal increase in the cost of the track if a wider gauge is adopted. In this connection, the following points are important.

(a) There is a proportional increase in the cost of acquisition of land, earthwork, rails, sleepers, ballast, and other track items when constructing a wider gauge.
(b) The cost of building bridges, culverts, and tunnels increases only marginally due to a wider gauge.
(c) The cost of constructing station buildings, platforms, staff quarters, level crossings, signals, etc., associated with the railway network is more or less the same for all gauges.
(d) The cost of rolling stock is independent of the gauge of the track for carrying the same volume of traffic.

Traffic considerations

The volume of traffic depends upon the size of wagons and the speed and hauling capacity of the train. Thus, the following points need to be considered.
(a) As a wider gauge can carry larger wagons and coaches, it can theoretically carry more traffic.
(b) A wider gauge has a greater potential at higher speeds, because speed is a function of the diameter of the wheel, which in turn is limited by the width of the gauge. As a thumb rule, diameter of the wheel is kept 75 per cent of gauge width.
(c) The type of traction and signalling equipment required are independent of the gauge.

Physical features of the country

It is possible to adopt steeper gradients and sharper curves for a narrow gauge as compared to a wider gauge.

Uniformity of gauge

The existence of a uniform gauge in a country enables smooth, speedy, and efficient operation of trains. Therefore, a single gauge should be adopted irrespective of the minor advantages of a wider gauge and the few limitations of a narrower gauge.

2.4 PROBLEMS CAUSED BY MULTI-GAUGE SYSTEM

The need for uniformity of gauge has been recognized by all the advanced countries of the world. A number of problems have cropped up in the operation of the Indian Railways because of the multi-gauge system (use of three gauges). The ill effects of change of gauge (more popularly known as *break of gauge*) are numerous; some of these are enumerated here.

Inconvenience to passengers

Due to change of gauge, passengers have to change trains mid-journey alongwith their luggage, which causes inconvenience such as the following:
(a) Climbing stairs and crossing bridges
(b) Getting seats in the compartments of the later trains
(c) Missing connections with the later trains in case the earlier train is late
(d) Harassment caused by porters
(e) Transporting luggage from one platform to another.
(f) Uncertainty and delay in reaching the destination

Difficulty in trans-shipment of goods

Goods have to be trans-shipped at the point where the change of gauge takes place. This causes the following problems:

(a) Damage to goods during trans-shipment

(b) Considerable delay in receipt of goods at the destination

(c) Theft or misplacement of goods during trans-shipment and the subsequent claims

(d) Non-availability of adequate and specialized trans-shipment labour and staff, particularly during strikes

Inefficient use of rolling stock

As wagons have to move empty in the direction of the trans-shipment point, they are not fully utilized. Similarly, idle wagons or engines of one gauge cannot be moved on another gauge.

Hindrance to fast movement of goods and passenger traffic

Due to change in the gauge, traffic cannot move fast which becomes a major problem particularly during emergencies such as war, floods, and accidents.

Additional facilities at stations and yards

Costly sheds and additional facilities need to be provided for handling the large volume of goods at trans-shipment points. Further, duplicate equipment and facilities such as yards and platforms need to be provided for both gauges at trans-shipment points.

Difficulties in balanced economic growth

The difference in gauge also leads to unbalanced economic growth. This happens because industries set up near MG/NG stations cannot send their goods economically and efficiently to areas being served by BG stations.

Difficulties in future gauge conversion projects

Gauge conversion is quite difficult, as it requires enormous effort to widen existing tracks. Widening the gauge involves heavy civil engineering work such as widening of the embankment, bridges and tunnels, as well as tracks; additionally, a wider rolling stock is also required. During the gauge conversion period, there are operational problems as well, since the traffic has to be slowed down and even suspended for a certain period in order to execute the work.

2.5 UNI-GAUGE POLICY OF INDIAN RAILWAYS

The problems caused by a multi-gauge system in a country have been discussed in the previous section. The multi-gauge system is not only costly and cumbersome but also causes serious bottlenecks in the operation of the Railways and hinders the balanced development of the country. Indian Railways therefore took the bold decision in 1992 of getting rid of the multi-gauge system and following the uni-gauge policy of adopting the broad gauge (1676 mm) uniformly.

2.5.1 Benefits of Adopting BG (1676 mm) as the Uniform Gauge

The uni-gauge system will be highly beneficial to rail users, the railway administration, as well as to the nation. Following are the advantages of a uni-gauge system:

No transport bottlenecks

There will be no transport bottlenecks after a uniform gauge is adopted and this will lead to improved operational efficiency resulting in fast movement of goods and passengers.

No trans-shipment hazards

There will be no hazards of trans-shipment and as such no delays, no damage to goods, no inconvenience to passengers of transfer from one train to another train.

Provisions of alternate routes

Through a uni-gauge policy, alternate routes will be available for free movement of traffic and there will be less pressure on the existing BG network. This is expected to result in long-haul road traffic reverting to the railways.

Better turnaround

There will be a better turnaround of wagons and locomotives, and their usage will improve the operating ratio of the railway system as a whole. As a result the community will be benefited immensely.

Improved utilization of track

There will be improved utilization of tracks and reduction in the operating expenses of the railway.

Balanced economic growth

The areas currently served by the MG will receive an additional fillip, leading to the removal of regional disparities and balancing economic growth.

No multiple tracking works

The uni-gauge project will eliminate the need for certain traffic facilities and multiple tracking works, which will offset the cost of gauge conversions to a certain extent.

Better transport infrastructure

Some of the areas served by the MG have the potential of becoming highly industrialized; skilled manpower is also available. The uni-gauge policy will help in providing these areas a better transportation infrastructure.

Boosting investor's confidence

With the liberalization of the economic policy, the uni-gauge projects of Indian Railways have come to play a significant role. This will help in boosting the investors' confidence that their goods will be distributed throughout the country in time and without any hindrance. This will also help in setting up industries in areas not yet exploited because of the lack of infrastructure facilities.

2.5.2 Planning of Uni-gauge Projects

The gauge-conversion programme has been accelerated on Indian Railways since 1992. In the eighth Plan (1993–97) itself, the progress achieved in gauge-conversion projects in five years was more than the total progress made in the past 45 years. The progress of gauge-conversion projects is briefly given in Table 2.3.

Table 2.3 Progress of gauge-conversion projects

Year	Progress in gauge conversion (kms)	Remarks
1947–1992	2500	Approx. figure
1993–1997	6897	Actual
1998–2004	3787	Actual
2005–2011	6564	Actual

The current position is that the gauge-conversion project still pending on Indian Railways is 8855 kms which is likely to be completed in next five years. Execution of a gauge conversion project is quite a tricky job and lot of planning is to be done for the same. For details refer Chapter 28.

2.6 LOADING GAUGE

Loading gauge represents the maximum width and height to which a rolling stock, namely a locomotive, coach, or wagon, can be built or loaded. Sometimes, a loading gauge is also used for testing loaded and empty vehicles as per the maximum moving dimensions prescribed for the section. On Indian Railways, the maximum height and width of rolling stock prescribed as per the loading gauge are given in Table 2.4.

Table 2.4 Maximum dimensions of rolling stock on Indian Railways

Gauge	Maximum height of rolling stock	Maximum width of rolling stock
BG	4140 mm (13'7")	3250 mm (10'8")
MG	3455 mm (11'4")	2745 mm (9'0")

In order to ensure that the wagons are not overloaded, a physical barrier is made by constructing a structure as per the profile of the loading gauge (see Fig. 2.2). This structure consists of a vertical post with an arm from which a steel arc is suspended from the top. The function of this structure is to ensure that the topmost and the widest portion of the load will clear all structures such as bridges and tunnels, along the route.

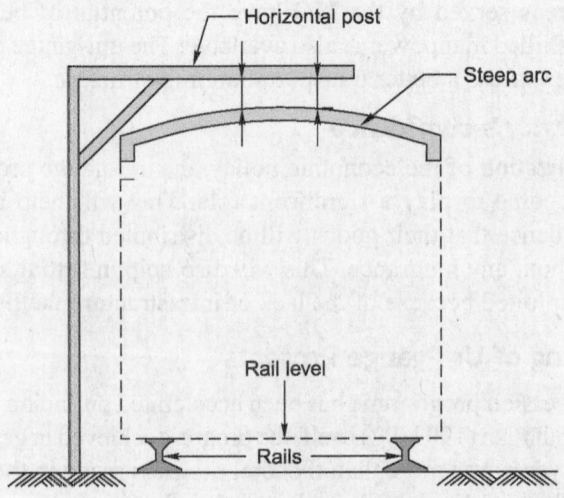

Fig. 2.2 Loading gauge

2.7 CONSTRUCTION GAUGE

The construction gauge is decided by adding the necessary clearance to the loading gauge so that vehicles can move safely at the prescribed speed without any infringement. The various fixed structures on railway lines such as bridges, tunnels, and platform sheds are built in accordance with the construction gauge so that the sides and top remain clear of the loading gauge.

SUMMARY

Three types of track gauges have been adopted on Indian Railways. The basic consideration behind the adoption of the metre gauge and narrow gauge was to provide access to undeveloped areas with low cost of construction. The multiple-gauge system has caused many problems and caused serious bottlenecks in the operation of the railways. The work on gauge conversion has been accelerated on Indian Railways since 1992. The uni-gauge system will be highly beneficial to rail users, the railway administration, and the nation.

REVIEW QUESTIONS

1. Define gauge problems with special reference to Indian Railways and bring out the effects of variations in the width of the gauge.
2. Why is it desirable to have, as far as possible, a uniform gauge for the railway network of a country?
3. List out the various gauges prevailing in India with their gauge widths. What factors govern the selection of a suitable gauge? Discuss.
4. What is the uni-gauge policy of Indian Railways? Describe the benefits of the uni-gauge system.
5. What is a loading gauge? How is it different from the construction gauge?
6. How many gauges exist on Indian Railways? Give their widths and route kilometres.
7. Write short notes on the following:
   (a) Break of gauge   (b) Standard gauge
   (c) Cape gauge   (d) Metre gauge

Choose the correct answer from the choices given.

8. The standard gauge has a width of:
   (a) 1365 mm   (b) 1435 mm
   (c) 1525 mm   (d) 1676 mm
9. Width of NG rail used in India is:
   (a) 762 mm   (b) 610 mm
   (c) both of (a) and (b)   (d) 877 mm
10. Gauge conversion refers to:
   (a) converting MG to BG   (b) converting NG to MG
   (c) converting NG to BG   (d) converting MG/NG to BG

Alignment of Railway Lines

INTRODUCTION

Alignment of railway line refers to the direction and position given to the centre line of the railway track on the ground in the horizontal and vertical planes. Horizontal alignment refers to the direction of the railway track in the plan including the straight path and the curves it follows. Vertical alignment refers to the direction it follows in a vertical plane including the level track, gradients, and vertical curves. The various factors affecting alignment, and the types of track alignments and their suitability in different terrains are discussed in this chapter.

3.1 IMPORTANCE OF GOOD ALIGNMENT

A new railway line should be aligned carefully after proper consideration, as improper alignment may ultimately prove to be more costly and may not be able to fulfil the desired objectives. Railway line constructions are capital-intensive projects; once constructed, it is very difficult to change the alignment of a railway line because of the costly structures involved, difficulty in getting additional land for the new alignment, and such other considerations.

3.2 BASIC REQUIREMENTS OF AN IDEAL ALIGNMENT

The ideal alignment of a railway line should meet the following requirements.

Purpose of the new railway line

The alignment of a new railway line should serve the basic purpose for which the railway line is being constructed. The purpose of constructing a new railway line may be one or more of the following:

(a) Strategic or political consideration   (Example Kashmir Rail Link)

(b) Development of backward areas   (Example Assam Rail link for development of Assam)

(c) Shortening of existing rail lines   (Konkan Rail-link was executed with a double purpose of shortening the existing rail route and also for development of East-Coast area)

(d) Connecting the trade centres

Integrated development

The new railway line should fit in with the general planning and form a part of the integrated development of the country.

Economic considerations

The construction of the railway line should be as economical as possible. The following aspects require special attention.

Shortest route   It is desirable to have the shortest and most direct route between the connecting points. The shorter is the length of the railway line, the lower would be the cost of its construction, maintenance, and operation. There can, however, be other practical considerations that can lead to deviation from the shortest route.

Construction and maintenance cost   The alignment of the line should be so chosen that the construction cost is minimum. This can be achieved by a balanced cut and fill of earthwork, minimizing rock cutting and drainage crossings by locating the alignment on watershed lines, and such other technical considerations. Maintenance costs can be reduced by avoiding steep gradients and sharp curves, which cause heavy wear and tear of rails and rolling stock.

Minimum operational expenses   The alignment should be such that the operational or transportation expenses are minimum. This can be done by maximizing the haulage of goods with the given power of the locomotive and traction mix. This can be achieved by providing easy gradients, avoiding sharp curves, and adopting a direct route.

Maximum safety and comfort

The alignment should be such that it provides maximum safety and comfort to the travelling public. This can be achieved by designing curves with proper transition lengths, providing vertical curves for gradients, and incorporating other such technical features.

Aesthetic considerations

While deciding the alignment, aesthetic aspects should also be given due weightage. A journey by rail should be visually pleasing. This can be done by avoiding views of borrow pits and passing the alignment through natural and beautiful surroundings with scenic beauty.

3.3 SELECTION OF A GOOD ALIGNMENT

Normally, a direct straight route connecting two points is the shortest and most economical route for a railway line, but there are practical problems and other compulsions which necessitate deviation from this route. The various factors involved in the selection of a good alignment for a railway line are as follows:

Choice of gauge

The gauge can be a BG (1676 mm), an MG (1000 mm), or even an NG (762 mm). All these gauges have been discussed in detail in Chapter 2. As per the latest policy of the Government of India, new railway lines are constructed on BG only.

Obligatory or controlling points

These are the points through which the railway line must pass due to political, strategic, and commercial reasons as well as due to technical considerations. The following are the obligatory or controlling points:

Important cities and towns   These are mostly intermediate important towns, cities, or places which are of commercial, strategic, or political importance.

Major bridge sites and river crossings   The construction of major bridges for large rivers is very expensive and suitable bridge sites become obligatory points for a good alignment.

Existing passes and saddles in hilly terrain   Existing passes and saddles should be identified for crossing a hilly terrain in order to avoid deep cuttings and high banks.

Sites for tunnels   The option of a tunnel in place of a deep cut in a hilly terrain is better from an economical viewpoint. The exact site of such a tunnel becomes an obligatory point.

Topography of the country

The alignment of a new railway line depends upon the topography of the country it traverses. The following few situations may arise:

Plane alignment   When the topography is plane and flat, the alignment presents no problems and can pass through obligatory points and yet have very easy gradients.

Valley alignment   The alignment of a railway line in a valley is simple and does not pose any problem. If two control points lie in the same valley, a straight line is provided between these points with a uniform gradient.

Cross-country alignment   The alignment of a railway line in such terrain crosses the watersheds of two or more streams of varied sizes. As the levels vary in cross-country, the gradients are steep and varying and there are sags and summits. The controlling or obligatory points for cross-country alignment may be the lowest saddles or tunnels. It may be desirable to align the line for some length along the watersheds so that some of the drainage crossings may be avoided.

Mountain alignment   The levels in mountains vary considerably, and if normal alignment is adopted, the grades would become too steep, much more than the ruling gradient (allowable gradient). In order to remain within the ruling gradient, the length of the railway line is increased artificially by the 'development process'. The following are the standard methods for the development technique:

Zigzag line method In this method, the railway line traverses in a zigzag alignment (Fig. 3.1) and follows a convenient side slope which is at nearly right angles to the general direction of the alignment. The line then turns about 180° in a horseshoe pattern to gain height.

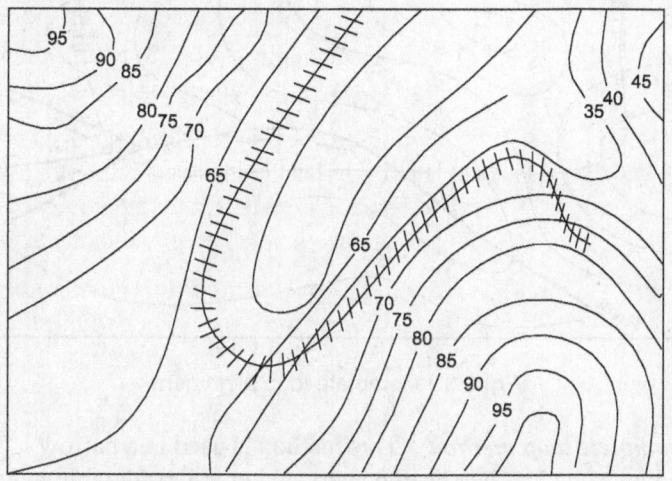

Fig. 3.1 Zigzag line alignment

Switch-back method In the case of steep side slopes, a considerable gain in elevation is accomplished by the switch-back method (Fig. 3.2). This method involves a reversal of direction achieved by a switch, for which the train has to necessarily stop. The switch point is normally located in a station yard. In Fig. 3.2, A and B are two switches and A_1 and B_1 are two buffer stops. A train coming from D will stop at B_1 and move in back gear to line BA. It will stop at A_1 again and then follow the line AC.

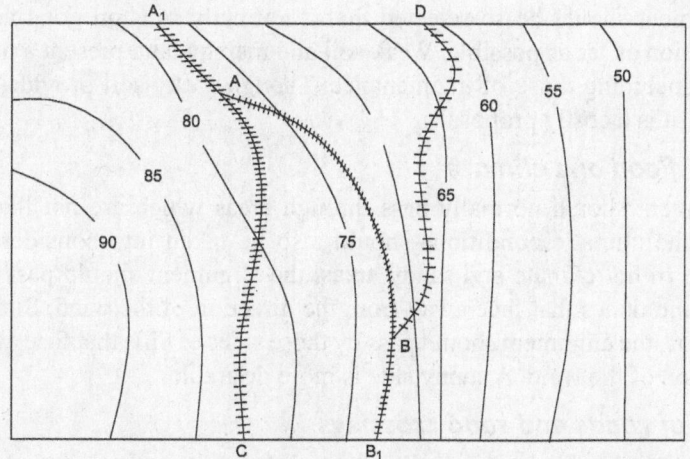

Fig. 3.2 Switch-back alignment

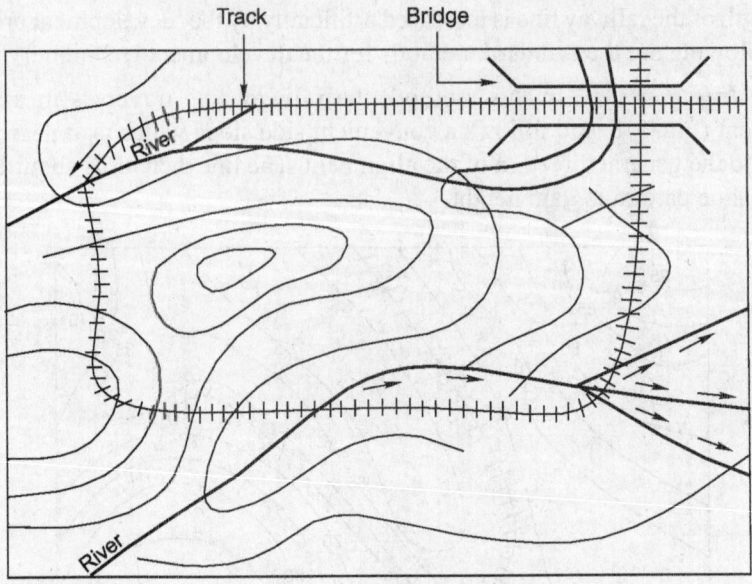

Fig. 3.3 Complete loop alignment

Spiral or complete loop method This method is used in a narrow valley where a small bridge or viaduct has been constructed at a considerable height to span the valley (Fig. 3.3). In this case, normally a complete loop of the railway line is constructed, so that the line crosses the same point a second time at a height through a flyover or a tunnel.

Geometrical standards

Geometrical standards should be so adopted as to economize as much as possible as well as provide safety and comfort to passengers. This can be done by adopting gradients and curves within permissible limits. Transition as well as vertical curves should be used to provide better comfort and safety.

Geological formation

The alignment should be so selected that it normally runs on good and stable soil formation as far as possible. Weak soil and marshy land present a number of problems including those of maintenance. Though rocky soil provides a stable formation, it is a costly proposal.

Effect of flood and climate

The alignment should normally pass through areas which are not likely to be flooded. The climatic conditions should also be taken into consideration for alignment. In hot climate and sandy areas, the alignment should pass by those sides of sand dunes that face away from the direction of the wind. Similarly, in cold regions, the alignment should pass by those sides of hills that face away from the direction of the wind. A sunny side is more desirable.

Position of roads and road crossings

A railway line should cross a road at right angles so as to have a perpendicular level crossing and avoid accidents.

Proximity of labour and material

The availability and proximity of local labour and good and cheap building material should also be considered when deciding the alignment.

Location of railway stations and yards

Railway stations and yards should be located on level stretches of land, preferably on the outskirts of a town or village so as to have enough area for the free flow of traffic.

Religious and historical monuments

The alignment should avoid religious and historical monuments, as it is normally not possible to dismantle these buildings.

Cost considerations

The alignment should be such that the cost of construction of the railway line is as low as possible. Not only the initial cost of construction but also the maintenance cost should be as low as possible. For this purpose, the alignment should be as straight as possible, with least earthwork, and should pass through a terrain with good soil.

Traffic considerations

The alignment should be so selected that it attracts maximum traffic. In this context, traffic centres should be well planned, so that the railway line is well patronized and the gross revenue arising out of traffic receipts is as high as possible.

Economic considerations

Keeping in mind the various considerations, it should be ensured that the alignment is overall economical. For this purpose, various alternate alignments are considered and the most economical one, which is cost-effective and gives the maximum returns is chosen.

The maximum annual (γ) return is calculated by the formula

$$\gamma = \frac{R - E}{I}$$

where R is the gross revenue earned by the railway line and E denotes the annual running expenses.

It may be noted here that R depends upon the route that proves to be advantageous when taking traffic into consideration and, therefore, should be given due weightage. The other way to maximize the annual return is to have sound and economical construction work so as to reduce the annual running expenses. A suitable balance has to be achieved between construction cost and operating expenses.

Political considerations

The alignment should take into account political considerations also. It should not enter foreign soil and should preferably be away from common border areas.

3.4 MOUNTAIN RAILWAYS

There are few hills or mountains that have been connected by railway lines. Some of the important features of these mountain railways are the following:

Gradients The gradients of mountain tracks are very steep. Normally tracks with gradients of 3 per cent or more are considered mountain tracks.

Gauge Normally narrow gauges with gauge widths of 762 mm or 610 mm are adopted for mountain railways.

Curvature The curvature of mountain tracks is very sharp; curves of up to 40° are normally adopted.

Alignment Mountain alignment is quite zigzag and not straight in order to gain heights easily. The type of alignments commonly followed are zigzag, switchback, and spiral.

Purpose Mountain railways have mostly been constructed for tourist traffic. In some cases, they may be constructed for exploiting hinterland or new areas.

Cost considerations The cost of construction of mountain railways is quite high because of the need for a large number of bridges and heavy earthwork. The cost, however, gets considerably reduced by adopting the narrow gauge (0.762 m or 0.61 m), and then it is possible to have very steep gradients and sharp curvature.

Some of the important mountain railways existing in India today are: (i) Kalka–Shimla Railway, (ii) Nilgiri Mountain Railway, (iii) Matheran Light Railway, (iv) Darjeeling Himalayan Railway, (v) The Kangra Valley Railway, and (vi) Haldwani–Kathgodam Railway.

3.5 RACK RAILWAYS

In the case of very steep gradients, much steeper than 3 per cent, it becomes difficult for a locomotive to pull the train load and hence is possible for the train to slide down or slip back along the down grade.

The rack railway system consists of three rails, i.e., one extra toothed rail in the middle in addition to the two normal rails. The locomotive also has a toothed pinion wheel whose teeth fit into the grooves of the central toothed rail. This locking arrangement helps to haul the train load and does not let the locomotive slip back. There are various such systems existing in the Indian Railways, such as the Fell system, Abt system, Riggenbach system, and Marsh system, each with its own characteristics.

Field photograph of the rack railway system
showing the third (toothed) rail, also see plate 1.

The Fell system was the first third-rail system for mountain railways with steep gradients. It uses a raised centre rail between the two running rails to provide extra traction and braking, or braking alone. Trains are propelled by wheels or braked by shoes pressed horizontally onto the centre rail, as well as by means of the normal running wheels. Extra brake shoes are fitted to specially designed or adapted Fell locomotives and brake vans, and for traction the locomotive has an auxiliary engine powering horizontal wheels which clamp onto the third rail. The Fell system was developed in the 1860s and was soon superseded by various types of rack railway for new lines. Brazil, France, and New Zealand are the only countries that have lines with Fell system.

The Riggenbach rack system was invented by Niklaus Riggenbach and was granted a French patent in 1863 based on a working model where the gear teeth were in the form of rollers arranged like the rungs of a ladder between two L-shaped wrought-iron rails. The Riggenbach system suffered from the problem that its fixed ladder rack was more complex and expensive to build than the other systems.

Abt rack railway system (Courtesy: A.M. Hurrell This file is licensed under the Creative Commons Attribution-Share Alike 3.0 Unported license; http://creativecommons.org/licenses/by-sa/3.0/deed.en), also see plate 1.

The Abt system was devised by Roman Abt, a Swiss locomotive engineer during the early 1880s. He worked to devise an improved rack system that overcame the limitations of the Riggenbach system. In 1882 Abt designed a new rack using solid bars with vertical teeth machined into them. Two or three of these bars are mounted centrally between the rails, with the teeth offset. The use of multiple bars with offset teeth ensures that the pinions on the locomotive driving wheels are constantly engaged with the rack. The first use of the Abt system was on the Harzbahn in Germany which opened in 1885.

With the help of this rack and pinion system, it is possible to move trains even on very steep gradients such as 1 in 5.

SUMMARY

The alignment of a railway line is extremely important as the subsequent costs of construction and operation depend heavily on it. The basic requirements and factors affecting alignment have been discussed in this chapter. The various types of engineering surveys required to mark the alignment of a railway line are described in the next chapter.

REVIEW QUESTIONS

1. State and discuss briefly the factors that control the alignment of a railway track.
2. In the process of selecting a suitable alignment for a railway line, what factors and parameters are kept in view? What analysis is done to assess the economic viability of the alignment?
3. Describe in brief the basic requirements of a good alignment. What are the factors that control the alignment of a railway line?
4. What do you understand by mountain railways? Describe in brief the various types of alignments used for mountain railways.

Choose the correct answer from the choices given.

5. Kashmir Rail link is being constructed with an objective, which is:
   (a) commercial reasons
   (b) development of backward areas
   (c) strategic and politcial considerations
   (d) any other reason
6. The maximum degree of the curve that can be used for mountain railways is:
   (a) 10°   (b) 16°
   (c) 30°   (d) 40°
7. The specialty of 'rack railway' is:
   (a) it has special locomotive
   (b) it has special track width
   (c) it uses three rails instead of conventional two rails
   (d) none of these

Engineering Surveys and Project Reports

INTRODUCTION

The construction of a new railway line is a capital-intensive project and each kilometre of a new railway line costs between ₹ 10 and 20 million depending upon the topography of the area, the standard of construction, and such other features. It is therefore natural that a lot of thought be given when making a final decision as to whether a new railway line is necessary or not.

4.1 NEED FOR A NEW RAILWAY LINE

The need for construction of a new railway line arises because of one or more of the following considerations.

Strategic reasons   It is sometimes necessary to extend the existing railway line to a new point of strategic importance so that the defence forces can move quickly to the same areas in case of any emergency such as a threat of war.

Political reasons   A new line sometimes becomes necessary to serve the political needs of the country, for example, the railway line from Pathankot to Jammu.

Development of backward areas   Railway lines are sometimes constructed to develop backward areas. Experience has shown that once railway connectivity is available, backward areas develop very fast. The Assam rail link can be classified in this category.

To connect new trade centres   Sometimes new trade centres are connected by railway lines for the quick transportation of goods between two trade centres or from the point of production to the point of consumption.

To shorten the existing rail link   The existing routes between two important points may be longer than required. New railway lines are constructed on a shorter alignment in such cases. A short route is not only economical, but also helps in the faster movement of goods and passengers. The Konkan Railway is a typical example.

Doubling of existing single railway lines is also done in a few cases to cope up with the additional requirement of traffic. Recently, a large number of projects have also been undertaken for converting the existing MG lines to BG lines in order to have a uniform gauge for the smooth flow of traffic.

4.2 PRELIMINARY INVESTIGATIONS FOR A NEW RAILWAY LINE

Whenever the construction of a new railway line is under consideration, preliminary investigations are done by the railway administration to determine how the proposed line will fit in with the general scheme of future railway development. The preliminary investigations are normally based on a careful study of the following:

  (a) Existing topo sheets and other maps of the area
  (b) Published figures of trade and population of the area to be served
  (c) Statistical data of existing railway lines in similar terrain in other areas

   As a result of these investigations, it becomes possible to decide whether or not the new railway line is required and surveys should then be undertaken to get more details of the new line being contemplated.

Note: For full details of construction of new lines, please refer to Chapter 28.

4.3 TYPES OF SURVEYS

Once a decision has been taken during preliminary investigations about the general feasibility and desirability of a railway line, surveys are undertaken before the construction of the new line. The following types of surveys are normally conducted:

  (a) Traffic survey
  (b) Reconnaissance survey
  (c) Preliminary survey
  (d) Final location survey

   The details of these surveys are discussed in the following sections.

4.4 TRAFFIC SURVEY

Traffic survey includes a detailed study of the traffic conditions in the area with a view to determine the following:

  (a) The most promising route for the railway in the area
  (b) The possible traffic the railway line will carry
  (c) The standard of railway line to be followed

   Traffic surveys are normally undertaken in conjunction with reconnaissance or preliminary engineering surveys so that the technical feasibility and relative costs of alternative proposals can be formulated. The traffic survey team should work in close cooperation with the engineering survey team. The survey team should visit all trade centres in the area and consult local bodies, state governments, and prominent citizens regarding trade and industry and propose the most suitable alignment for the new line.

Traffic survey consists of an economic study of the area keeping in mind the following considerations, information on which should be collected in detail:

(a) Human resources
(b) Agricultural and mineral resources
(c) Pattern of trade and commerce
(d) Industries located and projected
(e) Prospects of tourist traffic
(f) Existing transport facilities
(g) Locations of important government and private offices
(h) Planning for economic development of the area

The traffic survey team should make an assessment of the traffic likely to be carried by the new line. While carrying out the survey, details of traffic likely to be offered by various government organizations, public bodies, or private enterprises should be gathered.

At the end of the survey, a report should be formulated by the officer-in-charge of the survey. The formation of the report is governed largely by the nature of the terms of reference and the investigations made. The traffic survey report should normally contain the following information:

(a) History of the proposal and terms of reference
(b) General description
(c) Potentials and prospects
(d) Industrial and economic development and traffic projections
(e) Population projection and volume of passenger traffic
(f) Existing rates and rates to be charged
(g) Location of route or routes examined, alternate routes, and possible extensions
(h) Station sites and their importance
(i) Train services, section capacity, and various alternative ways of increasing capacity
(j) Coaching earnings
(k) Goods earnings
(l) Working expenses and net receipts
(m) Engineering features
(n) Telecommunication facilities
(o) Financial appraisal
(p) Conclusions and recommendations

4.5 RECONNAISSANCE SURVEY

Reconnaissance survey consists of a rapid and rough investigation of the area with a view to determine the technical feasibility of the proposal as well as the rough cost of one or more alternatives to the new line. The reconnaissance survey (RECCE) is normally based on contoured survey maps and other data already available without carrying out detailed investigations in the field. With the help of the maps, different alternative alignments of the new line are studied.

The general topography of the country is studied by the survey team and then field data is collected.

4.5.1 Survey Instruments

The reconnaissance survey is mostly conducted using survey instruments that rapidly measure approximate distances and heights. The survey instruments used are the following:

Prismatic compass   To get magnetic bearings of the proposed alignment
Aneroid barometer   To ensure relative heights of various points
Abney level or hand level or clinometer   To measure the gradients or angles of slopes
Binocular   To view the physical features
Pedometer   To get an idea of the total length traversed while walking

4.5.2 Modern Surveying Instruments

With the help of modern surveying instruments, it is possible to carry out fairly accurate surveying very expendiently and efficiently. Refer to Section 4.9 for further details.

4.5.3 Field Data

The following field data is collected during the reconnaissance survey:
  (a) General topography of the country
  (b) Approximate heights of the different points falling on the alignment
  (c) Positions of rivers, streams, and some hydrological details of the same
  (d) Positions of roads and highways
  (e) Nature of soil at different places
  (f) Rough location of various station sites
  (g) Controlling points on the alignment, through which the railway line must pass
  (h) Facilities for construction

4.5.4 Project Report for Reconnaissance Survey

Based on the above data, a report should be prepared by the engineer in charge of the project bringing out clearly, from the financial point of view, whether or not the prospects of the line surveyed are such as to make it worthwhile to undertake further investigations to construct the line. The project report should be accompanied by an abstract estimate of the cost of the line.

The report and estimate should be accompanied by a map of the area on a scale of 20 km to 1 cm and an index map of 2.5 km to 1 cm.

4.6 PRELIMINARY SURVEY

The preliminary survey consists of a detailed instrumental examination of the route to be selected as a result of the reconnaissance survey in order to estimate the cost of the proposed railway line. Based on the preliminary and traffic survey reports, the railway administration decides whether or not the proposed railway line is to be constructed.

4.6.1 Instruments for Preliminary Survey

The instruments to be used for a preliminary survey will depend on the topography of the country and its flora. The survey instruments normally used are the following:
  (a) Theodolite for traversing and pegging the centre line
  (b) Tacheometer for plotting the main features
  (c) Dumpy level for taking the longitudinal and cross levels
  (d) Plane table for getting details of various features
  (e) Prismatic compass for measuring the magnetic bearings of a particular alignment

4.6.2 Field Survey

The route selected is surveyed in greater detail in the preliminary survey. The survey normally covers a width of 200 m on either side of the proposed alignment. The following survey work is carried out.
  (a) An open traverse is run along the centre line of the proposed alignment with the help of a theodolite, tacheometer, or a compass.
  (b) Longitudinal and cross levels are taken on the proposed route for a width of 200 m on either side in order to make an accurate contour map.
  (c) Plane tabling of the entire area to obtain various geographical details is done.
  (d) Special survey of station sites, level crossings, and bridges using the plane table is conducted.

4.6.3 Data

The following information should normally be collected during a preliminary survey:
  (a) Geological information such as type of soil strata and the nature of rocks
  (b) Source of availability of construction materials such as sand, aggregate, bricks, cement, and timber
  (c) Facilities for construction such as the availability of labour and drinking water
  (d) Full details of the land and buildings to be acquired
  (e) Details of existing bridges and culverts along with information about proximity of tanks, bunds, etc., which may affect the design of bridges
  (f) Details of road crossings along with the angles of crossing and the traffic expected on the level crossings
  (g) High flood level and low water level of all the rivers and streams falling on the alignment
  (h) Full details of station sites along with the facilities required

4.6.4 Preparation of Project Report

A report based on the preliminary survey is prepared after obtaining an estimate of the cost. The project report should contain the following details.
  (a) Introduction
  (b) Characteristics of the project area

(c) Standard of construction

(d) Route selection

(e) Project engineering including cost estimate and construction schedule

(f) Conclusions and recommendations

4.6.5 Cost Estimate

The report should be accompanied by a cost estimate. The estimate based on the preliminary report should be sufficiently accurate to enable a competent authority to take a decision regarding the construction of the new line. The estimate should contain the following details:

(a) An abstract cost estimate of the line surveyed, accompanied by an abstract estimate of junction arrangements

(b) Detailed estimates of land, tunnels, major bridges, minor bridges, one kilometre of permanent way, rolling stock, and general charges

The report and estimate should also be accompanied by the following drawings:

(a) Map of the area (scale 20 km = 1 cm)

(b) Index plan and section (scale 0.5 km to 1 cm horizontal and 10 m to 1 cm vertical)

(c) Detailed plans and sections

(d) Plans of station yards

(e) Plans of junction arrangements

4.7 PRELIMINARY ENGINEERING-CUM-TRAFFIC SURVEY

In practice, and quite often, both the traffic survey and the preliminary engineering survey are carried out simultaneously in order to expedite the project. In such cases, techno-economic survey reports based on preliminary-cum-traffic surveys are compiled. Such techno-economic survey reports contain the following details.

(a) Introduction

(b) Traffic projections

(c) Analysis of alternatives

(d) Characteristics of project area

(e) Standards of construction (for new lines, multiple tracking schemes, gauge conversions)

(f) Route selection and project description

(g) Project engineering (for new lines, multiple tracking schemes, and gauge conversions)

(h) Cost, phasing, and investment schedules

(i) Financial appraisal

(j) Recommendations

4.8 FINAL LOCATION SURVEY

Once a decision has been taken to construct a railway line, a final location survey is done. The instruments used are generally the same as in the case of the preliminary survey. Final location survey is done to prepare working details and make accurate cost estimates in certain cases.

The principal differences between the preliminary survey and the final survey are as follows:

(a) In the final location survey, the alignment is fully staked with the help of a theodolite, whereas it is not obligatory to do so in the case of preliminary survey.

(b) In the final location survey, a more detailed project report is prepared and submitted.

(c) All working drawings are prepared in the final location survey.

The following tasks are carried out in the final location survey:

(a) The centre line is fully marked by pegs at 20 m. At each 100 m, a large peg should be used.

(b) Masonry pillars are built at tangent points of curves and along the centre line at intervals of 500 m.

(c) Longitudinal and cross levelling is done to ascertain the final gradient of the alignment. All gradients are compensated for curves.

(d) The sites for station yards are fully demarcated.

In the final location survey, the following set of drawings is prepared.

(a) General map of the country traversed by the project at a scale of about 20 km to 1 cm

(b) Index map, scale about 2.5 km to 1 cm

(c) Index plan and sections

(d) Detailed plans and sections

(e) Plans and cross section

(f) Plans of station yards

(g) Detailed drawings of structures

(h) Plans of junction arrangements

4.8.1 Objectives

The following broad objectives should be kept in mind when selecting the best possible alignment in the final location survey:

(a) Correct obligatory points

(b) Easy grades and flat curves

(c) Minimum cost of construction

(d) Minimum adverse effect on environment

(e) Ease of construction

(f) Potential for high speeds

(g) Avoidance of constraints for future expansion

(h) Minimum maintenance cost

4.8.2 Project Report

A final project report is prepared based on the final location survey. The report should consist of the following.

Introduction

The introduction of the report includes details of the following:

(i) Object of the investigation and background

(ii) Programme and methodology of the investigation

(iii) Special features of the investigation

Characteristics of the project areas

In this segment the topographical outline of the areas and geographical features of the country are given to the extent to which these are likely to affect the alignment, probable stability of the line, cost of construction, working expenses, or future prospects of the proposed line. Climatic and rainfall characteristics and environmental characteristics such as the presence of corrosive factors, pollution, etc., which may have an effect on the design and maintenance of structures and bridges, are also brought out.

Standard of construction

This segment gives details of the following.

Gauge This includes the gauge adopted for the proposed line and the reasons, if any, for adopting it.

Category of line This includes the category of the line, the maximum speed potential of the line, the maximum axle load, the loading standard of bridges, and the basis for adopting the same.

Ruling gradient This includes the gradient adopted and the basis for its selection.

Curves This includes the sharpest degree of curvature adopted, the basis for its adoption, and its impact on the projected speed compatible to the category of line.

Permanent way This includes the rail section adopted, the decision as to whether welding of the rails will be carried out or not, and the type and density of sleepers provided in the project estimate.

Ballast This includes the type and depth of ballast cushion provided.

Stations This includes spacing of stations in the case of new lines, provision for future intermediate stations, and the scale of facilities contemplated at stations.

Signalling and telecommunication This includes the standard of signalling adopted and the scale of communication facilities provided.

Traction This includes the type of traction proposed.

Other details This includes road crossings, station machinery, residential accommodation, service and maintenance facilities, etc.

Route selection

This segment provides relevant information and data related to the various alternative routes examined and gives an insight into the factors influencing the choice of the route adopted for the project.

Project engineering

This section furnishes information and data for the project manager to enable him to understand the scope and extent of the project and to assist him in formulating the strategy for the execution and management of the project. It must focus on the problems likely to be encountered, identify the areas requiring special attention, and place the knowledge and information gathered at the investigation stage for evolving optimal solutions.

Estimation of cost and construction schedule

This segment gives a cost estimate of preliminary expenses, land, formation, bridges, permanent way, station building and residential quarters, road crossings, station equipment, signalling and interlocking, rolling stock, etc. The schedule of construction as well as investment is also given. A network should be developed for projects costing above ₹ 5 million.

Project organization

This section details the organizational structure for the execution of the project, the proposed headquarters of the project manager and other construction officers, as well as the allocation of the various construction activities. Health and hygenic conditions provided to the staff as well as the provision of necessary medical establishments may be indicated, along with suggested plans providing for the housing of staff and labour and the construction of temporary office buildings. Comments regarding the availability of water for construction purposes and its suitability for drinking purposes may also be given. The purpose and final cost allocation of such plans may also be indicated.

Tabulated details

The report is accompanied by tabulated details of curve abstract, gradient abstract, bridge abstract, important bridges, stations, machinery, stations, and station sites.

Arrangement of documents in the report

All the documents pertaining to a final location survey report should be in the following order.
1. Covering note
2. Index
3. Report
4. A list of drawings accompanying the report
5. Appendices to the report
    (a) Historical and geographical aspects
    (b) Location report
    (c) Rates for construction work

4.9 MODERN SURVEYING INSTRUMENTS AND TECHNIQUES

Modern surveying instruments make extensive use of infrared beams, laser beams, as well as computers. Using these instruments, it is possible to carry out fairly accurate surveying efficiently at all times, eliminating human error.

Electromagnetic distance measurement (EDM) instruments

EDM instruments rapidly and automatically measure both horizontal and vertical distances. The readings can be displayed on built-in computer screens. Examples of such instruments are the geodimeter and tellurimeter, which have been used in the past for electronic distance measurement of up to 80 km during day or night. Modern EDM instruments are quite advanced and versatile.

Total station

Total station is an optical instrument used in modern surveying. It is a combination of an electronic theodolite (transit), an electronic distance measuring device (EDM), and software running on an external computer.

With this modern instrument, one may determine angles and distances from the instrument to points to be surveyed. With the aid of trigonometry, the angles and distances may be used to calculate the coordinates of actual position (X, Y, and Z or northing, easting, and elevation) of surveyed points, or the position of the instrument from known points, in absolute terms.

The data may be downloaded from the theodolite to a computer and application software will generate a map of the surveyed area.

Use of computers

The results of the field survey are recorded in the form of angles and distances in the normal field book or an electronic notebook. Using computers, it is possible to do all calculations as well as plot accurately. Thus, output from the EDM can be fed into the computer, which in turn can plot plans and sections.

Use of laser in surveying

Laser is an acronym for light amplification by stimulated emission of radiation. Its property of low diversion is used for alignment purposes. The invisible line of sight in ordinary survey instruments is replaced by the bright red beam of the laser. This beam is intercepted by the target composed of light-sensitive cells connected to the display panel. Its most important aspect is that the beam is in a perfect straight line. Distances up to 70 km can be measured using laser. For short distances infrared beams are used.

4.9.1 Modern Surveying Techniques for Difficult Terrain

In difficult terrains, particularly in hills, modern techniques and survey aids can be utilized for the preliminary survey of railway lines. Some of these survey aids are the following:

(a) Satellite imagery (remote sensing data)
(b) Aerial photographs
(c) Topographic maps/contour maps
(d) Digital terrain modelling (DTM)
(e) Photogrammetric plotted sheets
(f) Modern geotechnology in aid of preparation of subgrade formation

Modern techniques were utilized, e.g., for carrying out the survey of a proposed railway line in Kashmir valley from Udhampur to Qazigund. This area consists of difficult terrain, with numerous hills and valleys and is sensitive to terrorist activities, making fieldwork very difficult.

Satellite imagery

Satellite imagery provides a bird's eye view of large areas. Such maps are available from the Indian Space Research Organisation (ISRO) and are updated about once a month. Ground conditions can be well appreciated with a combination of satellite images and topographic maps. Using these, two or three promising alternative corridors can be marked. The corridor that satisfies the survey objectives best is then chosen for further analysis. The use of satellite data in the initial stages of planning has been greatest in those areas for which the existing map coverage and support data are inadequate, and where field evaluation is extremely difficult.

Aerial photographs

Aerial photographs of the entire country are taken once in every three to five years and these are available with the Survey of India. These photographs are used to gather further details of the chosen corridor. A critical examination of the corridor using the photographs helps in finalizing river crossings, tunnel locations, station sites, etc.

Digital terrain modelling

Further details of the railway line are worked out with the help of digital terrain modelling and contour maps. Computer aids are used for defining the most economical alignment. Ground stations are then fixed in the form of two mutually visible points about a kilometre apart in a 10 km stretch. All other details for the preliminary survey can be worked out with the help of contour maps, photogrammetric plotted sheets, and other computer aids.

4.9.2 Modern Geotechnology in Aid of Subgrade Preparation

Geotechnology has recently made great advancement. This technology can be used in efficient design of subgrade for railway lines to provide stable and trouble-free formation. Sometimes if the natural soil is not good, the technology can be used for providing right treatment to subgrade so as to get a sound and trouble-free formation.

In case of doubling and gauge conversion projects, if the existing formation is giving trouble, the same can be improved while executing these projects.

Some of these technologies or remedial measures are as follows:
 (i) Provision of pre-fabricated vertical drains
 (ii) Removal of soft soils of limited depth to be replaced by good material fill.
(iii) Provision of an inverted filter.
(iv) Cement grouting by a slurry of cement and sand to be pumped by pressure.
 (v) Provision of sand piles of about 30 cm diameter to a depth of 2 m to 3 m.
(vi) Soil stabilization by using geotextiles.

SUMMARY

The purpose of constructing a new railway line in an area may be strategic, political, or developmental. The new construction requires extensive thinking, planning, and investment. At this stage, preliminary investigations are carried out to determine the feasibility of the new project. Various types of surveys are conducted to this end. The money spent on surveys is non-recoverable and, therefore, it is extremely important to conduct survey work as precisely as possible. Various types of surveys are carried out using appropriate instruments. The outcome of these surveys is a detailed project report and elaborate working drawings.

REVIEW QUESTIONS

1. What are the various factors to be kept in mind when conducting a reconnaissance survey for a railway track?
2. Describe in detail the objectives and steps involved in the preliminary survey for a new railway alignment.

3. List the various surveys that need to be undertaken for the construction of a new railway line and outline their essential objectives.
4. Describe briefly the principal features of a preliminary survey.
5. Explain the following briefly (a) reconnaissance survey, (b) preliminary survey, (c) final location survey.
6. What is the basic difference between a preliminary survey and a final location survey? Describe briefly the instruments used in preliminary surveys and the details to be given in the project report of a final location survey.

Choose the correct answer from the choices given:

7. The purpose of constructing a new line from Jammu to Kashmir Valley was:
   (a) strategic reason
   (b) development of backward area
   (c) shorten existing rail link
   (d) none of the above

8. The instrument(s) used in preliminary survey:
   (a) theodolite
   (b) tacheometer
   (c) dumpy level
   (d) plan table
   (e) all of them

9. While conducting final location survey, the centre line is fully marked with pegs at a distance of:
   (a) 10 m
   (b) 20 m
   (c) 30 m
   (d) 50 m

10. Traffic survey of a new line should include the following:
    (a) The most promising route for the railway in the area
    (b) The possible traffic the railway line will carry
    (c) Standard of railway line to be followed
    (d) All of the above

11. Pedometer is used to measure:
    (a) distance
    (b) angle
    (c) height
    (d) none of the above

12. Aerial photographs of various areas are available with:
    (a) Indian Railways
    (b) CPWD
    (c) Air Authority of India
    (d) Survey of India

13. Satellite imageries give information about:
    (a) various satellites used by India
    (b) details of a particular satellite
    (c) large areas as bird's-eye view
    (d) none of these

14. The project report of reconnaissance survey is accompanied by a map to a scale of:
    (a) 1.0 km per cm
    (b) 2.0 km per cm
    (c) 2.5 km per cm
    (d) 4.00 km per cm

15. Field survey normally on either side of the track covers a width of:
    (a) 100 m
    (b) 200 m
    (c) 300 m
    (d) 500 m

Track and Track Stresses

INTRODUCTION

Track or permanent way is the railroad on which trains run. It consists of two parallel rails fastened to sleepers with a specified distance between them. The sleepers are embedded in a layer of ballast of specified thickness spread over level ground known as *formation*. The ballast provides a uniform level surface and drainage, and transfers the load to a larger area of the formation. The rails are joined in series by fish plates and bolts and these are fastened to the sleepers with various types of fittings. The sleepers are spaced at a specified distance and are held in position by the ballast. Each component of the track has a specific function to perform. The rails act as girders to transmit the wheel load of trains to the sleepers. The sleepers hold the rails in their proper positions, provide a correct gauge with the help of fittings and fastenings, and transfer the load to the ballast. The formation takes the total load of the track as well as of the train moving on it.

The permanent way or track, therefore, consists of the rails, sleepers, fittings and fastenings, the ballast, and the formation as shown in Fig. 5.1.

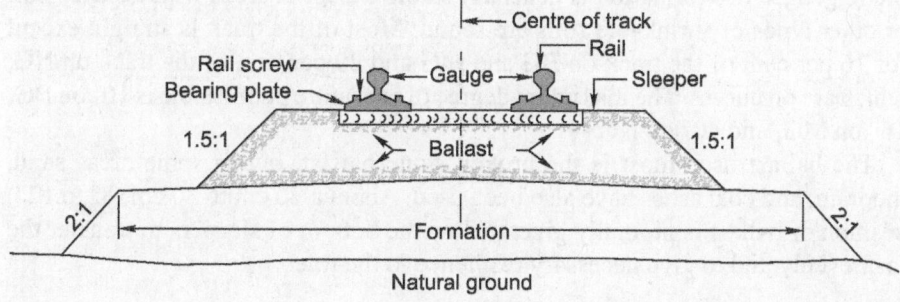

Fig. 5.1 Various components of a track

In the early days, a temporary track used to be laid for carrying earth and other building material for the construction of a railway line; this temporary track used to be removed subsequently. The track is also called the permanent way in order to distinguish the final track constructed for the movement of trains from the temporary track constructed to carry building material.

The specifications adopted by Indian Railways for various types of railway tracks are discussed in this chapter. The stresses developed in the different components of a railway track due to moving wheel load are also elaborated.

5.1 REQUIREMENTS OF A GOOD TRACK

A permanent way or track should provide comfortable and safe ride at the maximum permissible speed with minimum maintenance cost. To achieve these objectives, a sound permanent way should have the following characteristics:

(a) The gauge should be correct and uniform.
(b) The rails should have perfect cross levels. In curves, the outer rail should have proper superelevation to take into account the centrifugal force.
(c) The alignment should be straight and free of kinks. In the case of curves, a proper transition should be provided between the straight track and the curve.
(d) The gradient should be uniform and as gentle as possible. The change of gradient should be followed by a proper vertical curve to provide a smooth ride.
(e) The track should be resilient and elastic in order to absorb the shocks and vibrations of running trains.
(f) The track should have a good drainage system so that the stability of the track is not affected by waterlogging.
(g) The track should have good lateral strength so that it can maintain its stability despite variations in temperature and other such factors.
(h) There should be provisions for easy replacement and renewal of the various track components.
(i) The track should have such a structure that not only is its initial cost low, but also its maintenance cost is minimum.

5.1.1 Track Specifications on Indian Railways

Most of the railway lines on Indian Railways are single lines, generally with a formation of 6.85m (22'6") wide for broad gauge and 5.85m (19'1") wide for metre gauge. The formation is generally stable except in areas where clayey soil or other types of shrinkable soils are found. Most of the track is straight except for 16 per cent of the track on BG and MG and 20 per cent of the track on NG, which are on curves. The maximum degree of curvature permissible is 10° on BG, 16° on MG, and 40° on NG.

The ballast used most is the broken stone ballast, but in some areas sand, mooram, and coal ashes have also been used. About a 20 cm to 30 cm (8" to 12") cushion of ballast is normally given below the bottom of sleepers to transfer the load evenly and to give necessary resilience to the track.

The different types of sleepers used for BG are pre-stressed concrete (PSC) sleepers (94.2%), cast iron sleepers (4.0%), steel sleepers (1.5%), and wooden sleepers (0.3%). Similar figures for different type of sleepers for MG are PSC sleepers (9.8%), steel sleepers (14.7%), cast iron sleepers (69.0%), and wooden sleepers (6.5%). Experience has shown that cast iron sleepers are not suitable for high density routes. Prestressed concrete sleepers have been developed by Indian Railways and are being progressively laid on Group A and B routes. Sleepers are laid to various sleeper densities varying from M + 8 to M + 4 (1660 per km to 1310 per km) depending upon the weight and volume of traffic.

The rails standardized for Indian Railways are 60 kg and 52 kg for BG and 90 R, 75 R and 60 R, for MG. The rails are normally rolled in 13 m (42 ft) lengths for BG and 12 m (39 ft) lengths for MG. Rails are welded together to form longer rails and these are laid progressively in the track in order to reduce maintenance costs and the noise level, and providing a more comfortable travel. The rails (LWR) are welded in three rail panels in depots normally by the flash-butt welding method to form short welded rails (SWR). Long welded rails (LWR) are also being progressively introduced on various routes of Indian Railways. Thermit welding is normally done at the site to convert SWR into LWR. The fastenings used are mostly screw and rail spikes, keys, etc., but recently elastic fastenings like elastic clips (Pandrol Clip) and IRN202 clips have also been standardized on Indian Railways. The present position is that on BG track of main lines of Indian Railways about 85 per cent of the length is covered by LWR, about 95 per cent with PSC sleepers and about 88 per cent with 52 kg/60kg rails.

The turnouts used are normally 1 in 8.5 for goods trains and 1 in 12 as well as 1 in 16 for passenger trains. 1-in-20 turnouts were also designed and laid on Indian Railways for permitting higher speeds. Curved switches and thick web switches have also been introduced to permit higher speeds at turnouts. Currently, Fan Shaped Turnouts and Thick Web Switches and CMS crossing are being laid on PSC sleepers.

5.1.2 Recommended Track Structure for BG and MG Routes

Indian Railways (IR) has modernized its track structure in the recent past to meet the challenges of heavier loads and faster traffic. More than 10,000 km of the track is laid with 60-kg rails and more than 40,000 km is laid with concrete sleepers. The current IR standards for BG and MG routes are elaborated in the following sections.

Track structure for BG routes

The track on Indian Railways for BG sections is classified into five broad categories based on speed. These are group A, B, C, D, and E routes. Additionally, D spl and E spl routes have been subsequently added as explained in Section 1.8. The track standard on BG routes has been reviewed and revised from time to time based on the obtainable speeds and traffic conditions. The current BG track structure is detailed in Tables 5.1 to 5.4.

Track structure for MG routes

Tracks on MG routes have been classified based on speed and GMT in categories Q, RI, R2, R3, and S routes. The S route of MG has been further classified as S1, S2, and S3 routes as per the following details:

Table 5.1 Track structure for BG system of Indian Railways

Traffic density in GMT	A Route	B Route	C Route	D spl Route	D Route	E spl Route	E Route
> 20	60 kg	60 kg	60 kg	60 kg	60 kg	60 kg	60 kg
10–20	60 kg	60 kg	60 kg	60 kg	60 kg	60 kg	60 kg
5–10	60 kg	52 kg	52 kg	52 kg	52 kg	52 kg	52 kg
Under 5	52 kg	52 kg	52 kg	52 kg or 60 kg SH	52 kg or 60 kg SH	52 kg or 60 kg SH	52 kg or 60 kg SH

* Gross million tonnes per km per annum

On loop lines of A, B and C routes second hand 60 kg (SH) and second hand 52 kg (SH) can be used. For use of new rails in loop lines, the Railways Board's approval is required. Also see note (ii) below Table 5.2.

Table 5.2 Number of sleepers per km

Traffic density in GMT	A Route	B Route	C Route	D spl Route	D Route	E spl Route	E Route
>20	1660	1660	1660	1660	1660	1660	1660
10–20	1660	1660	1660	1660	1660	1660	1540
< 10	1540	1540	1540	1660	1540	1540	1540

Notes : (i) *On loop lines, 1540 sleepers can be laid on all routes of temperature Zone III & IV and A, B & C routes of Zonal I & II.*

(ii) *For routes identified for running of 22.3 tonne axle loads, rails of 60 kg 90 UTS and concrete sleepers with sleeper density of 1660 number per km. should be provided. As most of the routes are now cleared for 22.3 tonne axle loads, sleeper density on these should be generally 1660 sleepers per km. (Authority: Rly. Bd. letter No. 2007/CE-II/TS/6 dated 3.9.2007)*

* Fastenings used for wooden (W), steel (ST), and prestressed concrete (PSC) sleepers are given in Table 5.3. The track specification for loop lines and private sidings are given in Table 5.4.

Table 5.3 Type of sleepers fastening

Type of sleeper	PSC, W & ST	PSC, W & ST	PSC & W	PSC	PSC, W & ST	PSC	PSC, W & ST
Type of fastening	elastic	elastic	elastic	elastic	Existing std.	elastic	Existing std.

Table 5.4 Loop lines and other sidings

Permissible speed	Rail	Sleepers	Fitting and fastenings
Upto 50 kmph	52 kg (SH) 52 kg	1340 for Zone I & II 1540 for Zone 111 & IV	Elastic fastenings
> 50 kmph	60 kg	1660	Elastic fastenings

(a) Routes with a through movement of freight traffic are identified as S1 routes.

(b) Uneconomic branch lines are identified as S3 routes.

(c) Routes that are neither S1 nor S3 are identified as S2 routes.

The track standards being followed on MG routes are given in Table 5.5.

Table 5.5 Recommended track structures for MG lines

Item	Q routes for speeds > 75 kmph, GMT > 2.5	R routes for speeds up to 75 km/h			S routes for speeds < 75 kmph and GMT < 1.5	Remarks
		R1 routes, GMT > 5	R2 routes, GMT 2.5–5	R3 routes, GMT 1.5–2.5		
Rails	90 R new	90 R new	90 R (SS) or 75 R new	90 SS or 75 R new	SS rails of 90 R for S1 and 75 R for S2 and 60 R for S3	With elastic fastening
Sleepers	Concrete steel CST-9*	Concrete steel CST-9*	Concrete CST-9	Concrete CST-9	Concrete CST-9	*As an interim measure of up to 110 kmph
Sleeper density		M + 7 or 1540 per km			M + 4 or 1380 per km and M + 7* or 1540 per km	*Where LWR is contemplated
Ballast cushion	300 mm* or 250 mm	300 mm* or 250 mm	250 mm* or 200 mm	250 mm* or 200 mm	150 mm* or 200 mm with SWR or 250 mm*	* For speeds of 100 kmph or above

5.2 MAINTENANCE OF PERMANENT WAY

Permanent way is the backbone of any railway system, and the safety and comfort of the travelling public primarily rests on its proper maintenance. Till a decade ago, the tracks on Indian Railways were mostly manually maintained by beater packing as per a fixed timetable round the year. In recent years, however, on account of heavier and faster traffic and due to economic considerations, modern methods of track maintenance such as measured shovel packing, mechanized maintenance, and directed track maintenance have been tried and are in vogue in some sections on Indian Railways, particularly on high-speed routes.

Mechanical maintenance of the track has been introduced on about 14,500 km of high-density BG routes and the rest of the track is maintained through manual labour. The labour force directly employed for this task is about 190,000. Today, about 3000 km of track is being maintained by measured shovel packing, which is an improved method of manual packing. A need-based directed track maintenance system, which initiates maintenance work only when there is actual requirement, is being increasingly introduced in order to eliminate unnecessary maintenance work. It makes the labour force more productive. About 20,000 km of track is covered by this system.

A major portion of the track, however, continues to be maintained on a predetermined cyclic programme by the manual method of maintenance, i.e., beater packing. The full details of these methods of maintenance have been discussed in subsequent chapters.

5.2.1 Track Utilization

With the introduction of high-capacity bogie wagons and the replacement of steam locomotives with more powerful diesel and electric locomotives, the tracks have been subjected to heavier axle loads and higher operating speeds. During the period 1950–51 to 1994–95 the average density of traffic (in terms of net million tonne km per route km) has increased from 5.24 to 18.40 millions on BG and from 1.19 to 2.65 millions on MG. The increased track loading has necessitated improvement in the track structure and maintenance practices, especially over high-density and high-speed routes.

5.3 TRACK AS AN ELASTIC STRUCTURE

In 1888, Zimmerman propounded the theory that the track is an elastic structure. Rails are continuous beams carried on sleepers, which provide elastic support. The elastic nature of the rail supports affects the distribution of the wheel load over a number of sleepers in a rather complicated manner. The mode of distribution of load depends on the stiffness of the rails as well as the elasticity of the bed (sleepers and the ballast and formation taken together) on which the rail rests.

5.3.1 Track Modulus

Track modulus, like the modulus of elasticity, is an index of measurement of resistance to deformation. It is defined as the load in kilograms per unit rail length required to produce one unit depression in the rail bottom. The unit of track modulus is kg/cm^2.

The Research, Designs and Standards Organisation (RDSO) of Indian Railways has carried out a large number of investigations to determine the track modulus and vertical bending stresses in rails due to static loads on BG and MG tracks. These empirical studies reveal that the rail depression immediately below the load is not directly proportional to the load in the entire load range. Due to slacks and voids in the track structure, the track depression is disproportionately higher in the initial stages of loading. These slacks and voids get closed under the initial load and thereafter further depression per unit load is smaller and becomes proportionate to the increase in the load. It is found that an initial load of 4 tonnes for BG and 3 tonnes for MG gives the best results.

There are, thus, two well-designed load ranges, and a single value of the track modulus is not able to completely define the load–depression characteristics of a track. The complete relationship can be expressed by assuming that a linear load–depression relationship exists in the initial stage of the load and that there are two values of track modulus—one is the initial track modulus (K_i) and the other is the elastic track modulus (K_e).

The track modulus varies with the gauge as well as with the track standards, namely the type of rails, sleepers, sleeper density, and ballast cushion. The values of track modulus adopted on Indian Railways are given in Table 5.6.

Table 5.6 Details of track modulus

Gauge	Track standard	Initial track modulus (kg/cm²)	Elastic track modulus (kg/cm²)
BG	90 R rails, N + 3 Sleeper density and 200 mm ballast cushion	75	300
BG	52-kg rails, N + 6 Sleeper density 250 mm ballast cushion	120	380
MG	Rails 60 R and 75 R; sleeper density N + 3 and 200 mm ballast cushion	50	250

5.4 FORCES ACTING ON THE TRACK

A rail is subjected to heavy stress due to the following types of forces:
  (a) Vertical loads consisting of dead loads, dynamic augment of loads including the effect of speed, the hammer blow effect, the inertia of reciprocating masses, and so on.
  (b) Lateral forces due to the movement of live loads, eccentric vertical loading, shunting of locomotives, etc.
  (c) Longitudinal forces due to tractive effort and braking forces, thermal forces, etc.
  (d) Contact stresses due to wheel and rail contact
  (e) Stresses due to surface defects such as flat spots on wheels

5.4.1 Vertical Loads

The impact of vertical loads on rails is as follows.

Dead load of vehicles at rail-wheel contact

The value of dead load is usually taken from the axle-load diagram. It is, however, brought out that for various reasons the actual wheel loads, even in the static state on a level and perfect track, may be different from the nominal values. Cases have sometimes come to notice where a steam locomotive had a higher axle load than the nominal load or had different right and left wheel loads.

Dynamic augment of vertical loads

On account of vertical impact due to speed and rail vibrations, etc., the dynamic load is much more than the static load. The dynamic wheel load is obtained by increasing the static wheel load by an incremental amount given by the speed factor. Till 1965 Indian Railways used the 'Indian formula' for calculating the speed factor which was as follows:

$$\text{Speed factor} \quad = \frac{V}{18.2\sqrt{k}} \tag{5.1}$$

where V is the speed in km per hour and k is the track modulus in kg/cm².

After 1966, the 'German formula' given by Schram was adopted, which is as follows:
  (a) For speeds upto 100 kmph:

$$\text{Speed factor} = \frac{V^2}{30,000} \tag{5.2}$$

(b) For speeds above 100 kmph:

$$\text{Speed factor} = \frac{4.5\,V^2}{10^5} - \frac{1.5V^3}{10^7} \tag{5.3}$$

where V is the speed in kmph.

At a speed of 60 mph (96 kmph) and k value of 90 kg/cm², the Indian formula gives a speed factor of 55 per cent, whereas the German formula, as used by RDSO, gives a speed factor of 30 per cent. Investigations have been carried out by RDSO and different values of speed factors have been recommended for different types of vehicles running at different speeds.

Hammer blow effect

The centrifugal forces due to revolving masses in the driving and coupled wheels of a locomotive, such as crank pins, coupling rods, and parts of the connecting rod, are completely balanced by placing counterweights near the rim of the wheel, diametrically opposite to the revolving masses. The reciprocating masses of the piston, piston rod, cross head, and part of the connecting rod, by virtue of their inertia and oscillatory movement, produce alternating forces in the direction of the stroke and tend to cause the locomotive to oscillate sideways and nose across the track. In order to reduce this nosing tendency, a weight is introduced onto the wheels at the opposite side of the crank. The horizontal component of the centrifugal force of this added weight balances the inertial force in the line of stroke, but the vertical component throws the wheel out of balance in the plane perpendicular to the line of stroke. The vertical component of the centrifugal force of the weight introduced to balance the reciprocating masses causes variation in the wheel pressure on the rail, and is called the *hammer blow*. The heavier the weight added to balance the reciprocating masses, the greater the hammer blow.

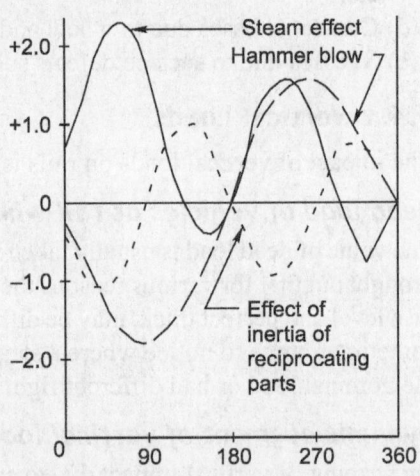

Fig. 5.2 Steam effect, hammer blow effect, and effect of inertia of reciprocating masses

The hammer blow effect occurs only in the case of steam locomotives. The hammer blow can be calculated as follows (see Fig. 5.2):

$$\text{Hammer blow} = \frac{M}{g}\,r(2\pi n)^2 \sin\theta \tag{5.4}$$

where M is the net weight in lbs, r is the crank pin diameter in ft, n is the number of revolutions of the wheel per second, and θ is the crank angle.

Steam effect

A steam locomotive works by converting coal energy into steam energy. Steam pressure acts on the piston and is transmitted to the driving wheels through the crank pins and connecting rod. The vertical component of the crank pins and connecting rod is at an angle to the piston rod. Its value is given by the formula

$$\text{Vertical component} = \frac{\pi}{4}d^2 p \frac{r \sin\theta \pm h}{L} \tag{5.5}$$

where L is the length of the connecting rod in inches, h is the height of the cross head above the centre line of the driving wheel in inches, and θ is the crank angle, i.e., the angle traversed by the crank since the beginning of the stroke.

The steam effect (Fig. 5.2) does not scynchronize with the hammer blow effect due to overbalance and is additional to the hammer blow only during some part of the revolution of the crank shaft.

Inertia of reciprocating masses

The reciprocating masses, due to their inertia and acceleration, alter the forces on the piston, and hence the force in the connecting rod is also affected during the revolution of the wheel. This is calculated as follows:

$$\text{Fv} = \frac{M}{g}r(2\pi n)^2\left(\cos\theta + \frac{r}{L}\cos 2\theta\right) \times \frac{r \sin\theta \pm h}{L} \tag{5.6}$$

where M is the mass of the reciprocating parts, L is the length of the connecting rod, n is the number of revolutions per second, h is the height of the cross head above the centre line of the driving wheel, and θ is the crank angle.

The maximum combined force of the hammer blow, the steam effect, and inertia for each driving wheel and the hammer blow effect of the coupled wheels do not act simultaneously due to the phase difference in the angular position of the counterweights in the coupled and driving wheels. The maximum combined effect of these forces is obtained by summing up the three curves for one complete revolution of the wheel.

Bending stresses on the rail due to vertical loads

The general theory of bending of rails is based on the assumption that the rail is a long bar continuously supported by an elastic foundation. Due to vertical loads, the rail is subjected to bending or flexural stresses. The bending stresses that a rail is subjected to as a result of vertical loads are illustrated in Fig. 5.3. The theory of stresses in rails takes into account the elastic nature of the supports. Based on this theory, the formula for bending moment is as follows:

$$M = 0.25\, pe^{-x/l}\left(\sin\frac{x}{l} - \cos\frac{x}{l}\right) \tag{5.7}$$

where M is the bending moment, p is the isolated vertical load, $l = (EI/k)^{1/4}$ is the characteristic length, EI is the flexural stiffnes of the rail, k is the track modulus, and x is the distance of the point from the load.

According to Eqn (5.7), the bending moment is zero at points where $x = \pi l/4$, $3\pi l/4$ and maximum where $x = 0$, $\pi l/2$, $3\pi l/2$, etc.

x	0	$\pi l/8$	$\pi l/4$	$3\pi l/8$	$\pi l/2$	$5\pi l/8$	$3\pi l/4$	$7\pi l/8$	πl	$9\pi l/8$	$5\pi l/4$
BM as a percentage of the maximum	$-$ 100.00	$-$ 36.67	$+$ 0.00	$+$ 16.50	$+$ 20.80	$+$ 18.40	$+$ 13.48	$+$ 8.60	$+$ 4.32	$+$ 1.59	$+$ 0.00

$$\text{BM} = \frac{Pl}{4} e^{x/l} \left(\sin\frac{x}{l} - \cos\frac{x}{l} \right)$$

$$\text{where } l = \left(\frac{EI}{\mu} \right)^{1/4}$$

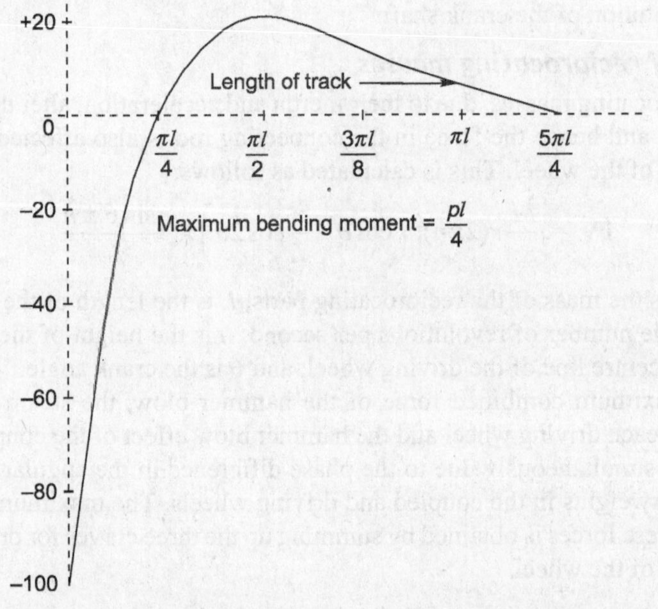

Fig. 5.3 Bending stresses due to vertical loads

For calculating the stresses acting on the rail, first the maximum bending moment caused due to a series of loads moving on the rail is calculated as per Eqn (5.7). The bending stress is then calculated by dividing the bending moment by the sectional modulus of the rail. The permissible value of bending stress due to a vertical load and its eccentricity is 23.5 kg/mm² for rails with a 72 UTS.

Virtual wheel load or Talbot Load (T_{lv})

As discussed in section 5.4, the rail is subjected to several forces such as dynamic effect of speed, hamer blow effect, steam effect in case of steam engine, and inertia of reciprocating masses, and the effect of adjoining wheels. Therefore, the instantaneous load applied on the rail is much higher than the static wheel load of the engine or the wagon. This is called the virtual wheel load or Talbot Load after the American scientist who calculated it first for a rail. This load is obtained by modifying the static wheel load taking into account all the factors to which the rail is subjected.

Effect of leading wheel

When a load is applied on a rail, it causes certain deflection and a wave-like effect is created in the rail with the passage of the train. As it will be explained in Chapter 11 (Section 11.1.1), the portion of the rail immediately under the wheel gets depressed due to wheel load and the portion ahead of the wheel gets lifted creating a wavy formation. The theory of elasticity, which is used to calculate stresses in a rail is based on the assumption that the rail is supported elastically on a number of sleepers and these supports can develop negative reactions completely. However, this is not practically the case as a small lift in the rail is always possible due to loose dog spikes, which create some gap between the spikes and the foot of the rail. This lift in the rail will increase the bending stress by about 10 per cent. The effect of the lift will be quite negligible if the wheels of the engine or the wagons are closely spaced. In that case, the lift will be suppressed by the lead wheel. Therefore, the effect of the lift in the rail on bending stress is considered only when the lead axle is at a distance greater than six times the distance of the point of contra flexure from the load. This is further explained in Example 5.1.

Stresses in the track

Stresses in a track due to the various types of forces are calculated based on the theory of double track modulus. The method requires determination of virtual wheel load or Talbot Load (T_{lv}) and is explained with the help of the following example.

Example 5.1 *Calculate the bending stresses in rail due to BOXN type wagon moving at a speed of 75 kmph on 52 kg rail load with M + 7 sleeper density and 250 mm ballast cushion. Use the following data.*

Wheel Load	=	*10.91 t*
Initial track modulus (K_i)	=	*125 kg/cm²*
Elastic track modulus (K_e)	=	*425 kg/cm²*
Moment of inertia of worn-out rail section (I_{xx})	=	*1942 cm⁴*
Section modulus in compression Z_c (worn)	=	*241.65 cm³*
Section modulus in tension Z_t (worn)	=	*256.95 cm³*

Solution

Step–1 Calculate the distance (X_1) from the load to the point of contra flexure of the rail using Eqn (5.8)

$$x_1 = 42.33 \sqrt[4]{\frac{I_{xx}}{k}} \tag{5.8}$$

For an initial load of 4 tonnes $x_{1i} = 42.33 \sqrt[4]{\dfrac{1942}{125}} = 84.04$ cm

For remaining load $x_{1e} = 42.33 \sqrt[4]{\dfrac{1942}{425}} = 61.88$ cm

Step–2 Speed Augmentation

The impact factor (dynamic augmentation) for a speed of 75 kmph is $\dfrac{v^2}{30000}$

$= 0.1875$ cm

The dynamic wheel load $= 10.91 \times 1.1875 = 12.96$ tonnes

Step–3 Effect of adjacement wheels
Wheel arrangement for BOXN wagon is shown in Fig. 5.4.

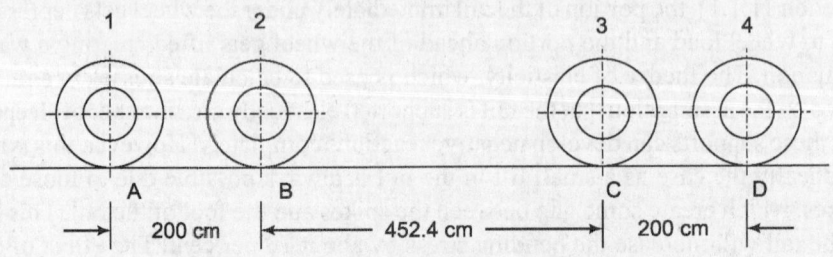

Fig. 5.4 Wheel arrangement for BOXN wagon

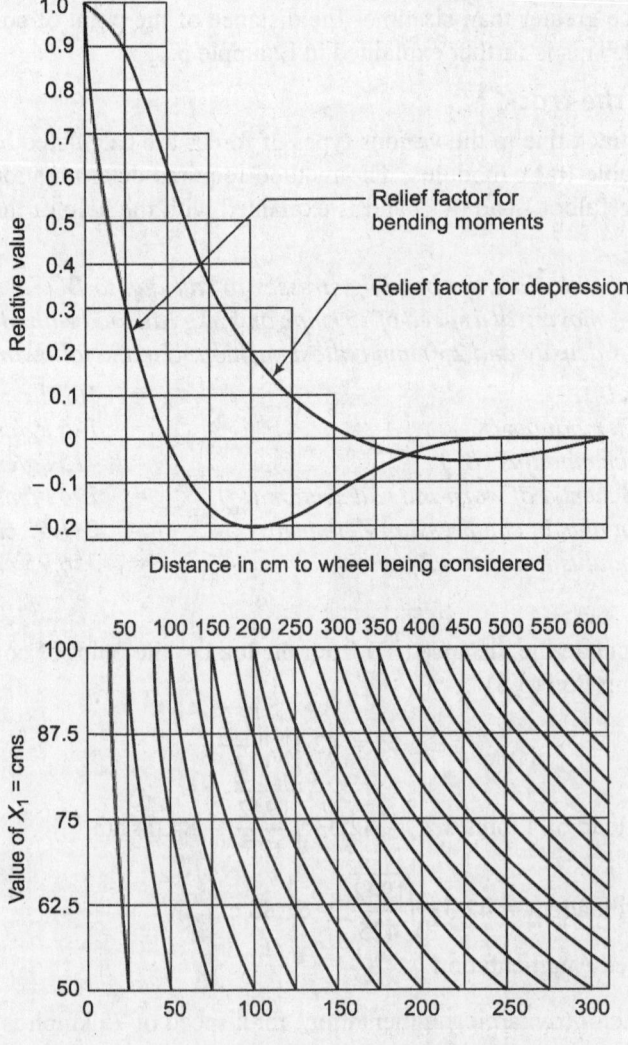

Fig. 5.5 Relief factors for BM and deflection in rail

Bending moment under the wheel load (P) is given by Eqn. (5.9)

$$BM = 0.318*P*X_1 \qquad (5.9)$$

The above equation gives BM at A due to a load at A. Its effect on any other point in the track is determined by multiplying the BM value by a factor, which is estimated from Fig. 5.5. This factor may be positive or negative depending upon the distance of the point from wheel.

The coefficients for bending moment from Fig. 5.5 are given in the following table.

	Coefficients for X_{1i} = 84.04 cm	Coefficients for X_{1e} = 61.88 cm
For a distance of 200 cm	– 0.19	– 0.10
For a distance of 452.4 cm	0	0
For a distance of 652.4 cm	0	0

Therefore the effect of adjacement wheels will be as below.

	For initial load of 4 tonnes	For remaining load of 8.96 tonnes
For a distance of 200 cm	$4 \times (–0.19)$ $= – 0.76$ tonnes	$8.96 \times (–0.10)$ $= – 0.896$ tonnes
For a distance of 452.4 cm	$4 \times 0.0 = 0$	$8.96 \times 0.0 = 0$
For a distance of 652.4 cm	$4 \times 0.0 = 0$	$8.96 \times 0.0 = 0$
Total	(–) 0.76 tonnes	(–) 0.896 tonnes

Step – 4 Determine Virtual Wheel Load (T_{lv})

(A) Virtual wheel load for an initial load of 4 tonnes

Wheel Number	1	2	3	4
Spacing (cm)	200	452.4	200	
Effect of Wheel 1	4.0	–0.76	0	0
Effect of wheel 2	–0.76	4.0	0	0
Effect of wheel 3	0	–0.76	4.0	0
Effect of wheel 4	0	0	–0.76	4.0
(T_{lv}) for part A	3.24	3.24	3.24	3.24

Effect of leading wheel is taken only if the distance between adjacent wheels (or axles) is more than 6 times X_1 (refer Section 5.4.3). In this case, $6.X_{1i}$ is 504.24 cm and no wheel except wheels 1 and 4 has an effect of leading axle, as the distance between all adjacent wheels is less than 504.24 cm.

Let us add 10% for the effect of leading axle.

T_{lv} for part A = 3.24 + 0.324 = 3.564 tonnes

(B) T_{lv} for remaining load of 8.96 tonnes

Wheel Number	1	2	3	4
Spacing (cm)	200	452.4	200	
Effect of wheel 1	8.96	−0.896	0	0
Effect of wheel 2	−0.896	8.96	0	0
Effect of wheel 3	0	0	8.96	−0.896
Effect of wheel 4	0	0	−0.896	8.96
(T_{lv}) for part B	8.064	8.064	8.064	8.064

$$6 \times X_{1e} = 6 \times 61.88 = 371.28 \text{ cm}$$

Spacing between wheels 2 and 3 is 452.4 cm, which is larger than 371.28 cm. Hence, wheels 2 and 3 will also have effect of leading axle.

Therefore T_{lv} for all wheels 1 to 4 = 8.064 + 0.8064 = 8.8704 t

For maximum value of virtual load, add T_{lv} for initial loading and elastic loading.

Wheel Number	1	2	3	4
(T_{lv}) for part A (initial loading)	3.564	3.240	3.240	3.240
(T_{lv}) for part B (elastic loading)	8.8704	8.8704	8.8704	8.8704
Total T_{lv} in tonnes	12.4344	12.1104	12.1104	12.4344
Maximum value of T_{lv} is tonnes	12.4344 (3.564 + 8.8704)			

Step–5 Calculate stress in rail due to vertical bending (Eqn 5.9)

$$\begin{aligned}
\text{Bending moment in the rail} &= 0.318 \times T_{lv} \times X_i + 0.318 \times T_{lv} \times X_e \\
&= 0.318 \times 3.564 \times 84.04 + 0.318 \times 8.8704 \\
&\quad \times 61.88 \\
&= 269.797 \text{ tonne. cm}
\end{aligned}$$

$$\text{Stress in rail head (compression)} = \frac{\text{Bending moment}}{Z_c} = \frac{269.796}{241.65}$$

$$= 1.116 \text{ tonnes/cm}^2$$

$$\text{Stress in rail foot (tensile)} = \frac{\text{Bending moment}}{Z_t} = \frac{269.796}{256.95}$$

$$= 1.05 \text{ tonnes/cm}^2$$

Dynamic overloading at joints

Due to modernization, Indian Railways is progressively using more and more number of diesel and electric engines. These engines have smaller diameter of wheels with higher unsprung (unsuspended) masses. This leads to dynamic overloading of joints resulting in the faster deterioration of the track near joints. The dynamic overloading at joints is estimated by using the following empirical formula.

$$F = F_o + 0.1188 \times V \times \sqrt{w} \tag{5.10}$$

Where, F = dynamic overload at dipped joints, tonnes

F_o = static wheel load, tonnes
V = speed of movement, km/hr
W = unsprung mass per wheel, tonnes

The unsprung mass for a WDM$_2$ locomotive is 1.985 tonnes. Therefoe, the dynamic overload at joint for a speed of 95 km/hr would be

$$F = 9.4 + 0.1188 \times 95 \times \sqrt{1.985} = 25.3 \text{ tonnes}$$

The permissible values on Indian Railways are given below.

Type of rolling stock	Total load for BG (tonnes)	Total load for MG (tonnes)
Locomotives	27	17
EMU stock	23	14
Coaches and wagons	19	11

5.4.2 Lateral forces

The lateral force applied to the rail head produces a lateral deflection and twist in the rail. Lateral force causes the rail to bend horizontally and the resultant torque causes a huge twist in the rail as well as the bending of the head and foot of the rail. Lateral deflection of the rail is resisted by the friction between the rail and the sleeper, the resistance offered by the rubber pad and fastenings, as well as the ballast coming in contact with the rail.

The combined effect of lateral forces resulting in the bending and twisting of a rail can be measured by strain gauges. Field trials indicate that the loading wheels of a locomotive may exert a lateral force of up to 2 tonnes on a straight track particularly at high speeds.

5.4.3 Longitudinal forces

Due to the tractive effort of the locomotive and its braking force, longitudinal stresses are developed in the rail. Temperature variations, particularly in welded rails, result in thermal forces, which also lead to the development of stresses. The exact magnitude of longitudinal forces depends on many variable factors. However, a rough idea of these values is as follows:

(a) Longitudinal forces on account of 30–40 per cent weight of locomotive of tractive effort for alternating current (ac).
(b) Longitudinal forces on account of 15–20 per cent of weight of braking force of the locomotive and 10–15 per cent weight of trailing load.

Tensile stresses are induced in winter due to contraction and compressive stresses are developed in summer due to compression. The extreme value of these stresses can be 10.75 kg/mm^2 in winter and 9.5 kg/mm^2 in summer.

5.4.4 Contact stresses between rail and wheel

Hertz formulated a theory to determine the area of contact and the pressure distribution at the surface of contact between the rail and the wheel. As per this theory, the rail and wheel contact is similar to that of two cylinders (the circular wheel and the curved head of the rail) with their axes at right angles to each other. The area of contact between the two surfaces is bound by an ellipse as shown in Fig. 5.6.

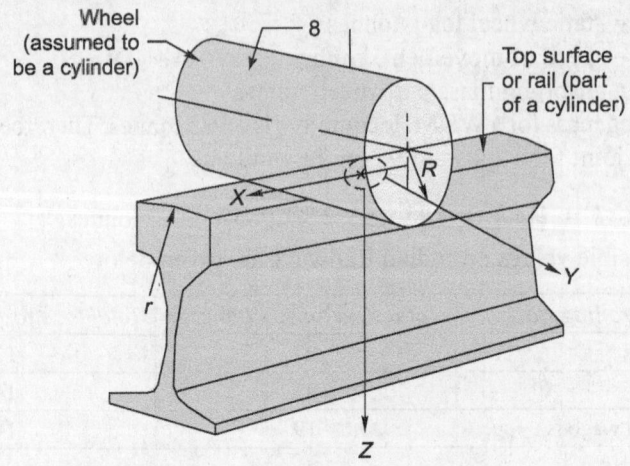

Fig. 5.6 Contact stresses between rail and wheel

The maximum contact shear stress (F) at the contact point between the wheel and the rail is given by the empirical formula

$$F = 4.13 \left(\frac{P}{R}\right)^{1/2} \tag{5.11}$$

where F is the maximum shear stress in kg/mm², R is the radius of the fully worn-out wheel in mm, and P is the static wheel load in kg + 1000 kg for on-loading on curves.

Contact stress for the WDM2 locomotive Static wheel load $(P) = 9400 + 1000 = 10,400$ kg. Radius of worn-out wheel for maximum wear of 76 mm (38 mm radius reduction):

$$R = \frac{1092}{2} - \frac{76}{2} = 508 \text{ mm}$$

$$F = 4.13 \left(\frac{P}{R}\right)^{1/2}$$

$$= 4.13 \left(\frac{10,400}{508}\right)^{1/2} = 18.7 \text{ kg/mm}^2$$

The contact stress for the WDM2 locomotive is 18.7 kg/mm². The maximum value is, however, limited to 21.6 kg/mm², which is 30 per cent of the ultimate tensile strength (UTS) value (72 kg/mm²) of the rail.

Surface defects

A flat on the wheel or a low spot on the rail causes extra stress on the rail section. Empirical studies reveal that an additional deflection of about 1.5 times the depth of the flat or low spot occurs at the critical speed (about 30 km/h). Additional bending moment is caused on this account with a value of about 370,000 for the BG group A route with the WDM4 locomotive.

5.4.5 Stresses on a sleeper

The sleepers are subjected to a large number of forces such as dead and live loads, dynamic components of tracks such as rails and sleeper fastenings, maintenance standards, and other such related factors.

Based on the elastic theory, the maximum load on a rail seat is given by the following formula:

$$\text{Maximum load on rail seat} = \frac{P}{Zkl} kS = \frac{PS}{Zl} \tag{5.12}$$

where P is the wheel load, k is the track modulus, S is the sleeper spacing, l is the characteristic length, and Z is the modulus of the rail section.

The maximum load on the rail seat is 30 per cent–50 per cent of the dynamic wheel load, depending on various factors and especially the packing under the sleeper.

The distribution of load under the sleeper is not easy to determine. The pattern of distribution depends on the sleeper as well as on the firmness of the packing under the sleeper. As the ballast yields under the load, the pressure under the sleeper is not uniform and varies depending on the standard of maintenance. The following two extreme conditions may arise:

End-bound sleeper   The newly compacted ballast is well compacted under the sleeper and the ends of the sleepers are somewhat hard packed. The deflection of the sleeper at the centre is more than that at the ends.

Centre-bound sleeper   As trains pass on the track, the packing under the sleeper tends to become loose because of the hammering action of the moving loads. The sleeper thus tends to be loose under the rail seat. Alternatively, due to defective packing, the sleeper is sometimes hard packed at the centre.

5.4.6 Stresses on ballast

The load passed onto the sleeper from the rail is in turn transferred to the ballast. The efficacy of this load transmission depends not only on the elasticity of the sleeper, but also on the size, shape, and depth of the ballast as well as the degree of compaction under the sleeper. Professor A.N. Talbot has analysed the pressure distribution in the ballast under the sleeper and investigations reveal that the pressure distribution curve under the sleeper would be shaped like bulbs as shown in Fig. 5.7.

The following are the important conclusions drawn from Fig. 5.7.
(a) The pressure on the sleeper is maximum at the centre of its width. This pressure decreases from the centre towards the ends.
(b) The vertical pressure under the sleeper is uniform at a depth approximately equal to the spacing between the sleepers.

5.6 PRESSURE ON FORMATION OR SUBGRADE

The live as well as dead loads exerted by the trains and the superstructure are finally carried by the subgrade. The pressure on the subgrade depends not only on the

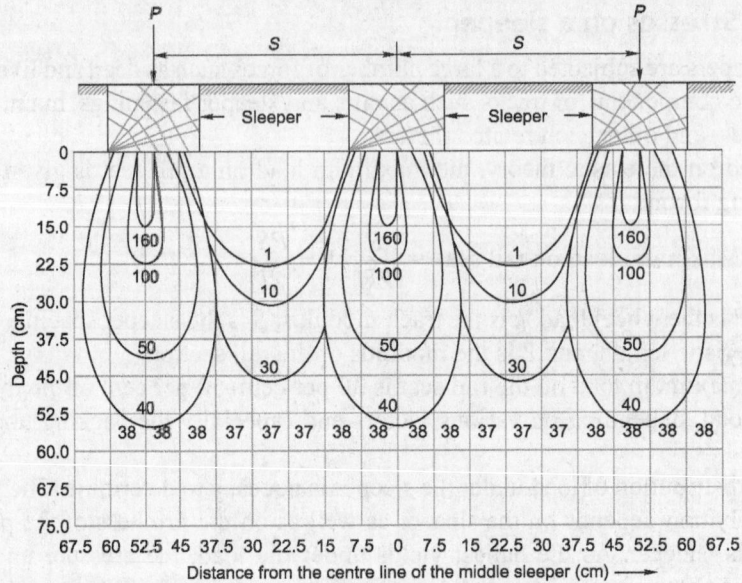

Fig. 5.7 Stresses in the ballast

total quantum of the load, but also on the manner in which it is transferred to the subgrade. The spacing between the sleepers; the size, depth, as well as compaction of the ballast under the sleeper; and the type of subgrade play an important role in the distribution of pressure on the subgrade.

The values of maximum formation pressure permitted on Indian Railways are the following:

For motive power	3.5 kg/cm² for BG
	2.5 kg/cm² for MG
For goods wagons	3.0 kg/cm² for BG
	2.3 kg/cm² for MG

5.9 RELIEF OF STRESSES

A train load consists of a number of wheel loads close to each other which act simultaneously on the rail. A single isolated wheel load creates much more bending moment in the rail as compared to a group of wheel loads, which on account of the negative bending moment under adjacent wheels provide what is known as a 'relief of stresses'. The rail stresses in this case are comparatively smaller. The value of relief of stresses depends upon the distance of the point of contraflexure of the rail and the spacing between the wheels, but its value can be as high as 50 per cent.

Permissible stresses on a rail section

The permissible bending stresses due to vertical load and its eccentricity, and lateral load on a rail section in Indian Railways is given in Table 5.7.

Table 5.7 Permissible bending stress

	kg/mm²	*t/in²*
Permissible stress due to bending	36.0	23.00
Minimum ultimate tensile strength	72.0	46.0

The stresses on a rail are measured by any of the following methods depending upon the facilities available.

(a) Photo-elastic method

(b) Electric resistance strain gauge method

(c) Method employed using special test frame

Today, Indian Railways mostly uses the electric resistance strain gauges for measuring rail stresses.

Whenever a new locomotive or rolling stock design is introduced in the railways, a detailed study is carried out followed by field trials to ensure that the permitted speed of the new locomotive or rolling stock does not cause excessive stresses on the track. The same stipulations are made whenever there is an increase in the speed or axle load of the existing locomotive or rolling stock design.

The various parameters and their limiting values required to be checked for BG are given in Table 5.8.

Table 5.8 Limiting values of stresses in BG

Parameter	*Permissible value*
Bending stress on the rail	36.0 kg/mm²
Contact stress between the rail and the wheel	21.6 kg/mm²
Dynamic overloads at rail joints due to unsuspended masses	27 t for locomotives and 19 t for wagons
Formation pressure	3.5 kg/cm² for locomotives and 3.0 kg/cm² for wagons
Fish plate stresses	30 kg/mm²
Bolt hole stresses	27 kg/mm²

5.7 CONING OF WHEELS

The tread of the wheels of a railway vehicle is not made flat, but sloped like a cone in order to enable the vehicle to move smoothly on curves as well as on straight tracks. The wheels are generally centrally aligned on a straight and level surface with uniform gauge, and the circumference of the treads of the inner and outer wheels are equal as can be seen in Fig. 5.8.

The problem, however, arises in the case of a curve, when the outer wheel has to negotiate more distance on the curve as compared to the inner wheel. Due to the action of centrifugal force on a curve, the vehicle tends to move out. To avoid this the circumference of the tread of the outer wheel is made greater than that of the inner wheel. This helps the outer wheel to travel a longer distance than the inner wheel.

The wheels of a railway vehicle are connected by an axle, which in turn is fixed on a rigid frame. Due to the rigidity of the frame, the rear axle has a tendency to move inward, which does not permit the leading axle to take full advantage of the

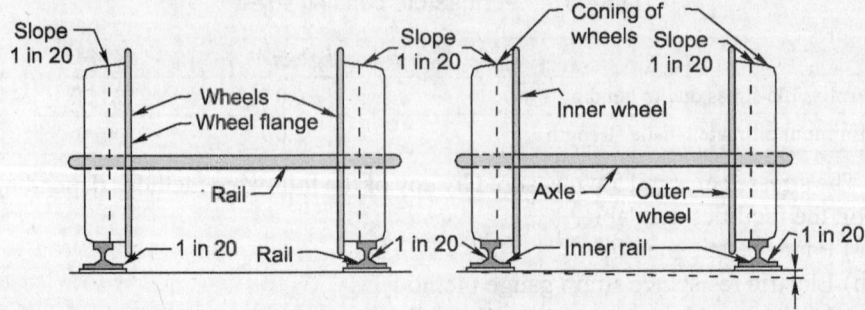

Fig. 5.8 Coning of wheels

coning. The rigidity of the frame, however, helps to bring the vehicle back into central alignment and thus works as a balancing factor.

The coning of wheels also helps to keep the vehicle centrally aligned on a straight and level track. Slight irregularities in the track do occur as a result of moving loads and the vagaries of the weather. The wheels, therefore, move from side to side and therefore the vehicles sway. Due to the coning of wheels, this side movement results in the tread circumference of one wheel increasing over the other. As both the wheels have to traverse the same distance, this causes one wheel to slide. Due to the resistance caused by the sliding, any further side movement is prevented. If there was no coning, the side movement would have continued and the flange of the wheel would have come in contact with the side of the rail, causing jerks and make the ride uncomfortable.

Coning of wheels causes wear and tear due to the slipping action. It is, however, useful in the following ways:

(a) It helps the vehicle to negotiate a curve smoothly.

(b) It provides a smooth ride.

(c) It reduces the wear and tear of the wheel flanges.

The slip can be mathematically calculated as follows:

$$\text{Slip} = \frac{2\pi\theta}{360} G \tag{5.13}$$

where θ is the angle at the centre of the curve fixed by the rigid wheel box and G is the gauge in metres.

The approximate value of the slip for broad gauge is 0.029 metre per degree of the curve.

5.8 TILTING OF RAILS

Rails are tilted inward at an angle of 1 in 20 to reduce wear and tear on the rails as well as on the tread of the wheels. As the pressure of the wheel acts near the inner edge of the rail, there is heavy wear and tear of the rail. Lateral bending stresses are also created due to eccentric loading of rails. Uneven loading on the sleepers is also likely to cause them damage. To reduce wear and tear as well as lateral stresses, rails are tilted at a slope of 1 in 20, which is also the slope of the wheel cone. The rail is tilted by 'adzing' the wooden sleeper or by providing canted bearing plates.

SUMMARY

Permanent way consists of rails, sleepers, the ballast, sleeper fittings, and the subgrade. The strength of each of these components is essential for the safe running of trains. The stresses developed in each component due to the movement of wheel loads should be within permissible limits as specified for different types of tracks. The concept of load distribution in a railway track is based on the elastic theory, but most of the equations used to calculate stresses in the different components lack a theoretical background. The coning of wheels helps reduce the wear and tear of the wheel flanges, providing a smooth ride. The ill effects of coning on horizontal curves are reduced by laying the rails at a slope of 1 in 20.

REVIEW QUESTIONS

1. What do you understand by a railway track or a permanent way? Mention the requirements of an ideal permanent way.
2. What are the component parts of a permanent way?
3. What is 'track modulus'? Indicate its usual range of values for a broad gauge track.
4. How is track modulus expressed? State the factors affecting it and give the values of at least one of these factors for the tracks in our country.
5. Draw a typical cross section of a permanent way. Explain briefly the functions of the various components of the railway track.
6. Discuss the necessity and effects of coning of wheels.
7. What are the various types of stresses induced in a rail section? Explain briefly how these are evaluated.

   Choose the coorect answer from the choices given.
8. Three wheels of an engine are equi-spaced with centre-to-centre distance of 142 cm. The central wheel carries a static load of 10 tonnes, while the outer wheels carry 8.8 tonnes each. Calculate the bending moment under the first wheel using the following data.

Moment of inertia of rail section	=	1950 cm^4
Initial track modulus	=	90 kg/cm^2
Elastic track modulus	=	300 kg/cm^2
Speed of movement	=	80 kmph

Choose the correct answer from the choices given.

9. The rail section for Group A route having traffic density of above 20 GMT is:

   (a) 52 kg   (b) 60 kg

   (c) 90 R   (d) none of these
10. Sleeper density in terms of number of sleepers per km for traffic density of 10 to 20 GMT is:

    (a) 1660   (b) 1540

    (c) 1310   (d) none of these

11. The rail section prescribed for MG Q routes is
    - (a) 52 kg
    - (b) 90 RNew
    - (c) 90 R (SH)
    - (d) none of these

12. The value of maximum formation presence permitted on Indian Railways for motive power for BG is:
    - (a) 3.0 kg/cm^2
    - (b) 3.5 kg/cm^2
    - (c) 2.5 kg/cm^2
    - (d) none of these

13. The permissible value of dynamic overloading at joints on Indian Railways for BG locomotive is:
    - (a) 23 tonnes
    - (b) 25 tonnes
    - (c) 27 tonnes
    - (d) none of these

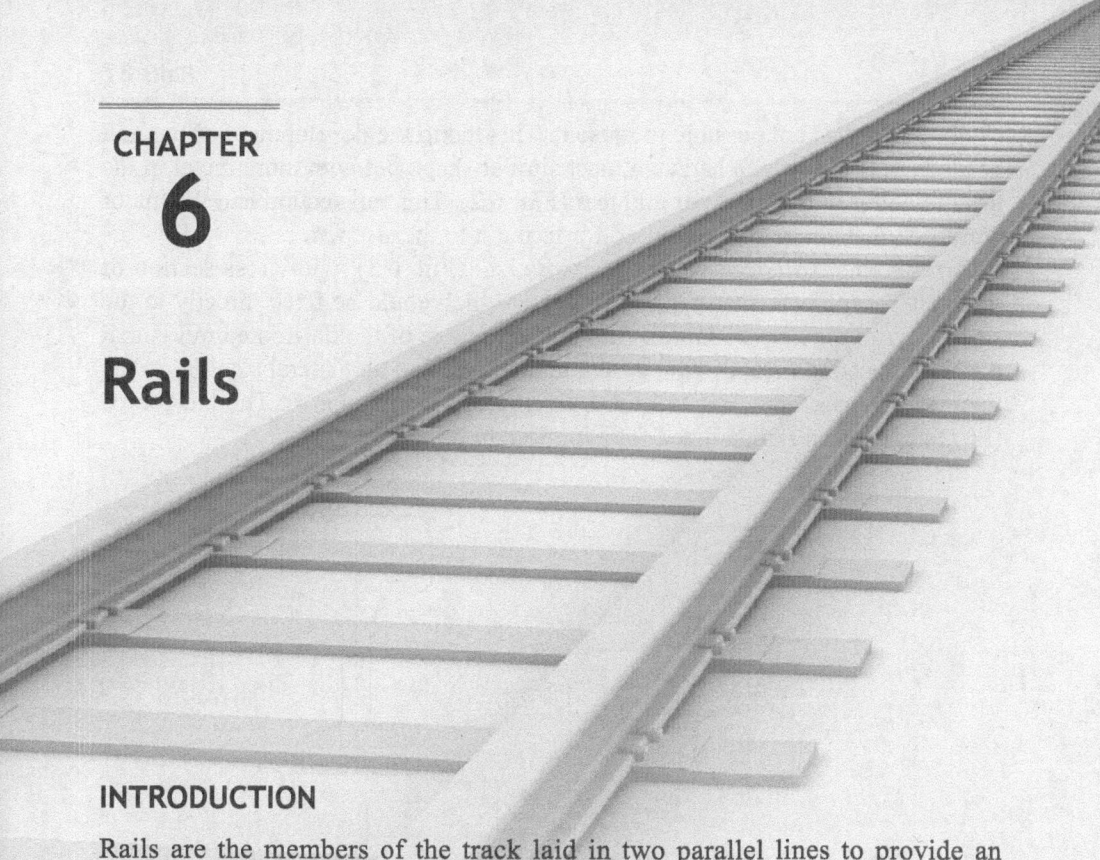

Rails

INTRODUCTION

Rails are the members of the track laid in two parallel lines to provide an unchanging, continuous, and level surface for the movement of trains. To be able to withstand stresses, they are made of high-carbon steel. Standard rail sections, their specifications, and various types of rail defects are discussed in this chapter.

6.1 FUNCTION OF RAILS

Rails are similar to steel girders. They perform the following functions in a track:

(a) Rails provide a continuous and level surface for the movement of trains.
(b) They provide a pathway which is smooth and has very little friction. The friction between the steel wheel and the steel rail is about one-fifth of the friction between the pneumatic tyre and a metalled road.
(c) They serve as a lateral guide for the wheels.
(d) They bear the stresses developed due to vertical loads transmitted to them through axles and wheels of rolling stock as well as due to braking and thermal forces.
(e) They carry out the function of transmitting the load to a large area of the formation through sleepers and the ballast.

6.2 TYPES OF RAILS

The first rails used were double headed (DH) and made of an I or dumb-bell section (Fig. 6.1). The idea was that once the head wore out during service, the rail could be inverted and reused. Experience, however, showed that the bottom table of the rail was dented to such an extent because of long and continuous contact with the chairs that it was not possible to reuse it. This led to the development of the bull headed (BH) rail, which had an almost similar shape but with more metal in the head to better withstand wear and tear (Fig. 6.2). This rail section had the major drawback that chairs were required for fixing it to the sleepers.

A *flat-footed rail*, also called a *vignole rail* (Fig. 6.3), with cross section of inverted T- type was, therefore, developed, which could be fixed directly to the sleepers with the help of spikes. Another advantage of the flat-footed rail is that it is a more economical design, giving greater strength and lateral stability to the track as compared to a BH rail for a given cross-sectional area. The flat-footed (FF) rail has been standardized for adoption on Indian Railways.

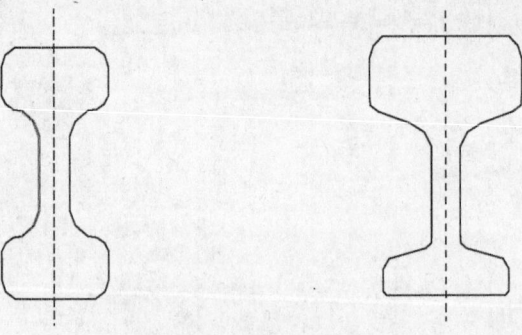

Fig. 6.1 Double headed rail   **Fig. 6.2** Bull headed rail

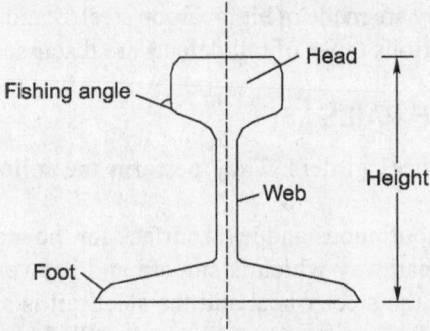

Fig. 6.3 Flat-footed rail

6.3 REQUIREMENTS OF AN IDEAL RAIL SECTION

The requirements of an ideal rail section are as follows:
  (a) The rail should have the most economical section consistent with strength, stiffness, and durability.
  (b) The centre of gravity of the rail section should preferably be very close to the mid-height of the rail so that the maximum tensile and compressive stresses are equal.
  (c) A rail primarily consists of a head, a web, and a foot, and there should be an economical and balanced distribution of metal in its various components so that each of them can fulfil its requirements properly.

The requirements, as well as the main considerations, for the design of these rail components are as follows:

Head　The head of the rail should have adequate depth to allow for vertical wear. The rail head should also be sufficiently wide so that not only is a wider running surface available, but also the rail has the desired lateral stiffness.

Web　The web should be sufficiently thick so as to withstand the stresses arising due to the loads bore by it, after allowing for normal corrosion.

Foot　The foot should be of sufficient thickness to be able to withstand vertical and horizontal forces after allowing for loss due to corrosion. The foot should be wide enough for stability against overturning. The design of the foot should be such that it can be economically and efficiently rolled.

Fishing angles　These must ensure proper transmission of loads from the rails to the fish plates. The fishing angles should be such that the tightening of the plate does not produce any excessive stress on the web of the rail.

Height of the rail　The height should be adequate so that the rail has sufficient vertical stiffness and strength as a beam.

6.3.1 Standard Rail Section

The rail is designated by its weight per unit length. In FPS units, it is the weight in lbs per yard and in metric units it is in kg per metre. A 52 kg/m rail denotes that it has a weight of 52 kg per metre.

The weight of a rail and its section is decided taking into consideration the following:

(a) Heaviest axle load
(b) Maximum permissible speed
(c) Depth of ballast cushion
(d) Type and spacing of sleepers
(e) Other miscellaneous factors

The standard rail sections in use in Indian Railways are 60 kg, 52 kg, 90 R, 75 R, 60 R, and 50 R. The two heavier rail sections, 60 kg and 52 kg, were recently introduced and are designated in metric units. Other rails are designed as per the revised British Standard specifications and are designated in FPS units though their dimensions and weight are now in metric units. In the nomenclature 90 R, 75 R, etc., R stands for revised British specifications.

Taking into consideration, the requirement of future traffic, Indian Railways in its new specification (IRS T-12-2009) has standardized two heavier rails also, viz. 68 kg and ZU-1-60.

Branding of Rail

Every rail rolled has a brand on its web, which is repeated at an interval of three metres or less.

As per IRS-T-12-2009, brand marks should be rolled on one side of the web of each rail at least every three metres. The letters should be at least 20 mm in height and at least one metre above the surface of the web of the rail.

The brand marks should be as follows:

IRS–52kg – 880 – SAIL X 10 → OB

The definitions for the various abbreviations are as follows:
 (a) IRS-52-kg: Number of IRS rail section, i.e., 52 kg
 (b) 880: Grade of rail section, i.e., 880 (or 710)
 (c) SAIL: Manufacturer's name, e.g., Steel Authority of India Ltd (SAIL)
 (d) X 10: Month and year of manufacture, 10th month of year 2010 (X 10)
 (e) →: An arrow showing the direction of the top of the ingot
 (f) OB: Process of steel making, e.g., open hearth basic (OB)

The brand marks on the rails are to be rolled in letters at least 20 mm in size and 1.5 mm in height at intervals of 1.5 to 3.0 m.

The standard rail sections and standard rail length prescribed on Indian Railways are given in Table 6.1.

Table 6.1 Standard rail sections

Gauge	Rail section	Type of section	Rail length
Broad gauge	60 kg/m	UIC	13 m (42 ft as per old standards)
	52 kg/m	IRS	
	90 lb/yd	RBS	
Metre gauge	90 lb/yd	RBS	12 m (39 ft as per old standards), except 90 R rails, which are of 13 m length
	75 lb/yd	RBS	
	60 lb/yd	RBS	
Narrow gauge	50 lb/yd	RBS	12 m (39 ft as per old standards)

UIC—International Union of Railways, IRS—Indian Railway Standard, RBS—Revised British Standard.

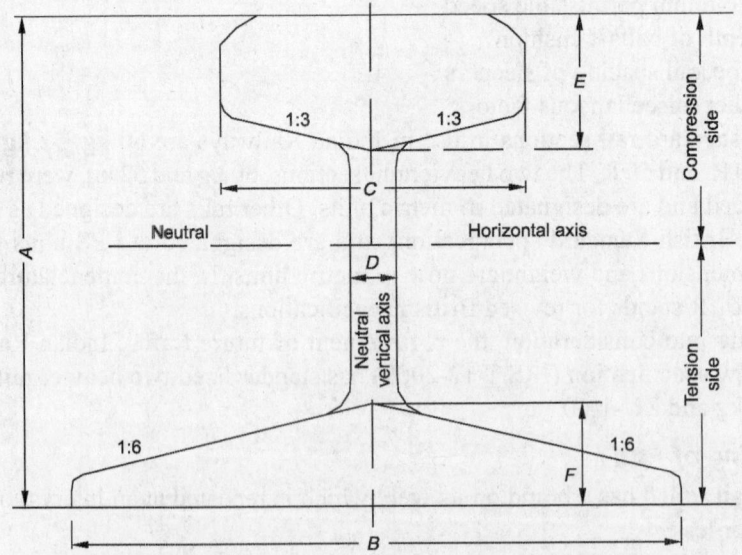

Fig. 6.4 Standard flat-footed rail section

Detailed dimensions of standard rail sections are shown in Fig. 6.4 and Table 6.2.

Table 6.2 Details of standard rail sections

Rail section	Wt/ metre (kg)	Area of section (mm²)	Dimensions (mm)					
			A	*B*	*C*	*D*	*E*	*F*
50 R	24.80	3168	104.8	100.0	52.4	9.9	32.9	15.1
60 R	29.76	3800	114.3	109.5	57.2	11.1	35.7	16.7
75 R	37.13	4737	128.6	122.2	61.9	13.1	39.7	18.7
90 R	44.61	5895	142.9	136.5	66.7	13.9	43.7	20.6
52 kg (IRS)	51.89	6615	156.0	136.0	67.0	15.5	51.0	29.0
60 kg (UIC)	60.34	7686	172.0	150.0	74.3	16.5	51.0	31.5

It may be mentioned here that the 90 R rail section is adequate only for an annual traffic density of about 10 GMT (gross million tonnes per km/annum), speeds of up to 100 km per hour, axle loads up to main line (ML) standard, and a service life of about 20–25 years. Realizing these limitations, the Indian Railways, in 1959, designed a heavier rail section of 52 kg/m to meet the requirements of heavier and faster traffic. This rail section was recommended for use on all BG main line routes with future speeds of up to 130 km per hour and traffic density of 20–25 GMT. The important dimensions of 52 kg and 60 kg rails are shown in Fig. 6.5.

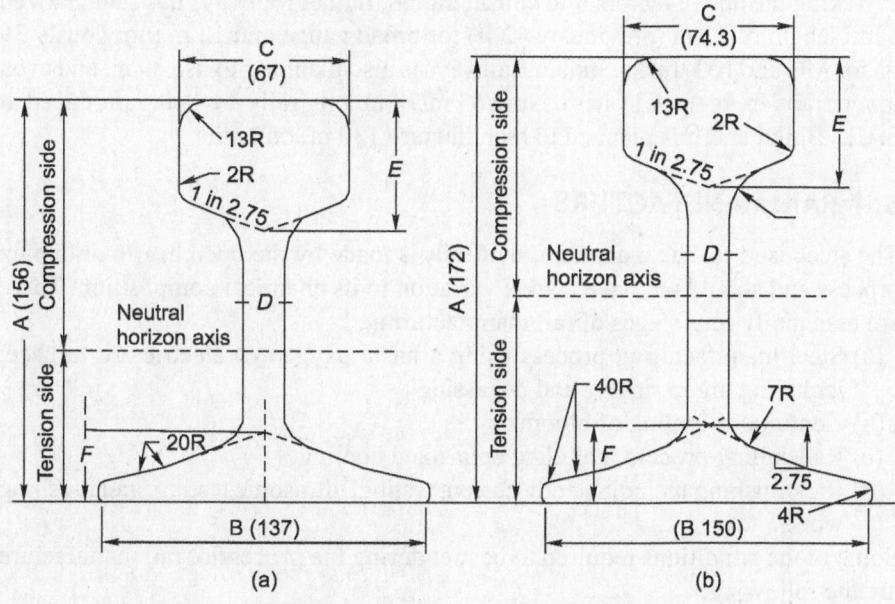

Fig. 6.5 (a) 52 kg rail (b) 60 kg rail

The traffic density on the BG track routes of Indian Railways is increasing very fast. Accordingly, to meet the future requirements of traffic, a new design has been finalized for the 60 kg UIC section rail. The rail section has been designed for speeds of up to 160 km per hour and a traffic density of about 35 GMT.

Weight of rails

Though the weights of a rail and its section depend upon various considerations, the heaviest axle load that the rail has to carry plays the most important role. The following is the thumb rule for defining the maximum axle load with relation to the rail section:

Maximum axle load = 560 × sectional weight of rail in lbs per yard
or kg per metre

- For rails of 90 lbs per yard,
  Maximum axle load = 560 × 90 lbs = 50,400 lbs or 22.5 tonnes
- For rails of 52 kg per m,
  Maximum axle load = 560 × 52 kg = 29.12 tonnes
- For rail of 60 kg per m,
  Max. axle load for 60 kg/m rail = 560 × 60 kg = 33.60 tonnes

Length of rails

Theoretically, the longer is the rail, the lesser would be the number of joints and fittings required and the lesser the cost of construction and maintenance. Longer rails are economical and provide smooth and comfortable rides. The length of a rail is, however, restricted due to the following factors:

(a) Lack of facilities for transport of longer rails, particularly on curves
(b) Difficulties in manufacturing very long rails
(c) Difficulties in acquiring bigger expansion joints for long rails
(d) Heavy internal thermal stresses in long rails

Taking the above factors into consideration, Indian Railways has standardized a rail length of 13 m (previously 42 ft) for broad gauge and 12 m (previously 39 ft) for MG and NG tracks. Indian Railways is also planning to use 39 m, and even longer rails in its track system. Now 65 m/78 m long rails are being produced at SAIL, Bhilai and it is planned to manufacture 130 m long rails.

6.4 RAIL MANUFACTURE

The steel used for the manufacture of rails is made by the open hearth or duplex process and should not have a wide variation in its chemical composition. There are essentially four stages of rail manufacturing.

(a) Steel manufacturing process using a basic oxygen or electric arc furnace, including argon rinsing and degassing
(b) Continuous casting of blooms
(c) Rail rolling process including controlled cooling
(d) Rail finishing including eddy current testing, ultrasonic testing, and finishing work

Some of the conditions required to be met during the process of rail manufacture are the following:

(i) The steel used for the manufacture of rail shall be made by basic oxygen or electric are furnace process and continuously cast. Any other method of casting should have prior approval of the purchaser. For molten steel secondary ladle refining is mandatory.
(ii) The cross-sectional area of the bloom shall not be less than ten times that of the rail section to be produced.
(iii) For head hardening, rails should be suitably heat treated to meet the requirements of this specification.

A typical flow chart for the manufacture of rails at the Bhilai steel plant in India is given in Fig. 6.6.

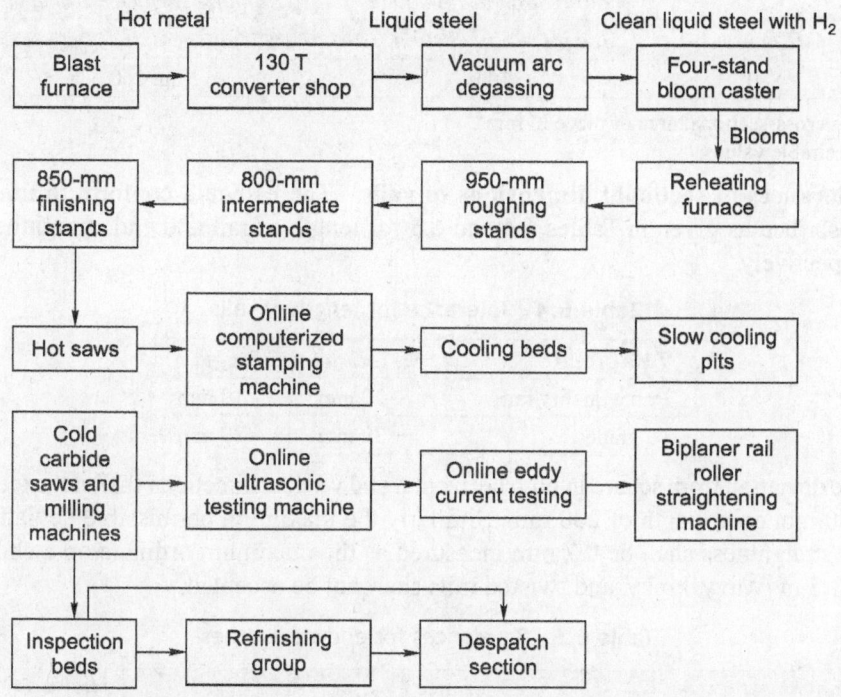

Fig. 6.6   Process of rail manufacture at Bhilai steel plant, India

6.4.1 Salient Features of Rail Specification (T-12-2009)

The Rail Specification (T-12-2009) was initially adopted in 1934 and subsequently revised in 1939, 1950, 1953, 1955, 1958, 1960, 1964, 1988, 1996, and in 2009.

The present version has been adopted in 2009 specifying the requirements of the prime rail and IU rails having ultimate tensile strength (UTS) of 880 MPa, 1080 MPa CR, and 1080 MPa HH. This sepecification also specifies the requirements of special classes of rail steel such as Niobium (NB), Vanadium (VN), corrosion resistant rail steel such as Copper Molybdenum (CM), Nickel Chromium Copper (NC).

This specification applies to flat bottom rails. It specifies the quality of steel, manufacturing process, chemical compostion, acceptance tests/retests, qualifying criteria and other technical conditions of supply.

Chemical composition of rails   The chemical composition of rails should be as per limits of various chemicals as given in Table 6.3.

Table 6.3   Desirable properties of steel for rails

(a) Chemical composition

Grade	C	Mn	Si	S (max.)	P (max.)	Al (max.)	Liquid hydrogen
880	0.6–0.8	0.8–1.3	1.3–0.5	0.035[*]	0.035[*]	0.02	3.00

[*] The maximum value for finishing is 0.040.

Table 6.3 (*Contd.*)

(b) Mechanical properties

UTS (MPa; min.)	Elongation % on gauge length = 5.65/S* (min.)	Running surface hardness
880	10.0	Min. 260[†]

* S = cross-sectional area of piece in mm^2.
[†] Desirable values.

Tolerances in sectional dimensions of rails These should conform to limits prescribed as given in Tables 6.4 and 6.5 for length of rail and end straightness respectively.

Table 6.4 Tolerance for length of rails

Type of rail	Tolerance in length	
Prime quality rail	+ 20 mm	– 10 mm
IU grade	+ 30 mm	– 30 mm

The deviation from square in both horizontal and vertical directions shall not exceed 0.60 mm on a length of 200 mm. Similarly the maximum permissible deviation for straightness shall be 0.7 mm measured as the maximum ordinate on a chord of 1.5 m. Wavy, kinky, and twisted rails shall not be accepted.

Table 6.5 Tolerances for end straightness

Straightness	Tolerance			Remarks
	Class A Rails	**Class B rails**	**Class 1 A grade rails**	
Horizontal	Deviation of 0.5 mm	Deviation of 0.7 mm	Deviation of 1.5 mm	Measured as max. ordinate from a chord
Vertical (up sweep)	Deviation of 0.4 mm	Deviation of 0.5 mm	Deviation of 1.5 mm	of 2 m for class A rails, 1.5 m for class B & 1 A rails.

Ultrasonic testing of rails The limits of permissible defects for ultrasonic testing of rails shall be as follows:

Head	:	1.5 mm dia through hole
Web	:	2.00 mm dia through hole
Web & foot junction	:	2.0 mm dia through hole
Foot	:	0.5 mm deep, 12.5 mm long and 1.0 wide notch (inclined at 20″ with vertical axis)

Surface quality Tables 6.6 and 6.7 present the limits of surface defects in primary rails and class 1A grade rails.

Table 6.6 Limits of surface defects in primary rails

Item	Details of hot mark
Hot marks	(i) For 0.5 mm depth, hot mark width from 1.5 mm to 4.0 mm (ii) For 0.4 mm depth, hot mark width for 1.2 to 4.0 mm (iii) For 0.3 mm depth, hot mark width from 0.9 to 4.0 mm
Cold marks	Cold formed scratches anywhere on the rail should not exceed 0.5 mm.
Seams	Not greater than 0.2 mm in depth.

Table 6.7   Limits of surface defects in 1A rails

Type of defect	Location	Permissible dimensions of defects
Seams	(a) Table of rails, side of the head of rail, bottom and side of the foot of rail (excepting, middle third of the foot)	Up to 3 mm in depth
	(b) Middle third of the bottom surface of the foot on the rail	Up to 2 mm in depth
Scabs	Table of rail and side of the head of the rail	75 mm × 25 mm not to exceed 3 mm in depth

6.4.2 Qualifying Criteria of Rails

The following tests should by done for each rail section, grade, and class after any change in the process of manufacture, which may affect the results annually for the first three years after adoption of the revised specification. If results of these three years are consecutively found satisfactory, this frequency may be relaxed to three years by the purchaser. The types of measurement are as follows:

 (i)   Residual stress measurement
 (ii) Fracture toughness measurement
 (iii) Fatigue test

In case any sample fails to meet the requirement laid in qualifying criteria, the manufacturer shall review its process of manufacturing within six months to eliminate any shortcomings and fresh qualifying criteria test shall be undertaken under intimation to the purchaser.

6.4.3 Acceptance Tests of Rails

The following acceptance tests shall be conducted for Grade 880, 1080 CR, 880 CM, 880 NC, 880 VN, and 880 NB Rails:

 (i)   Chemical analysis
 (ii) Tensile test
 (iii) Sulphur print
 (iv) Hardness test (for information and record)
 (v)   Falling weight test
 (vi) Hydrogen content
 (vii) Inclusion rating level

The deatails of some of the important tests are given here.

Falling weight or tup test   A rail piece of 1.5 m (5 ft) is cut. The rail is supported between the bearers at a prescribed distance. A tup of specified weight (1000 kg for a 90 R rail) is dropped from a height of 7.2 m on the centre of the test piece. The specimen should withstand the blow without any fracture. One falling weight test is done for every cast of 100 metric tonnes. The weight of the tup, distance between the centres of the bearings, and the weight of the drop for different rail sections are given in Table 6.8.

Table 6.8 Details of the tup test

Rail section	Wt of tup (kg)	Distance between centres of bearings (m)	Height of drop (m)
60 kg/m	1270	1.07	8.4
52 kg/m	1270	1.20	7.6
90 R	1000	1.10	7.2
75 R	1000	1.10	6.0

Tensile test   A test piece is taken from the head of a rail section and subjected to the tensile test. The tensile strength of the rail should not be less than 72 kg/mm^2, with a minimum elongation of 14 per cent for medium manganese rails and 12 per cent for carbon quality rails. This test is optional and is to be carried out when required by the inspecting official.

Hammer test   The foot of the test rail piece is rigidly gripped in a vertical position and the head of the rail is struck with a 4.5 kg hammer. Sufficient number of blows are given till the web bends and the dimensional value of A given in Table 6.9 (Fig. 6.7) is achieved. No fracture should occur or a lap be disclosed, otherwise the batch is rejected. This test is no longer required as per IRS/T- 18-88. The values of A for various rail sections are listed in Table 6.9.

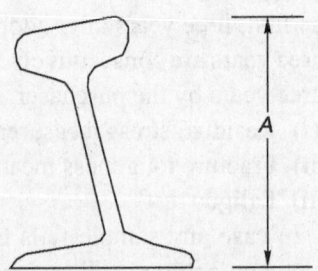

Fig. 6.7   Hammer test

Table 6.9   Values of *A* for different rail sections

Rail section	52 kg/m	90 R	75 R	60 R	50 R
Dimension A in mm	157	144.5	130	117.7	108

6.4.4 90 UTS Rails

Indian Railways has mostly been using medium manganese rails with an ultimate tensile strength (UTS) of 72 kg/mm^2 manufactured at the Bhilai steel plant. The service life of 52 kg (72 UTS) rails is only about 350 GMT. On a section with an annual traffic density of about 20 GMT, the renewal cycle is just about 17–18 years, which is rather short as compared to the service life of 50 years of a concrete sleeper. Moreover, such rails wear faster on curves and gradient sections.

In view of the above considerations, Indian Railways has been importing 52-kg and 90 R, 90 UTS rails for some time. These rails have the following main advantages:

1. The service life of 90 UTS rails is about 50 per cent more than that of conventional medium manganese 72 UTS rails.
2. The total GMT that 72 and 90 UTS rails can carry during their primary service life is as follows:

        52 kg (72 UTS): 350 GMT
        52 kg (90 UTS): 525 GMT
        60 kg (90 UTS): 900 GMT

3. 90 UTS rails are more resilient against wear and have a hardness of about 270 BHN (Brinell hardness number) as against that of 220 BHN of medium manganese rails with 72 UTS.

4. The allowable shear stress of 90 UTS rails is much higher, as can be seen from the following comparative figures shown below.

Rails	Allowable shear stress
Medium manganese rails (72 UTS)	18.0 kg/mm^2
Wear-resistant rails (90 UTS)	22.5 kg/mm^2

Studies have shown that the maximum shear stress due to BOXN wagons could be of the order of 20.0 kg/mm^2, which is in excess of the permissible shear stress for medium manganese 72 UTS rails. Therefore, for routes on which BOXN wagons are running, it is desirable to have 90 UTS rails.

6.4.5 110 UTS Rails

Due to the introduction of BOXN stock, there is not only higher shear stress in rails but the wear of rails is also faster and excessive as the BOXN stock have cast wheels with higher hardness. This has lead to the introduction of even higher UTS rails of 110 kg per sq. mm.

Indian Railways has imported about 45,000 tonnes of 110 UTS rails in 1990 which have been laid on busy corridors like Mumbai Division of Central Railways, which have predominantly EMU traffic. These 110 UTS rails have given so far better performance and their further behaviour is still under watch.

While handling 90 UTS and higher UTS rails, following points require special consideration:
  (i) Protection of straightness
  (ii) Protection of rail surface
  (iii) Prevention of metallurgical damage
  (iv) Protection from contact with injurious substances
  (v) Precautions to minimize danger to personnel handling rails
  (vi) Proper stacking to avoid damages to rails
  (vii) Handling of rails in flash butt welding plants and production units in proper manner by magnetic lifting devises, etc., to avoid damages.
  (viii) Adequate precautions for handling of rails in electrified areas.

End-hardened rails

These are rails with ends that are hardened by oil or water quenching. The wear and tear and end batter of such rails is considerably less.

Head-hardened rails

These are rails with heads that have been hardened by passing them through a thermal treatment plant. The head is hardened for a depth of about 12 mm from the surface. Head hardened rails have a longer service life that extends up to two to three times more compared to ordinary medium manganese rails.

The chemical composition of head-hardened steel (grade 1080) is prescribed as given in Table 6.10.

Table 6.10 Chemical composition of head-hardened rail

Item	Carbon	Manganese	Silicon	Sulphur	Phosphorus
Limit of values	0.72–0.82	0.75–1.05	0.05–0.30	0.035 max.	0.035 max.

6.5 RAIL WEAR

Due to the passage of moving loads and friction between the rail and the wheel, the rail head gets worn out in the course of service. The impact of moving loads, the effect of the forces of acceleration, deceleration, and braking of wheels, the abrasion due to rail–wheel interaction, the effects of weather conditions such as changes in temperature, snow, and rains, the presence of materials such as sand, the standard of maintenance of the track, and such allied factors cause considerable wear and tear of the vertical and lateral planes of the rail head. Lateral wear occurs more on curves because of the lateral thrust exerted on the outer rail by centrifugal force. A lot of the metal of the rail head gets worn out, causing the weight of the rail to decrease. This loss of weight of the rail section should not be such that the stresses exceed their permissible values. When such a stage is reached, rail renewal is called for.

In addition, the rail head should not wear to such an extent that there is the possibility of a worn flange of the wheel hitting the fish plate.

6.5.1 Types of Wear on Rails

A rail may face wear and tear in the following positions:
  (a) On top of the rail head (*vertical wear*)
  (b) On the sides of the rail head (*lateral wear*)
  (c) On the ends of the rail (*battering of rail ends*)

Wear is more prominent at some special locations of the track. These locations are normally the following:
  (a) On sharp curves, due to centrifugal forces
  (b) On steep gradients, due to the extra force applied by the engine
  (c) On approaches to railway stations, possibly due to acceleration and deceleration
  (d) In tunnels and coastal areas, due to humidity and weather effects

6.5.2 Measurement of Wear

Wear on rails can be measured using any of the following methods:
  (a) By weighing the rail
  (b) By profiling the rail section with the help of lead strips
  (c) By profiling the rail section with the help of needles
  (d) By using special instruments designed to measure the profile of the rail and record it simultaneously on graph paper

6.5.3 Methods to Reduce Wear

Based on field experience, some of the methods adopted to reduce vertical wear and lateral wear on straight paths and curves are as follows:

(a) Better maintenance of the track to ensure good packing as well as proper alignment and use of the correct gauge
(b) Reduction in the number of joints by welding
(c) Use of heavier and higher UTS rails, which are more wear resistant
(d) Use of bearing plates and proper adzing in case of wooden sleepers
(e) Lubricating the gauge face of the outer rail in case of curves
(f) Providing check rails in the case of sharp curves
(g) Interchanging the inner and outer rails
(h) Changing the rail by carrying out track renewal

6.5.4 Rail End Batter

The hammering action of moving loads on rail joints batters the rail ends in due course of time. Due to the impact of the blows, the contact surfaces between the rails and sleepers also get worn out, the ballast at places where the sleepers are joined gets shaken up, the fish bolts become loose, and all these factors further worsen the situation, thereby increasing rail end batter.

Rail end batter is measured as the difference between the height of the rail at the end and at a point 30 cm away from the end. If the batter is up to 2 mm, it is classified 'average', and if it is between 2 and 3 mm, it is classified as 'severe'. When rail end batter is excessive and the rail is otherwise alright, the ends can be cropped and the rail reused.

Rail lubricators are provided on sharp curves, where lateral wear is considerable. The function of lubricators is to oil the running face of the outer rail in order to reduce the friction. It has been noticed that this considerably reduces the wear, by up to 50 per cent. There are many mechanical devices that can be attached to the wheels to provide such lubrication. In these mechanical arrangements, the wheels of moving trains normally cause the lubricant to flow on the side of the rail either by the action of the wheels pressing the plunger up and down or by ramps on account of the rails being depressed by wheels. Sometimes the movement of trains also causes lubricants to flow. Based on the principle of the plunger being pressed by moving wheels, P and M type lubricators have been provided on curves in some sections such as the 'Ghat section' of Central Railway and these are working very satisfactorily. For more details on wear, including limit of wear, refer to Chapter 23.

6.6 OTHER DEFECTS IN RAILS

Rail wear and battering of rail ends are the two major defects in rails. However, some other types of defects may also develop in a rail and necessitate its removal in extreme cases. These are as follows:

Hogging of rails

Rail ends get hogged due to poor maintenance of the rail joint, yielding formation, loose and faulty fastenings, and other such reasons. Hogging of rails causes the quality of the track to deteriorate. This defect can be remedied by measured shovel packing. (For details, refer to Chapter 20.)

Flange oilers lubricate wheel flanges to reduce rail wear in tight curves, Middelburg, Mpumalanga, South Africa. (Courtesy: Col André Kritzinger. This file is licensed under the Creative Commons Attribution-Share Alike 3.0 Unported license; http://creativecommons.org/licenses/by-sa/3.0/deed.en)

Scabbing of rails

The scabbing of rails occurs due to the falling of patches or chunks of metal from the rail table. Scabbing is generally seen in the shape of an elliptical depression, whose surface reveals a progressive fracture with numerous cracks around it.

Wheel burns

Wheel burns are caused by the slipping of the driving wheel of locomotives on the rail surface. As a consequence, extra heat is generated and the surface of the rail gets affected, resulting in a depression on the rail table. Wheel burns are generally noticed on steep gradients or where there are heavy incidences of braking or near water columns.

Shelling and black spots

Shelling is the progressive horizontal separation of metal that occurs on the gauge side, generally at the upper gauge corner. It is primarily caused by heavy bearing pressure on a small area of contact, which produces heavy internal shear stresses.

Corrugation of rails

Corrugation consists of minute depressions on the surface of rails, varying in shape and size and occurring at irregular intervals. The exact cause of corrugation is not yet known, though many theories have been put forward. The factors which help in the formation of rail corrugation, however, are briefly enumerated here.

  (a) Metallurgy and age of rails

    (i) High nitrogen content of the rails

    (ii) Effect of oscillation at the time of rolling and straightening of rails

(b) Physical and environment conditions of track
  (i) Steep gradients
  (ii) Yielding formation
  (iii) Long tunnels
  (iv) Electrified sections
(c) Train operations
  (i) High speeds and high axle loads
  (ii) Starting locations of trains
  (iii) Locations where brakes are applied to stop the train
(d) Atmospheric effects
  (i) High moisture content in the air particularly in coastal areas
  (ii) Presence of sand

The corrugation of rails is quite an undesirable feature. When vehicles pass over corrugated rails, a roaring sound is produced, possibly due to the locking of air in the corrugation. This phenomenon is sometimes called 'Roaring of rails'. This unpleasant and excessive noise causes great inconvenience to the passengers. Corrugation also results in the rapid oscillation of rails, which in turn loosens the keys, causes excessive wear to fittings, and disturbs the packing.

Corrugation can be removed by grinding the rail head by a fraction of a millimetre. No method has been standardized on Indian Railways to grind rail surfaces. The problem of corrugation, however, has been tackled in great detail on German Railways, where two types of equipment are normally used for rail grinding.

  (i) Hand or motor-driven trollies that move on the rails at slow speeds and grind the individual rails one by one
  (ii) Rail grinding train, which moves at a speed of 30 km per hour and grinds both rails simultaneously

6.7 RAIL FAILURE

A rail is said to have failed if it is considered necessary to remove it immediately from the track on account of the defects noticed on it. The majority of rail failures originate from the fatigue cracks caused due to alternating stresses created in the rail section on account of the passage of loads. A rail section is normally designed to take a certain minimum GMT of traffic, but sometimes due to reasons such as an inherent defect in the metal, the section becomes weak at a particular point and leads to premature failure of the rail.

6.7.1 Causes of Rail Failures

The main causes of failure of rails are as follows:

Inherent defects in the rail These are due to manufacturing defects in the rail, such as faulty chemical composition, harmful segregation, piping, seams, laps, and guide marks.

Defects due to fault of the rolling stock and abnormal traffic effects Flat spots in tyres, engine burns, skidding of wheels, severe braking, etc.

Excessive corrosion of rails This generally takes place due to weather conditions, the presence of corrosive salts such as chlorides and constant exposure of the rails to moisture and humidity in locations near water columns, ashpits, tunnels, etc. Corrosion normally leads to the development of cracks in regions with a high concentration of stresses.

Badly maintained joints Poor maintenance of joints such as improper packing of joint sleepers and loose fittings.

Defects in welding of joints These defects arise either because of improper composition of the thermit weld metal or because of a defective welding technique.

Improper maintenance of track Ineffective or careless maintenance of the track or delayed renewal of the track.

Derailments The rails are damaged during derailment.

6.7.2 Classification of Rail Failures

The classification of rail failures on Indian Railways has been codified for easy processing of statistical data. The code is made up of two portions—the first portion consisting of three code letters and the second portion consisting of three or four code digits.

First portion of the code The three code letters make up the first portion and denote the following.
   (i) Type of rail being used (O for plain rail and X for points and crossing rails)
  (ii) Reasons for withdrawal of rail (F for fractured, C for cracked, and D for defective)
 (iii) Probable cause failure (S for fault of rolling stock, C for excessive corrosion, D for derailment, and O for others)

Second portion of code The second portion consisting of three or four digits gives the following information.
   (i) First digit indicate the location of the fracture on the length of the rail (1 for within fish plate limits and 2 for other portions on the rail).
  (ii) Second digit indicate the position in the rail section from where the failure started (0 for unknown, 1 for within rail head, 2 for surface of rail head, 3 for web, and 4 for foot).
 (iii) Third digit indicate the direction of crack or fracture (0 to 9).
 (iv) Any other information about the fracture, where it is necessary to provide further subdivision. No specific system is recommended for this code.

6.7.3 Metallurgical Investigation

The following types of defective rails should normally be sent for metallurgical investigation.
   (i) Rails that have been removed from the track as a result of visual or ultrasonic detection
  (ii) Rail failures falling in categories in which cracks or surface defects develop at specified locations

6.8 RAIL FLAW DETECTION

A defect in a rail which will ultimately lead to the fracture or breakage of the rail is called a *flaw*. From the point of view of the ultimate consequence of the flaw resulting in a fracture, it is necessary to detect these flaws and take timely action to remove them. Rail flaws can be detected either by visual examination of the rail ends or by rail flaw detection equipment.

Visual examination of rail ends In this method, the joint is first opened after removing the fish plates. The rail ends are then cleaned using kerosene oil and visually examined in detail with the help of a magnifying glass for any hairline crack, etc. White chalk is sometimes rubbed on the rail ends so as to identify the flaw clearly. A mirror is used to reflect light on the joint in case sufficient light is not available.

Ultrasonic rail flaw detectors Ultrasonic rail flaw detectors (USFDs) have been progressively used in recent years on Indian Railways for detecting flaws. This method is also known as the non-destructive method of testing rails.

6.8.1 Theory of Ultrasonic Rail Flaw Detectors

Vibration waves of a frequency of more than 20,000 cycles per second are termed as ultrasonic waves. These waves have the property of being able to pass through materials and following the normal principles of light waves of refraction, reflection, and transmission. Whenever there is a change of medium, some of the ultrasonic energy gets reflected and the rest gets transmitted. The amount of energy reflected depends upon the physical properties of the two media. When travelling through steel, if these waves come across air either from the bottom of the steel or from any flaw inside the steel, the reflection is almost 100 per cent. This property has been found most useful for detecting flaws in rails. Thus, when ultrasonic waves are fed from a location on a rail, they pass through the rail metal and are normally reflected only from the foot. However, if a discontinuity exists in the rail metal due to some flaw, the ultrasonic waves get reflected back from the location of the flaw, which can be picked up and the defect located.

Production of ultrasonic waves

There are several methods of producing ultrasonic energy. The most common and simple method of producing ultrasonic frequency is by using '*crystal transducers*', which normally produce ultrasonic waves of a frequency of up to 15 MHz. The crystals generally used for this purpose are made either of quartz or of barium titanate, cut to special size, shape, and dimensions. These crystals have the peculiar property of changing dimensions and generating vibrations in a particular direction when an oscillating electric charge is applied to the crystal faces. Also, when these crystals are made to vibrate, they produce an oscillating electric current. The crystals, as such, have the potential of generating ultrasonic vibrations, as also of converting the waves received after reflection into electric current. They also possess reversible properties.

These crystals are housed in metal holders protected by superior quality Perspex and are then called probes. There are two types of probes used for ultrasonic testing.

Normal probe   This probe consists of two semi-cylindrical thin crystals with a vertical separating layer through the crystals and Perspex. These probes transmit ultrasonic waves vertically downwards when put on the rail and are suitable for detecting horizontal or inclined flaws, including bolt hole cracks.

Angle probe   In this probe, crystals are mounted on an angular surface capable of transmitting pulses at an outward angle, which may be forward or backward or both, with separate transmitting and receiving crystals. The waves emitted by these probes follow an inclined path and are suited to detect inclined and vertical defects.

Techniques of ultrasonic testing

A number of techniques have been used in ultrasonic testing to suit the design of different types of equipment. Some of these techniques are as follows:
(a) Frequency modulation
(b) Pulse echo
(c) Transmission
(d) Resonance
(e) Acoustic range

Indian Railways uses the frequency modulation and pulse echo techniques which have been explained in the following:

Frequency modulation technique   In instruments utilizing frequency modulation ultrasonic waves are created with the help of a probe crystal and transmitted continuously into the rail at rapidly changing frequencies. It is necessary for the rail head to be wet to enable the ultrasonic waves to pass efficiently from the crystal to the rail. The waves that get reflected from the opposite face are received continuously by the crystal. There is interference between the transmitted waves and the reflected waves, which causes resonance. As the frequencies of the transmitted waves are changing constantly, such resonance takes place at regular intervals. When the position of reflection is changed due to a flaw in the metal, the resonance gets affected, which can be easily detected by the operator. Instruments manufactured on this principle such as the audi-gauge are light, portable, and simple in mechanism. However, these instruments have certain limitations.
(a) Fine vertical cracks are not readily detected because the single vertical probe does not find any surface defect from which it can be reflected.
(b) Cracks wholly below bolt holes are also not detected, as the vibrations are interrupted by the hole.

Pulse echo system   In the pulse echo technique, a pulsed ultrasonic beam of very high frequency is produced by a pulse generator and sent in to the rail. At the opposite face, the ultrasonic waves are reflected and the echo is picked up by the crystal transducers. A discontinuity or defect in the rail will also produce the echo. The time interval between the initial pulse and the arrival of the echoes is measured with the help of a cathode ray tube. There may be multiple reflections of the echo but the one arising due to a fault can easily be determined by its relative position and amplitude.

The more sophisticated types of instruments that are based on the pulse echo system are being manufactured by the firm Kraut Kramer at present.

6.8.2 Kraut Kramer Multi Probe Rail Testing Trolley

Kraunt Kramer multi proble rail testing trolley is the most common type of equipment used on Indian Railways for detecting flaws in the rail (Fig. 6.8). The equipment is fitted on a hand trolley that is carried on the rails. There are two probes: a normal probe and an angle probe, both act independently. The probe material used for the production of ultrasonic waves is barium titanate, which produces and transmits vertical ultrasonic waves of four megacycles frequency through the vertical probe and two megacycles frequency through the angle probe. The cylindrical probe is mounted on a knuckle jointed holder frame and has a renewable bakelite wear plate at the base. As the probe is worked over the rail, the bakelite piece takes the wear completely. The height of the probe above the rail surface can be adjusted in the holder assembly. The normal probe is powerful enough to scan the entire rail depth for defects. It can detect longitudinal discontinuities in the head at the junction of either the web and the foot or the web and the head as well as cracks from bolt holes. It cannot, however, detect vertical cracks. The defects detected by the normal probe are represented on the oscilloscope screen in the form of firm echoes protruding from a baseline. Ordinarily, two echoes are visible on the screen, the initial echo due to partial reflection of the waves from the rail top and back echo from the bottom of the rail. Any echo between the initial and back echo with a corresponding reduction in the height of the back echo is termed a *flaw echo* and is indicative of a flaw. The position of the flaw can be known by reading the distance of this intermediate echo from the initial echo, which will be the distance of the flaw from the rail top.

Fig. 6.8   Ultrasonic rail testing trolley

The ultrasonic rail testing trolley can be used without any block protection, but one has to be vigilant about movement of trains. Progress depends upon the experience and efficiency of the operator. The work is quite strenuous in nature and a single operator cannot observe the screen continuously for a long time. Work can also not be done during the middle of the day during the summer, because the operator will not be able to pick up the signals clearly. On account of these limitations, the work progresses rather slowly and two operators can cover approximately two to three kilometers of rails per day.

6.8.3 Classification of Rail Flaws

Depending upon the nature and extent of internal flaws, traffic density, and speed in the section, the defects noticed by rail flaw detection methods have been classified into three major categories, i.e., IMR, REM, and OBS.

IMR defects A defect that is serious in nature and can lead to sudden failure is classified as IMR. Immediately after detection, clamped fish plates should be provided for the defective portions and a speed restriction of 30 km per hour imposed till the IMR rail is removed by a sound-tested rail piece of not less than six metres. A watchman should be posted till the clamped fish plates are provided to avoid any mishaps. IMR stands for immediate removal.

REM defects These are the type of defects that warrant early removal of the rail from the track. These defects are marked with red paint. REM stands for remove.

OBS defects These are defects that are not so serious. The rail need not be removed in such cases but should be kept under observation. These defects are marked with yellow paint. OBS stands for observe. OBS defects have been further classified as OBS (E) and OBS (B). An OBS defect located within the fish-plated zone is designated OBS (E). Similarly, an OBS defect on major bridges and up to 100 m on their approaches is designated OBS (B).

In the case of the need-based concept of ultrasonic testing (explained in Section 6.8.6), there are only two types of defects, IMR and OBS. However, if the defects occur at welded joints, they are called IMR (R) and OBS (W) defects.

6.8.4 Ultrasonic Rail Flaw Testing Car

As the portable ultrasonic rail flaw detector can test only two to three kilometers of rails everyday, advanced railways such as German Railways have developed a *rail testing car* that can test a much longer length of track much more effectively. The test car tests the track at a speed of 30 km per hour and each of the two rails is tested ultrasonically by means of five probes (0°, ±35°, ±70°) and an airborne sound assembly. The test results are recorded photographically on a tape, a subsequent examination of which can reveal all the flaws. The flaws can then be properly classified. Their position in the track can also be pinpointed with respect to kilometrage to an accuracy of 1 m. This special car tests the track during the day and covers 100–200 km per day. The car covers approximately 20,000 km of track every year on German Railways. The average cost of testing is ₹ 300–500 per km of the track. With the present trend in increase in speed, the need for ultrasonic inspection of rails is felt all the more to avoid hazards due to rail fractures, and in this context, the use of the test car in Indian Railways is considered a technical and operational necessity.

6.8.5 Self-propelled Ultrasonic Rail Testing Car

Indian Railways has recently procured a self-propelled ultrasonic rail testing (SPURT) car of make MATRIX-VUR-404 from Sa Matrix Industries, Paris, at a total cost of approximately ₹ 25 million, inclusive of ancillary equipment. The MATRIX car can detect, measure, record, and simultaneously analyse the internal defects of rails using a non-destructive method of rail flaw detection. The rail testing car has been designed for simultaneous examination of rails, points, and crossings at a maximum speed of 40 km per hour. It consists of the following parts.

(a) An ultrasonic detection lorry
(b) An electronic unit including all circuit transmitters, receivers, selectors, and auxiliaries

(c) Two multi-track recorders installed at either end of the car

(d) A real-time automatic defect analyser

The various components of rail flaw detecting equipment are shown in the schematic diagram in Fig. 6.9. Testing is done using five probes at different angles, which are able to detect rail flaws.

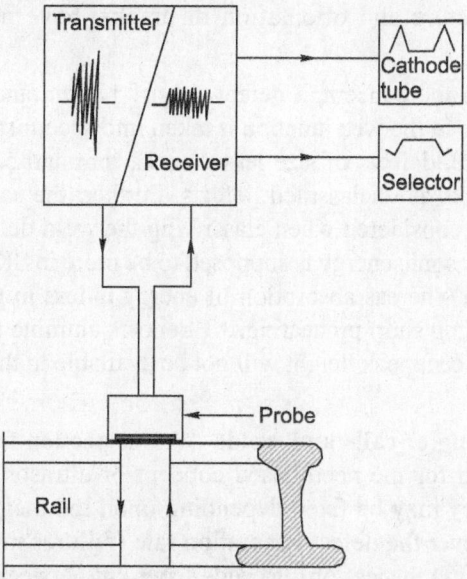

Fig. 6.9 Various components of ultrasonic rail flaw detecting equipment

The SPURT car can detect most rail defects that normally develop under traffic during service. The type of defect, its size, and its position in the rail section is automatically determined. The SPURT car is able to screen the rail section completely in the web and almost completely in the head and the zone of the foot below the web. The flange of the foot and the top corners of the head, however, are not screened. The defects recorded are automatically analysed. The results are given in a synthesized form in a prescribed manner.

It is expected that the SPURT car will be used intensively in Indian Railways and that it will be possible to control rail fractures/failures in Indian Railways to a considerable extent.

6.9 NEED-BASED CONCEPT OF USFD TESTING

Indian Railways has decided to introduce a need-based concept of USFD testing based on a Russian concept. As per the present policy, need-based rail inspection is being progressively introduced on A (not covered by the SPURT car), B, and D routes of Indian Railways. The introduction of this concept will require changing the present classification of defects, frequency of inspection, detection equipment, organization, etc.

The following are the important features of the need-based scheme of USFD testing:

Traffic density and periodicity In the need-based concept, the stipulated frequency of ultrasonic inspection is one after the passage of every 8 GMT, with periodicity varying from two to six months, depending on the sectional GMT.

Important related parameters The system has been evolved based on the consideration of two important related parameters: permissible condemning defects size and inspection frequency. Other important factors such as microstructure of rail steel and nature and orientation of cracks, have not been taken into consideration.

Defect size Under the concept, a defect size of 12 mm and above in the head and 5 mm and above in the web junction is taken into account for the classification of defects. Therefore, defects of size less than 12 mm and 5 mm are allowed to continue in the tracks as unclassified defects. Further, the same size of artificial flaw, i.e., 12 mm, is considered when classifying the weld defect too. Attenuation or absorption of ultrasonic energy is supposed to be more in SKV/AT welded joints due to coarse grain, whereas absorption of energy is less in rail steel. (SKV is a German word meaning short preheating. AT denotes alumino thermit.) Obviously, the specified 60 per cent peak height will not be available in the same grain setting in the case of welds.

Frequency of testing of rails and welds An inspection frequency of 8 GMT has been prescribed for the need-based concept of ultrasonic testing. A higher inspection frequency may be fixed depending upon the incidence of defects. In view of this, whenever the defect generation rate (failures in service and defects detected during USFD inspection) exceeds 1 per km between successive tests in a stretch, the inspection frequency should be doubled in that particular stretch. When calculating the defect generation rate, only rail defects (IMR) or fractures with an apparent origin other than the bolt area and detectable by USFD should be considered.

Testing of rails After the initial testing of rails in the rail manufacturing plant, the first re-test normally need not be done until the rails have covered 15 per cent of their service life in GMT as given in Table 6.11. For rails rolled in April 1999 and later, the test-free period will be 25 per cent instead of 15 per cent.

Table 6.11 Service life of rails

Gauge	Rail section	Assessed GMT service life for T-12-72 UTS rails	Assessed GMT service life for T-12-90 UTS rails
BG	60 kg	550	800
	52 kg	350	525
	90 R	250	375
MG	75 R	150	225
	60 R	125	–

Whenever rails are not tested in the rail manufacturing plant, the test-free period will be applicable, and rail testing will be done at the periodic interval given in Table 6.12 right from the day of its laying in the field. This table gives the frequency of ultrasonic testing after the passage of 8 GMT, subject to a maximum interval of one year.

Table 6.12 Frequency of ultrasonic testing of rails for all BG and MG routes

BG routes		MG routes	
GMT	*Frequency*	*GMT*	*Testing frequency*
Less than 5	Once in 2 years	Less than 2.5	Once in 5 years
5 to 8	12 months	2.5 ot 5.0	Once in 3 years
8 to 12	9 months	Above 5.0	Once in 2 years
12 to 16	6 months		
16 to 24	4 months		
24 to 40	3 months		
above 40	2 months		

Notes   (i)   Frequency of testing of welded joints by 0° and 70° probes shall be as per table given above.

(ii)   Testing of flange portion of AT welded joints by 70°, 1.25 MHz probe by hand probing, shall be done after GMT for first testing for all AT welds (conventional as well as SKV), after initial acceptance test. Subsequent tests for conventional welds only are required to be done after passage of 40 GMT. In case of SKV welds subsequent flange testing is required to be done when weld failures due to defects in flange exceeds 1 percent in a year in a block section. A round of testing be also carried out after 8 years life incidence of 1 per cent gross failure in a block section since the second test, i.e., the next test after final acceptance test.

In the need-based concept, the actions suggested in Table 6.13 are taken when defective rails are detected.

Table 6.13 Actions to be taken after detection of rail/weld defects

Classifi-cation	*Painting on both faces of the weld*	*Action to be taken*	*Interim action*
IMR/ IMR (W)	Three crosses made with red paint.	The flawed portion should be replaced by a sound-tested rail piece of not less than 6 m length within three days of detection.	PWI/USFD (ultrasonic flaw detecting officer) shall impose speed restriction of 30 kmph immediately and this to be continued till the flawed rail weld is replaced. He should communicate the location of the flaw to the sectional PWI, who shall ensure that a clamped, joggled fish plate is provided within 24 hrs.
OBS/ OBS (W)	One cross made with red paint.	Rail/weld to be provided with clamped, joggled fish plate within three days. PWI/USFD to specifically record his observations of the location in his register in subsequent rounds of testing.	• PWI/USFD to advise sectional PWI within 24 hrs about flaw location. • Keyman to watch during daily patrolling till it is joggled fish plated.

(Contd.)

Table 6.13 (*Contd.*)

Classifi-cation	Painting on both faces of the weld	Action to be taken	Interim action
Defective weld 'DFW' with 0°, 2MHz 70° 2MHz, 45° 2MHz or 70° 2MHz, (8mm × 8 mm) probe	Two cross-es made with red paint.	i) In case protection of weld has been done using joggled fish-plates with clamps, the defective weld shall be replaced within 15 days. However, in case the protection has been done using joggled fish plates with 2 far end tight bolts, the speed re-striction imposed shall continue till the defec-tive weld is replaced which should not be later than 3 months.	PWI USFD shall impose speed restriction of 30 kmph or stricter immediately. He should com-municate to sectional PWI about the flaw location who shall ensure following: Protection of defective weld by joggled fishplates using minimum two tight clamps/2 far end tight bolts one on each side after which speed restriction can be relaxed up to 75 kmph for goods train and 100 kmph for passenger trains on BG and 30 kmph for train and 60 kmph for passenger trains on MG

SUMMARY

Various types of rail sections and their specifications have been discussed in this chapter. Flat-footed rails are commonly used on Indian Railways. A rail section may develop different types of defects during its service life. The defect in a rail should immediately be attended to; otherwise, it will lead to the failure of the rail. The various types of defects in a rail and the remedial measures to be adopted have been highlighted in this chapter.

REVIEW QUESTIONS

1. What are the functions of rails? Name the various types of rails in use. Which type is widely used now? How is the weight of a rail section usually determined?
2. It is observed that at present tracks are mostly laid with flat-footed rails. Give reasons for this preference in relation to other types of rail sections.
3. Defects in rails can be divided into the following three ctaegories:
   (a) Defective rail steel
   (b) Surface defects
   (c) Service defects
   Explain these defects clearly.
4. What is meant by wear of rails? Categorize the types of rail wear and enumerate the methods by which wear in rails can be measured.
5. What factors govern the permissible limit of rail wear?
6. Determine the suitable rail section for a locomotive carrying an axle load of 22.5 tonnes. (*Ans*: 90 R)

Choose the correct answer from the chocies given:

7. The area of section in mm^2 of 52 kg (IRS) rail is:
   - (a) 5250
   - (b) 5895
   - (c) 6615
   - (d) 7686

8. The total GMT which 60 kg (90 UTS) rail can carry is:
   - (a) 350
   - (b) 450
   - (c) 525
   - (d) 800

9. The total service life of 52 kg (72 UTS) rail in terms of GMT is:
   - (a) 350
   - (b) 250
   - (c) 200
   - (d) none of these

10. In need-based system of USFD testing of rails, rail with IMR defect should be replaced within:
    - (a) 7 days
    - (b) 10 days
    - (c) 3 days
    - (d) 1 day

11. The correct sequence of development of rails is:
    - (a) Bull headed, double headed, flat footed
    - (b) Double headed, bull headed, flat footed
    - (c) Bull headed, flat footed, double headed

12. A rail has brand mark IRS-52kg:- 710-TISCO-II 1991-OB; here OB represents:
    - (a) brand of rail
    - (b) type of boiler used
    - (c) process of steel making
    - (c) none of these

13. UTS is the abbreviation of:
    - (a) ultimate tensile strength
    - (b) universal turnout standards
    - (c) under tension stress
    - (d) none of these

14. The total GMT which 52 kg (90 UTS) can carry is:
    - (a) 350
    - (b) 450
    - (c) 525
    - (d) 800

CHAPTER 7

Sleepers

INTRODUCTION

Sleepers are the transverse ties that are laid to support the rails. They have an important role in the track as they transmit the wheel load from the rails to the ballast. Several types of sleepers are used on Indian Railways. The characteristics of these sleepers and their suitability with respect to load conditions are described in this chapter.

7.1 FUNCTIONS AND REQUIREMENTS OF SLEEPERS

The main functions of sleepers are as follows:
  (a) Holding the rails in their correct gauge and alignment
  (b) Giving a firm and even support to the rails
  (c) Transferring the load evenly from the rails to a wider area of the ballast
  (d) Acting as an elastic medium between the rails and the ballast to absorb the blows and vibrations caused by moving loads
  (e) Providing longitudinal and lateral stability to the permanent way
  (f) Providing the means to rectify the track geometry during their service life
   Apart from performing these functions the ideal sleeper should normally fulfil the following requirements.
  (a) The initial as well as maintenance cost should be minimum.
  (b) The weight of the sleeper should be moderate so that it is convenient to handle.
  (c) The designs of the sleeper and the fastenings should be such that it is possible to fix and remove the rails easily.
  (d) The sleeper should have sufficient bearing area so that the ballast under it is not crushed.
  (e) The sleeper should be such that it is possible to maintain and adjust the gauge properly.

(f) The material of the sleeper and its design should be such that it does not break or get damaged during packing.

(g) The design of the sleeper should be such that it is possible to have track circuiting.

(h) The sleeper should be capable of resisting vibrations and shocks caused by the passage of fast moving trains.

(i) The sleeper should have anti-sabotage and anti-theft features.

7.2 SLEEPER DENSITY AND SPACING OF SLEEPERS

Sleeper density is the number of sleepers per rail length. It is specified as $M + x$ or $N + x$, where M or N is the length of the rail in metres and x is a number that varies according to factors such as (a) axle load and speed, (b) type and section of rails, (c) type and strength of the sleepers, (d) type of ballast and depth of ballast cushion, and (e) nature of formation.

If the sleeper density is $M + 7$ on a broad gauge route and the length of the rail is 13 m, it implies that $13 + 7 = 20$ sleepers will be used per rail length of the track on that route. The number of sleepers in a track can also be specified by indicating the number of sleepers per kilometre of the track, for example, 1540 sleepers/km. This specification becomes more relevant particularly in cases where rails are welded and the length of the rail does not have much bearing on the number of sleepers required. This system of specifying the number of sleepers per kilometre exists in many foreign countries and is now being adopted on Indian Railways as well.

The spacing of sleepers is fixed depending upon the sleeper density. Spacing is not kept uniform throughout the rail length. It is closer near the joints because of the weakness of the joints and impact of moving loads on them. There is, however, a limitation to the close spacing of the sleepers, as enough space is required for working the beaters that are used to pack the joint sleepers. The standard spacing specifications adopted for a fish-plated track on Indian Railways are given in Table 7.1. The notations used in this table are explained in Fig. 7.1.

Table 7.1 Spacing of sleepers for a fish-plated track

Spacing of sleepers	BG centre-to-centre spacing (mm)		MG centre-to-centre spacing (mm)	
	Wooden	Metal	Wooden	Metal
Between joint sleepers (a)	300	380	250	330
Between joint sleepers and the first shoulder sleeper (b)	610	610	580	580
Between first shoulder sleeper and second shoulder sleeper (c) for sleeper density $M + 4$	700 (640)*	720 (630)	700 (620)	710 (600)
Between intermediate sleepers (d) for sleeper density $M + 4$	840 (680)	830 (680)	820 (720)	810 (640)

* Values within parentheses are those for sleeper density $M + 7$.

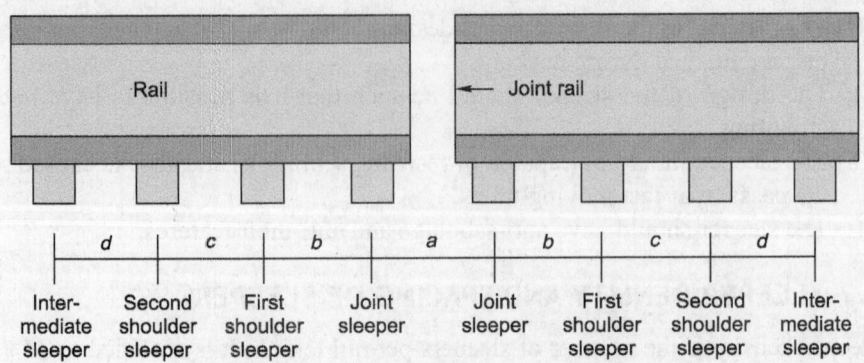

Fig. 7.1 Spacing of sleepers on a fish-plated track

Nowadays sleeper density is also indicated in terms of the number of sleepers/km. The sleeper spacing required for various sleeper densities is given in Table 7.2.

Table 7.2 Spacing of sleepers for welded track

No. of sleepers per km	Exact centre-to-centre spacing required as per calculation (mm)	Centre-to-centre spacing to be provided in the field (mm)	
		LWR track	SWR track
1660	602.4	600	–
1540	649.3	650	660
1310	763.3	–	780

7.3 TYPES OF SLEEPERS

The sleepers mostly used on Indian Railways are (i) wooden sleepers, (ii) cast iron (CI) sleepers, (iii) steel sleepers, and (iv) concrete sleepers. Table 7.3 compares the important characteristics of these types.

Table 7.3 Comparison of different types of sleepers

Characteristics	Type of sleeper			
	Wooden	Steel	CI	Concrete
Service life (years)	12–15	40–50	40–50	50–60
Weight of sleeper for BG (kg)	83	79	87	267
Handling	Manual handling; no damage to sleeper while handling	Manual handling; no damage to sleeper while handling	Manual handling; liable to break by rough handling	No manual handling; gets damaged by rough handling
Type of maintenance	Manual or mechanized	Manual or mechanized	Manual	Mechanized only
Cost of maintenance	High	Medium	Medium	Low

Contd.

Table 7.3 *(Contd.)*

Characteristics	Type of sleeper			
	Wooden	*Steel*	*CI*	*Concrete*
Gauge adjustment	Difficult	Easy	Easy	No gauge adjustment possible
Track circuiting	Best	Difficult; insulating pads are necessary	Difficult; insulating pads are necessary	Easy
Damage by white ants and corrosion	Can be damaged by white ants	No damage by white ants but corrosion is possible	Can be damaged by corrosion	No damage by white ants or corrosion
Suitability for fastening	Suitable for CF* and EF†	Suitable for CF and EF	Suitable for CF only	Suitable for EF only
Track elasticity	Good	Good	Good	Best
Creep	Excessive	Less	Less	Minimum
Scrap value	Low	Higher than wooden	High	None

* CF stands for conventional fastening.
† EF stands for elastic fastening.

7.4 WOODEN SLEEPERS

The wooden sleeper is the most ideal type of sleepers and its utility has not decreased with the passage of time. However, due to the overriding need to preserve forests and also the need to use timber for domestic use and for architectural purpose, it was felt that alternative materials should be explored for sleepers. Accordingly steel, cast iron, and cement concrete have been extensively used for manufacture of sleepers. The different types of sleepers for BG on Indian Railways is presented in Table 7.4.

Table 7.4 Different types of sleepers for BG on Indian Railways

Type of sleeper	Percentage use
Concrete sleeper	94.2
Cast iron sleepers	4.0
Steel sleepers	1.5
Wooden	0.3
All sleepers	100.0

Due to economical, technical, and environmental reasons the use of new wooden sleepers on Indian Railways has been prohibited.

7.4.1 Advantages and Disadvantages of Wooden Sleepers

Wooden sleepers have the following main advantages and disadvantages.

Advantages

(a) Cheap and easy to manufacture
(b) Absorbs shocks and bears a good capacity to dampen vibrations; therefore, retains the packing well

(c) Easy handling without damage
(d) Suitable for track-circuited sections
(e) Suitable for areas with yielding formations
(f) Alignment can be easily corrected
(g) More suitable for modern methods of maintenance
(h) Can be used with or without stone ballast
(i) Can be used on bridges and ashpits also
(j) Can be used for gauntleted track

Disadvantages
(a) Lesser life due to wear, decay, and attack by vermin
(b) Liable to mechanical wear due to beater packing
(c) Difficult to maintain the gauge
(d) Susceptible to fire hazards
(e) Negligible scrap value

Specifications of wooden sleepers The size of a wooden sleeper should be economical. It should provide the desired strength as a beam as well as adequate bearing area. The depth of a sleeper governs its stiffness as a beam and its length as well as its width control the necessary bearing area. The bearing length under each rail seat is 92 cm (3 ft) for a BG wooden sleeper, thereby giving an area of 2325 cm^2 under each rail seat. The sizes of sleepers used for BG, MG, and NG as well as the bearing area per sleeper are given in Table 7.5.

Table 7.5 Sizes of wooden sleepers and bearing areas

Gauge	Size (cm)	Bearing area per sleeper (m^2)
BG	275 × 25 × 13	0.465
MG	180 × 20 × 11.5	0.3098
NG	150 × 18 × 11.5	0.209

Wooden sleepers required for bridges, and points and crossings are of a thicker section—25 cm × 15 cm or 25 cm × 18 cm.

7.4.2 Composite Sleeper Index

The composite sleeper index (CSI), which evolved from a combination of the properties of strength and hardness, is an index used to determine the suitability of a particular timber for use as a sleeper from the point of view of mechanical strength.

CSI is given by the formula

$$\text{CSI} = \frac{S + 10H}{20} \tag{7.1}$$

where S is the figure for the general strength for both green and dry timber at 12 per cent moisture content and H is the figure for the general hardness for both green and dry timber at 12 per cent moisture content. The minimum CSI prescribed on Indian Railways is given in Table 7.6.

Table 7.6 Sleeper type and corresponding CSI

Type of sleeper	Minimum CSI
Track sleeper	783
Crossing sleeper	1352
Bridge sleeper	1455

7.4.3 Durable and Non-durable Types of Wooden Sleepers

Wooden sleepers may be classified into two categories, durable and non-durable.

Durable type of wooden sleepers

Durable sleepers do not require any treatment and can be laid directly on the track. These are produced from timbers such as *teak, sal, nahor, rosewood, anjan, kongu, crumbogam kong, vengai, padauk, lakooch, wonta, milla, and crul.*

Non-durable type of wooden sleepers

Non-durable sleepers require treatment before being put on the track. Non-durable sleepers are made of wood of trees such as *chir, deodar, kail, gunjan, and jamun.*

The primary service life of these wooden sleepers is approximately as follows:

	BG	MG
Durable	19 years	31 years
Non-durable	12.5 years	15.5 years

Treated and untreated wooden sleepers

These include the following:
 (a) 'U' or untreated sleepers comprising all the sleepers made of wood from naturally durable species
 (b) 'T' or treated sleepers consisting of the rest of the sleepers

7.4.4 Treatment of Sleepers

Earlier when wooden sleepers were used extensively a number of treatment plants were set up by Indian Railways such as Dhilwan (Punjab), Naharkatia (Assam), Clutterbuckgang (UP), and Olvakot (Kerala).

All these plants utilize the pressure treatment process and the preservative is forced into the wood under pressure using any one of the following three methods.
 (i) Full cell (Bethell) process
 (ii) Empty cell (Ruefling process)
 (iii) Empty cell (Lowry process)

Note: None of these plants are currently working and as such it is not important to discuss the same in the book.

7.4.5 Seasoning of Sleepers

Wooden sleepers are seasoned to reduce the moisture content so that their treatment is effective. The Indian Standard code of practice for preservation of timber lays down that the moisture content in the case of sleepers to be treated by pressure treatment should not be more than 25 per cent.

The seasoning of sleepers can be done by any one of the following processes.

Artificial seasoning in kiln This is a controlled method of seasoning the timber, normally used in USA and other advanced countries, under conditions of temperature and relative humidity, which are in the range of natural air seasoning.

Boulton or boiling under vacuum process This is a process in which unseasoned wood is treated with hot preservative to remove the moisture content. This is adopted in the Naharkatia depot.

Note: Nowadays Indian Railways does not do artificial seasoning of wooden sleepers.

Air seasoning This is the method adopted extensively for the seasoning of wooden sleepers in India. The sleepers are stacked in the timber yard and a provision is made for enough space for the circulation of air in between the sleepers. The sleepers are stacked in any one of the following ways:
 (a) One and nine method (Fig. 7.2)
 (b) Close crib method
 (c) Open crib method (Fig. 7.3)

Normally, the one and nine method is adopted on Indian Railways for stacking the sleepers. About six months are required to air season the timber fully by this method.

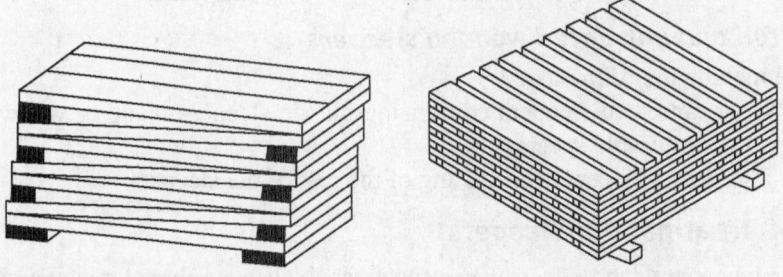

Fig. 7.2 One and nine method   **Fig. 7.3** Open crib method

7.4.6 Laying of Wooden Sleepers

Great care should be taken in laying wooden sleepers. Untreated wooden sleepers should be laid with the sapwood side upwards and the heartwood side downwards so as to ensure minimum decay due to fungus, etc., attacking from below. More moisture would also percolate into the sleepers if laid otherwise. In the case of treated sleepers, however, the heartwood side is kept upwards and the sapwood side downwards. This is done because the sapwood side contains more creosote and is liable to less damage from vermin and fungus.

7.4.7 Adzing of Wooden Sleepers

In order to enable the rails to be slightly tilted inwards at a cant of 1 in 20, wooden sleepers are required to be cut to this slope at the rail seat before laying. This process of cutting the wooden sleeper at a slope of 1 in 20 is known as 'adzing of the wooden sleeper'.

It may be pointed out that adzing or cutting of a wooden sleeper at a slope of 1 in 20 is done with great care, otherwise the slope will vary from sleeper to sleeper resulting in a rough ride. The adzed surface of a wooden sleeper is treated with coal tar or creosote to ensure proper protection of the surface. Normally, adzing of a wooden sleeper is done only when bearing plates are not provided.

7.5 STEEL CHANNEL SLEEPERS

In view of the great shortage of wooden sleepers, steel channel sleepers have been developed by Indian Railways particularly for use on girder bridges. Steel channel sleepers can be used for welded plates, riveted plates, as well as open web girders.

Composite sleepers have been developed indigenously in India as a replacement for wooden sleepers. These are made from waste products such as used rubber tyres, and the manufacturers claim a lifespan of about 40 years for these sleepers. The Patel Group of Industries is one such firm that has developed these composite sleepers.

Composite sleepers are similar to wooden sleepers and use similar fittings. These sleepers are under trial and the results so far have been quite encouraging.

7.6 STEEL TROUGH SLEEPERS

About 4 per cent of the track on Indian Railways is laid on steel trough (ST) sleepers (Fig. 7.4). The increasing shortage of timber in the country and other economical factors are mainly responsible for the use of steel sleepers in India. Steel sleepers have the following main advantages/disadvantages over wooden sleepers.

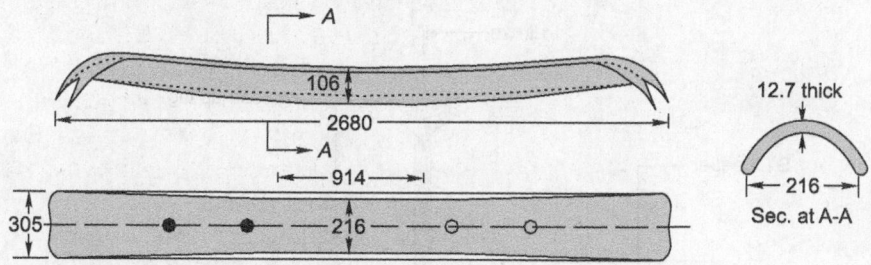

Fig. 7.4 Steel trough sleeper (BG 90 R)

Advantages
  (a) Long life
  (b) Easy to maintain gauge and less maintenance problems
  (c) Good lateral rigidity
  (d) Less damage during handling and transport
  (e) Simple manufacturing process
  (f) Very good scrap value
  (g) Free from decay and attack by vermin
  (h) Not susceptible to fire hazards

Disadvantages
  (a) Liable to corrode
  (b) Unsuitable for track-circuited areas
  (c) Liable to become centre-bound because of slopes at the two ends

(d) Develops cracks on rail seats during service

(e) Design is rail specific

7.6.1 Design Features

The steel trough sleeper essentially consists of a rolled steel plate of about 2 mm thickness pressed into a suitable trough shape and the rail seat canted to 1 in 20. The ends of the rolled section are flattened out in the shape of a spade to retain the ballast. Two alternative types of sleepers have been designed for each rail section as per the following details.

1. In one type, the lugs or jaws are pressed out of the plate itself to accommodate the foot of the rail and the key (Fig. 7.5). There are several maintenance problems with these pressed up lugs, as they give way due to the movement of the keys as well as due to the vibrations and impact of the moving loads.

2. In order to obviate this defect, another sleeper design has been adopted. In this design, two holes are punched into either side of the plate to accommodate specially designed 'loose jaws' (Fig. 7.6). The rails are held with the help of two standard keys driven either into the pressed up lugs or into the loose jaws.

The gauge is adjusted to the extent of ± 3 mm by properly driving in the keys. In the double-line section, the keys are driven in the direction of the traffic. The approximate weight of a standard BG trough sleeper is 81 kg and that of an MG sleeper is 35 kg. The ST sleeper has an average life of about 50 years. It is an acceptable type of sleeper for use with long-welded rails because of its lateral stability and its adaptability for use along with elastic fastenings.

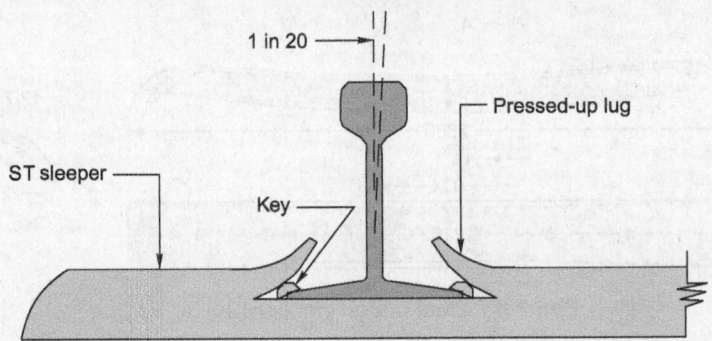

Fig. 7.5 ST sleeper with pressed-up lugs

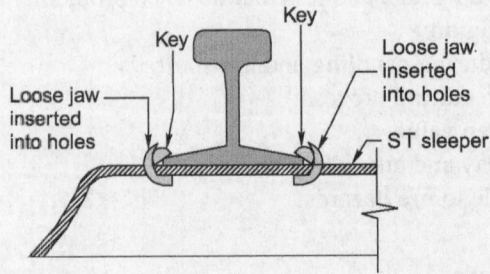

Fig. 7.6 Sleeper with loose jaws inserted into holes

7.6.2 Classification of ST Sleepers

All steel sleepers conforming to Indian Railways specifications T-9 are classified as first quality sleepers. The sleepers not accepted as first quality but free from the following defects are termed second quality steel trough sleepers.

(a) Inward tilt at rail seat beyond the limits of 1 in 15 to 1 in 25
(b) Sleepers with a twist
(c) Heavy scale fitting or deep grooves or cuts
(d) Deep guide marks at heads, blisters, etc.

All first quality sleepers are normally marked by a green dot. Sleepers that have been rejected as first quality sleepers on account of pipes, seams, and laps but are free from the defects indicated above are marked with a cross (×) in yellow paint at the centre. All other second quality steel trough sleepers are marked distinctly with a 15-cm-wide strip of yellow paint at one end. Sleepers that are unfit as second quality are given a distinct red paint mark to avoid mixing them up with first and second quality sleepers during loading.

7.6.3 Maintenance Problems of ST Sleepers

It has been noticed that the keys used to fix rails on steel sleepers tend to become loose due to the bending of the pressed up lugs or due to wear at the rail seat. The holes also get elongated during service. Special types of shims and liners are provided in these cases to hold the gauge well. Mota Singh Liner is a very effective type of liner used for holding the correct gauge for oblong holes with loose jaws. Another maintenance problem with ST sleepers is that these tend to become centre-bound if due care is not taken while packing. The ballast is normally removed from the centre of the sleepers after packing so as to ensure that centre binding of the sleepers does not take place. Sometimes the alignment of steel sleeper tracks also gets affected by the overdriving of the keys.

7.7 CAST IRON SLEEPERS

Cast iron (CI) sleepers are also used on Indian Railways and currently about 8.5 per cent of the track consists of CI sleepers, which may be either pot type or plate type. The main advantages and disadvantages of CI sleepers over steel trough sleepers are the following.

Advantages of CI Sleepers
(a) They undergo less corrosion.
(b) There is less probability of cracking at rail seat.
(c) They are easy to manufacture.
(d) They have high scrap value.

Disadvantages
(a) Gauge maintenance is difficult as the bars get bent.
(b) Provides less lateral stability.
(c) Unsuitable for track-circuited lines.
(d) Not very suitable for mechanical maintenance and/or MSP because of rounded bottom.
(e) Susceptible to breakage.

7.7.1 CI Pot Sleepers

Cast iron pot sleepers (Fig. 7.7) consist of two hollow bowls or pots of circular or elliptical shape placed inverted on the ballast section. The two pots are connected by a tie bar with the help of cotters and gibs; the gauge can be adjusted slightly [± 3 mm (1/8″) by changing their positions. The rail is placed on top of the pots in a rail seat provided with a cant of 1 in 20 and is held in position with the help of a key. The pot sleeper suffers from the drawback that is cannot be used on curves sharper than 4° on BG. Most of the fittings are hidden and their inspection and maintenance is quite difficult. These sleepers have become obsolete now and are not being procured by the Indian Railways any more.

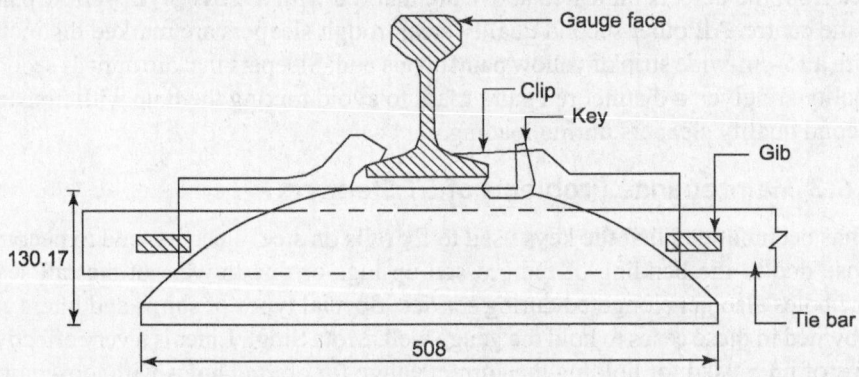

Fig. 7.7 CI pot sleeper (dimensions in mm)

7.7.2 CST-9 Sleepers

The CST-9 sleeper is a standard sleeper and is being most extensively used on Indian Railways (IR). It is called CST-9 (Central Standard Trial-9) (Fig. 7.8) because it is the ninth of the series produced by the Central Standard Office. The sleeper is a combination of pot, plate, and box sleepers. It consists of two triangular inverted pots on either side of the rail seat, a central plate with a projected keel, and a box on top of the plate. The two CI plates are connected by a tie bar with the help of four cotters. The rails are held to the sleeper by two-way keys provided at each rail seat on the side of the gauge face. The gauge is adjusted to a value of ± 5 mm by altering the relative positions of the four cotters.

Table 7.5 Details of CST-9 sleeper (Fig. 7.9)

Rail	Gauge	RDSO drawing number	Wt (kg)	A (mm)	B (mm)	C (mm)	D (mm)
52kg	BG	T-478 (M)	43.55	800	330	140	89
90R	BG	T-478 (M)	43.55	800	330	140	89
90R	MG	T-2366	–	700	300	132	85
75R	MG	T-498 (M)	24.50	650	270	114	77
60R	MG	T-10257	20.07	650	270	114	77
50R	NG	T-438	–	533	228	108	69

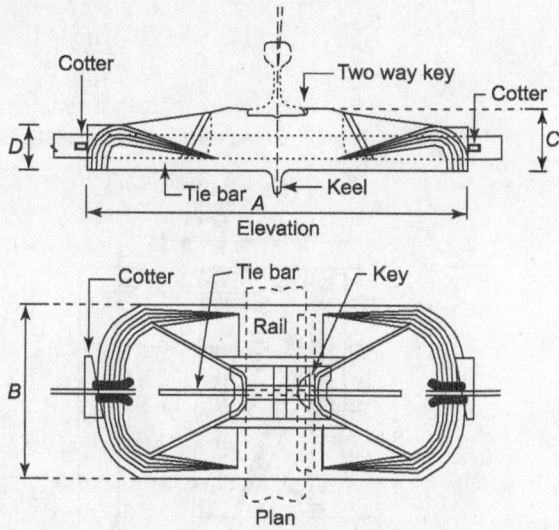

Fig. 7.8 CST-9 sleeper

The rail seat of a CST-9 sleeper is 115 mm wide along the length, and this marrow bearing tends to reduce the rocking of the sleeper under the wave motion of the rail. The sleeper is designed to provide a firm support to the rail and provides fairly good lateral and longitudinal stability to the rails. The dimensions of CST sleepers in use on IR are given in Table 7.5 The sleeper provides a bearing area approximately equal to the effective bearing area of a standard BG wooden sleeper, i.e., 5 sq. ft, for the plates. CST-9 plates are also available with reverse jaws (T-443 type) to serve as an anti-sabotage measure; a few of these are provided in each rail length. Normally, three reverse jaw CST-9 sleepers are provided per rail to serve anti-sabotage purposes. The weight of a CST-9 sleeper assembly along with fastenings for BG is 102 kg and for MG is 58 kg.

The CST-9 sleeper is one of the most popular sleepers on Indian Railways at present. The sleeper has, however, certain limitations when combined with the modern track a mentioned in the following.

Disadvantages of CI Sleepers

(a) As the sleeper does not have a flat bottom, it is not quite suitable for MSP and mechanical maintenance with tie tamers.

(b) The suitability of a CST-9 sleeper on LWRs, particularly on the breathing lengths, is doubtful because of rigid fastenings and the inability of the fastenings to hold the rail with a constant toe load.

(c) The rail seat wears out quickly causing the keys to become loose.

(d) The sleeper has only limited longitudinal and lateral strength to hold LWRs, particularly in the breathing length.

(e) Due to the use of less metal under rail seat, the shocks and vibrations are directly transmitted to the ballast, resulting in poor retention of packing (loose packing) and hence an increased frequency of attention.

7.7.3 CST-9 Sleeper for MG

A new design of the CST-9 sleeper (T-2366) has recently been developed on Indian Railways for 90 R rails on MG lines as shown in Fig. 7.9.

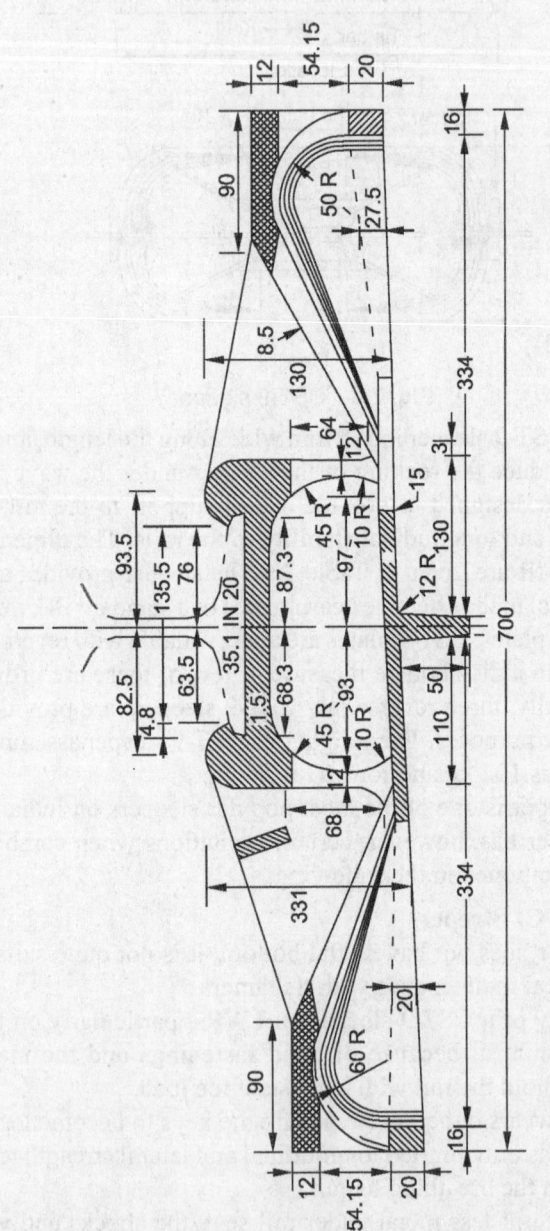

Fig. 7.9 CST-9 sleeper for MG (units in mm)

7.7.4 CST-10, CST-11, CST-12, and CST-13 Sleepers

These sleepers were subsequently designed as modification of CST-9 sleepers, but have not been found to be successful in actual field trials.

7.8 CONCRETE SLEEPERS

The evolution of concrete sleepers has been mainly due to economic considerations coupled with the elements of changing traffic pattern.

7.8.1 Evolution and History of Development of Concrete Sleeper

In the early period of history of railways, wood was the only material used for making sleepers in Europe. Even in those days, occasional shortage of wooden sleepers and their increasing price posed certain problems and this gave a fillip to the quest for alternative material for sleepers.

With the development of concrete technology in the nineteenth century, cement concrete had established its place as a versatile building material that could be adopted suitably to meet the requirements of a railway sleeper. In 1877, Monnier, a French gardener and inventor of reinforced concrete, suggested that cement concrete could be used for making sleepers for railway tracks. Monnier in fact designed a concrete sleeper and obtained a patent for it, but his design did not work successfully. The design was further developed and the railways of Austria and Italy produced the first concrete sleepers with a promising design around the turn of the nineteenth century. This was closely followed by other European railways, where large-scale trials of concrete sleepers were done mostly due to economic considerations.

However, not much progress could be achieved till World War II when wooden sleepers practically disappeared from the European market and their prices shot up. Almost at the same time, as a result of extensive research carried out by French Railways and other European railways, the modern track was born. Heavier rail sections and long-welded rails came into existence. The necessity of a heavier and better type of sleeper that could fit the modern track was felt. These conditions gave a spurt to the development of concrete sleepers and countries such as France, Germany, and Britain went a long way in developing concrete sleepers to perfection.

7.8.2 Development of Concrete Sleepers

The development of concrete sleepers that took place on various railway systems was mainly based on the following concepts of design.

(a) RCC or prestressed sleepers similar in shape and size to wooden sleepers
(b) Block-type RCC sleepers connected by a steel tie bar
(c) Prestressed concrete blocks and a steel or an articulated concrete tie bar
(d) Prestressed (pre-tensioned or post-tensioned) type of concrete sleepers

These four concepts of design are the basis of the development of present-day concrete sleepers.

7.8.3 Advantages and Disadvantages of Concrete Sleepers

Concrete sleepers have the following advantages and disadvantages.

Advantages

(a) Concrete sleepers, being heavy, lend more strength and stability to the track and are specially suited to LWR due to their great resistance to buckling of the track.

(b) Concrete sleepers with elastic fastenings allow a track to maintain better gauge, cross level, and alignment. They also retain packing very well.

(c) Concrete sleepers, because of their flat bottom, are best suited for modern methods of track maintenance such as MSP and mechanical maintenance, which have their own advantages.

(d) Concrete sleepers can be used in track-circuited areas, as they are poor conductors of electricity.

(e) Concrete sleepers are neither inflammable nor subjected to damage by pests or corrosion under normal circumstances.

(f) Concrete sleepers have a very long lifespan, probably 40–50 years. As such rail and sleeper renewals can be matched, which is a major economic advantage.

(g) Concrete sleepers can generally be mass produced using local resources.

Disadvantages

(a) Handling and laying concrete sleepers is difficult due to their heavy weight. Mechanical methods, which involve considerable initial expenditure, have to be adopted for handling them.

(b) They are heavily damaged at the time of derailment.

(c) They have no scrap value.

(d) They are not suitable for beater packing.

(e) They should preferably be maintained by heavy 'on track' tampers.

7.8.4 Design Considerations of Concrete Sleepers

Two different concepts are being adopted by German and French engineers in designing the section of a concrete sleeper. The Germans, having adopted a beam type sleeper, consider the sleeper as a rigid, stiff, and continuous beam supported on a firm and unyielding bed. The French engineers however, consider the sleeper as two separate blocks connected by a tie bar and resting on a resilient ballast bed. The former design is based on static loading, while the latter theory caters for a slightly differential settlement of ballast support. As the calculations based on the latter theory are quite complicated and difficult, the sleeper design based on this concept has been evolved mostly on an empirical basis.

The forces and factors considered in the design of concrete sleepers are the following:

(a) Forces acting on a sleeper

(b) Effects of the geometric form including shape, size, and weight

(c) Effect of the characteristics of fastenings used

(d) Provision of failure against derailments

7.8.5 Need for Concrete Sleepers in India

In India there has been a chronic shortage of wooden sleepers over the last few decades. Wooden sleepers of various species in India have a short lifespan of about 15–20 years. In view of this drawback of wooden sleepers, cast iron and steel trough sleepers have been used extensively. The consumption of these metal sleepers at present is quite high and Indian Railways consumes about 40 per cent of the entire pig iron production in the country. There is a need to reduce pig iron consumption by the Railways so that the iron can be made available in large quantities for defence purposes and other heavy engineering industries. In addition, higher speeds, welding of rails, and installation of long-welded rails have recently been introduced on Indian Railways. A sleeper for a long welded track has to be heavy and sturdy and should be capable of offering adequate lateral resistance to the track. Wooden and steel sleepers were found to be totally lacking in these requirements. Both these considerations led to investigations for selecting a suitable concrete sleeper for use on Indian Railways.

7.8.6 Loading Conditions Adopted on Indian Railways

Concrete sleepers have been designed by the Research Design and Standard Organization (RDSO) wing of Indian Railways for the following different loading conditions:

BG sleeper

(a) 15 tonnes vertical loads at the rail seat.
(b) Vertical load of 13 tonnes at rail seats plus a reaction at the centre of the sleeper equal to half of the load under the rail seat.
(c) A vertical load of 13 tonnes and a lateral load of 7 tonnes directed towards the outside of one rail only.

The sleeper is designed to resist a bending moment of 1.33 tonnes m at the rail seat and 0.52 tonnes m at the centre of the sleeper.

MG sleeper

(a) Vertical loads of 10 tonnes at the rail seats plus a reaction at the centre of sleeper equal to half of that under the rail seat.
(b) Vertical loads of 8 tonnes at the rail seats with 4.5 tonnes lateral force directed towards the outside of one rail only.

7.8.7 Types of Concrete Sleepers

The various types of concrete sleepers (prestressed, pre-tension, post-tension, and two-block) being manufactured on Indian Railways have been described in Table 7.8.

Table 7.8 Types of concrete sleepers manufactured on Indian Railways

Gauge	Type of sleeper	Rail section	Standard drawing number	Sleeper design number
BG	Mono block	60 kg	RDSO/T-2496	PDS-14
BG	Mono block	52 kg	RDSO/T-2495	PDS-12
BG	Mono block	60 kg/52 kg	RDSO/T-3602	Post-tension type
BG	Mono block	90 R/75 R	RDSO/T-2521	RCS-6
BG	Mono block	90 R	RDSO/T-2503	PCS-17
MG	Twin block	75 R/60 R	RDSO/T-3518	PCS-12
BG	Twin block	75 R	RDSO/T-153	PCS-11

Mono-block prestressed concrete sleepers with Pandrol clips

The mono-block prestressed concrete sleeper (Fig. 7.10), which is similar to the German B-58 type of sleeper, has an overall length of 2750 mm and a weight of 270 kg approximately. The sleeper has a trapezoidal cross section with a width of 154 mm at the top and 250 mm at the bottom and a height of 210 mm at the rail seat. A cant of 1 in 20 is provided on the top surface of the sleeper for a distance of 175 mm on either side of the centre line of the rail to cover the area of rail fittings. The sleeper is prestressed with 18 high tensile steel (HTS) strands of 3 × 3 mm diameter and 126 mm-diameter mild steel links. The initial prestressing of the steel is 100 kg/cm^2. The 28-day crushing strength of the concrete is normally not less than 525 kg/cm^2.

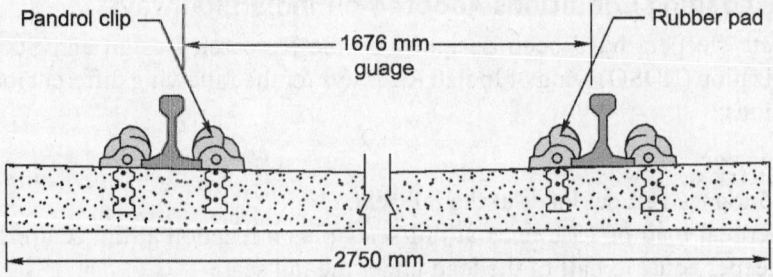

Fig. 7.10 Mono-block prestressed concrete sleeper

The rail rests on a grooved 130 × 130 mm rubber pad, with the grooves lying parallel to the axis of the rail. The fastenings provided for the 52 kg rail are Pandrol clips, which are held in malleable cast iron inserts as shown in Fig. 7.11.

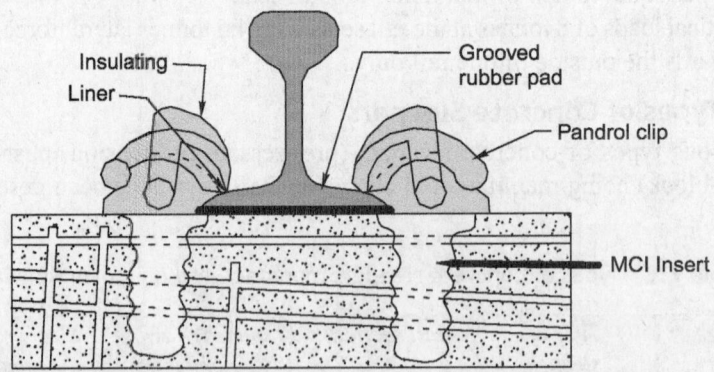

Fig. 7.11 Details at rail seat of a prestressed concrete sleeper

PCS-12 and PCS-14

PCS-12 is the latest type of prestressed concrete (PRC) sleeper for use on BG routes with 52-kg rails and elastic rail clips. For use with 60-kg rails and elastic rail clips, the PCS-14 sleeper has been standardized on Indian Railways.

The important dimensions of both of these types of sleepers are shown in Fig. 7.12 and listed as follows:

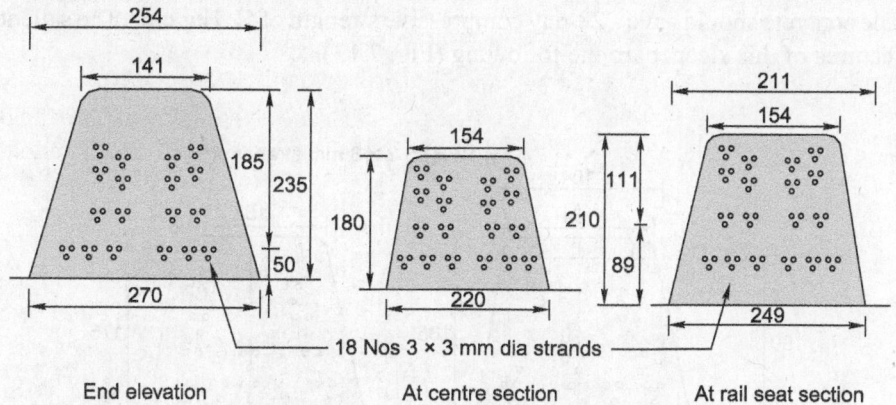

End elevation   At centre section   At rail seat section

Fig. 7.12   PCS-12 mono-block concrete sleeper (units in mm)

- Length = 2750 mm
- Weight = 267 kg
- Reinforcement: Eighteen 3 × 3 mm diameter strands
- Concrete is to be of controlled quality with a minimum 28-day crushing strength of 525 kg/cm^2
- Each strand to be tensioned with an initial tensile force of 2730 kg

Mono-block post-tension type of concrete sleepers for BG

The first factory in India for the manufacture of post-tension type of mono-block concrete sleepers was set up by Northern Railways at Allahabad in collaboration with M/s Dyckerhoff and Widmann (D&W) of West Germany. The factory, which started production in 1981, has a planned capacity of manufacturing 300,000 concrete sleepers per year.

The salient features of post-tension type of concrete sleepers are the following:

Size of sleeper
- Length = 2750 mm
- Width at centre = 160 mm (top)
  - 200 mm (bottom)
- Depth at centre = 180 mm
- Weight = 295 kg

Design features
- Initial prestressing force = 37 tonnes
- Final prestressing force = 31 tonnes
- Minimum concrete strength in 28 days = 550 kg/cm^2
- Minimum strength of concrete at the time of applying prestress = 450 kg/cm^2

The use of concrete sleepers using the post-tension method has not been successful on Indian Railways and its manufacture has since been stopped.

Mono-block PRC sleepers for MG (PCS-17)

A design for mono-block PRC sleepers (PCS-17) has recently been standardized for MG. The sleeper has a trapezoidal cross section similar to that of a BG sleeper. The concrete should have a 28-day compressive strength of 525 kg/cm². The salient features of this sleeper are the following (Fig. 7.13).

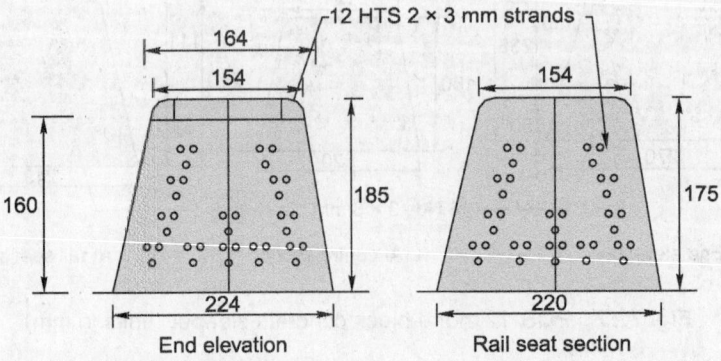

Fig. 7.13 PCS-17 concrete sleeper for MG (units in mm)

- Length = 2000 mm
- Weight = 158.5 kg
- Reinforcement: Twelve 3 × 3 mm diameter strand of HTS wire tensioned to initial force of 2730 kg

PRC sleepers can be used for 90 R rails with elastic rail clips and glass filled nylon liners (GFN 66) and on sole plates.

Two-block RCC sleeper for BG yards

A design for a two-block RCC sleeper for BG yards has been standardized by RDSO as per drawing number RDSO/T-2521 for extensive trials on Indian Railways. There is a general scarcity of wooden and CST-9 sleepers for use in BG yards and the new RCC sleepers will ease the situation in a big way.

Some of the salient features of this sleeper are as follows:

- Considering low speeds in yard lines and less impact effect, the rail seat design load has been taken only as 10 tonnes without any lateral thrust.
- Size at rail seat (top width × bottom width × depth) = 22 cm × 30 cm × 17 cm
- Overall length of the sleeper = 247.5 cm
- Weight of the sleeper = 170 kg
- Main reinforcement in each block
    - At top: Five 8-mm-diameter steel bars
    - At bottom: Two 8-mm-diameter steel bars
- The fastenings used are steel clips and a spring washer with screw fitted to a polythene dowel.

Two-block concrete sleeper for MG yards

Two-block concrete sleepers for use in MG yards have recently been developed. The sleeper consists of two cement concrete blocks, each weighing about 36 kg and consisting of an MS reinforcement of about 7 kg. The two RCC sleeper blocks are connected by an angle tie bar of $55 \times 50 \times 6$ mm section and 1.5 m length. The rail is fixed to the sleeper block either by a clip and bolt arrangement or by polythene dowels and rail screws. A pad is provided below the rail seat to provide cushioning.

Mono-block versus two-block concrete sleepers

There are relative advantages and disadvantages of mono-block and two-block concrete sleepers. Some of these are enumerated below.

(a) Mono-block sleepers give better longitudinal and lateral stability to the track compared to two-block concrete sleepers.

(b) The mono-block concrete sleeper, being a monolithic concrete mass, is likely to have a longer working life compared to the two-block concrete sleeper connected with a tie bar. In the latter case, a tie bar is weak and has a comparatively shorter life due to corrosion, etc.

(c) The mono-block concrete sleeper requires heavy capital expenditure for its manufacture, being a prestressed reinforced concrete unit, compared to the two-block sleeper, which is an ordinary reinforced concrete sleeper.

(d) In a mono-block prestressed concrete sleeper, a crack that develops because of overstressing is likely to close down upon return to normal condition, whereas in a two-block sleeper, such a crack will continue to remain open.

(e) Mono-block sleepers are likely to become centre-bound unlike two-block sleepers.

(f) During derailments and rough handling the tie bars of a two-block sleeper get deformed, thereby affecting the gauge.

(g) In a two-block sleeper, the two blocks are not likely to rest on the ballast in a way that each rail is properly inclined to the vertical, a feature which could affect the alignment and gauge of the track.

7.8.8 Sleepers for Turnouts

A railroad turnout is a mechanical installation that enables trains to be guided from one line of rail tracks to another. In this section we will discuss sleepers and sleeper designs for turnouts.

Prestressed concrete sleepers for turnouts

Due to the acute shortage of wood, especially of long timbers required for points and crossings, it was felt necessary to develop PRC sleepers for use on turnouts in track-circuited areas. RDSO developed a PRC sleeper design with a rectangular cross section in July 1986 for 1-in-12 left-hand turnouts with a 7730-mm curved switch for use with 52-kg rails. These PRC sleepers for turnouts have been manufactured in the PRC sleeper factory at Khalispur, and currently these sleepers are on trial on Northern Railway.

The salient features of these sleepers are the following:

(a) The sleepers have a rectangular cross section.

(b) There are 74 sleepers comprising 21 sleepers in switch assembly, 3 in intermediate sub-assembly and 18 in crossing sub-assembly.

(c) The sleepers are of varying lengths and design. There are 16 different turnout sleeper designs.

(d) These sleepers require the use of a number of fittings different from the existing standard fittings. The grooved rubber pads are of a standard 4.5 mm thickness, but of varying size.

New fan-type concrete sleeper for turnouts

The prestressed concrete sleepers discussed above are suitable only for 1-in-12 turnouts. RDSO has developed a new fan-type sleeper that can be used for 1-in- 8½ as well as 1-in-12 turnouts.

The new design of concrete sleepers has the following characteristics.

(a) The cross section of the sleeper in the new design is trapezoidal instead of rectangular as in the earlier design.

(b) The layout of the sleepers is fan shaped and the same design of sleepers can be used for right-hand as well as left-hand turnouts by rotating them 10° in a horizontal plane.

(c) Apart from approach sleepers, 54 concrete sleepers are used for 1-in-8½ turnouts and 83 concrete sleepers are used for 1-in-12 turnouts.

(d) The concrete used has a 28-day crushing strength of 600 kg/cm^2.

(e) The sleepers are laid perpendicular to the main line on the switch portion. In the lead portion, sleepers are laid equally inclined to the straight and turnout tracks. In the crossing portion, the sleepers are laid perpendicular to the bisecting line of the crossing.

(f) The sleepers under the switch portion have dowels for fixing slide chairs with the help of screws. These sleepers are laid perpendicular to the main line and, therefore, can be used for both left-hand and right-hand turnouts.

(g) The mark 'RE' is provided on the fan-shaped PRC turnout sleepers at one end. The sleepers should be so laid that the end with the RE mark is always laid on the right-hand side.

7.8.9 Laying of the Concrete Sleepers on Turnouts

The turnout locations where concrete sleepers are to be laid should have a clean ballast cushion of 30 cm thickness. Extra ballast should be available on the cess and the area should have good drainage. Depending upon the availability of space and various other site conditions, one of the following three methodologies or their combinations can be adopted for laying concrete sleeper turnouts.

(i) Assembling the turnout at the site and replacing it during the block period by means of either cranes or rollers.

(ii) Carrying parts of the assembled turnout on dip lorries and replacing them during the block period.

(iii) Replacing the existing turnout sleepers by new concrete sleepers except for the switch portion, which can be assembled as one unit.

The assembling and laying should normally be done using a crane of suitable capacity. After removing old turnout sleepers, the ballast bed at the level of the bottom of the concrete sleepers for turnouts should be evened out. Vibrating rollers should be employed to the extent possible for compaction of ballast bed.

Turnouts with concrete sleepers can be maintained in any one of the following ways:

(a) Using points and crossing tamper

(b) Using off-track tampers with lifting jacks

(c) Measured shovel packing

In case of emergencies such as derailments, when the sleepers may be damaged, temporary repairs should be carried out by interlacing wooden sleepers for permitting traffic with restricted speed. The damaged concrete sleepers are replaced by a fresh lot of turnout concrete sleepers as a permanent measure as early as possible. The wooden sleepers and any other damaged sleepers are replaced one by one with new turnout sleepers.

7.8.10 Manufacture of Concrete Sleepers

Prestressed concrete sleepers can be of the pre-tensioned or post-tensioned type. In the case of pre-tensioned sleepers, the force is transferred to the concrete through bonds or through a combination of bonds and positive anchors. Bond transmission lengths and the losses in prestress vitally affect the design and determine the quality of manufacture. In the post-tensioned type of sleeper, the force is transferred only through positive anchors.

Mono-block prestressed concrete sleepers

Mono-block concrete sleepers are generally manufactured by the 'long line method'. In this method, at a time, 30–40 moulds for casting concrete sleepers are kept in about 100–120-m-long casting beds. High tensile steel wires with diameters of 5 mm are anchored at the end block between the tension towers and moulds, and stretched by a specially designed tensioning method. The tensile stress in the wires should not exceed 70 per cent of the specified minimum UTS (ultimate tensile stress). High-quality concrete, with a pre-designed mix, is then filled into the moulds. The newly laid cement concrete is thoroughly mixed and consolidated by means of high-frequency vibrators. The concrete is then cured after about three hours, preferably by steam. The wires are then destressed by Hover's method of destressing. The wires are cut and the line is released. The sleepers are further cured by submerging them into a water tank for a period of 14 days. Alternatively, the sleepers can also be steam cured.

Another method adopted sometimes for the manufacture of prestressed mono-block concrete sleepers is the short line method or 'stress bench method'. This process involves the use of short stress benches that accommodate four to five sleepers. The ends of the benches serve as anchor plates and comprise an iron frame to bear the initial prestressing force. The benches are on wheels and are mobile. The prestressing is done as in the case of the long line method. The concreting, vibrating, etc., is however, done at a fixed place, the stress benches being moved into position one after another. This leads to better quality control in concrete mixing and compaction. Generally, after casting, the benches are taken into steam chambers for curing with an overall turnround period of about 24 hours and a steam curing cycle of about 16 hours. This method of manufacture gives qualitatively better results and has been adopted by M/s Daya Engineering Works Pvt. Ltd, Gaya, and M/s Concrete Products and Construction Co., Chennai.

Prestressed mono-block concrete sleepers can also be manufactured by the individual mould method. This method is generally used when prestressing is transferred to concrete through bonds and positive anchorages in the case of pre-tensioned sleepers or only by positive anchors in the case of post-tensioned sleepers. The mould for the pre-tensioned type is designed to take the initial prestressing force and hence has to be sturdier than the moulds used in other systems. The moulds can adjust one to three sleepers, and as they move along the assembly line, various tasks, such as cleaning of moulds, insertion of high tensile stress wires, prestressing of wires, fixing inserts, concreting, vibrating, steam curing, and remoulding, are carried out on the manufacturing belt. This system involves a greater degree of automation, yields qualitatively better results, and requires the least amount of workforce. In India, factories utilizing this technique have currently gone into production at Secunderabad and Bharatpur.

Two-block type of concrete sleepers

The manufacture of two-block concrete sleepers is simple and similar to that of any other ordinary precast RCC unit. These sleepers are manufactured in a mould in which the necessary reinforcement and tie bar are placed in position. Concrete of designed mix is then poured into the mould and vibrated. The mould is removed after the concrete is set and the blocks are cured in water for a period of 14 days.

Post-tension type of concrete sleepers

Post-tension type of concrete sleepers were earlier manufactured in the concrete sleeper plant at Allahabad as per the design submitted by D&W of Germany, approved by the Railway Board. The specialty of this patent design of D&W lies in the use of high tensile steel rods bent into the U shape known as hair pins, slits, and nuts. This process also involved the instantaneous demoulding of the products.

The technology of post-tension concrete sleepers has become outdated over time. The sleepers manufactured in the concrete sleeper plant (CSP) at Allahabad have been quite uneconomical and their rejection rate has also been quite high. In view of this, the manufacture of concrete sleepers by the post-tension method has been stopped in the CSP at Allahabad since July 1995.

7.8.11 Testing of Concrete Sleepers

In addition to the control checks exercised on the material and manufacturing process, the concrete and the finished sleepers are subjected to the following periodical checks and tests.

(a) The minimum 28-day compressive strength of the test cube should not be less than 525 kg/cm^2. Sleepers from occasional batches in which the minimum crushing strength falls below 525 kg/cm^2 but not below 490 kg/cm^2 may be accepted subject to their passing the increased frequency of testing for static bending strength.

(b) The minimum compressive strength of the test cube of concrete at detensioning should not be less than 370 kg/cm^2.

(c) The modulus of rupture should be as specified in the Concrete Bridge Code.

(d) The dimensional tolerance and surface finish of the sleepers should be checked using suitable templates and gauges.

(e) The cracking and failure moments of the sleepers should be tested at the following sections by applying suitable loads:

- Positive cracking moment at rail seat bottom
- Negative cracking moment at centre section top
- Positive cracking moment at centre section bottom
- Failure moment at rail seat bottom

(f) For the abrasion resistance test, the concrete sleeper is subjected to a vibrating load under specified conditions. After 300 hours of operating time, the loss in weight due to abrasion should not be more than 3 per cent.

7.8.12 Handling of Concrete Sleepers

Concrete sleepers weigh about 215 to 270 kg and about six to eight persons are required to handle one sleeper. The mechanical handling of concrete sleepers is, therefore, desirable for safety purposes.

7.8.13 Prohibited Locations for Use of Concrete Sleepers

Concrete sleepers, because of their heavy weight and rigidity of structure, are not suited to yielding formations, fish-plated joints, and places where uniform packing cannot be achieved. Concrete sleepers as such are normally laid at only those locations where LWRs are permissible. These should be used only with long welded rails. Fish-plated joints on concrete sleeper tracks, where unavoidable, should have wooden sleepers at joints. These sleepers should not be laid at the following locations:

(a) New formation in banks unless specially compacted
(b) Any rock cuttings, except where a minimum depth of 300 mm of ballast cushion has been provided
(c) Un-ballasted lines in yards
(d) Curves of radius less than 500 m
(e) Troublesome formations
(f) Near ashpits and other locations where drivers habitually drop ash
(g) At locations where excessive corrosion is expected
(h) On un-ballasted bridges and on arch bridges, where the height between the arch and the bottom of the ballast section is less than 1 m, and on slab bridges, where the ballast cushion between the bottom of the sleepers and the top of the slab is less than 300 mm
(i) With fish-plated tracks

7.8.14 Laying of Concrete Sleepers

Concrete sleepers are heavy, and as such manual handling of concrete sleepers is not only difficult, but may generally damage the sleeper as well. In exceptional cases, however, manual handling, including manual laying of concrete sleepers, is resorted to after taking adequate precautions.

In the case of mechanical relaying system, normally two portal cranes are used on Indian Railways and relaying is done using prefabricated panels. The existing rail panels are removed by gantry cranes, the ballast is levelled up, and prefabricated panels are then laid with the help of portal cranes. The following operations are involved.

(a) Preparation work at the site of relaying
(b) Pre-assembly of panels in base depots

(c) Actual relaying operation

(d) Post-relaying work

The full details of the manual relaying method as well as of the mechanical relaying system are given in Chapter 21.

7.8.15 Maintenance of Concrete Sleepers

The following points need attention in the maintenance of concrete sleepers.

(a) Concrete sleepers should normally be maintained with heavy on-track tampers. For spot attention, MSP or off-track tampers may be used. The size of chips for MSP should be 8 mm–30 mm as required.

(b) Only 30 sleeper spaces are to be opened out at a time between two fully boxed track stretches of 30 sleepers length each in case a LWR track exists.

(c) Concrete sleepers should be compacted well and uniformly to give a good riding surface. Centre binding of mono-block concrete sleepers should be avoided, for which the central 800 mm of the sleeper should not be hard packed.

(d) Both ends of the concrete sleepers should be periodically painted with anticorrosive paint to prevent corrosion of the exposed ends of prestressing wires. In the case of two-block sleepers, the tie bars should be examined every year, and if any sign of corrosion is noticed, the affected portion should be painted with an approved paint.

(e) Mechanical equipment should be used for laying and maintaining concrete sleepers as far as possible.

(f) Wherever casual renewal of concrete sleepers is to be done, the normal precautions followed for LWR tracks should be taken.

(g) The elastic rail clip should be driven properly to ensure that the leg of the clip is flush with the end face of the insert. Overdriving and underdriving should be guarded against, as these cause eccentric loading on the insulations, resulting in their displacement and in the variation of load.

(h) A vigilant watch should be kept to ensure that no creep occurs in any portion of the concrete sleeper track or there is no excessive movement near the switch expansion joint (SEJ).

(i) It must be ensured that the rubber pads are in their correct positions. Whenever it is found that the rubber pads have developed a permanent set, these should be replaced by new ones. Such examinations can be done at the time of destressing. Toe load can also be lost due to ineffective pads.

(j) Nylon or composite insulating liners used with Pandrol clips should be examined periodically for signs of cracking and breakage. Adequate care should be exercised when driving the clip at the time of installation to prevent damage.

(k) One of the biggest problems regarding the maintenance of a concrete sleeper track is that the elastic rail clips get seized with malleable cast iron (MCI) inserts not only during regular maintenance, but also during destressing, other incidental works, and derailments. The following remedial measures are suggested.

    (i) At the base depot, all the elastic rail clips and MCI inserts should be thoroughly cleaned. Grease should then be applied on the central leg of the elastic rail clip (ERC) and the eye of the MCI insert. These should then be driven into place at the time of assembly of the service pan.

(ii) During service, all the elastic rail clips must be taken out from the MCI inserts and cleaned with a wire brush and emery paper, especially on the central leg. The eyes of the MCI inserts must also be cleaned of any debris or rusted material. The central·leg of the ERC should then be covered with good quality grease. The eyes of the MCI inserts should be smeared with the same grease before the treated ERCs are driven back. This has to be repeated every year in corrosion prone areas. A maintenance checklist for concrete sleepers is given in Table 7.9.

Table 7.9 Maintenance checklist for concrete sleepers

Item	*Points for checking*
Location of concrete sleepers	Concrete sleepers should normally be laid on a LWR/ CWR track, first preference being given to high-speed routes and then to other routes. The track standard for the use of a concrete sleeper has been specified in Chapter 5.
	Concrete sleepers should be used only at permitted locations. See Section 7.8.13
Sleeper spacing	Spacing should be uniform, 60 cm for a sleeper density of 1660/ km and 65 cm for a sleeper density of 1540/km.
Ballast section	The specified ballast section for LWR should be followed.
	In two-block RCC sleepers, a 1033-mm-wide central trough should be provided to avoid corrosion of the tie bar.
Handling of concrete sleepers	Preferably mechanized means such as gentry cranes should be used. In exceptional cases, manual handling may be done using sleeper slings and rail dollies, taking proper precautions to avoid damage to the sleeper.
Laying concrete sleepers	Mechanical means, i.e., portal cranes with a pre-assembled panel should be adopted.
	Manual laying should be adopted only in exceptional conditions and that too with proper precautions.
Maintenance of concrete sleepers	On-track tampers should be used for regular maintenance of long stretches.
	Off-track tampers such as Chinese tampers or measured shovel packing should be used for isolated or short stretches.
	In emergencies, a blunt end beater should be used for packing.
Maintenance of fastenings used with concrete sleeper track	Overdriving or underdriving of Pandrol clips should be guarded against.
	It should be ensured that the rubber pad is in its correct position and renewed when these develop permanent set.
	Care should be taken while driving the clip into position to avoid damage to liners. Cracked liners should be replaced.
	At the time of initial laying as well as during service, all the MCI inserts and ERCs should be thoroughly cleaned and then grease applied on the central leg of ERC and the eye of the MCI insert.

7.8.16 Derailment on a Concrete Sleepers Track

Derailment is a type of accident that occurs when the wheels of a vehicle mount the rail head. It causes excessive damage to the track in general and sleepers in particular.

The following actions should be taken in the eventuality of derailment on a track with concrete sleepers.

(a) When the damage to concrete sleepers is not extensive and it is possible to allow the traffic to pass at a restricted speed, suitable speed restriction should be imposed after assessing the damage to the track. Sleepers should be replaced as in the case of casual renewals while taking all precautions. After all the damaged sleepers are replaced, the affected portion as well as the portions 100 m on either side adjacent to it should be distressed, and normal speed should be restored after consolidation.

(b) When the damage to the concrete sleeper is extensive and the track is distorted in such a way that it is not possible to allow traffic to pass even at a restricted speed, the affected portion should be isolated by introducing buffer rails on either end of it. The distorted track should be removed and replaced by the track laid on single-rail panels using the available rails and sleepers. The section should then be converted into long welded rails using concrete sleepers, taking the usual precautions laid down in the LWR manual.

7.8.17 Concrete Sleepers for Heavier Axle Load

The existing PSC sleeper, which was designed for 22.5 tonne axle load, has been in service since three decades and has given good service.

(i) To keep pace with the increased freight traffic, Indian Railways is contemplating to increase throughput by allowing higher axle load on existing track. Moreover, heavy haul operations on new Dedicated Freight Corridor planned on eastern and western sections are also on the anvil.

(ii) Dedicated freight corridor has been planned for 30 tonnes axle load operation and feeder routes from existing network are planned to carry 25 tonnes axle load. Therefore, 25 tonnes and 30 tonnes axle load will coexist and these are Indian Railways' business requirements.

A detailed study has been carried out for examining the feasibility of the existing PSC sleepers for 25 and 30 tonnes axle loads. The results of the study are as follows:

(i) 25 tonnes axle load can be run on existing PSC sleepers laid to a density of 1660/km.

(ii) A slight modification of existing design with initial pre-stress level as 75 per cent of breaking load can also be tried for 25 tonnes axle load operation.

(iii) For 30 tonnes axle load operations, new sleepers design should be used.

It is brought out that a new sleepers design (EDO/T-2255) has been developed using 18 number, 3mm × 3 high-tension steel strands. In the design, rail seat load has been taken as 20 tonnes, which is derived from the existing practices.

This sleeper is under trial. The trial casting of the new design of PSC sleeper for 30 tonnes axle is being done at Concrete Sleepers Plant, Anwargang, Kanpur.

For laboratory testing, static bend test shall be performed at Kanpur while fatigue test shall be conducted at Bridge & Structure lab of RDSO.

7.8.18 Planning of Concrete Sleepers on Indian Railways

Indian Railways is modernizing its track in a big way to meet the challenges of heavier traffic at faster speeds. The modern track consisting of long welded 52-kg/60-kg rails, concrete sleepers, and elastic fastenings can meet the above requirements.

Prestressed concrete sleepers are most economical and technically best suited for high speeds and heavy traffic density. They provide a stable track structure, which requires less maintenance efforts. Maintenance of concrete sleeper track should, however, be done using track machines only.

It has been proposed that concrete sleepers should be provided on all important routes on Indian Railways. Adequate capacity has been developed for the production of these sleepers to meet all the requirements of Indian Railways. During 2010–11 about 8 million concrete sleepers (for BG) and 6500 sets of concrete turnout sleepers were produced. The intake of wooden sleepers for main lines has been completely stopped and emphasis is being laid on using concrete sleepers on turnouts.

Indian Railways is the world leader in the manufacture of concrete sleepers and currently manufactures about 60 per cent of the total concrete sleepers in the world.

Indian Railways has plans to further boost up the manufacture of concrete sleepers in order to take the challenge of moving faster and heavier traffic. It is planning to set up more factories in near future. It is estimated that Indian Railways will be manufacturing more than 15 million concrete sleepers per year in the near future.

SUMMARY

Sleepers support rails and transfer the live load of moving trains to the ballast and formation. Wooden sleepers are the best, as they satisfy almost all the requirements of an ideal sleeper. Scarcity of timber has led to the development of metal and concrete sleepers. Concrete sleepers have high strength and a long life, and are most suitable for modern tracks. Indian Railways has developed designs for prestressed concrete sleepers and these are being extensively used on all important routes.

REVIEW QUESTIONS

1. What are the requirements of sleepers used in a railway track? Give a neat sketch of a typical BG mono-block prestressed concrete sleeper. What are its advantages and drawbacks?
2. List the various types of sleepers used on Indian Railways. Which type would you consider to be the best for modern tracks and why?
3. Enumerate the loading conditions adopted by RDSO for the design of mono-block prestressed concrete sleepers in India.

4. List the various types of metal sleepers in use on Indian Railways. Describe mono-block prestressed concrete sleepers with a neat sketch. What are the reasons for their ever-increasing adoption the world over?

5. Discuss the factors on which sleeper density depends. How is sleeper density expressed?

6. Compare the characteristics of the different types of sleepers used in India.

7. Explain the functions of sleepers and ballast in a railway track. Explain how the spacing of sleepers is determined. Give specific reasons for the necessity of regular maintenance of the ballast.

8. Draw a neat sketch of the prestressed concrete sleeper used on Indian Railways for BG tracks. Give details of the location of wires and the seating and fastening arrangements.

9. What are the different types of sleepers used in the track on Indian Railways? Write down in brief the advantages and disadvantages of each type.

10. What are the advantages and disadvantages of steel trough sleepers? What is the function of tie bars in the case of cast iron pot sleepers? What is the relation between sleeper density and the depth of ballast?

11. What are the loading conditions adopted by Indian Railways for the design of concrete and? Discuss briefly the relative advantages and disadvantages of mono-block and two-block sleepers.

12. What are the various methods of manufacture of concrete sleepers? Discuss briefly one of these methods used in India.

13. What is the future scope of concrete sleepers on Indian Railways? Discuss briefly the planning being done for the production of concrete sleepers in India.

Choose the correct answer from the choices given.

14. The weight of a CST-9 sleeper assembly alongwith fastening of MG is:
    (a) 58 kg   (b) 60.5 kg
    (c) 65 kg   (d) None of these

15. For a BG route with M + 7 sleeper density, the number of sleepers per rail length is:
    (a) 18   (b) 19
    (c) 20   (d) 21

16. On either side of the centre line of rails, a cant of 1 in 20 in the sleeper is provided for a distance of:
    (a) 50 mm   (b) 165 mm
    (c) 175 mm   (d) 185 mm

17. A mono-block sleeper has:
    (a) square section   (b) rectangular section
    (c) trapezoidal section   (d) semi-circular section

18. Pick up the incorrect statement from the following:
    (a) Sleepers transfer the load of moving locomotive to the girders of the bridges.
    (b) Sleepers act as a non-elastic medium between the rails and the ballast.
    (c) Sleepers hold the rails at 1-in-20 tilt inward.
    (d) Sleepers hold the rails loose on curve.

19. Minimum CSI for wooden sleepers used over bridge girders is:
    (a) 1455   (b) 135
    (c) 1255   (d) 1155

20. The sleeper density of a BG track is (M + 6) in metric units. The number of sleepers for 1.04 km length of track is:
    (a) 1520   (b) 1630
    (c) 1720   (d) 1800

21. In fan-shaped layout of PRC sleepers the special type of concrete sleepers used for BG 1-in-8 1/2 turnout are:
    (a) 45   (b) 54
    (c) 65   (d) 83

22. Overall length of mono-block concrete sleepers is:
    (a) 2500 mm   (b) 2625 mm
    (c) 2750 mm   (d) 2900 mm

23. Minimum CSI prescribed on Indian Railways for wooden crossing sleeper:
    (a) 652   (b) 783
    (c) 1352   (d) 1455

24. Using a sleeper density of M + 6 for a BG track, determine the number of sleepers required for construction of 100 panels of 13 metres each:
    (a) 1000   (b) 1500
    (c) 1900   (d) 2000

25. The weight of a concrete sleeper is about:
    (a) 50 kg   (b) 100 kg
    (c) 200 kg   (d) 300 kg

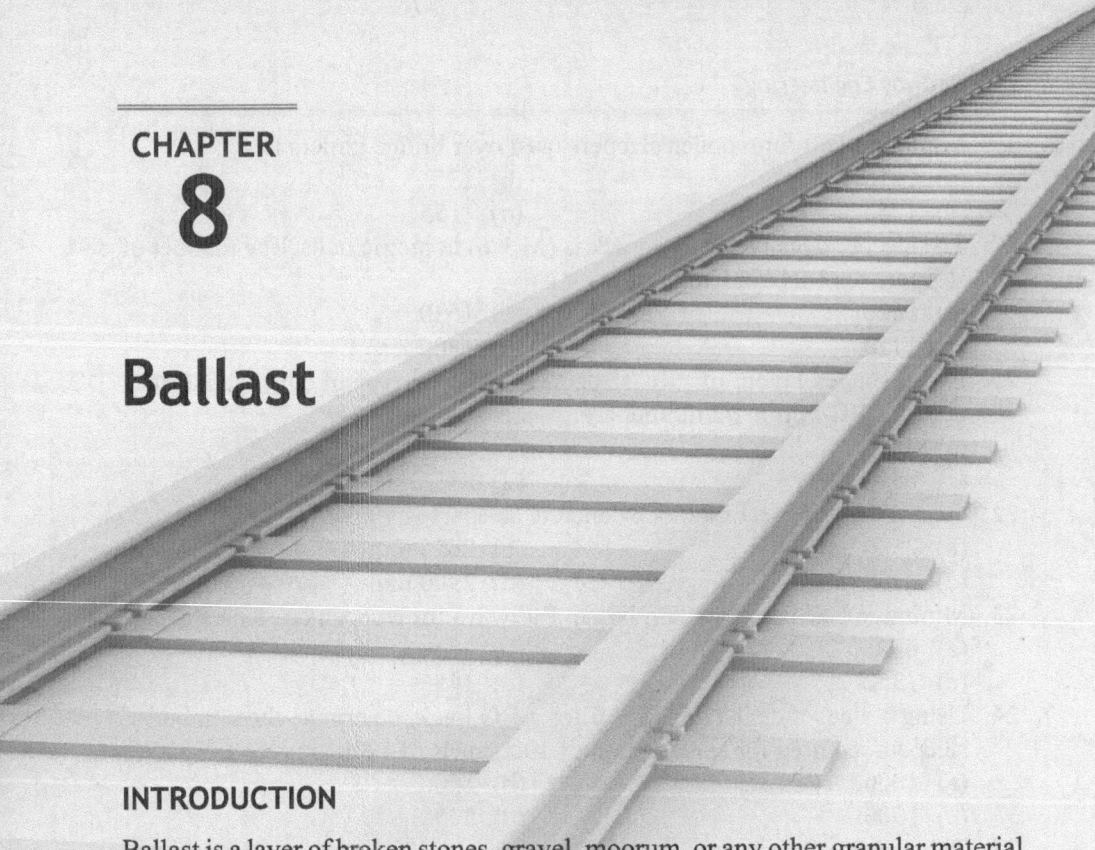

Ballast

INTRODUCTION

Ballast is a layer of broken stones, gravel, moorum, or any other granular material placed and packed below and around sleepers for distributing load from the sleepers to the formation. It provides drainage as well as longitudinal and lateral stability to the track. Different types of ballast materials and their specifications are discussed in this chapter.

8.1 FUNCTIONS OF BALLAST

The ballast serves the following functions in a railway track.
- It provides a level and hard bed for the sleepers to rest on.
- It holds the sleepers in position during the passage of trains.
- It transfers and distributes load from the sleepers to a large area of the formation.
- It provides elasticity and resilience to the track for proper riding comfort.
- It provides the necessary resistance to the track for longitudinal and lateral stability.
- It provides effective drainage to the track.
- It provides an effective means of maintaining the level and alignment of the track.

8.2 TYPES OF BALLAST

The different types of ballast used on Indian Railways are described here.

Sand ballast

Sand ballast is used primarily for cast iron (CI) pots. It is also used with wooden and steel trough sleepers in areas where traffic density is very low. Coarse sand is preferred in comparison to fine sand. It has good drainage properties, but has the drawback of blowing off because of being light. It also causes excessive wear of the rail top and the moving parts of the rolling stock.

Moorum ballast

The decomposition of laterite results in the formation of moorum. It is red, and sometimes yellow, in colour. The moorum ballast is normally used as the initial ballast in new constructions and also as sub-ballast. As it prevents water from percolating into the formation, it is also used as a blanketing material for black cotton soil.

Coal ash or cinder

This type of ballast is normally used in yards and sidings or as the initial ballast in new constructions since it is very cheap and easily available. It is harmful for steel sleepers and fittings because of its corrosive action.

Broken stone ballast

This type of ballast is used the most on Indian Railways. A good stone ballast is generally procured from hard stones such as granite, quartzite, and hard trap. The quality of stone should be such that neither is it porous nor does it flake off due to the vagaries of weather. Good quality hard stone is normally used for high-speed tracks. This type of ballast works out to be economical in the long run.

Other types of ballast

There are other types of ballast also such as the brickbat ballast, gravel ballast, kankar stone ballast, and even earth ballast. These types of ballast are used only in special circumstances.

The comparative advantages, disadvantages, and suitability of different types of ballast are given in Table 8.1.

Table 8.1 Comparison of different types of ballast

Type of ballast	Advantages	Disadvantages	Suitability
Sand ballast	Good drainage properties	Causes excessive wear	Suitable for CI pot sleeper tracks
	Cheap	Blows off easily	Not suitable for high-speed tracks
	No noise produced on the track	Poor retentivity of packing	
	Good packing material for CI sleepers	Track cannot be maintained to high standards	
Moorum ballast	Cheap, if locally available	Very soft and turns into dust	Used as a sub-ballast
	Prevents water from percolating	Maintenance of track is difficult	Initial ballast for new construction
	Provides good aesthetics	Quality of track average	
Coal ash or cinder	Easy availability on railways	Harmful for steel sleepers	Normally used in yards and sidings

Contd.

Table 8.1 *Contd.*

Type of ballast	Advantages	Disadvantages	Suitability
	Very cheap	Corrodes rail bottom and steel sleepers	Suitable for repairs of formations during floods and emergencies
	Good drainage	Soft and easily pulverized	Not fit for high-speed tracks
		Maintenance is difficult	
Broken stone ballast	Hard and durable when procured from hard rocks	Initial cost is high	Suitable for packing with track machines
	Good drainage properties	Difficulties in procurement	Suitable for high-speed tracks
	Stable and resilient to the track	Angular shape may injure wooden sleepers	
	Economical in the long run		

8.3 SIZES OF BALLAST

Previously, 50 mm (2″) ballasts were specified for flat-bottom sleepers such as concrete and wooden sleepers, and 40 mm (1.5″) ballasts for metal sleepers such as CST-9 and trough sleepers. Now, to ensure uniformity, 50 mm (2″) ballasts have been adopted universally for all types of sleepers.

Points and crossings are subjected to heavy blows of moving loads and hence are maintained to a higher degree of precision. A small sized, 25 mm (1″) ballast is, therefore, preferable because of its fineness for slight adjustments, better compaction, and increased frictional area of the ballast.

For uniformity sake, the Indian Railways has adopted the same standard size of ballast for the main line as well as for points and crossings.

This standard size of ballast should be as per Indian Railways specification. The specification provides grading of ballast from 25 mm to 65 mm, maximum quantity of ballast being in the range of 40 mm to 50 mm size.

8.4 REQUIREMENTS OF A GOOD BALLAST

Ballast material should possess the following properties.
   (a) It should be tough and wear resistant.
   (b) It should be hard so that it does not get crushed under the moving loads.
   (c) It should be generally cubical with sharp edges.
   (d) It should be non-porous and should not absorb water.
   (e) It should resist both attrition and abrasion.
   (f) It should be durable and should not get pulverized or disintegrated under adverse weather conditions.

(g) It should allow for good drainage of water.

(h) It should be cheap and economical.

8.5 DESIGN OF BALLAST SECTION

The design of the ballast section includes the determination of the depth of the ballast cushion below the sleeper and its profile. These aspects are discussed as follows.

8.5.1 Minimum Depth of Ballast Cushion

The load on the sleeper is transferred through the medium of the ballast to the formation. The pressure distribution in the ballast section depends upon the size and shape of the ballast and the degree of consolidation. Though the lines of equal pressure are in the shape of a bulb as discussed in Chapter 5 (Fig. 5.7), yet for simplicity, the dispersion of load can be assumed to be roughly 45° to the vertical. In order to ensure that the load is transferred evenly on the formation, the depth of the ballast should be such that the dispersion lines do not overlap each other.

For the even distribution of load on the formation, the depth of the ballast is determined by the following formula (refer to Fig. 8.1):

Sleeper spacing = width of the sleeper + 2 × depth of ballast   (8.1)

For example, if a BG track is laid with wooden sleepers with a sleeper density of $N + 6$, then the average sleeper spacing would be 68.4 cm. If the width of the sleeper is 25.4 cm, then the depth of the ballast cushion would be

$$d = \frac{68.4 - 25.4}{2} = 21.5 \text{ cm}$$

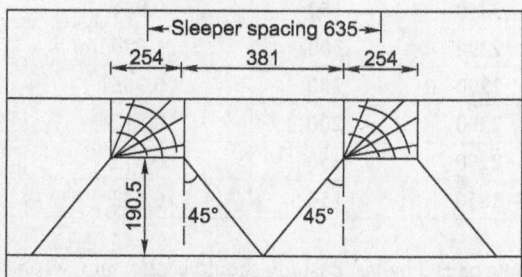

Fig. 8.1   Minimum depth of ballast cushion (dimensions in mm)

A minimum cushion of 15–20 cm of ballast below the sleeper bed is normally prescribed on Indian Railways.

8.5.2 Ballast Profile for Fish-plated Track

The ballast profile for a fish-plated track is shown in Fig. 8.2. The requirements of ballast for different groups of railway lines as adopted on Indian Railways are given in Table 8.2.

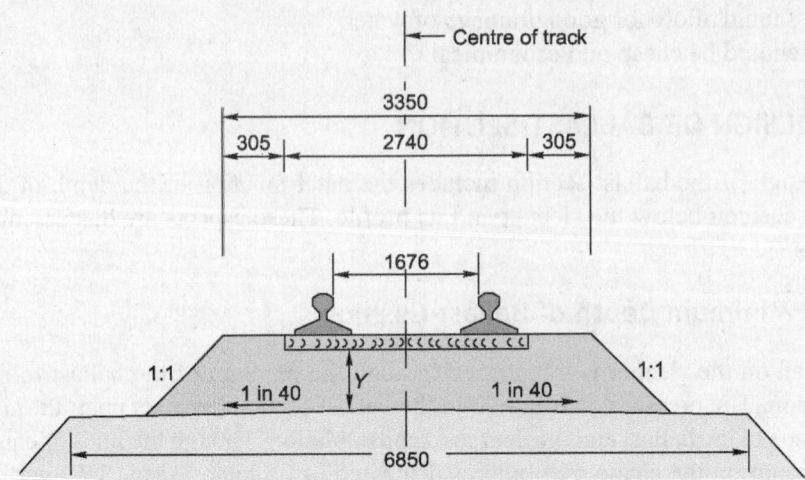

Fig. 8.2 Standard ballast profile for BG (other than LWR/CWR)

Table 8.2 Recommended depth of ballast cushion and ballast profile for fish-plated track

Gauge and route	Top width of ballast (mm)	Ballast cushion (mm)	Quantity of ballast required per metre	
			On straight and curves flatter than 600 m radius (m³)	On curves sharper than 600 m radius (m³)
BG (A)	3350	300	1.588	1.634
BG (B & C)	3350	250	1.375	1.416
BG (D)	3350	200	1.167	1.202
BG (E)	3350	150	0.964	0.996
MG (Q)	2290	300	1.070	1.145
MG (R1)	2290	250	0.965	1.033
MG (R2)	2290	200	0.817	0.905
MG (S)	2290	150	0.673	0.725
NG	1850	150	0.543	0.584

Notes:
1. For group D route on BG ballast cushion should be 250 mm, when speed is 100 kmph.
2. For Q routes on MG ballast cushion should be 300 mm for speed of 100 kmph and above and 250 mm for speeds less than 100 kmph.
3. For R2 route on MG the ballast cushion should be 250 mm where LWR is contemplated and 200 mm otherwise.
4. In case of SWR track, the minimum depth of ballast cushion should be 200 mm for BG as well as for MG.
5. For fish plated track on the outside of curves sharper than 600 m radius on BG and MG and 300 m on NG, width of shoulder ballast should be increased to a maximum of 400 mm on BG and MG 300 mm on NG (Extra width is 100 mm for BG and MG) to give additional lateral support.
6. For short-welded track, the ballast width to be increased on outside of curves is as follows: BG: 400 mm for curves flatter than 875 m radius and 450 mm for curves sharper than

875 m radius. MG: 350 mm for curves flatter than 600 m radius and 380 mm for curves sharper than 600 m radius.

7. For ballasting on turn in curves of turnouts negeotiated by passenger trains in station yards, the width of the outside shoulder should be increased to a maximum of 550 mm for BG and MG and 450 mm for NG.
8. The slopes of the ballast section should be normally (horizontal: vertical) 1:1 or 1.5:1, but can vary according to the type of ballast used.
9. A cross slope of 1 in 400 is given on the formation to have good drainage of the ballast section.
10. The ballast below sleeper is called ballast cushion, that outside the sleeper is called shoulder ballast and that between the sleepers is called crib ballast.
11. For ballast section for LWR track, please refer next section.
12. The increase in ballast cushion is carried out during complete track renewal.

8.5.3 Ballast Profile for Long-welded Rail Tracks

The ballast profile for a long-welded rail (LWR) track is shown in Fig. 8.3. The requirements of ballast for different types of sleepers on a BG railway line are given in Table 8.3.

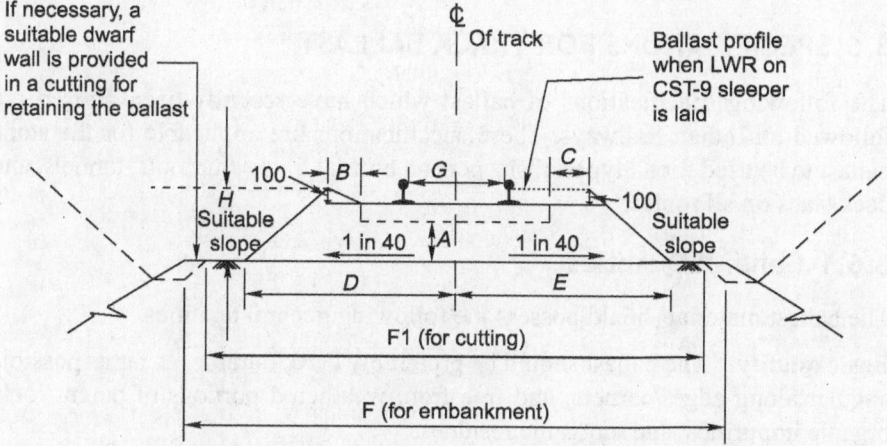

Fig. 8.3   Ballast profile for LWR track (single-line BG)

Table 8.3   Ballast requirement (mm) for single-line BG LWR tracks

Type of sleeper	A	B	C*	D	E*	F	F1	H	Quantity of ballast per metre (m³)	
									Straight track	Curved track
Wooden	250	350	500	2270	2420	6850	6250	540	1.682	1.646
	300	350	500	2270	2420	6850	6250	590	1.782	1.853
	†	350	500	2270	2420	6850	6250	640	1.982	2.060
Steel trough	250	350	500	2270	2430	6850	6250	550	1.762	1.827
	300	350	500	2270	2440	6850	6250	600	1.962	2.035
	†	350	500	2270	2430	6850	6250	650	2.162	2.242
Prestressed concrete	250	350	500	2270	2510	6850	6250	630	1.954	2.032
	300	350	500	2270	2510	6850	6250	680	2.158	2.243
	†	350	500	2270	2510	6850	6250	730	2.362	2.455

(Contd.)

Table 8.3 Contd.

Type of sleeper	A	B	C*	D	E*	F	F1	H	Quantity of ballast per metre (m³)	
									Straight track	Curved track
Two-block concrete	250	350	500	2270	2510	6850	6250	630	2.110	2.193
	300	350	500	2270	2510	6850	6250	680	2.314	2.405
	†	350	500	2270	2510	6850	6250	730	2.518	2.616

* On the outer side of the curves only. Cess may be widened where required depending on local conditions.
† 200 mm over 150 mm sub-ballast.

The minimum clean stone ballast cushion below the bottom of sleeper (A) is 250 mm. For routes where speeds are to be more than 130 kmph, A is 300 mm–200 mm along with 150 mm of sub-ballast. Suitable dwarf walls should be provided in the case of cuttings, if necessary, for retaining the ballast.

8.6 SPECIFICATIONS FOR TRACK BALLAST

The following specifications of ballast which have recently been revised are followed on Indian Railways. These specifications are applicable for the stone ballast to be used for all types of sleepers on normal tracks, turnouts, tunnels, and deck slabs on all routes.

8.6.1 General Qualities

The ballast material should possess the following general qualities.

Basic quality   The ballast should be preferably hard, durable, as far as possible angular along edges/corners, and free from weathered portions of parent rock, organic impurities, and inorganic residues.

Particle shape   Ballasts should be preferably cubical in shape. Individual pieces should not be flaky and should have flat faces generally with not more than two rounded/sub-rounded faces.

Mode of manufacture   Ballasts for all BG main lines and running lines, except on E routes, but including E special routes, should be machine crushed. For other BG lines and MG/NG routes planned or sanctioned for conversion, the ballast should preferably be machine crushed. Hand-broken ballast can be used in exceptional cases with the prior approval of the chief track engineer or the chief administrative officer (CAO). Such approval should be obtained prior to the invitation of tenders. Hand-broken ballasts can be used without any formal approval on MG and NG routes not planned or sanctioned for conversion.

8.6.2 Physical Properties

All ballast samples should possess the physical properties given in Table 8.4 when tested in accordance with IS:2386 (IV)–1963.

Table 8.4 Physical requirements of ballast

Characteristics	BG, MG, and NG (planned/ sanctioned for conversion)	NG and MG (other than those planned for conversion)
Aggregate abrasion	30% maximum*	35% maximum
Aggregate impact	20% maximum*	30% maximum
Water absorption	1%	–

* In exceptional cases, relaxable on technical and/or economic grounds up to 35 per cent and 25 per cent, respectively, by the chief track engineer (CTE) in open lines and the chief administration officer (construction) (CAO/C) for construction projects. Relaxation in abrasion and impact values is given prior to the invitation of tender and should be incorporated in the tender document.

8.6.3 Size and Gradation

The ballast should satisfy the size and gradation requirements given in Table 8.5.

Table 8.5 Ballast gradation

Size of sieve, mm	% retained
65	5% maximum
40	40% to 60%
20	Not less than 98% for machine crushed and not less than 95% for hand broken

8.6.4 Oversized Ballast

Retention on 65 mm square mesh sieve   A maximum of 5 per cent ballast retained on a 65 mm sieve is allowed without deduction in payment. In case the ballast retained on a 65 mm sieve exceeds 5 per cent but is less than 10 per cent, payment at a 5 per cent reduction of 5 per cent in the contracted rate is made for the full stack. Stacks retaining more than 10 per cent of ballast on a 65 mm sieve are rejected.

Retention on 40 mm square mesh sieve   In case the ballast retained on a 40 mm square mesh sieve (machine crushed only) exceeds the 60 per cent limit prescribed above, payment at the following reduced rates is made for the full stack in addition to the reduction as worked out above.

  (i) 5 per cent reduction in contracted rates if the retention on a 40 mm square mesh sieve is between 60 per cent (excluding) and 65 per cent (including).
  (ii) 10 per cent reduction in contracted rates if retention on a 40 mm square mesh sieve is between 65 per cent (excluding) and 70 per cent (including).
  (iii) In case the retention on a 40 mm square mesh sieve exceeds 70 per cent, the stack is rejected.
  (iv) In the case of hand-broken ballast supply, 40 mm sieve analysis may not be carried out. The executive may, however, ensure that the ballast is well graded between 65 mm and 20 mm.

8.6.5 Undersized Ballast

The ballast is treated as undersized and rejected if
  (a) retention on a 40 mm sieve is less than 40 per cent and
  (b) retention on a 20 mm sieve is less than 98 per cent (for machine crushed ballast) or 95 per cent (for hand-broken ballast).

8.6.6 Shrinkage Allowance

Payment is made for the gross measurements either in stacks or in wagons without any deduction for shrinkage/voids. However, when ballast is supplied in wagons, up to 8 per cent shrinkage is permitted at the destination by the consignee verifying the booked quantities.

8.6.7 Sampling and Testing

The following procedure is specified for the sampling and testing of the ballast.
  (a) A minimum of three samples of ballast should be taken for sieve analysis for measurement done on any particular date, even if the number of stacks to be measured is less than three.
  (b) The tests for abrasion value, impact value, and water absorption should be done in approved laboratories or in the Railways' own laboratories (A list of these laboratories should be given in the tender document).
  (c) In order to ensure the supply of a uniform quality of ballast, the specifications given in Table 8.6 should be followed with respect to sampling, testing, and acceptance of the ballast. The tests given in this table may be carried out more frequently if warranted at the discretion of the Railways.

Table 8.6 Frequency of tests for ballast supply

Item	Supply in stacks		Supply in wagons
	For a stack of volume less than 100 m³	*For a stack of volume more than 100 m³*	
Number of size and gradation tests	One for each stack	One for each stack	One for each wagon
Size of one sample* (m³)	0.027	0.027 for every 100 m³ or part thereof	0.027
Abrasion value, impact value, and water absorption test †	One test for every 200 m³ of ballast	One test for every 2000 m³ of ballast	One test for every 200 m³ of ballast

* This sample should be collected using a wooden box of internal dimensions 0.3 m × 0.3 m × 0.3 m from different parts of the stack/wagon.

† These tests should be done for the purpose of monitoring the quality during supply. In case the test results are not as per the prescribed specifications at any stage, further supplies should be suspended till suitable corrective action is taken and supply as per specifications is ensured.

  (d) On supply of the first 100 m³, tests for size gradation, abrasion value, impact value, and water absorption (if prescribed) should be carried out by the Railways. Further supply should be accepted only after the first batch satisfies these tests. The Railways reserves the right to terminate the contract as per the general conditions of contract (GCC) at this stage itself in case the ballast supply fails to meet any of these specifications.

8.7 COLLECTION AND TRANSPORTATION OF BALLAST

The collection and transportation of ballast can be done by either of the following methods:

- Collecting the ballast at ballast depots and transporting it to the site in ballast trains
- Collecting the ballast along the cess and putting the same on the track directly

The mode of collection is decided taking into account the proximity of the quarry, availability of good stone ballast, serving roads alongside the railway line for the carriage of ballast, availability of ballast trains, turnround of ballast trains, and availability of traffic blocks for unloading.

8.7.1 Collection at Ballast Depots

The following procedure is adopted when ballasts are collected at ballast depots:

(a) The space alongside the railway line meant for stacking is divided into a convenient number of zones and demarcated.

(b) For each depot, a diagram indicating the site details of all the measured stacks is maintained.

(c) All stacks in each zone are serially numbered.

(d) Operations of collecting and training out materials should not be carried out at the same time in any zone.

(e) The ground on which the stacks are made should be selected and levelled.

(f) Measurements should be taken for complete stacks. The measured stack should be identified suitably by lime sprinkling or any other method.

(g) As soon as a stack is lifted, it should be recorded on the depot diagram, which should always be kept up-to-date. Challans should be prepared after loading the ballast into wagons.

8.7.2 Collection of Ballast Along the Cess

In case the ballast is collected along the side of the cess, the inspector in charge should maintain a separate register showing the measurement of stacks as well as their disposition (from km to km). The stacks should be serially numbered between successive posts. Entries should be made in a register whenever stacks are removed and ballast is put onto the track. The record should state the place where the removed ballast has been used as well as the date of removal of the stack. Materials passing through a 6 mm square mesh are classified as 'dust' (limited to 1 per cent).

8.8 METHODS OF MEASUREMENT

The quantity of ballast can be measured either in a stack or in a wagon. Both methods are explained below.

8.8.1 Stack Measurement

Stacking should be done on an almost plane and firm ground with good drainage. The height of the stack should not be less than 1 m except in hilly areas, where it may be 0.5 m. The top width of the stack should not be less than 1 m and should be kept parallel to the ground plane. The side slopes of the stack should not be flatter than 1.5:1 (horizontal: vertical). The volume of each stack should normally not be less than 30 m^3 in plain areas and 15 m^3 in hilly areas.

8.8.2 Wagon Measurement

In the case of the ballast supply being directly loaded into wagons, a continuous white line should be painted inside the wagon to indicate the level up to which the ballast can be loaded. The volume in cubic metres corresponding to the white line should also be painted outside the wagon on both sides. In addition to the painted line mentioned above, short pieces of flats (cut pieces of tie bars or otherwise) punched with the volume should be welded in the centre of all the four sides of the wagon.

8.9 LABORATORY TESTS FOR PHYSICAL PROPERTIES OF BALLAST

The following tests are recommended to judge the suitability of the ballast material for a railway track.

8.9.1 Aggregate Abrasion Value

To check for aggregate abrasion, a test sample of 10 kg of clean ballast conforming to the following grading is taken:

Passing the 50-mm sieve and retained on the 40 mm square mesh sieve: 5000 g

Passing the 40-mm and retained on the 25 mm square mesh sieve: 5000 g

The sample, along with the abrasive charge, is placed in the Los Angeles machine, which is rotated at a speed of 30–33 rpm for 1000 revolutions. The sample is sieved and material coarser than the 1.70 mm sieve is washed, dried, and weighed. The difference between the original weight (A) and the final weight of the sample (B) is expressed as a percentage of the original weight of the test sample. This value is reported as the abrasion value and can be calculated as follows:

$$\text{Aggregate abrasion value} = \frac{A - B}{A} \times 100 \tag{8.2}$$

8.9.2 Aggregate Impact Value

To check for aggregate impact, the test sample is prepared out of the track ballast in such a way that it has a grading that passes the 12.5 mm sieve and is retained on the 10 mm sieve. The ballast sample is oven dried and placed duly tamped in the different stages in a cylindrical metal container with 75 mm diameter and 50 mm depth (weight A). The cup of the impact testing machine is fixed firmly in position on the base of the machine and the entire test sample is placed in it and compacted by 25 strokes of the tamping rod. The test hammer weighing about 14 kg is raised 380 mm above the upper surface of the cup and dropped. The test sample is subjected to a total of 15 such blows. The sample is then removed and sieved using a 2.36 mm sieve and the weight of quantity retained is measured (weight *B*) as follows:

$$\text{Aggregate impact value} = \frac{A-B}{A} \times 100 \tag{8.3}$$

8.9.3 Flakiness Index

Flakiness index of an aggregate is the percentage by weight of the particles with a least dimension (thickness) less than three-fifths of their mean dimension. The test is not applicable to sizes smaller than 6.3 mm.

Track ballast sample of sufficient quantity is taken to provide a minimum of 200 pieces, which is weighed (weight *A*). The sample consisting of aggregates is sieved as per the prescribed procedure in a series of sieves. The flaky material is separated and weighed (weight *B*). The flakiness index is then determined by the total weight of the material passing the various sieves, expressed as a percentage of the total weight of the sample gauged.

$$\text{Flakiness index} = \frac{B}{A} \times 100 \tag{8.4}$$

8.9.4 Specific Gravity and Water Absorption Test

A sample consisting of at least 2000 g of aggregate is washed thoroughly to remove finer particles and dust. The whole material is then drained, placed in a wire basket, and immersed in distilled water at a temperature between 22°C and 32°C. The sample is shaken, jolted, and dried as per specific procedure. The sample is then placed in an oven in a shallow tray at a temperature between 100°C and 110°C. It is then removed from the oven, cooled in the container, and weighed (weight *C*). The specific gravity and water absorption are calculated as follows:

$$\text{Specific gravity} = \frac{C}{B-A} \tag{8.5}$$

$$\text{Water absorption (\% by weight)} = \frac{100(B-C)}{C} \tag{8.6}$$

where *A* is the weight in grams of saturated aggregate in water, *B* is the weight in grams of saturated dry aggregate in air, and *C* is the weight in grams of oven-dried aggregate in air.

8.10 ASSESSMENT OF BALLAST REQUIREMENTS

The requirements of the ballast should be assessed separately for:
  (a) correcting the deficiencies existing in the track as well as those arising out of overhauling, through packing and deep screening,
  (b) providing adequate cushion for mechanical tamping, and
  (c) providing extra cushion while converting into LWR.

The ballast required for maintenance purposes is estimated by assessing the quantity approximately, if necessary by a survey, over every 1 km of rail length.

Care should be taken to ensure that the cores under the sleepers are not disturbed.

In case of deep screening, the ballast required for recoupment and providing a standard section should be assessed by deep screening the ballast on a trial basis. For this, full depth screening is done for a length of two to three sleepers at every 0.5 to 1 km interval. In this case, the screening is done under the sleepers as well. The quantity of ballast required for deep screening is roughly taken as 1.5 per cent of the existing quantity of ballast based on field trials.

The quantities assessed above will be the net quantities of ballast required to recoup the deficiencies or to provide required profiles/sections. These net quantities of ballast may be enhanced suitably (say by 8 per cent) to arrive at the gross quantities of ballast needed for the purpose of procurement in case the measurements are proposed to be taken in stacks or wagons at the originating station.

8.11 GUIDELINES FOR PROVISION OF SUB-BALLAST

The sub-ballast is normally made of granular material and is provided between the formation and the ballast in order to distribute the load evenly over the formation. The following points should be kept in mind while selecting a material for sub-ballasts.

(a) The material should consist of coarse granular substances such as river gravel, stone chips, quarry grit, and predominantly coarse sand. Ash, cinder, and slag containing predominantly fine and medium sand should not be used.
(b) The material should be non-cohesive and graded. The uniformity coefficient should be more than 4 to ensure that the sub-ballast is well graded.
(c) The material should not contain more than 15 per cent of fines that measure less than 75 microns.
(d) The thickness of the sub-ballast should not be less than 150 mm.

Table 8.7 Ballast required for points and crossings

Type of sleeper	Angle of crossing	Ballast cushion (mm)	Total quantity of ballast (m^3)
PSC	1 in 8 1/2	300	79.14
PSC	1 in 12	300	113.97
ST/Wooden	1 in 8 1/2	300	68.68
ST/Wooden	1 in 12	300	93.38

8.12 SELF-STABILIZING TRACK

The railways use rails fitted on sleepers with double elastic fastenings and concrete sleepers spread over ballast bed, the ballast providing the necessary self-draining, load distributing, as well as geometry correction medium. The ballast under the action of moving trains gets subjected to disturbing vibrations and the geometry is disturbed as the ballast loses its compactness and yields.

In the new concept of *self-stablizing track* running trains create conditions so that the track gets stabilized with the help of ballast. This is done by suitable control of the inertia of mass ballast that vibrates in track. It is thus possible to stabilize the response of track under running trains rather than destabilizing the track.

8.12.1 Details of Self-stabilizing Track

In the self-stabilzing track, predetermined quantities of ballast are pre-compacted and firmly constrained. Further, an elastic medium in the form of reinforced rubber or polyethylene layer of the order of a few millimetres thick, inserted between the sleeper bottom and the top of the constrained ballast block, modifies the coefficient of restitution and also helps in controlling the vibration profile below the sleepers.

The overall effect is to reduce the frequency and amplitude of vibrations being transmitted below the sleeper because of moving train with ballast not getting disturbed and the formation protected from destabilizing vibrations. The constrained ballast in pre-formed wire mesh cages of specific shapes arranged in a particular manner, combine with each other to provide elastic vertical support to the sleeper as well as improved lateral resistance to the track.

There are pre-formed cages which hold ballast such as U-shaped gabions at the sleeper end to develop lateral resistance.

8.12.2 Advantages

(i) By using these modular ballast filled cages, a lot of material in ballast used for conventional track is saved, which could be around 70 per cent because in the middle one third of the sleeper no ballast will be needed.

(ii) The lateral strength of track also is considerably augmented because of positive interlocking of the U-shaped gabion containing well-compacted ballast coupled with the PSC sleepers providing more resistance as compared to a loose ballast. This improves safety of track against buckling under thermal loads.

(iii) The modular steel cages holding pre-compacted ballast, may provide better stability to track in the lateral direction as well as in the vertical direction, because of positive interlocking between the two modules.

SUMMARY

Ballast is a layer of granular material provided below and around sleepers to distribute load from the sleepers over a larger area of the formation. Any granular material can be used as ballast if it satisfies certain requirements of strength, size, and gradation. The ballast gets crushed because of the dynamic action of the wheel load and, therefore, requires regular maintenance. The thickness of the ballast cushion under the sleepers depends upon the axle load, type of sleepers, sleeper density, and other related factors.

REVIEW QUESTIONS

1. Mention the properties required of a good ballast for a railway track.
2. Explain briefly how the pressure created by wheel loads is transmitted through the ballast. What factors of the ballast influence the intensity of pressure on the formation?
3. What is ballast? Why is it used in railway tracks? Briefly explain the various types of ballasts.

4. Explain the following with respect to the ballasts used on Indian Railways.
   (i) Functions
   (ii) Requirements of ideal ballast material
   (iii) Different materials used for ballast and their relative merits
5. Name six materials commonly used as ballast on Indian Railways. Write down the specifications of an ideal stone ballast.
6. Determine the optimum thickness of the stone ballast required below sleepers of density $M + 7$ on a BG track. (***Ans:*** 19.8 cm)
7. Sketch a typical section of an MG line on wooden sleepers and show the ballast cushion and side slopes for a sleeper density of $M + 3$.

Choose the correct answer from the choices given.

8. The recommended width of ballast for BG(A) route is:
   (a) 3650 mm
   (b) 3350 mm
   (c) 2750 mm
   (d) 2290 mm
9. The ballast requirement for BG track having ST sleeper with 250 mm ballast cushion is:
   (a) 0.965 m^3
   (b) 1.375 m^3
   (c) 1.588 m^3
   (d) 1.750 m^3
10. For over-size ballast, ballast retained on 60 mm sieve should not be more than:
    (a) 5%
    (b) 10 to 20%
    (c) 20 to 30%
    (d) 40 to 60%
11. Under-size hand-crusted ballast will be rejected if retention of the ballast on 40 mm square sieve is less than:
    (a) 25%
    (b) 40%
    (c) 80%
    (d) 95%
12. The minimum depth of ballast below the sleeper on any guage adopted on Indian Railways is:
    (a) 150 mm
    (b) 200 mm
    (c) 300 mm
    (d) 250 mm
13. To prevent percolation of water into formation, moorum is used as a blanket for:
    (a) black cotton soil
    (b) sandy soil
    (c) clayey soil
    (d) all of the above
14. Maximum width of ballast shoulder on outside of turn in curve on BG is:
    (a) 550 mm
    (b) 450 mm
    (c) 500 mm
    (d) 350 mm
15. The standard width of ballast for BG track on Indian Railways is:
    (a) 3.35 m
    (b) 3.53 m
    (c) 2.35 m
    (d) 2.53 m
    (e) none of these

CHAPTER
9

Subgrade and Formation

INTRODUCTION

Subgrade is the naturally occuring soil which is prepared to receive the ballast. The prepared flat surface, which is ready to receive the ballast, along with sleepers and rails, is called the formation. The formation is an important constituent of the track, as it supports the entire track structure. It has the following functions:
  (a) It provides a smooth and uniform bed for laying the track.
  (b) It bears the load transmitted to it from the moving load through the ballast.
  (c) It facilitates drainage.
  (d) It provides stability to the track.

9.1 GENERAL DESCRIPTION OF FORMATION

The formation can be in the shape of an embankment or a cutting. When the formation is in the shape of a raised bank constructed above the natural ground, it is called an *embankment*. The formation at a level below the natural ground is called a *cutting*. Normally, a cutting or an excavation is made through a hilly or natural ground for providing the railway line at the required level below the ground level.

The formation (Fig. 9.1) is prepared either by providing additional earthwork over the existing ground to make an embankment or by excavating the existing ground surface to make a cutting. The formation can thus be in the shape of either an embankment or a cutting. The height of the formation depends upon the ground contours and the gradients adopted. The side slope of the embankment depends upon the shearing strength of the soil and its angle of repose. The width of the formation depends upon the number of tracks to be laid, the gauge, and such other factors. The recommended widths of formation as adopted on Indian Railways for BG, MG, and NG are given in Table 9.1.

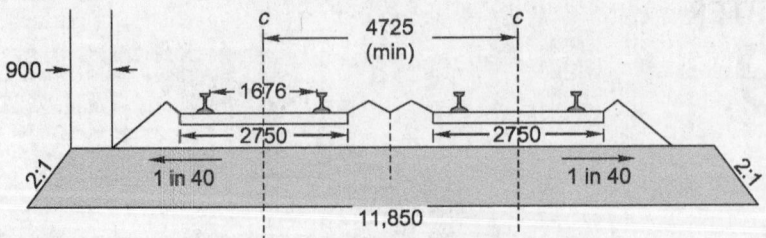

(a) Cross section of bank

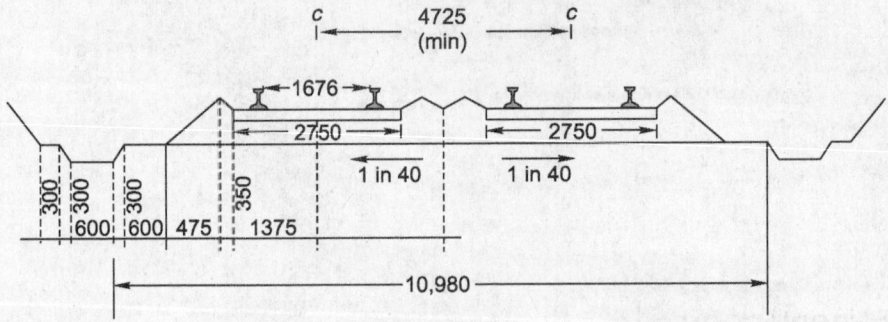

(b) Cross section of cutting

Fig. 9.1 Typical cross section of bank and cutting for BG double line (dimensions in mm)

Table 9.1 Width of formation for different tracks

Gauge	Type of sleepers	Single-line section		Double-line section	
		Bank width (m)	Cutting width (m)	Bank width (m)	Cutting width (m)
BG	W,* ST,† and concrete	6.85	6.25	12.155	11.555
MG	W, ST, CST-9, and concrete	5.85	5.25	9.81	9.21
NG	W, ST, and CST-9	3.70	3.35	7.32	7.01

*   W stands for wooden sleepers.
†   ST stands for steel trough sleeper.

The following points are relevant with respect to the dimensions given in Table 9.1.

(a) The widths have been calculated for a minimum depth of 900 mm in banks and 600 mm in cuttings and a ballast profile slope of about 1:1.

(b) The width of a double-line section has been calculated with a track centre of 5.30 m on BG and 3.96 m on MG. These dimensions are based on a ballast cushion of 300 mm.

(c) The side drain should have a minimum of 0.30 m horizontal berm on the side (i.e., other than the track side) in order to be fully effective.

9.1.1 Slopes of Formation

The side slopes of both the embankment and the cutting depend upon the shearing strength of the soil and its angle of repose. The stability of the slope is generally determined by the *slip circle method*. In actual practice, average soil such as sand or clay may require a slope of 2:1 (horizontal: vertical) for an embankment and 1:1 or 0.5:1 or even steeper particularly when rock is available for cutting.

To prevent erosion of the side slopes due to rain water, etc., the side slopes are turfed. A thin layer of cohesive soil is used for this purpose. Alternatively, the slopes are turfed with a suitable type of grass. Sometimes the bank also gets eroded due to standing water in the adjoining land. A toe and pitching are provided in such cases.

9.2 EXECUTION OF EARTHWORK

The stability of the formation depends, apart from other factors, upon the subgrade material and the methods of construction. Experience has shown that many of the problems in the maintenance of the track are due to incorrect methods of execution of earthwork. In order to have a certain uniformity in practices, guidelines have been laid down on Indian Railways for the execution of earthwork in embankments and cuttings in new constructions, doubling, and conversion projects. These guidelines, given briefly in the following sections, are required to be modified to suit local conditions and prevailing circumstances.

Mechanical compaction of earthwork

For mechanical compaction, earthwork should be done in layers not exceeding a thickness of 300 mm to 650 mm in the loose state using static and vibratory rollers, respectively. The layers should be compacted preferably at or near the optimum moisture content with suitable rollers so as to achieve the dry density that is 98 per cent of laboratory density.

The top of the formation should be finished to a slope of 1 in 30 away from the centre. An extra wide bank of 50 cm should be rolled on either side and then dressed to size to avoid any loose earth at the shoulder.

Proper quality control should be exercised during mechanical compaction. Coarse-grained soil which contains fines up to 5 per cent passing through a 75-micron sieve should be compacted to get the relative density of a minimum of 70 per cent. However, all other types of soil, when compacted, should normally have at least 98 per cent of the maximum dry density determined by using Proctor's compaction test in the laboratory.

9.2.1 Soil Classification

Soil exploratory surveys are carried out in the beginning by taking soil samples from the site. The soil is then classified as 'good' or 'other-than-good' depending upon its grain size and consistency limits. Generally, coarse-grained soils come under the category of good soils. Fine-grained soils such as inorganic clay, silts, sandy soils, and clayey soils are grouped under the category of other-than-good soils. However, the Indian Standard method of soil classification, which is also based on grain size distribution and consistency limits of the soil, is more scientific and elaborate. The grain size is determined by mechanical analysis of the soil. The soil is screened through a set of sieves, and based on the sieve analysis, the soil is classified into gravel (coarser than 2.00 mm), coarse sand (2.00–0.60 mm), medium sand (0.60–0.20 mm), fine sand (0.20–0.06 mm), silt (0.06–0.002 mm), or clay (finer than 0.002 mm).

Black cotton soil

Black cotton soil is a type of shrinkable soil which changes its properties considerably with change in moisture content. With the addition of water, this type of soil swells, thereby losing its strength. Loss of moisture may result in cracks in the soil. During dry seasons, the ballast penetrates into these cracks causing the track to sink. The situation worsens during the rainy season, when water entering these cracks makes the soil soft, and with the hydrostatic pressure and impact of moving loads, deeper ballast pockets are formed. This undesirable property of swelling and shrinkage of black cotton soil presents a lot of problems in the maintenance of proper levels of subgrade.

Remedial measures

The suggested remedies for the problems discussed above with respect to black cotton soil are as follows:
  (a) Treating the top layer of the soil with quick lime so as to reduce the harmful effects of the soil
  (b) Providing a blanket of a graded inverted filter at the top of the embankment
  (c) Consolidating the soil at optimum moisture content
  (d) Providing a bituminous carpet or other similar intercepting material such as polythene sheets to intercept the surface water from getting into the formation
  (e) Improving the drainage conditions of the formation at surface and sub-surface levels

9.2.2 Specifications for Embankments in Good Soil

The following guidelines are followed on Indian Railways for embankment construction on good soil.

For embankments up to 6 m high

The earthwork should be carried out manually in layers not exceeding 30 cm in thickness. All clods of earth should be broken. Earthwork should be carried out in this manner for a height of up to 1 m below the formation. The earthwork is then exposed to rains for one season before taking up the remaining work. The remaining earthwork is carried out by mechanical compaction of the soil in layers not exceeding 30 cm at optimum moisture content in order to obtain at least 90 per cent of the maximum dry density.

For embankments more than 6 m high

In the first working season, up to 6 m or less of earthwork should be done and exposed to the rains. In the second working season, earthwork should be progressed further up to a distance of 1 m lower than the formation level and exposed to rains. The remaining earthwork should be done in the third working season by mechanical compaction. The work can also be completed in the second working season if mechanical compaction is used.

On high-speed and heavy-density routes, a blanket of suitable material or a sub-ballast of 30 cm thickness may be provided. The formation should be given a cross slope of 1 in 40 or 1 in 30 from the centre towards the cess.

9.2.3 Specifications for Cuttings in a Good Soil

The following guidelines are followed by Indian Railways for cuttings in good soil.
 (a) If the normal dry density of the top 30 cm of soil is less than 90 per cent of the maximum dry density, the formation should be rolled to obtain the desired density.
 (b) The road bed should be given a cross slope of 1 in 40 or 1 in 30 from the centre towards the drains on either side.

9.2.4 Embankments in Other-than-good Soils

The guidelines adopted by Indian Railways for embankment construction in soil that is not categorized as good are as follows:
 (a) The earthwork should be compacted to full height at optimum moisture content in layers not exceeding 30 cm in thickness in order to obtain 90 per cent of the maximum dry density.
 (b) A blanket of suitable material of height not less than 30 cm should be provided on the road bed and should be compacted.
 (c) A cross slope of 1 in 30 should be provided from the centre towards the cess.

9.2.5 Cuttings in Other-than-good Soils

The guidelines listed here are followed by Indian Railways for digging cuttings in soil that is not categorized as good.
 (a) The cutting should be provided with drainage.
 (b) A 30-cm-thick blanket of suitable material should be provided in two layers at optimum moisture content and duly compacted.
 (c) A cross slope of 1 in 30 should be provided.

Track drainage

Track drainage is defined as the interception, collection, and disposal of water from, upon or under the track. It is accomplished by a surface and sub-surface drainage system. Proper drainage of the subgrade is very vital, as excess water reduces the bearing capacity of the soil as well as its resistance to shear. The full details about track drainage can be obtained from Chapter 19 where this subject is dealt with in depth.

9.3 BLANKET AND BLANKETING MATERIAL

A blanket can be defined as an intervening layer of superior material that is provided in the body of the bank just underneath the ballast cushion. It is different from the sub-ballast, which is provided above the formation. The functions of the blanket are two fold:

(a) To-minimize the puncturing of the stone ballast into the formation soil

(b) To reduce the ingress of rainwater into the formation soil

The blanket should generally cover the entire width of the formation from the shoulder, except in the case of sand or similar erodable material, where it should be confined within berms of a width of 60–75 cm. The depth of the blanket should normally be about 30 cm in ordinary clayey soil. However, if the formation soil is particularly weak, a thicker layer of up to 60 cm may be necessary, depending on the shear properties of the formation soil. The blanket material should have the following properties.

For sand, quarry grit, gravel, and other non-cohesive materials

(a) The blanket material should be coarse and granular.

(b) If the material contains plastic fines, the percentage of fines (particles measuring up to 75 microns) should not exceed 5. If the fines are non-plastic, then they can be allowed up to a maximum of 12 per cent.

(c) The material should be properly graded and its particle size distribution curve should lie within the standard enveloping curves.

For Macadam

(a) The liquid limit should not exceed 35 and the plasticity index should be below 10.

(b) The uniformity coefficient should be above 4, preferably above 7. The coefficient of curvature, which is $\dfrac{D_{30}^2}{D_{60} \times D_{10}}$, should be between 1 and 3.

(c) When macadam is used as the blanketing material, it should be compacted in a suitable number of layers, at or near the optimum moisture content so as to achieve not less than 90 per cent of the maximum dry density as determined by Proctor's test using heavy compaction.

(d) If an erodable material is used as a blanket, it should be confined in a trench and sand drains should be provided across the track and the blanket. These cross sand drains with adequate slope should be 5–30 cm below the bottom of the blanket and spaced 2–4 m apart.

9.3.1 Depth of Blanket Layer

The depth of blanket layer of a specified material depends primarily on the type of subgrade soil and the axle load of the traffic.

The depth of blanket to be provided for an axle load of upto 22.5 tonnes, and different types of subgrade soils (describing top one metre thickness) is described below. In case more than one type of soil exists in the top one metre, then the soil requiring higher thickness of blanket will govern.

Soil not requiring blanket

- Rocky beds, except those, which are very susceptible to weathering, e.g., rocks consisting of shales and other soft rocks, which become muddy after coming into contact with water
- Well-graded gravel (GW)
- Well-graded sand (SW)
- Soils-conforming to specifications of blanket material

Table 9.2 lists the depth of blanket for different types of soils.

Table 9.2 Depth of blanket for different types of soils

Type of soil	*Minimum thickness of blanket*
Poorly graded gravel (GP) having uniformity coefficient more than 2	45 cm
Poorly graded sand (SP) having uniformity coefficient more than 2	
Silty gravel (GM)	
Silty gravel-clayey gravel (GM-GC)	
Clayey gravel (GC)	60 cm
Silty sand (SM)	
Clayey sand (SC)	
Clayey silty sand (SM-SC)	
Silt with low plasticity (ML)	100 cm
Silty clay of low plasticity (ML-CL)	
Clay of low plasticity (CL)	
Silt of medium plasticity (MI)	
Clay of medium plasticity (CI)	
Rocks which are very susceptible to weathering	

9.4 FAILURE OF RAILWAY EMBANKMENT

A railway embankment may fail due to the following causes:
   (a) Failure of the natural ground
   (b) Failure of the fill material in the embankment
   (c) Failure of the formation top

   Whatever may be the cause of failure, the normal symptoms are as follows:
   (a) Variation in cross levels
   (b) Loss of ballast
   (c) Upheaval of the ground beyond the toes of the embankment
   (d) Slips in bank slopes
   These failures are discussed in the subsequent sections.

9.4.1 Failure of Natural Ground

The natural ground on which the embankment is made can fail either due to shear failure or due to excessive settlement. Failure of this kind is generally associated with the upheaval of the ground beyond the toes of the embankment. Shear failure of natural ground generally takes place when construction is in progress or immediately after construction. Once the ground stabilizes, it hardly fails under existing embankments.

   The following remedial measures are generally adopted to improve the load-carrying capacity of natural ground and expedite the process of settlement.
   (a) Provision of suitably spaced sheet or ordinary piles on either side of the embankment, which will check shear failure by obstructing the slipping mass.

(b) Provision of a balancing embankment to increase the load on the natural ground to check its heaving tendency.

(c) Provision of sand drains to help quicker consolidation.

9.4.2 Failure of the Fill Material of Embankments

Sometimes shear failure and excessive settlement of an embankment take place due to the failure of the fill material of the embankment. This can easily be avoided by judicious selection of the fill material, better construction procedures, and adopting a suitably designed section for a new embankment. The main reasons for this type of failure are the following:

(a) Heavy traffic causing excessive stress in the soil, beyond its safe limit

(b) Inadequate side slopes of the bank

(c) Percolation of water in the embankment, thereby increasing the weight of the soil on one hand and reducing its bearing capacity and shear resistance on the other. Shear failure of existing embankments is quite common and occurs due to slips. Other causes of failure are the weights of the embankment and the moving loads on it. The forces resisting the failure are the cohesion and internal friction of the fill material.

The following types of slip failures may occur along different planes, as shown in Fig. 9.2.

- A slip passing through the toe of the bank known as *toe failure*.
- A slip passing below the toe of the bank through its base known as *base failure*.
- A slip passing above the toe of the bank through its slope known as *slope failure*.

The remedies effective for such failures are listed below.

(a) Providing vertical piles on the slope on either side of the track, spaced at suitable intervals. These piles, which may be of scrap rail, bullies, etc., help check shear failure by causing an obstruction for the slip mass. This method was adopted on Ganga Bridge, Mokameh, and was found to be quite successful.

(b) Providing balancing embankments on either side of the embankment as shown in Fig. 9.2.

(c) Flattening the side slopes.

(d) Reducing the height of the embankment.

(e) Providing a lighter material at the top of the embankment, replacing the older material.

(f) Providing proper surface and sub-surface drainage.

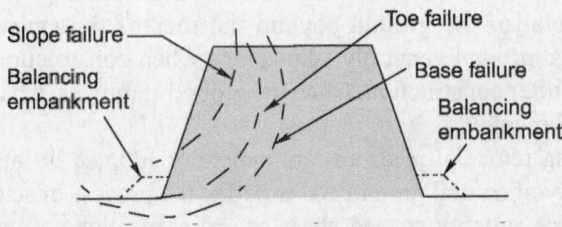

Fig. 9.2 Slope failure, toe failure, and base failure

9.4.3 Failure of Formation Top

Failure of the formation top is very common in clayey soils during or just after monsoons. Some locations may trouble throughout the year. The main causes for such failures are the following.

Low bearing capacity of the soil   Sinking of the ballast and the track, and the heaving up of cesses and bulging of side slopes as a consequence. The ballast punches into the formation causing ballast pockets.

Action of water and moving loads   The top soil becomes soft and gets pumped up, resulting in the sinking of the ballast. The ballast also gets clogged and looses its drainage property.

Effect of weather   Cracks develop on the formation during the summer months and the ballast sinks through the cracks, resulting in the settlement of the track. The situation gets further worsened during the monsoons when water seeps through these cracks, turning the upper layers of the formation to slush and resulting in the formation of deeper ballast pockets.

The impact of moving loads and the development of hydrostatic pressure further deepens the ballast on the side slopes as well and can lead to slips in extreme cases. These failures present considerable problems in the maintenance of the track. Not only is the track geometry affected thereby requiring frequent attention, but also huge quantities of ballast are lost every year, making its maintenance difficult and expensive.

The following remedial measures can be adopted depending upon the situation.

Provision of an inverted filter

The bearing capacity of the soil is improved by the provision of a blanket of adequate thickness (inverted filter, Fig. 9.3) between the ballast and the weak formation. The blanket should be of a non-cohesive material with adequate bearing capacity to withstand the load thereon. The blanketing material should conform to the following specifications.

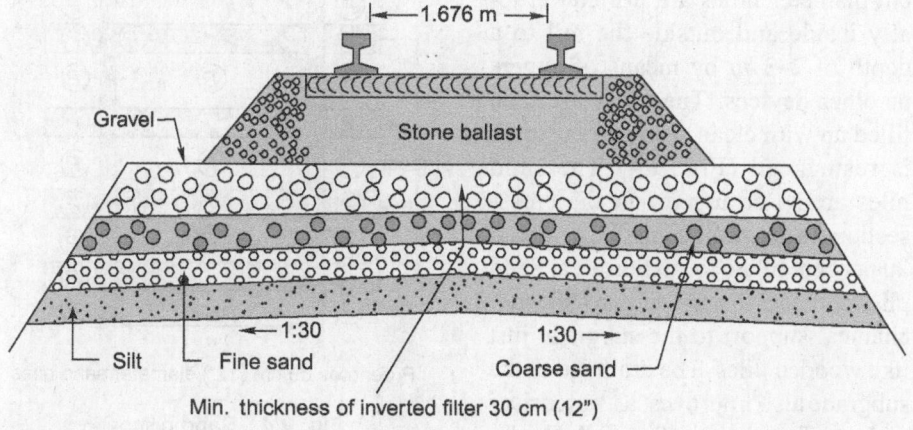

Fig. 9.3   Inverted filter and underground drainage

(a) The liquid limit of the blanketing material should not be greater than 35 and the plasticity index should not be greater than 15.

(b) The blanketing material should have such a grain size that the fine soil from the bottom does not mix up with the water. The material should, therefore, conform to the specifications of an inverted filter.

(c) The blanketing material should be well graded, starting from a fine size to a size slightly smaller than stone ballast, the finer size lying at the bottom.

The inverted filter blanket is a very effective method of improving the bearing capacity of the soil. It serves as a barrier for the upward movement of the clay. It also provides a porous medium to drain off the surface water. The blanket also works as a capillary cut-off layer. The blanket can be inserted by imposing a traffic block of four to five hours or by temporarily operating only one line.

Improvement of surface drainage

Surface drainage can be improved by diverting ground water, providing catch water drains, etc., as well as draining the sub-surface structure.

Cement grouting

For cement grouting, a slurry or grout of cement and sand is pumped into the embankment by pneumatic injections. A 25 mm-diameter steel pipe is coupled with a rubber hose pipe of the same diameter, and through this grout of cement and sand in the ratio of 1:2 to 1:6 is injected under a pressure of 60 psi (4.2 kg/cm^2) with the help of a pneumatic hammer. The injection points are kept close to both ends of the sleeper in a staggered position at an interval of 1.5 m or so. Pumping is continued till the grout appears through the ballast and reaches its top surface.

Cement grouting is considered to be a very effective method of treating the subgrade. It fills the cracks, preventing the water from flowing into the subgrade, and seals off the moisture entering it. The soil is stabilized and develops better properties and strength.

Sand piling

In sand piling (Fig. 9.4), a series of 30-cm-diameter holes are drilled vertically inside and outside the rail to a depth of 2–3 m by means of augers or other devices. The holes are then filled up with clean sand and the track is resurfaced (Fig. 9.4). The sand piles are so arranged that the cross-sectional area of the sand piles is about 20 per cent of the formation area. Sand piles compact the soil and provide mechanical support to the subgrade just like wooden piles. The drainage of the subgrade also improves, as water rises to the surface through the sand piles by capillary action and evaporates.

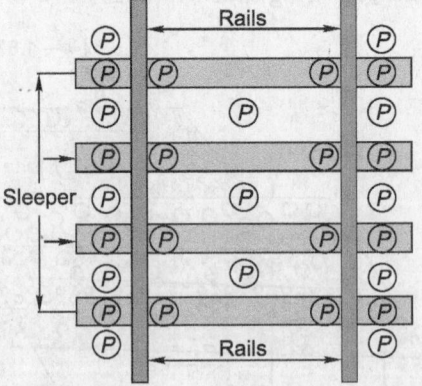

P denotes 30 cm (12") diameter sand piles

Fig. 9.4 Sand piling

Deep screening of ballast and drainage of water pockets

The problem of ballast pockets can be tackled by assessing the depth of the penetration of the ballast in the bank. For this purpose, about five vertical trial bores are drilled to get a complete picture of the drainage condition of the sub-surface structure. Water pockets can then be removed by any of the following methods, depending upon the situation.

(a) If the problem has just started, it can be remedied by deep screening and the provision of a pervious layer of 30–60 cm on the cess. If necessary, the water pockets can be drained using a perforated pipe drain inserted with the help of a jack, as shown in Fig. 9.5.

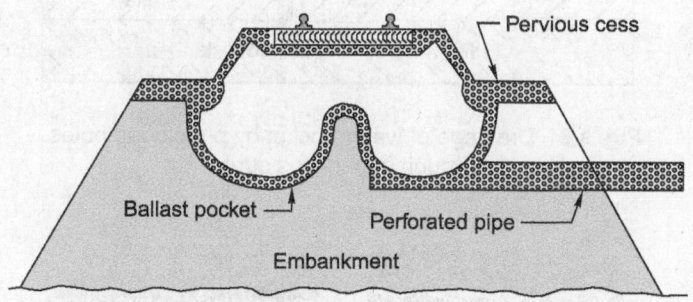

Fig. 9.5 Drainage of ballast pockets

(b) Cement grouting can also be done to seal the water pockets in case the problem is in a very small stretch. This method is, however, very expensive.

(c) If the problem is extensive, a geological survey should be done to assess the type of soil strata available. In case there is impervious soil under the water pocket, it can be drained out using a perforated pipe (Fig. 9.5)—deep screening of the ballast is done, the water pockets are drained out using a perforated pipe, and then an inverted filter of about 30 cm thickness is provided.

(d) Counterfeit drains are sometimes provided to drain the water pockets. Such drains are generally 60 cm wide and spaced at intervals of 10 m or so, depending upon the extent of the problem.

(e) Water may also be held up in the ballast pockets by an impervious layer of soil over a good pervious layer of soil of fissured strata. In this case the remedy lies in drilling a tap hole in the thin impervious strata, allowing the water to go into the pervious subsoil, where it gets drained automatically (Fig. 9.6).

9.4.4 Soil Stabilization by Geotextiles

A new method of stabilization of soil using geotextiles has recently been developed in many countries. Geotextiles are made up of polymers and have the unique property of allowing water, but not soil fines, to pass through. Geotextiles not only work as separators and filters, but also help drain the water and provide reinforcement to the soil bed (Fig. 9.7).

A layer of geotextile is normally either laid directly below the ballast or sandwiched between layers of sand. On Indian Railways, the geotextile is proposed

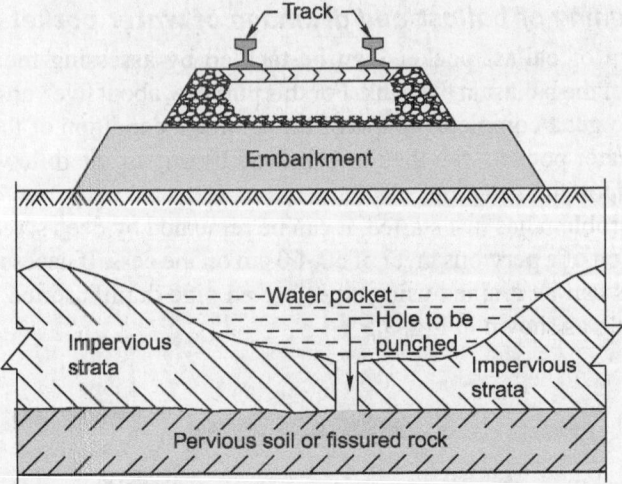

Fig. 9.6 Drainage of water pocket by puncturing holes through impervious strata

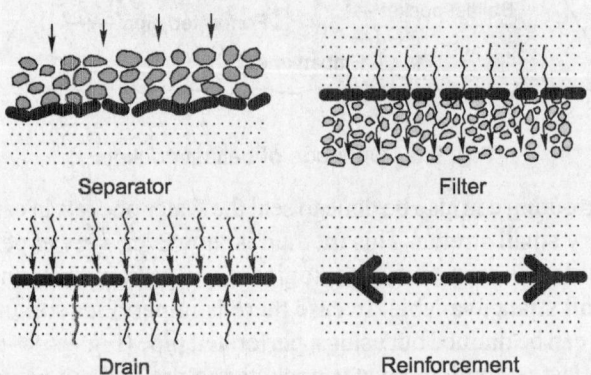

Fig. 9.7 Functions of geotextiles

to be sandwiched between a 50 mm layer of sand on top and a 25 mm layer of sand below so that the ballast does not rest directly on the geotextile and incidences of tear and puncture are reduced. Figure 9.8 illustrates how geotextiles are laid.

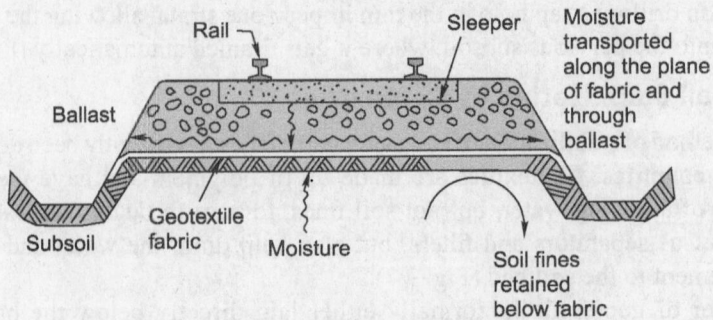

Fig. 9.8 Laying of geotextiles

9.5 SUBGRADE MATERIAL

Subgrade material is normally obtained from the borrow pits in the adjoining land. The selection of soils for subgrade is, therefore, quite difficult unless either the alignment is changed and the subgrade material is imported from an outside place.

The requirements of subgrade material are as follows:

 (i) It should be able to bear the load transferred to it by the ballast section.
 (ii) It should prevent the ballast from puncturing into it.
(iii) It should drain off entering from top.
(iv) It should not change its volume due to variation in moisture.

9.6 SITE INVESTIGATIONS

The following data should be collected for determining the type of treatment to be given to the formation.

History of the affected section   This includes information about the period of construction, method of construction, date of opening to traffic, sub-soil, bank settlement slips, and speed restrictions.

Site details   These include various details of the site such as bank heights, depth of cuttings, nature of existing slopes, drainage conditions, stagnation of water, condition, and proximity of borrow pits, signs of movement and bulging of the slopes, and groundwater level and its position during rains.

Number of visits to track   The particulars of the number of visits to the track should be obtained from the gang charts for the last five years to get an idea of track maintainability. Man-days utilized for maintenance per kilometre should also be determined vis-à-vis the total number of men required for normal maintenance.

Ballast penetration profiles   These should be obtained at regular intervals to indicate the extent of ballast penetration and the condition of the ballast.

Exact nature of present problem   The exact nature of the present problem should be identified.

Remedial measures

The relevant remedial measures should be formulated based on the site investigations and soil testing. Some of the suggested remedial measures for the formation problems generally encountered are listed in Table 9.3.

Table 9.3   Remedial measures for foundation problems

Nature of problem	Remedial measure
Track level variations due to:	
Inadequate drainage due to high cess, dirty ballast	Improve side drainage by lowering the cess and screening the ballast
Weakening of soil at the top of the formation upon contact with rain water, resulting in mud pumping	Cationic bituminous emulsion below ballast; provision of a moorum/sand blanket of depth 20–30 cm below ballast; laying of geotextiles

Contd.

Table 9.3 Contd.

Nature of problem	Remedial measure
Strength failure below ballast or between sleepers causing heaving of cess	Provision of a 30–60-cm-deep blanket below ballast; provision of sub-ballast
Seasonal variations in the moisture content of the top of the formation in expansive soils, causing alternate heaving and shrinkage of formation	Treatment with lime slurry pressure injection; use of a macadam blanket of depth 30–45 cm with macadam lining
Gradual subsidence of the bank core under live loads due to inadequate initial compaction/consolidation of embankment	Cement grouting of ballast pockets if ballast pockets are permeable; sand or boulder drains
Gradual consolidation of earth below embankment	Lime piling in subsoil; sand drains in subsoil
Creeping away of formation soil	Easing of side slopes
Coal ash pockets due to treatment of previous slips	Sand drains below deepest level of coal ash; cement pressure grouting
Instability of bank/cutting slopes due to:	
Inadequate side slopes causing bank slips after prolonged rains	Flattening of slopes and provision of berms; improvement in drainage
Consolidation/settlement of subsoil causing bank slips	Providing of sand drains to expedite consolidation
Hydrostatic pressure built up because of live loads in ballast pockets containing water, causing bank slips	Draining out ballast pockets by sand or boulder drains; Cement sand pressure grouting of ballast pockets
Creeping of soil	Reducing stresses by the provision of side berms or by flattening the slopes
Swelling of over-consolidated clay slopes in cuttings causing loss of shear strength and slipping	Flattening side slopes
Erosion of banks	Providing turfing, mats, etc.

SUMMARY

Subgrade is the natural soil which is prepared and compacted at its maximum density to receive the ballast and the track. It is also called the formation. The track should be so designed that the stresses transmitted to the formation do not exceed the permissible limits. The bearing capacity of the soil depends upon the type of soil and the degree of compaction. The formation may fail in different modes, and remedial measures should be taken in time to safeguard the track.

REVIEW QUESTIONS

1. Briefly describe the normal and special measures adopted to ensure the stability of railway embankments.
2. Define good soil and other-than-good soil. Give the specifications of embankment in good soil and other-than-good soil.
3. Prepare a neat sketch of a typical cross section of an embankment with the ballast section for a double-line BG track and indicate its dimensions and salient features.

4. Discuss the requirements for the stability of an embankment and the precautionary measures commonly adopted against failure. What are the requirements of the subgrade material for a railway track?

5. What are the main causes of failure of a railway embankment? Discuss the remedial means for each one of them.

6. What is formation width? Give the standard formation width for a BG track in cutting and embankments. Illustrate your answer with suitable sketches.

7. What are the main functions of formation? Give the width of embankment and cutting for BG and MG tracks. If concrete sleepers are provided, what will their dimensions be?

Choose the correct answer from the choices given.

8. Formation width for cutting for single BG railway line is:
   (a) 5.25 m   (b) 5.85 m
   (c) 6.25 m   (d) 6.85 m

9. Centre-to-centre spacing of formation for double railway line for BG is:
   (a) 4250 mm   (b) 4725 mm
   (c) 5050 mm   (d) 5350 mm

10. While providing inverted filter to improve the bearing capacity of soil, the liquid limit of blanketing material should not be greater than:
    (a) 15   (b) 20
    (c) 27   (d) 35

11. To have good drainage of the ballast section, cross slope of the formation should be:
    (a) 1 in 30   (b) 1 in 40
    (c) 1 in 20   (d) 1 in 50

12. For yard drainage Pucca drains should be designed for the velocity of:
    (a) 0.05 to 1 m/sec   (b) 1.0 to 2.0 m/sec
    (c) 2.0 to 2.5 m/sec

13. For yard drainage, Kutcha drains should be designed for the velocity of:
    (a) 2.0 to 2.5 m/sec   (b) 1.0 to 2.0 m/sec
    (c) 0.5 to 1 m/sec

10

Track Fittings and Fastenings

INTRODUCTION

The purpose of providing fittings and fastenings in railway tracks is to hold the rails in their proper position in order to ensure the smooth running of trains. These fittings and fastenings are used for joining rails together as well as fixing them to the sleepers. They serve their purpose so well that the level, alignment, and gauge of the railway track are maintained within permissible limits even during the passage of trains. The important fittings and fastenings commonly used on Indian Railways are listed in Table 10.1.

Table 10.1 Types of track fittings

Purpose and type	Details of fittings and fastenings
Joining rail to rail	Fish plates, combination fish plates, bolts, and nuts
Joining rail to wooden sleepers	Dog spikes, fang bolts, screw spikes, and bearing plates
Joining rail to steel trough sleepers	Loose jaws, keys, and liners
Joining rail to cast iron sleepers	Tie bars and cotters
Elastic fastenings to be used with concrete, steel, and wooden sleepers	Elastic or Pandrol clip, IRN 202 clip, HM fastening, MCI insert, rubber pads, and nylon liners

The number of various fittings and fastenings required per sleeper for ordinary or conventional fastening as well as elastic fastenings for different types of sleepers are summarized in Table 10.2.

All these fittings and fastenings together with other ancillary features, are discussed in this chapter.

Table 10.2 Number of fastenings

Type of sleeper	Ordinary fastenings per sleeper	Number	Elastic fastenings per sleeper	Number
Wooden	Dog spikes or	8	CI bearing plates	2
	Screw spikes	8	Plate screws	8
	Keys for CI bearing plates	4	Pandrol clips	4
			Rubber pads	2
Concrete	No ordinary fastening	–	Pandrol clips	4
			Nylon liners	4
			Rubber pads	2
			MCI inserts	4
Steel trough	Keys	4	Modified loose jaws	4
	Loose jaws	4	Pandrol clips	4
			Rubber pads	2
CST-9	Plates	2	Pandrol clips	4
	Tie bar	1	Rubber pads	2
	Cotters	4		
	Keys	4		

10.1 RAIL-TO-RAIL FASTENINGS

Rail-to-rail fastenings involve the use of fish plates and bolts for joining rails in series. A detailed description of these are given in the following sections.

10.1.1 Fish Plates

The name 'fish plate' derives from the fish-shaped section of this fitting (Fig. 10.1). The function of a fish plate is to hold two rails together in both the horizontal and vertical planes. Fish plates are manufactured using a special type of steel (Indian Railways specification T-1/57) with the following composition:

Carbon: 0.30–0.42 per cent

Manganese: not more than 0.6 per cent

Courtesy: Photograph by Les Chatfield (This work is licensed under the Creative Commons Attribution 2.0 License; http://creativecommons.org/licenses/by/2.0/)

Silicon: not more than 0.15 per cent

Sulphur and phosphorous: not more than 0.06 per cent

The steel used for fish plates should have a minimum tensile strength of 5.58 to 6.51 tonnes/cm^2 with a minimum elongation of 20 per cent. Fish plates are designed to have roughly the same strength as the rail section, and as such the section area of two fish plates connecting the rail ends is kept about the same as that of the rail section. As fish plates do not go as deep as the rail, the strength of a pair of fish plates is less than that of the rail section, about 55 per cent, when only vertical bending is taken into consideration. Fish plates are so designed that the fishing angles at the top

Fish plate between two sections of jointed bullhead rail

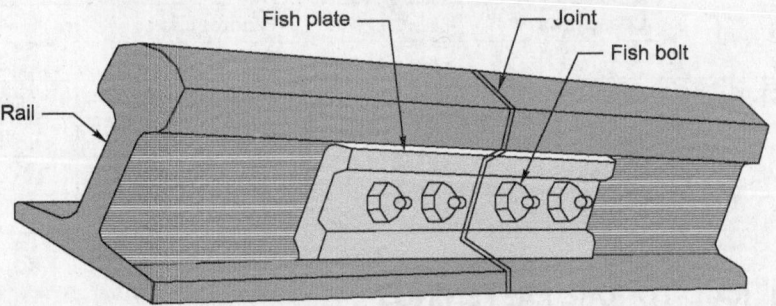

Fig. 10.1 Fish plate

and bottom surfaces coincide with those of the rail section so as to allow perfect contact with the rail as shown in Fig. 10.2. The details of standard fish plates used on Indian Railways for different rail sections are given in Table 10.3.

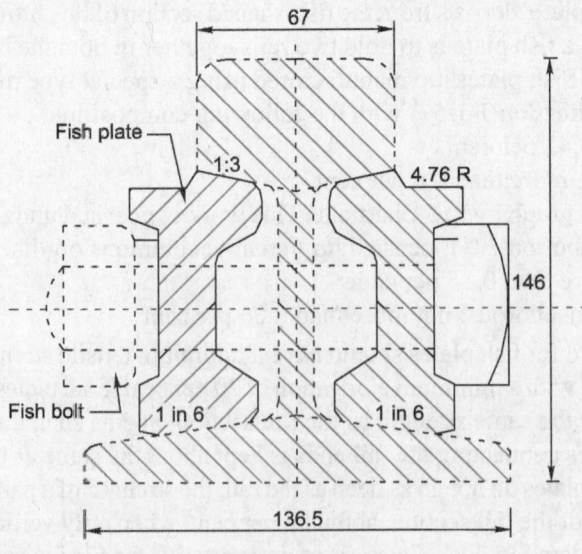

Fig. 10.2 Fish plate for 90 R rails

Table 10.3   Details of standard fish plates

Rail section	Drg. No.	Weight per pair in (kg)	Total length of fish plate in mm (inches)	Length from centre to centre of hole in mm (inches)	Diameter of fish bolt holes in mm
60 kg	T-1898	34.90	610(24)	166(6.5)	27
52 kg	090 M	28.71	610 (24)	166 (6.5)	27
90 R	071 M	26.11	610 (24)	166 (6.5)	27
90 R	059 M	19.54	460 (18)	114 (4.5)	27
75 R	060 M	13.58	420 (16.5)	102 (4)	27
60 R	961 M	9.98	410 (16)	102 (4)	24
50 R	1898 M	8.31	410 (16)	102 (4)	20

10.1.2 Combination Fish Plates

Combination or junction fish plates (Fig. 10.3) are used to connect rails of two differential sections. These are designed to cover the rail section at either end adequately up to the point in the centre where the rail section changes. Another

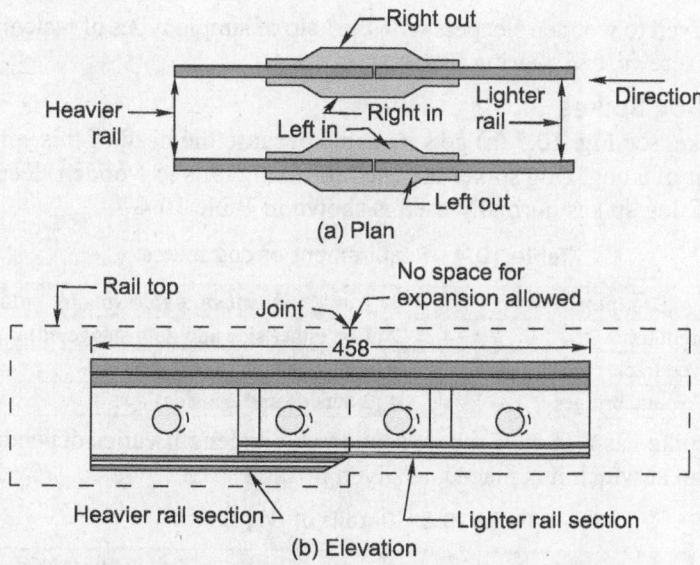

(a) Plan

(b) Elevation

Fig. 10.3   Combination fish plate (dimensions in mm)

design feature in these junction fish plates is the elimination of the expansion gap in order to give them more strength. In spite of the varying depths of the combination fish plates used in the fitting of 52 kg/90 R, 90 R/75 R, 75 R/60 R, etc., rail sections, the use of junction fish plates provides a common top table for the two rail sections they join. A uniform system of marking and exact nomenclature is adopted for each junction fish plate for proper identification. Fish plates are marked right in, right out, left in, and left out depending upon their position with respect to the direction from the lighter rail to the heavier rail (as shown in Fig. 10.4).

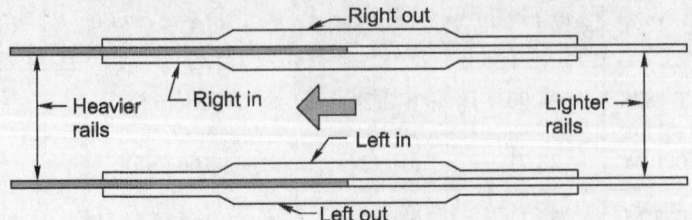

Fig. 10.4 Marking of combination fish plates

In case of any difficulty in obtaining a combination fish plate, the following alternate arrangement can be made.

1. First the composite rail, normally of a length not less than 4 m, is prepared by welding together two rail pieces of different rail sections.
2. This composite rail piece is then inserted at the joint in lieu of the combination fish plate.
3. Normal fish plates are then used to join the composite rail piece to the rail lengths on either side, which have a rail section identical to that of the composite rail piece.

10.2 FITTINGS FOR WOODEN SLEEPERS

Rails are fixed to wooden sleepers with the help of simple types of fastenings such as spikes, screws, and bearing plates.

10.2.1 Dog Spikes

A dog spike, see Fig. 10.5 (a) gets its name because the head of this spike looks like the ear of a dog. Dog spikes are used for fixing rails to wooden sleepers. The number of dog spikes normally used is shown in Table 10.4.

Table 10.4 Requirement of dog spikes

Location	Number of dog spikes
On straight track	2 (1 on either side and duly staggered)
On curved track	3 (2 outside and 1 inside)
Joint sleepers, bridges	4 (2 outside and 2 inside)

The dog spike has a 16-mm square section and its length varies depending upon the location at which it is placed, as given in Table 10.5.

Table 10.5 Details of dog spikes

Location of dog spike	Length of dog spike	
	mm	in.
BG points and crossings	160	6.5
BG track with canted bearing plates; MG points and crossings	135	5.375
MG track with canted bearing plates; NG points and crossings	120	4.75
MG track without bearing plates; NG track with or without bearing plates	110	4.5
BG track without bearing plates	120	4.75

10.2.2 Round Spikes

Round spikes, see Fig. 10.5(b) are used along with anticreep bearing plates for fixing rails to sleepers. These are also used for fixing assemblies of switches onto wooden sleepers. The round spike has a round section of diameter 18 mm, and its length depends upon the purpose it serves. Round spikes have become obsolete now.

10.2.3 Fang Bolts

Fang bolts, see Fig. 10.5(c) are employed under the switches for fastening slide chairs to the sleepers. These are used in locations where the gauge is to be preserved.

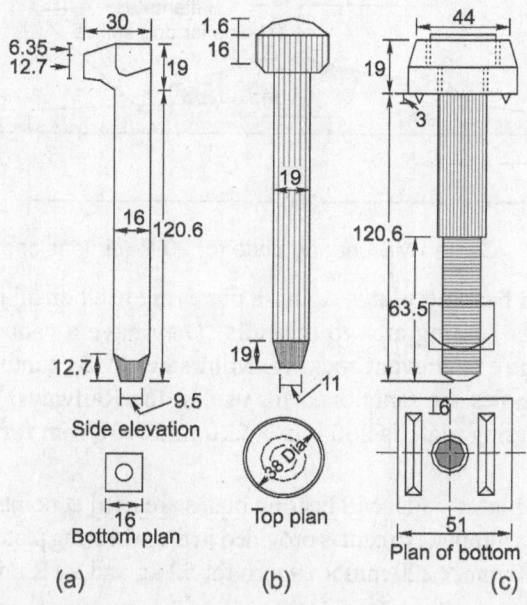

Fig. 10.5 (a) Dog spike, (b) round spike, (c) fang bolt

10.2.4 Screw Spikes

Indian Railways has developed screw spikes with diameters of 20 mm and 22 mm (Fig. 10.6) to be used on high-speed, main, and trunk routes in order to increase the lifespan of wooden sleepers. Screw spikes with a diameter of 20 mm are called 'plate screws' and are used in place of round spikes for fixing rails to sleepers with the help of anticreep bearing plates. Screw spikes with a diameter of 22 mm are called 'rail screws' and are used to directly fasten the rails to the sleepers with or without the use of bearing plates. They are also used on bridges and platform lines. Plate and rail screws should be preferred to round and dog spikes in order to conserve the life of wooden sleepers.

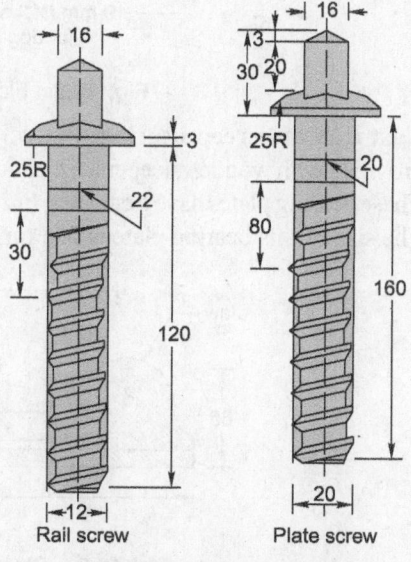

Fig. 10.6 Screw spikes

10.2.5 Bearing Plates

Bearing plates are used for fixing wooden sleepers to rails. The different types of bearing plates used on Indian Railways are described here.

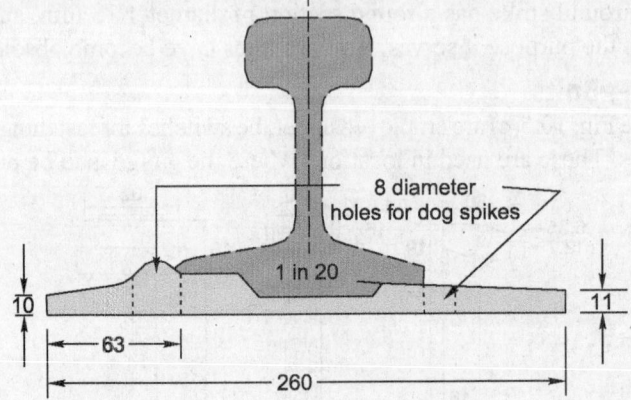

Fig. 10.7   Canted MS bearing plate for 90 R (dimensions in mm)

Mild steel canted bearing plates   These plates are used on all joints and curves to provide a better bearing area to the rails. They have a cant of 1 in 20 and a groove in the centre to prevent rocking. Mild steel (MS) canted bearing plates with only round holes are sanctioned for use on the Railways. The normal size of this type of bearing plate is 260 mm × 220 mm × 18 mm for 52 kg and 90 R rails (Fig. 10.7).

Flat MS bearing plates   Flat MS bearing plates are used at points and crossings in the lead portion of a turnout. No cant is provided in these bearing plates. The size of this bearing plate is 260 mm × 220 mm × 19 mm for 52 kg and 90 R rails (Fig. 10.8).

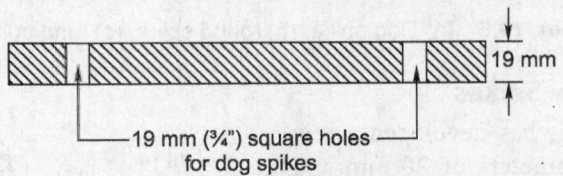

Fig. 10.8   Flat MS bearing plate

Cast iron anticreep bearing plates   Cast iron (CI) anticreep bearing plates are provided with wooden sleepers at locations where the rails are likely to develop creep. These bearing plates have a cant of 1 in 20 and can be fixed using normal round spikes. The size of this bearing plate is 285 mm × 205 mm for BG tracks (Fig. 10.9).

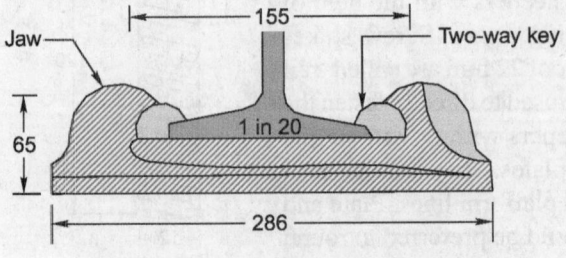

Fig. 10.9   CI anticreep bearing plate

Special CI bearing plates for BH rails   Special CI bearing plates are used for fixing bull headed (BH) rails. The rail is held in position with the help of a spring key (Fig. 10.10).

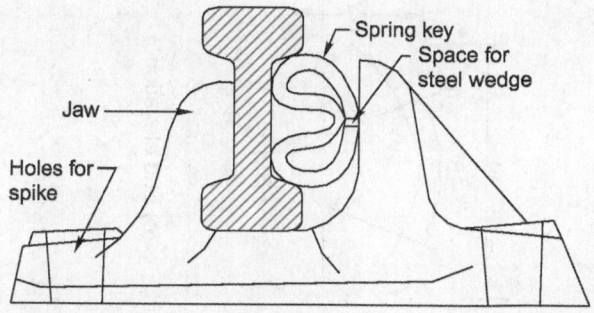

Fig. 10.10   CI bearing plate for BH rail

10.3 FITTINGS OF STEEL TROUGH SLEEPERS

The fittings required for metal sleepers are different from those used for wooden sleepers. Loose jaws, keys, and rubber pads are used to fix rails to steel sleepers.

Loose jaws

Loose jaws (Fig. 10.11) and keys are used for holding the rail and the steel trough sleeper together. The older type of trough sleepers were easily damaged, cracked, or deformed due to the provision of pressed-up lugs. These problems have been solved by introducing spring steel loose jaws, which have been standardized on Indian Railways. These jaws can be easily replaced whenever necessary. They are manufactured using spring steel and the weight of 100 loose jaws is approximately 28.8 kg.

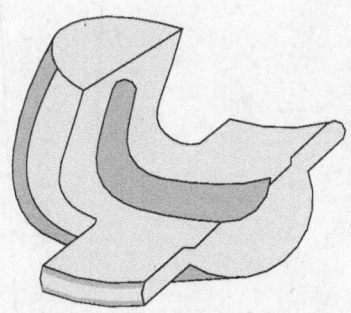

Fig. 10.11   Spring steel loose jaw

Two-way keys

Two-way keys (Fig. 10.12) are universally used for fixing trough sleepers, pot sleepers, and CST-9 sleepers. A two-way taper is provided at both ends of a two-way key and as such the key can be driven in either direction. These keys are manufactured using a special rolled section. The length of the keys for BG is about 190 mm with a taper of 1 in 32. A gauge variation of ±3 mm can be adjusted by altering the extent to which these keys are driven in.

Rubber-coated and epoxy-coated fish plates

Earlier, rubber-coated fish plates were used at insulated joints on Indian Railways on a trial basis. The results indicated that these fish plates get damaged early in service, thereby limiting their life. Therefore, epoxy-coated fish plates are now being used.

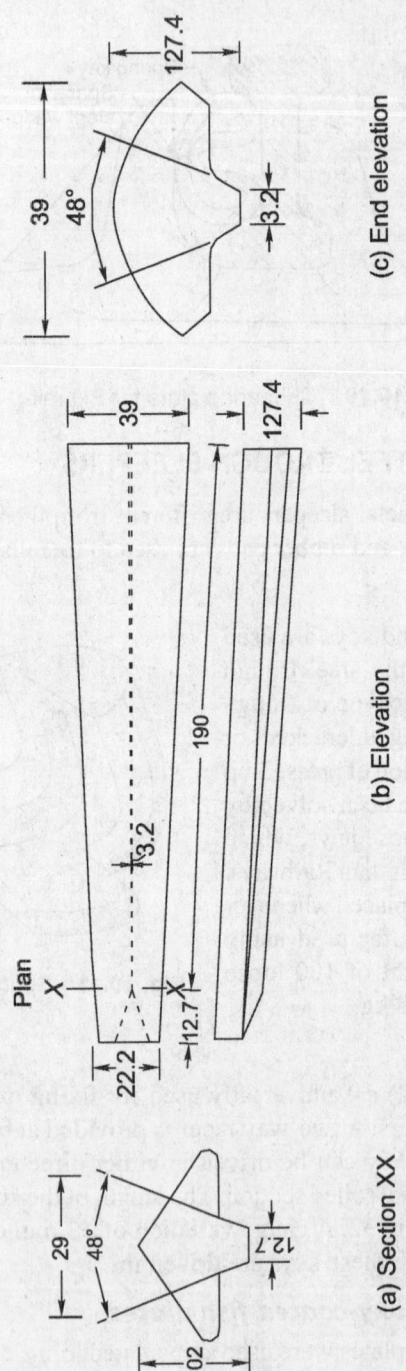

Fig. 10.12 Two-way keys (dimemions in mm)

10.3.4 Mota Singh Liner

The holes in trough sleepers get elongated during service due to the wear and tear caused on account of moving loads. The Mota Singh liner (Fig. 10.13) is liner used effectively with loose jaws for overcoming the problem of elongated holes.

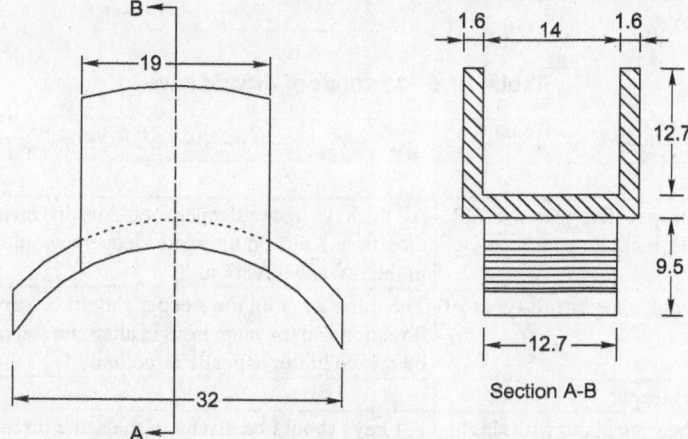

Fig. 10.13   Mota Singh liner (all dimensions are in mm)

10.4 FITTINGS OF CI SLEEPERS

Rails are fixed to cast iron sleepers using cotters and tie bars. These fittings are described below.

Cotters

Cotters (Fig. 10.14) are used for fixing tie bars to CI sleepers. Cotters are classified according to their methods of splitting. The four different types of cotters being used on Indian Railways are as follows.

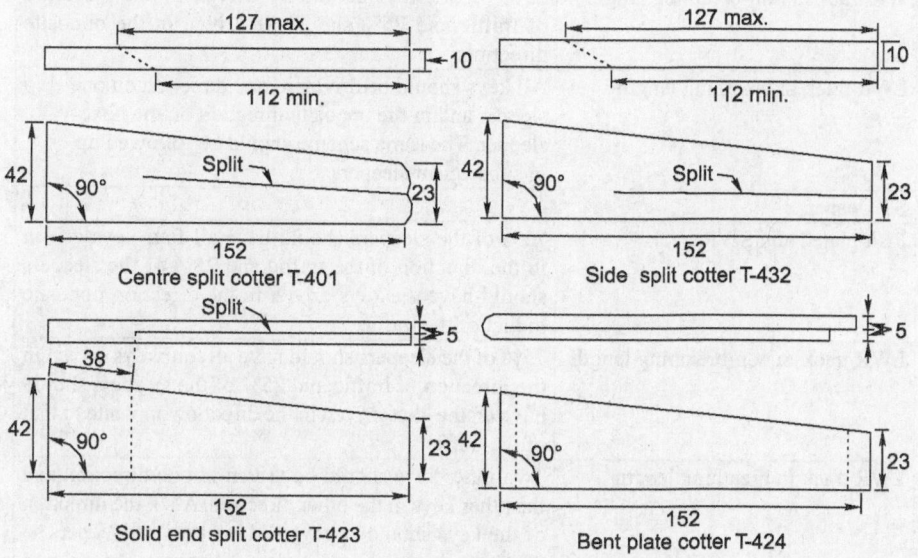

Fig. 10.14   MS cotters (all dimensions are in mm)

(a) Centre split cotter
(b) Side split cotter
(c) Solid end split cotter
(d) Bent plate cotter

The various methods of driving keys for different types of sleepers are listed in Table 10.6.

Table 10.6 Methods of driving keys

Type of sleeper and track	Direction of driving
Single line	
CST-9 sleeper (fish-plated, SWR, and LWR tracks)	All the keys in one sleeper should be driven in the same direction. Keys on alternate sleepers should be driven in the reverse direction.
Steel trough sleeper (all types of tracks)	The outer keys on the sleeper should be driven in one direction and the inner keys in alternate sleepers should be driven in the opposite direction.
Wooden sleeper	
Anticreep bearing plates with single-key configuration	All keys should be driven in the same direction. Keys should be driven in reverse direction in alternate sleepers.
Anticreep bearing plates with double-key configuration	The outer keys on a sleeper should be in one direction and the inner keys in the opposite direction. The pattern of driving keys should be reversed in alternate sleepers.
Double line	
CST-sleeper	
Fish-plated and SWR track	The direction of 75% of the keys should be in the direction of the traffic and that of 25 per cent should be in the opposite direction.
WR track in non-breathing length	75% of the keys should be driven in the direction of traffic and 25% should be driven in the opposite direction.
LWR track in breathing length	All keys should be driven in one direction on one sleeper and in the opposite direction on the next sleeper. The same scheme should be followed up in subsequent sleepers.
ST sleeper	
Fish-plated and SWR track	75% of the sleepers should have all four keys driven in the direction of the traffic and 25% of the sleepers should have the keys driven in the direction opposite to that of the traffic.
LWR track in non-breathing length	75% of the sleepers should have all four keys driven in the direction of traffic and 25% of the sleepers should have all the keys driven in the direction opposite to that of the traffic.
LWR track in breathing length	Two inner keys should be driven in one direction and the other keys in the other direction. Also, the direction of the keys should be reversed in alternate sleepers so as to prevent relative movement between the rail and the sleeper.

The overall dimensions, taper, etc., of these four cotters are by and large identical; they only differ in their methods of splitting. The length of a cotter is 152 mm and the approximate weight is 350g per piece.

MS tie bars

MS tie bars are used for holding the two plates of CST-9 sleepers together. The normal length of a tie bar is 2720 mm for BG and 1870 mm for MG. The section of a BG tie bar measures 50 mm × 13 mm and that of an MG tie bar measures 45 mm × 10 mm.

10.5 ELASTIC FASTENINGS

The primary purpose of a fastening is to fix the rail to the sleeper. The rail may be fixed either directly or indirectly with the help of fastenings. In the process, the fastening gets subjected to strong vertical, lateral, and longitudinal forces. The forces, which are predominantly dynamic, increase rapidly with increasing loads and speeds. In addition, vibrations are generated by moving loads mainly on account of geometrical irregularities in the track and due to the forces set up by the imbalance in the rolling stock. The traditional rigid fastening, which so far has fulfilled its task to a certain extent, is no longer able to effectively meet the present challenge of heavy dynamic forces and, therefore, becomes loose under the impact of high-frequency vibrations of the order of 800 to 1000 cycles per second, even at a moderate speed of 100 km per hour. In fact, this type of fastening is unable to hold the rail to the sleeper firmly for a satisfactory length of time because of the constant pressure exerted by moving loads. Due to the shocks and vibrations caused by moving loads, the rigid fastenings become loose, an interplay between the components of the track develops, track parameters get affected, and rapid deterioration of the track begins. To solve these problems a fastening which could safeguard track parameters and dampen the vibrations is required. This has led to the development of the elastic fastening.

10.5.1 Requirements of an Elastic Fastening

An ideal elastic fastening should meet the following requirements:
 (a) It should hold the gauge firmly in place.
 (b) It should have an adequate toe load which should not reduce under service.
 (c) It should provide sufficient elasticity to absorb the vibrations and shocks caused by moving loads.
 (d) It should help in keeping the track well maintained.
 (e) It should offer adequate resistance to lateral forces in order to maintain the stability of the track.
 (f) It should provide adequate resistance to the longitudinal forces that are a result of the acceleration of moving loads and other miscellaneous factors. These longitudinal forces tend to cause the development of creep in the track.
 (g) It should be of the 'fit and forget' type so that it requires least maintenance.
 (h) It should not lose its properties even when it is used over and over.
 (i) It should have as few parts as possible and these parts should be easy to manufacture, lay, and maintain.

(j) It should be irremovable so that once fitted it cannot be taken out and as such it should not be vulnerable to sabotage or theft.

(k) It should be universally applicable so that it can be used with wooden, steel, or concrete sleepers.

(l) It should be cheap and long lasting.

10.5.2 Types of Elastic Fastenings

An elastic fastening is usually in the form of a clip. Various types of clips have been developed over the years; these are discussed here in detail.

Pandrol clip or elastic rail clip

The Pandrol PR 401 clip (also known as an elastic rail clip) (Fig. 10.15) is a standard type of elastic fastening used on Indian Railways. It is a 'fit and forget' type of fastening that requires very little attention towards its maintenance. The clip is made of a silico–manganese spring steel bar with a diameter of 20.6 mm and is heat treated. It exerts a toe load of 710 kg for a nominal deflection of 11.4 mm. The toe load is quite adequate

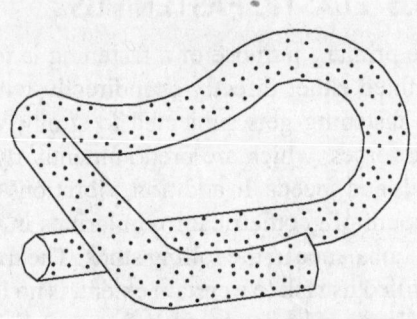

Fig. 10.15 Pandrol clip or elastic rail clip

to ensure that no relative movement is possible between the rail and the sleeper. Pandrol clips can be driven with the help of an ordinary 1.8 kg hammer and require no special tools. In order to ensure that the correct toe load is exerted, the Pandrol clip should be driven to such an extent that the outer leg of the clip is flush with the outer face of the CI insert. Figure 10.16 shows an isometric view of the clip fixed on the rail while Fig. 10.17 shows the clip fixed to a rail seat.

The Pandrol or elastic clip can be fixed on wooden, steel, cast iron, and concrete sleepers with the help of a base plate and some other ancillary fittings. Pandrol clips are the most widely used clips with concrete sleepers on Indian Railways. Therefore, it becomes imperative that a detailed account of the same be given.

Concrete sleepers with Pandrol/elastic clips In the case of concrete sleepers, malleable cast iron (MCI) inserts are punched directly into the sleepers during manufacture. The Pandrol clip is fixed in the holes of the CI insert. A 4.5 mm-thick grooved rubber pad is provided under the rail seat to make it doubly elastic. Insulated liners provide the necessary insulation (Fig. 10.18).

Drawbacks of Pandrol clip

The Pandrol clip suffers from the following drawbacks
- Its use makes the adjustment of the gauge impossible.
- It has a point contact and this causes indentation on the foot of the rail due to a heavy toe load and a small contact area.
- It does not provide enough safeguard from theft or sabotage because it can easily be taken out using an ordinary hammer.
- It gets caught inside the malleable cast iron (MCI) insert during service.

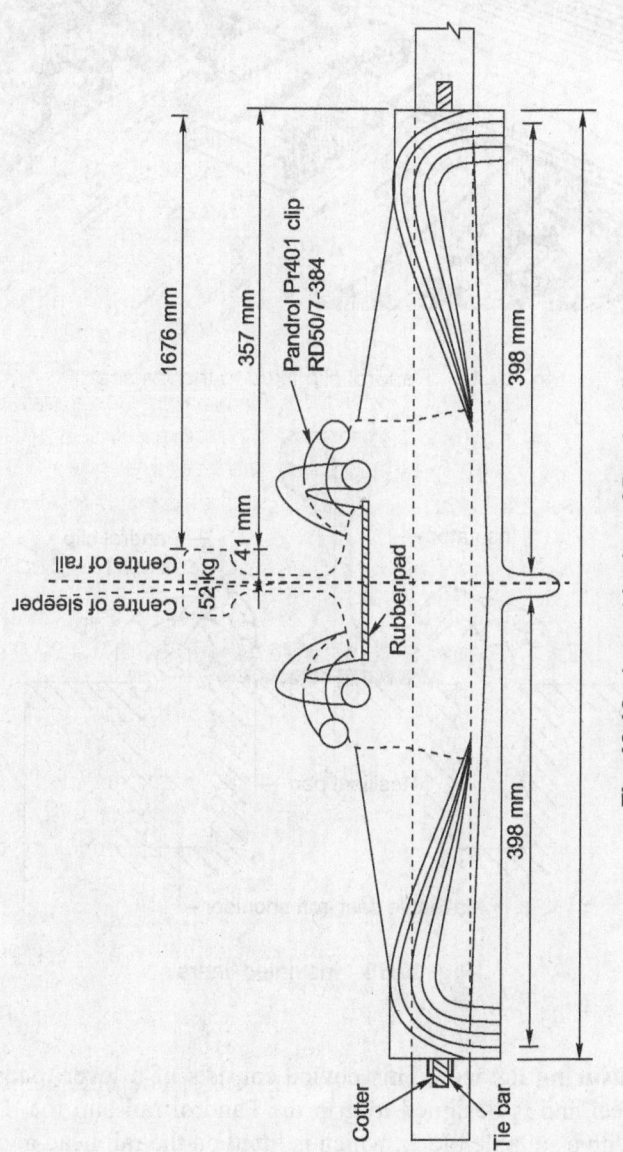

Fig. 10.16 Isometric view of Pandrol clip

Fig. 10.17   Pandrol clip fixed to the rail seat

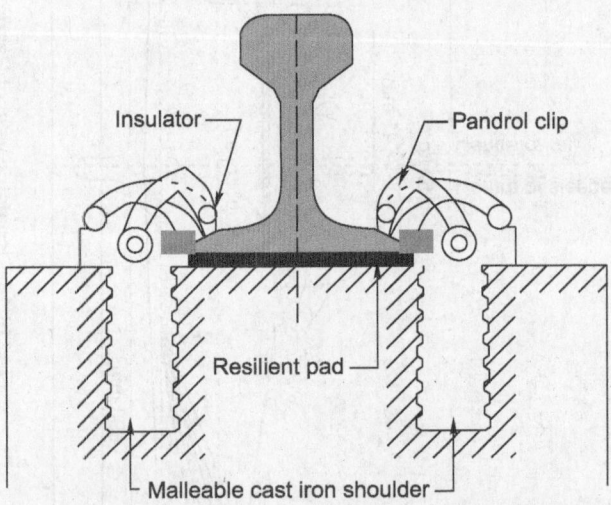

Fig. 10.18   Insulated liners

Toe load measuring device   This device consists of a lever made of silico–manganese steel and is designed to grip the Pandrol rail slip toe. It is used in conjunction with a suitable block, which is fitted on the rail head and acts as the fulcrum. To operate the device, a force is gradually applied to the handle and the reading of the dial gauge at which the Pandrol clip toe is just lifted above the rail seat is noted. The reading of the dial gauge indicates the toe load, which is pre-calibrated in the laboratory.

IRN 202 clip

The IRN 202 clip (Fig. 10.19) is an elastic fastening designed by RDSO to suit two-block reinforced cement concrete (RCC) sleepers. The clip is manufactured using a silico–manganese spring steel bar of diameter 18 mm, suitably heat treated to Brinel hardness number (BHN) 375–415. The assembly is designed for a toe load of 1000 kg and a toe deflection of 18.5 mm.

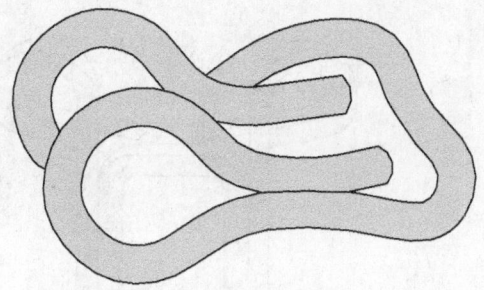

Fig. 10.19 IRN 202 clip

The assembly has a creep resistance generally equal to 50 per cent of the total toe load of the rail.

The clip essentially consists of outer legs connected by means of two coils. It is held in position by a bolt and clamp arrangement. The clamp is made up of the same material as the clip. The bolt, which has a diameter of 19 mm, is made of mild steel. The clip holds the track gauge easily and effectively. The inner legs rest against the bottom flange of the rail to provide an elastic gauge check. After the clip is placed in position, the nut is tightened to depress the inner legs with respect to the toe till these touch the sleeper surface. This stage depends on the designed toe load and the toe load deflection. At this stage, it is not possible to tighten the nut any further. The nut will remain in position for quite some time, as the tension in the bolt does not vary much during summer.

The advantage of the IRN 202 clip is that the rail can be changed without removing the fastening simply by loosening the bolt and pushing the rail out. However, the IRN 202 clip suffers from the following drawbacks.

- Corrosion of the highly stressed part (heel) of the assembly can lead to the development of cracks.
- It is not a fit and forget type of fastening and requires frequent attention such as oiling and tightening of the nuts to maintain the required toe load.
- It is a comparatively costlier and heavier clip.

Lock spikes

Lock spikes (Fig. 10.20) are used with wooden sleepers. The lock spike type LG-20 has been tried on Indian Railways. It is a 165 mm- (6.5″.) long spike with a round section of diameter 16 mm (5/8″). The sizes of the holes bored into the sleeper are 14 mm and 12 mm for hard wood and soft wood respectively. The spike, which appears to have a good future, is still under trial.

Spring steel clip

A large number of spring steel clips (Fig. 10.21), have been tried on Indian Railways. The assembly consists of a double elastic fastening used on a prestressed concrete sleeper. In this assembly, the rail rests on a grooved rubber pad and is held vertically by a pair of spring clips at each rail seat. The clip is pressed with the help of a nut tightened on a 22 mm bolt, which is inserted from the underside of the sleeper. The clip is manufactured using EN-48 steel. The nut is tightened

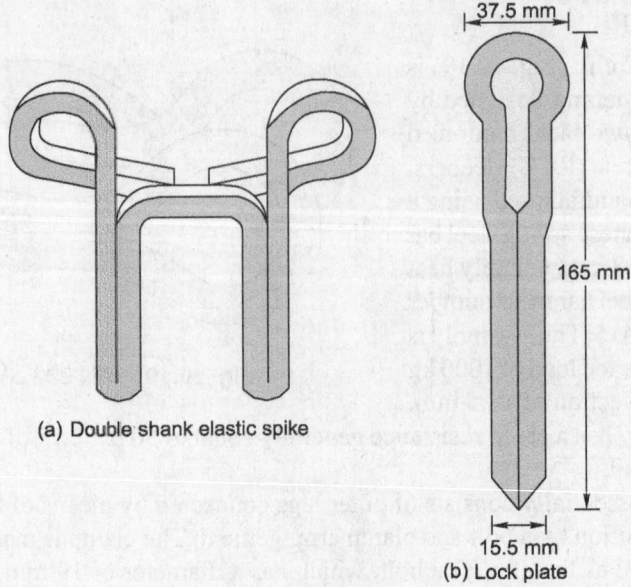

37.5 mm

165 mm

15.5 mm

(a) Double shank elastic spike

(b) Lock plate

Fig. 10.20 Lock spike

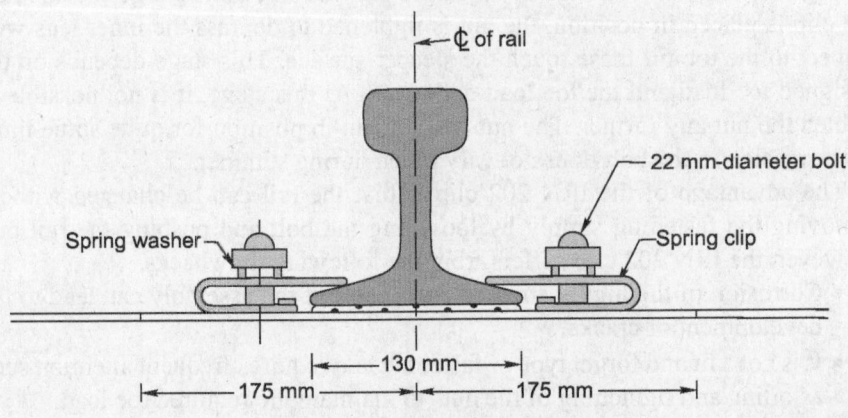

₵ of rail

22 mm-diameter bolt

Spring washer

Spring clip

130 mm

175 mm

175 mm

Fig. 10.21 Spring steel clip

to a torque of 13.5 kg. cm to obtain a resistance of 1 tonne per pair of clips. The clip is still in its experimental stage on Indian Railways.

Elastic rail clip MK-III

RDSO has designed a new type of elastic rail clip known as ERC MK-III (Fig. 10.22) which suits both 52 kg and 60 kg rails along with 6 mm thick rubber pads. In the case of 60 kg rails, two liners of 16 mm thickness are used, whereas in the case of 52 kg rails, one liner of 16 mm thickness is used on the non gauge side and another liner of 10 mm thickness is used on the gauge side. The clip can also be used with 6 mm thick rubber pads in place of the usual 4.5 mm thick rubber pads on the existing 52 kg PRC sleeper.

The ERC MK-III clip has been modified from the standard elastic rail clip to the extent that the distance of the toe of the clip has increased with respect to the

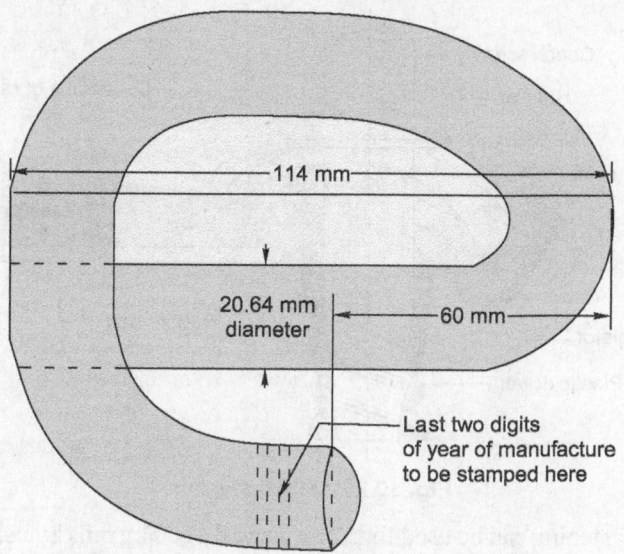

114 mm

20.64 mm
diameter

60 mm

Last two digits
of year of manufacture
to be stamped here

Fig. 10.22   Pandrol clip (ERC rail clip) MK III

centre leg. The space curves of the clip have also been modified to achieve a higher toe load. The diameter of the standard ERC has been retained, that is 20.6 mm.

The new ERC MK-III has a toe load of 900–1100 kg with a toe deflection of 15.5 mm. The clip is still under trial.

Limitations of elastic rail clip   The elastic clip (or Pandrol clip) presently being used on Indian Railways has the following limitations.
  (a) The elastic rail clip (ERC) does not permit 52 kg rails to be interchangd with 60 kg rails. Therefore, whenever traffic requirements demand the replacement of 52 kg rails with 60 kg rails, the sleepers also have to be replaced or costlier special steel alloy rails have to be used.
  (b) ERC can be easily removed from the track.
  (c) ERC gets jammed/rusted in the insert and tends to lose the designed toe load.

The only fastenings in the world proven to permit the interchangeability of 52 kg rails with 60 kg rails are the HM fastenings of German design and the NABLA fastenings of French design. RDSO has recently designed an elastic rail clip (mark III) which also permits interchangeability of the two types of rails, but it is still in its trial stage.

Herbert Meir fastening

The Herbert Meir (HM) fastening (Fig. 10.23) basically consists of four coach screws which are tightened against the plastic dowels of the PRC sleepers and press the HM clip assembly to give the desired toe load. Each clip weighs about 510 g and can give a toe load of about 1 tonne. The gauge is maintained with the help of an angled guide plate. A thin insulated shim is placed between the angled plate and the concrete sleepers. A grooved rubber pad is placed below the seat to provide the necessary dampening effect and resistance to the lateral movement of rails.

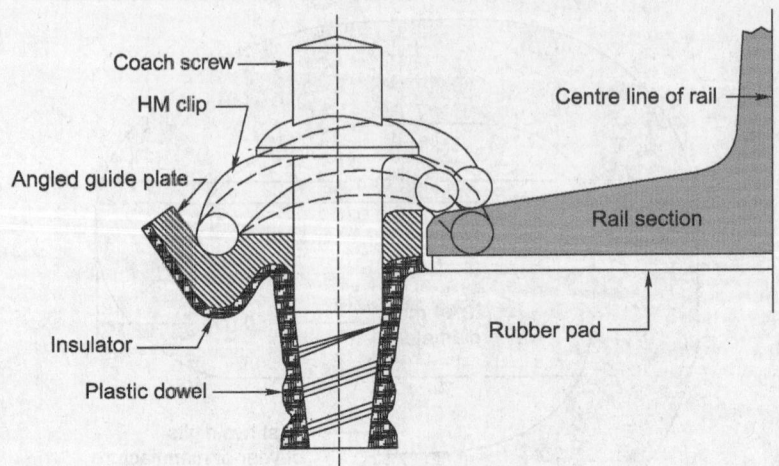

Fig. 10.23 HM fastening

The HM fastening can be used for 52 kg as well as 60 kg rails by using a suitable size of angled guide plates and insulating shims.

New elastic fastening (G Clip)

A new elastic fastening named G clip has recently been developed by an Indian Company (Logwell Forge Ltd.), which the firm claims is an improved version of Pandrol E-Clip.

The G clip has a toe load of 1000 kg to 1300 kg with a deflection of 11.5 mm. The G clip is manufactured from special steel 25 A 58 BS-970 and has a hardness of 44 to 48 HRC. The G clip has a diameter of 20.64 mm and weight of about 825 gms. Flat bearing area of the G clip is 15 mm × 36 mm. Figure 10.24.(a) illustrates the Logwell G clip.

The manufacturers claim that the installation of G clip with concrete sleepers is very easy and they have developed suitable design for all types of rails and sleepers. The isometric view of Logwell G clip is given in Figure 10.24(b).

After carying out a series of trials, the Indian Railway Board has recently approved the adoption of G clips.

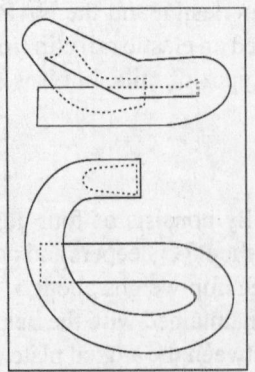

Fig. 10.24 (a) Logwell G clip   **Fig. 10.24 (b)** Isometric view of G clip assembly

10.6 OTHER FITTINGS AND FASTENINGS

Some of the other fittings and fastenings are malleable cast iron inserts, rubber pads, composite liners, and pilfer-proof elastic fastenings. These are discussed in detail here.

10.6.1 Malleable Cast Iron Inserts

Malleable cast iron (MCI) inserts are directly fixed onto concrete sleepers during manufacture. These are manufactured according to the Indian Railway Standard (IRS) specification T-32-76. These inserts are of the following two types:

1. Stem-type MCI insert for use in normal pre-tension concrete sleepers. This insert is provided in concrete sleepers being manufactured in all the concrete sleeper factories in India except the one located at Allahabad. The weight of the stem-type insert is about 1.6 kg per piece.
2. Gate-type MCI insert for use in the post-tension concrete sleepers being manufactured at Allahabad. The approximate weight of the gate-type MCI insert is 1.7 kg per piece.

10.6.2 Rubber Pads

A rubber pad (Fig. 10.25) is an integral part of an elastic fastening. It is provided between the rails and the sleepers and has the following functions:

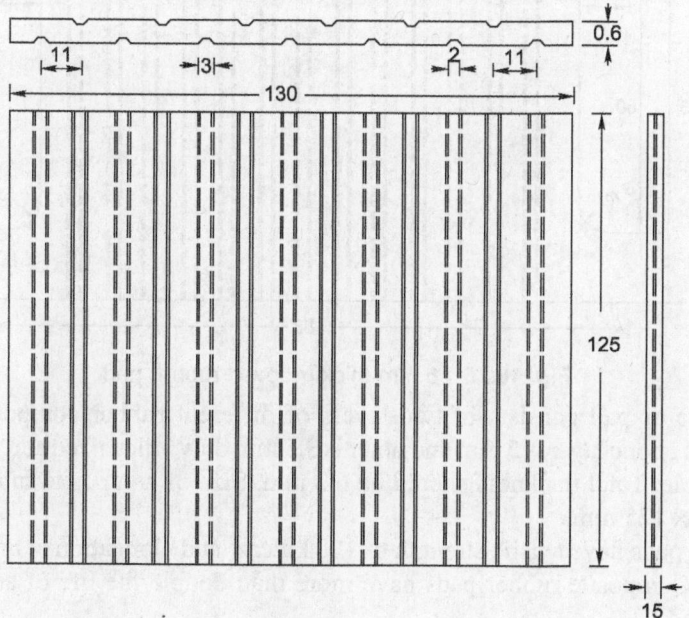

Fig. 10.25 Rubber pad

- It absorbs shocks.
- It dampens and absorbs vibrations.
- It resists the lateral movements of the rails.
- It prevents the abrasion of the bottom surface of the rail, which would otherwise come in direct contact with the sleepers.
- It provides electrical insulation between the rails in an electrified area.

4.5 mm-thick grooved rubber pad   Indian Railways uses grooved rubber pads of 4.5 mm thickness made of special quality rubber. The grooves aid in the uniform distribution of the load on sleepers and help to limit the lateral expansion of the rubber under the pressure of dynamic loads.

6 mm0-thick grooved rubber pad   RDSO has recently designed 6 mm thick grooved rubber pads with horns (Drg. No. RDSO/T-37) for use on 60 kg rails (Fig. 10.26). It was noticed that normal 4.5 mm-thick rubber pads (IRST-37-1982) got crushed wihtin six to seven years and, therefore, thicker, grooved rubber pads with a service life of 15–20 years were designed particularly for use on 60 kg UIC rails. These rubber pads are still under trial.

Composite grooved rubber sole plate for 60 kg. UIC Rail   The composite rubber sole plates are manufactured using natural rubber, Ribbed Smoked Sheet (RSS) of grade 1 to 4 or blended with styrene butadiene rubber. The composite

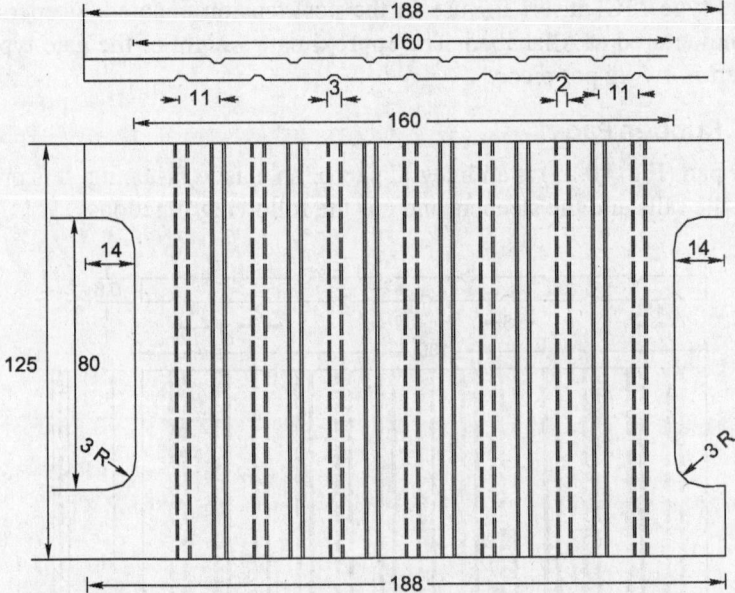

Fig. 10.26   6 mm-thick grooved rubber pad

sole plate or pad consists of two layers of different rubber compounds. The thickness of one layer is 3 mm and other is 3.5 mm duly vulcanized and physically inseperable. Total thickness should be 6.2 mm. Size of composite rubber pad is 188 mm × 125 mm.

These pads have tensile strength of 120 kg/cm^2 and elongation at break is 250 kg/cm^2. Composite rubber pads have more than double the life of an ordinary rubber pad.

10.6.3 Composite Liners

The Indian Railways mostly uses nylon insulating liners. These liners, however, get crushed under the toe load exerted by Pandrol clips. To eliminate such premature failure, the following two types of composite liners have been evolved by RDSO.

(a) Composite liner with malleable cast iron and nylon components (Drg. No. RDSO/T653/1)

(b) Composite liners with MS and nylon components (Drg. No. RDSO/T-1895)

These liners have been developed on the basis of the designs of the liners adopted by British Railways, which have been reported to provide trouble-free service. Composite liners have been used on Indian Railways for the last few years and are serving the purpose well.

10.6.4 Glass-filled Nylon Liners

RDSO has developed glass-filled nylon liners (Fig. 10.27) (GFN-66) of 4 mm thickness particularly for track-circuited areas and sections subject to server corrosion. These glass-filled nylon liners are considered to be technically superior to other liners because they are single piece, have a longer life, and are free from corrosion. These liners are used extensively on Indian Railways particularly with the ERC clip assembly on 60 kg and 52 kg rails and PRC sleepers.

Fig. 10.27 Glass-filled nylon liner

It has been noticed that the GFN-66 liners tend to break, particularly in yards where these liners have been fitted in the ERC clip assembly on concrete sleepers. This happens due to the rusting of the rail surface and uneven seating. To avoid breakage of GFN-66 liners, it is necessary that proper precautions be taken during initial laying to ensure that the rail surface is free from rust, etc., and that the liners are fitted evenly on the 1 in 6 sloping surface of the rail flange.

A new design of GFN-66 liners with a thickness of 6 mm (Drg. No. DSO/T-2505 Alt II) has recently been developed and is expected to be sturdier and provide a better service life.

10.6.5 Pilfer-proof Elastic Fastenings for Concrete Sleepers

The current design of elastic fastenings (Pandrol clips) is such that they can be easily removed by a single stroke of a hammer. A new type of elastic rail clip, which is pilfer-proof, has been recently developed by RDSO. A pilfer-proof elastic fastening is easy to fit in the assembly but is difficult to remove without damaging the system.

The design of a pilfer-proof elastic rail fastening consists of a clip of almost the same design as that of the normal elastic fastening as well as a new fitting known as the *pilfer-proof circlip*. The circlip is a standard mechanical component manufactured according to IS specifications and is generally used for restraining the axial movements of the components mounted on shafts.

10.7 TESTING OF FASTENINGS

Both elastic and rigid fastenings are tested in the laboratory for their suitability in the field. The vibrogir and pulsator are used to test these fastenings.

The vibrogir is used in laboratories for checking the effectiveness of various fastenings. With the help of this equipment it is possible to produce high-frequency vibrations in the laboratory, very similar to those produced on a real track. By applying a frequency of 50 Hz, the rail and sleeper are made to vibrate at a rate of 700–800 Hz with an acceleration of 70 g and an amplitude of the order of 0.1–0.3 mm. One hour of working of a vibrogir corresponds to almost 4.05 GMT of traffic and 300 hours of its working creates the same effect on a fastening as 20 years of service under normal track conditions.

The pulsator not only simulates vibrations just like the vibrogir, but also applies vertical and lateral pressure on the rail fastening at a frequency of 250–500 cycles/minute.

Tests carried out with the help of vibrogirs and pulsators clearly establish the superiority of elastic fastenings over rigid fastenings.

SUMMARY

Fittings hold rails in position and thus help provide a smooth ride. Fish plates and bolts are used to join the rails in series while different types of fastenings are used to fix the rails to the sleepers. The traditional rigid types of fastenings are not able to meet the challenges posed by heavy dynamic forces and become loose under high-frequency vibrations. Elastic fastenings are found to be very suitable for high-speed tracks. New design of elastic fastenings are being developed to overcome as many drawbacks as possible.

REVIEW QUESTIONS

1. With the help of a suitable sketch explain the assembly of the Pandrol clip in elastic fastenings for concrete sleepers.
2. Illustrate with sketches the various fastenings used to fasten rails to sleepers. Discuss their merits and demerits.
3. Explain briefly the functions of the following in a railway track:
    (a) Dog spike   (b) Fish plate
    (c) Tie bar   (d) Cotters
    (e) Screw spike
4. What do you understand by anchors and what are their functions in railways? What are the advantages and disadvantages of bearing plates?
5. Name the different types of track fittings. Name the different types of spikes generally used and draw a sketch of any one of them.
6. Draw the details of a rail held to a wooden sleeper by the following.
    (a) Dog spikes on an MS bearing plate
    (b) Anticreep keys on a CI bearing plate
    (c) Pandrol clip on a CI bearing plate
    What are the advantages and disadvantages of these fastenings?
7. What is the difference between an ordinary fish plate and a combination fish plate?
8. What are the requirements of an elastic fastening? Briefly describe the various elastic fastenings being used on Indian Railways.
9. Describe the various type of fittings used for wooden sleepers and steel trough sleepers.
10. Draw a sketch of an elastic rail clip and explain how it is fixed to a concrete sleeper.
11. Describe the functions of a rubber pad. Draw a dimensioned sketch of the same.

12. Differentiate between the following:
    (a) Flat bearing plate and canted bearing plate
    (b) Dog spike and screw spike
    (c) Ordinary fish plate and combination fish plate
    (d) Cotters and liners
    (e) Elastic rail clip and spring steel clip
    (f) Glass-filled nylon liners and Mota Singh liners

Choose the correct answer from the choices given.

13. Spring steel loose jaws are used to hold the rail with steel trough sleeper with the help of:
    (a) one way key   (b) two way key
    (c) anti sabotage key   (d) thicker keys
14. The pitch of hole in fish plate of 52 kg rails is:
    (a) 166 mm   (b) 168 mm
    (c) 167 mm   (d) 169 mm
15. The diameter of a rail screw is:
    (a) 20 mm   (b) 22 mm
    (c) 24 mm   (d) none of these
16. The longitudinal movement of the rail with respect to sleepers in track is:
    (a) temperature variation   (b) creep
    (c) SD   (d) none of these
17. The Mota Singh liner is an effective liner used with loose jaws steel trough sleeper for overcoming the problems of:
    (a) circular holes   (b) elongated holes
    (c) hoxogand holes   (d) none of these
18. The normal length of MS tie bar for BG is:
    (a) 2720 mm   (b) 2730 mm
    (c) 1870 mm   (d) none of these
19. The length of the BG tie bar length is:
    (a) 50 mm × 12 mm   (b) 45 mm × 10 mm
    (c) 55 mm × 12 mm   (d) none of these
20. The length of dog spikes used in BG turnout is:
    (a) 160 mm   (b) 135 mm
    (c) 120 mm   (d) 110 mm
21. The number of keys used in a CST-9 sleeper is:
    (a) 2   (b) 3
    (c) 4   (d) none of these
22. The overall depth of a dog spike for a BG track with anti creep bearing plate is:
    (a) 160 mm   (b) 155 mm
    (c) 135 mm   (d) 120 mm
23. The number of dog-spikes on sleepers over the bridge per rail seat is normally:
    (a) 2 per joint   (b) 6 per joint
    (c) 4 per joint   (d) 8 per joint
24. The number of dog spikes normally used per rail seat on curved track is:
    (a) one on either side   (b) two outside and one inside
    (c) one outside and two inside   (d) two outside and two inside

25. Rails are fixed on steel sleepers:
    (a) by bearing plates
    (b) by dog spikes
    (c) by keys in lugs or jaws
    (d) none of these
26. The main function of a fish plate is:
    (a) to join the two rails together
    (b) to join rails with the sleeper
    (c) to allow rail to expand and contract freely
    (d) none of the above
27. Canted bearing plates:
    (a) have a cant of 1 in 20
    (b) are flat bottom
    (c) are used on turnout
    (d) all of the above
28. Fish plate is named so because:
    (a) it is manufactured out of special steel
    (b) the fitting looks like a fish
    (c) it is named after the inventor
    (d) none of these
29. The number of bolts in fish-plate is generally:
    (a) 2
    (b) 3
    (c) 4
    (d) 8
30. All the holes in a fish-plate are:
    (a) circular
    (b) oval
    (c) circular and Square
    (d) oval and circular
31. Loose jaw is used with:
    (a) wooden sleeper
    (b) steel trough sleeper
    (c) CI sleeper
    (d) concrete sleeper
32. The horizontal taper at the tip of a two-way key is:
    (a) 1 in 24
    (b) 1 in 20
    (c) 1 in 32
    (d) 1 in 60
33. The weight of a pair of fish plates per pair is for a 52-kg rail is:
    (a) 28.71
    (b) 26.11
    (c) 19.54
    (d) none of these
34. The length of dog spikes in a BG track without bearing plates is:
    (a) 110 mm
    (b) 120 mm
    (c) 135 mm
    (d) none of these
35. The size of MS tie bar on BG is:
    (a) $2720 \times 55 \times 13$ mm
    (b) $1870 \times 45 \times 10$ mm
    (c) $2720 \times 50 \times 12$ mm

Creep of Rails

INTRODUCTION

Creep is defined as the longitudinal movement of the rail with respect to the sleepers. Rails have a tendency to gradually move in the direction of dominant traffic. Creep is common to all railway tracks, but its magnitude varies considerably from place to place; the rail may move by several centimetres in a month at few places, while at other locations the movement may be almost negligible.

11.1 THEORIES FOR THE DEVELOPMENT OF CREEP

Various theories have been put forward to explain the phenomenon of creep and its causes, but none of them have proved to be satisfactory. The important theories are briefly discussed in the following subsections.

11.1.1 Wave Motion Theory

According to wave motion theory, wave motion is set up in the resilient track because of moving loads, causing a deflection in the rail under the load. The portion of the rail immediately under the wheel gets slightly depressed due to the wheel load. Therefore, the rails generally have a wavy formation. As the wheels of the train move forward, the depressions also move with them and the previously depressed portion springs back to the original level. This wave motion tends to move the rail forward with the train. The ironing effect of the moving wheels on the wave formed in the rail causes a longitudinal movement of the rail in the direction of traffic resulting in the creep of the rail (Fig. 11.1).

11.1.2 Percussion Theory

According to percussion theory, creep is developed due to the impact of wheels at the rail end ahead of a joint. As the wheels of the moving train leave the trailing

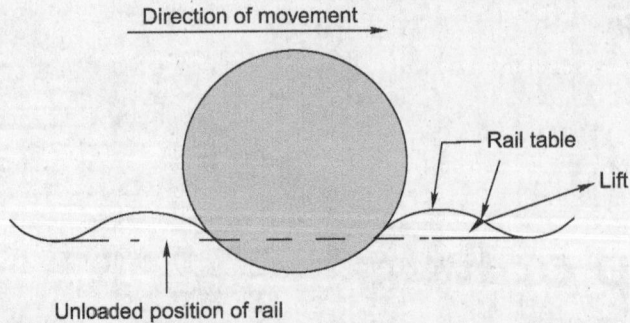

Fig. 11.1 Wave motion theory for development of creep

rail at the joint, the rail gets pushed forward causing it to move longitudinally in the direction of traffic, and that is how creep develops. Though the impact of a single wheel may be nominal, the continuous movement of several wheels passing over the joint pushes the facing or landing rail forward, thereby causing creep (Fig. 11.2).

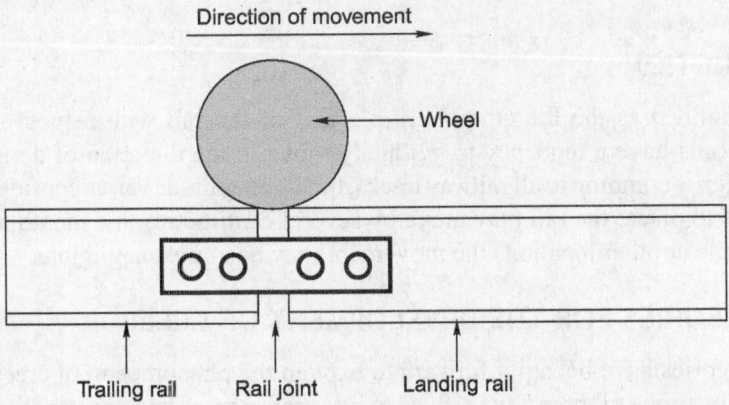

Fig. 11.2 Percussion theory for development of creep

11.1.3 Drag Theory

According to drag theory, the backward thrust of the driving wheels of a locomotive has the tendency to push the rail backwards, while the thrust of the other wheels of the locomotive and trailing wagons pushes the rail in the direction in which the locomotive is moving. This results in the longitudinal movement of the rail in the direction of traffic, thereby causing creep.

11.2 CAUSES OF CREEP

The main factors responsible for the development of creep are as follows.

Ironing effect of the wheel The ironing effect of moving wheels on the waves formed in the rail tends to cause the rail to move in the direction of traffic, resulting in creep.

Starting and stopping operations When a train starts or accelerates, the backward thrust of its wheels tends to push the rail backwards. Similarly, when the train slows down or comes to a halt, the effect of the applied brakes tends to push the rail forward. This in turn causes creep in one direction or the other.

Changes in temperature Creep can also develop due to variations in temperature resulting in the expansion and contraction of the rail. Creep occurs frequently during hot weather conditions.

Unbalanced traffic In a double-line section, trains move only in one direction, i.e., each track is unidirectional. Creep, therefore, develops in the direction of traffic. In a single-line section, even though traffic moves in both directions, the volume of traffic in each direction is normally variable. Creep, therefore, develops in the direction of predominant traffic.

Poor maintenance of track Some minor factors, mostly relating to poor maintenance of the track, also contribute to the development of creep. These are as follows:
- Improper securing of rails to sleepers
- Limited quantities of ballast resulting in inadequate ballast resistance to the movement of sleepers
- Improper expansion gaps
- Badly maintained rail joints
- Rail seat wear in metal sleeper track
- Rails too light for the traffic carried on them
- Yielding formations that result in uneven cross levels
- Other miscellaneous factors such as lack of drainage, and loose packing, uneven spacing of sleepers

11.3 EFFECTS OF CREEP

The following are the common effects of creep.

Sleepers out of square The sleepers move out of their position as a result of creep and become out of square. This in turn affects the gauge and alignment of the track, which finally results in unpleasant rides.

Expansion in gaps get disturbed Due to creep, the expansion gaps widen at some places and close at others. This results in the joints getting jammed. Undue stresses are created in the fish plates and bolts, which affect the smooth working of the switch expansion joints in the case of long welded rails.

Distortion of points and crossings Due to excessive creep, it becomes difficult to maintain the correct gauge and alignment of the rails at points and crossings.

Difficulty in changing rails If, due to operational reasons, it is required that the rail be changed, the same becomes difficult as the new rail is found to be either too short or too long because of creep.

Effect on interlocking The interlocking mechanism of the points and crossings gets disturbed by creep.

Possible buckling of track   If the creep is excessive and there is negligence in the maintenance of the track, the possibility of buckling of the track cannot be ruled out.

Other effects   There are other miscellaneous effects of creep such as breaking of bolts and kinks in the alignment, which occur in various situations.

11.4 MEASUREMENT OF CREEP

Creep can be measured with the help of a device called creep indicator. It consists of two creep posts, which are generally rail pieces that are driven at 1 km intervals on either side of the track. For the purpose of easy measurement, their top level is generally at the same level as the rail. Using a chisel, a mark is made at the side of the bottom flange of the rail on either side of the track. A fishing string is then stretched between the two creep posts and the distance between the chisel mark and the string is taken as the amount of creep.

According to the prescribed stipulations, creep should be measured at intervals of about three months and noted in a prescribed register, which is to be maintained by the permanent way inspector (PWI). Creep in excess of 150 mm (6 in.) should not be permitted on any track and not more than six consecutive rails should be found jammed in a single-rail track at one location. There should be no creep in approaches to points and crossings.

11.5 ADJUSTMENT OF CREEP

When creep is in excess of 150 mm resulting in maintenance problems, the same should be adjusted by pulling the rails back. This work is carried out after the required engineering signals have been put up and the necessary caution orders given. The various steps involved in the adjustment of creep are as follows:
  (i) A careful survey of the expansion gaps and of the current position of rail joints is carried out.
 (ii) The total creep that has been proposed to be adjusted and the correct expansion gap that is to be kept are decided in advance.
(iii) The fish plates at one end are loosened and those at the other end are removed. Sleeper fittings, i.e., spikes or keys, are also loosened or removed.
 (iv) The rails are then pulled back one by one with the help of a rope attached to a hook. The pulling back should be regulated in such a way that the rail joints remain central and suspended on the joint sleepers.

The pulling back of rails is a slow process since only one rail is dealt with at a time and can be done only for short isolated lengths of a track. Normally, about 40–50 men are required per kilometre for adjusting creep. When creep is required to be adjusted for longer lengths, five rail lengths are tackled at a time. The procedure is almost the same as the preceding steps except that instead of pulling the rails with a rope, a blow is given to them using a cut rail piece of a length of about 5 m.

11.6 CREEP ADJUSTER

A creep adjuster is normally used when extensive work is involved. The creep adjuster is set at the centre of the length of the track, to be tackled, with the wide joints behind it and the jammed joints ahead of it.

The following steps are adopted while using a creep adjuster:

  (i) Expansion liners of the correct size are put in all the expansion gaps.

 (ii) All the keys on the side (with wide joints) of the creep adjuster are removed and all fish bolts loosened.

(iii) The creep adjuster is then used to close up the gaps to the required extent by pushing the rails forward. A gap of a few inches is left between the rail ends opposite the adjuster.

(iv) The corrected rails are then fastened with keys. After that, the rails on the other side of the adjuster are tackled.

 (v) The operation leaves some of the expansion gaps too wide which are tackled by the creep adjuster when it is set in the next position.

(vi) The corrected rails are then fastened and the adjuster is shifted to the new position.

(vii) The whole process is repeated again and again till the requisite attention has been paid to the entire length of the rail. In the end it may be necessary to use a rail with the correct size of closure (bigger or smaller) to complete the work.

11.7 PORTIONS OF TRACK SUSCEPTIBLE TO CREEP

The following locations of a track are normally more susceptible to creep.

- The point where a steel sleeper track or CST-9 sleeper track meets a wooden sleeper track
- Dips in stretches with long gradients
- Approaches to major girder bridges or other stable structures
- Approaches to level crossings and points and crossings
- Steep gradients and sharp curves

11.8 MEASURES TO REDUCE CREEP

To reduce creep in a track, it should be ensured that the rails are held firmly to the sleepers and that adequate ballast resistance is available. All spikes, screws, and keys should be driven home. The toe load of fastenings should always be slightly more than the ballast resistance. Creep anchors can effectively reduce the creep in a track. At least eight of these creep anchors must be provided per panel. Out of the large number

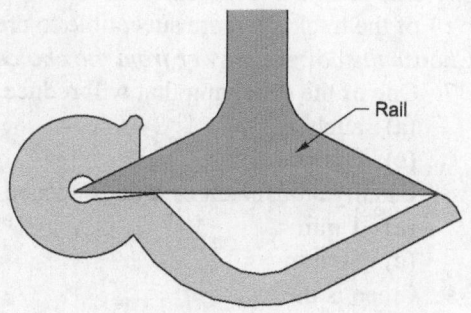

Fig. 11.3 Fair V anchor

of creep anchors tried on Indian Railways, the 'fair T' and 'fair V' anchors, have been standardized for use. The fair 'V' anchor, which is more popular, is shown in Fig. 11.3. The creep anchor should fit snugly against the sleeper for it to be fully effective. The following measures are also helpful in reducing creep.

(a) The track should be well maintained—sleepers should be properly packed and the crib and shoulder ballast should be well compacted.

(b) A careful lookout should be kept for jammed joints that exist in series. In the case of a fish-plated track, more than six consecutive continuously jammed joints should not be permitted. In the case of SWR tracks, more than two consecutive jammed joints should not be permitted at rail temperatures lower than the maximum daily temperature (T_m) in the case of zones I and II and lower than ($T_m - 5°C$) in the case of zones III and IV. Regular adjustment may be necessitated on girder bridges.

(c) Anticreep bearing plates should be provided on wooden sleepers to arrest creep, but joints sleepers should have standard canted bearing plates with rail screws.

SUMMARY

Creep is the longitudinal movement of rails with respect to sleepers. It is common in all tracks and is a severe type of track defect. In severe cases, it can result in the buckling of the track and can eventually derail the train. Therefore, it is very important to attend to creep immediately after it has been detected. Certain preventive measures can be taken to reduce creep, but it cannot be eliminated completely.

REVIEW QUESTIONS

1. Define 'creep of rail'. What are its effects?
2. What are the effects of creep of rails? Discuss the factors influencing the creep. What are the limits of creep that call for immediate remedial measures?
3. Discuss briefly the causes and effects of creep in the railway track.
4. What are the causes of creep? How can creep be adjusted?
5. What are the various theories that have been put forward to explain the development of creep? Explain wave motion theory.
6. Explain various measures that can be adopted to reduce creep. Which portions of the track are more suceptible to creep?

Choose the correct answer from the choices given.

7. One of the following that will reduce creep of rails is:
   (a) anchors
   (b) spikes
   (c) chairs
   (d) bearing plates
8. Usually adjustment of rails is needed whenever creep exceeds:
   (a) 1 mm
   (b) 10 mm
   (c) 50 mm
   (d) 150 mm
9. Creep is the:
   (a) longitudinal movement of rail
   (b) lateral movement of rail
   (c) vertical movement of rail
   (d) difference in level of two rails
10. Read the following statement regarding creep.
   (a) Creep is greater on curves than on rails tangents.
   (b) Creep in new rail is more than that in old rail.
   (c) Creep is more on steep gradients than on level track.

The correct statements is/are:

(a) only (a)   (b) only (c)

(c) (b) and (c)   (d) (a), (b) and (c)

11. The number of creep anchors normally provided per rail to arrest creep is:

(a) 4   (b) 6

(c) 8   (d) 10

12. Creep indicators are placed at an interval of:

(a) 1 km   (b) 2 km

(c) 3 km   (d) 4 km

CHAPTER
12

Geometric Design of Track

INTRODUCTION

Geometric design of a railway track includes all those parameters which determine or affect the geometry of the track. These parameters are as follows:

1. Gradients in the track, including grade compensation, rising gradient, and falling gradient
2. Curvature of the track, including horizontal and vertical curves, transition curves, sharpness of the curve in terms of radius or degree of the curve, cant or superelevation on curves, etc.
3. Alignment of the track, including straight as well as curved alignment

It is very important for tracks to have proper geometric design in order to ensure the safe and smooth running of trains at maximum permissible speeds, carrying the heaviest axle loads. The speed and axle load of the train are very important and sometimes are also included as parameters to be considered while arriving at the geometric design of the track.

12.1 NECESSITY FOR GEOMETRIC DESIGN

The need for proper geometric design of a track arises because of the following considerations:

(a) To ensure the smooth and safe running of trains
(b) To achieve maximum speeds
(c) To carry heavy axle loads
(d) To avoid accidents and derailments due to a defective permanent way
(e) To ensure that the track requires least maintenance
(f) For good aesthetics

12.2 DETAILS OF GEOMETRIC DESIGN OF TRACK

The geometric design of the track deals with various aspects, which are as follows:

Alignment of railway track   The subject of railway track alignment has been covered in Chapter 2.

Curves   Details regarding curves and their various aspects are discussed in Chapter 13.

Gradients   Details regarding gradients are discussed in the following section.

12.3 GRADIENTS

Gradients are provided to negotiate the rise or fall in the level of the railway track. A rising gradient is one in which the track rises in the direction of movement of traffic and in a down or falling gradient the track loses elevation the direction of movement of traffic.

A gradient is normally represented by the distance travelled for a rise or fall of one unit. Sometimes the gradient is indicated as per cent rise or fall. For example, if there is a rise of 1 m in 400 m, the gradient is 1 in 400 or 0.25 per cent, as shown in Fig. 12.1.

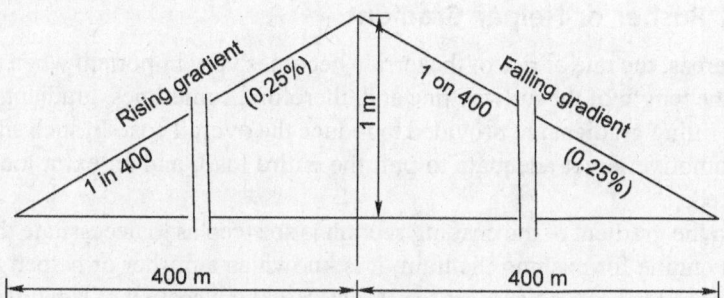

Fig. 12.1   Rising and falling gradient

Gradients are provided to meet the following objectives:
(a) To reach various stations at different elevations
(b) To follow the natural contours of the ground to the extent possible
(c) To reduce the cost of earthwork

The following types of gradients are used on the railways:
(a) Ruling gradient
(b) Pusher or helper gradient
(c) Momentum gradient
(d) Gradients in station yards

12.3.1 Ruling Gradient

The ruling gradient is the steepest gradient that exists in a section. It determines the maximum load that can be hauled by a locomotive on that section. While deciding the ruling gradient of a section, it is not only the severity of the gradient, but also its length as well as its position with respect to the gradients on both sides that have to be taken into consideration. The power of the locomotive to be put into service on the track also plays an important role in taking this decision, as the locomotive should have adequate power to haul the entire load over the ruling gradient at the maximum permissible speed.

The extra force P required by a locomotive to pull a train of weight W on a gradient with an angle of inclination θ is as follows:

$$P = W \sin \theta$$
$$= W \tan \theta \text{ (approximately, as } \theta \text{ is very small)}$$
$$= W \times \text{gradient}$$

Indian Railways does not specify any fixed ruling gradient owing to enormous variations in the topography of the country, the traffic plying on various routes, and the speed and type of locomotive in use on various sections. Generally, the following ruling gradients are adopted by Indian Railways when there is only one locomotive pulling the train.

In plain terrain: 1 in 150 to 1 in 250

In hilly terrain: 1 in 100 to 1 in 150

Once a ruling gradient has been specified for a section, all other gradients provided in that section should be flatter than the ruling gradient after making due compensation for curvature.

12.3.2 Pusher or Helper Gradient

In hilly areas, the rate of rise of the terrain becomes very important when trying to reduce the length of the railway line and, therefore, sometimes, gradients steeper than the ruling gradient are provided to reduce the overall cost. In such situations, one locomotive is not adequate to pull the entire load, and an extra locomotive is required.

When the gradient of the ensuing section is so steep as to necessitate the use of an extra engine for pushing the train, it is known as a pusher or helper gradient. Examples of pusher gradients are the Budni–Barkhera section of Central Railway and the Darjeeling Himalayan Railway section.

12.3.3 Momentum Gradient

The momentum gradient is also steeper than the ruling gradient and can be overcome by a train because of the momentum it gathers while running on the section. In valleys, a falling gradient is sometimes followed by a rising gradient. In such a situation, a train coming down a falling gradient acquires good speed and momentum, which gives additional kinetic energy to the train and allows it to negotiate gradients steeper than the ruling gradient. In sections with momentum gradients there are no obstacles provided in the form of signals, etc., which may bring the train to a critical juncture.

12.3.4 Gradients in Station Yards

The gradients in station yards are quite flat due to the following reasons:

(a) It prevents standing vehicles from rolling and moving away from the yard due to the combined effect of gravity and strong winds.

(b) It reduces the additional resistive forces required to start a locomotive to the extent possible.

It may be mentioned here that generally, yards are not levelled completely and certain flat gradients are provided in order to ensure good drainage. The maximum gradient prescribed in station yards on Indian Railways is 1 in 400, while the recommended gradient is 1 in 1000.

12.4 GRADE COMPENSATION ON CURVES

Curves provide extra resistance to the movement of trains. As a result, gradients are compensated to the following extent on curves:

(a) On BG tracks, 0.04 per cent per degree of the curve or 70/R, whichever is minimum

(b) On MG tracks, 0.03 per cent per degree of curve or 52.5/R, whichever is minimum

(c) On NG tracks, 0.02 per cent per degree of curve or 35/R, whichever is minimum

where R is the radius of the curve in metres. The gradient of a curved portion of the section should be flatter than the ruling gradient because of the extra resistance offered by the curve.

Example 12.1   Find the steepest gradient on a 2° curve for a BG line with a ruling gradient of 1 in 200.

Solution

(i) Ruling gradient = 1 in 200 = 0.5%

(ii) Compensation for a 2° curve = 0.04 × 2 = 0.08%

(iii) Compensated gradient = 0.5 − 0.08 = 0.42% = 1 in 238

The steepest gradient on the curved track is 1 in 238.

SUMMARY

Geometric design of a railway track is the scientific method of laying the various components of the track. It ensures the safety of the trains when they run at the maximum permissible speed and carry heavy axle loads. Gradient is a feature of the geometric design that is controlled by the hauling capacity of the engine. Other design features are horizontal curves, superelevation, and vertical curves. These features are discussed in Chapter 13.

REVIEW QUESTIONS

1. Define (a) ruling gradient, (b) pusher gradient, and (c) momentum gradient.
2. What is meant by grade compensation for curvature?
3. What do you understand by the geometric design of a track? Enumerate the parameters which affect the geometric design.
4. Write short notes on the following:
   (a) Gradient in station yard
   (b) Objectives for gradients
   (c) Momentum gradient

Choose the correct answer from the choices given.

5. The gradient for a BG track where the grade resistance together with curve resistance due to a 2° curve is equal to the resistance due to a ruling gradient, of 1 in 200 is:
   (a) 1 in 150
   (b) 1 in 200
   (c) 1 in 238
   (d) 1 in 250

6. The extent that a ruling gradient of 1 in 150 on a BG line should be downgraded to accommodate a 3° curve is:
   (a) 1 in 135
   (b) 1 in 150
   (c) 1 in 183
   (d) 1 in 200

7. A falling gradient followed by a rising gradient is known as:
   (a) ruling gradient
   (b) momentum gradient
   (c) pusher gradient
   (d) angular gradient

Curves and Superelevation

INTRODUCTION

Curves are introduced on a railway track to bypass obstacles, to provide longer and easily traversed gradients, and to pass a railway line through obligatory or desirable locations. Horizontal curves are provided when a change in the direction of the track is required and vertical curves are provided at points where two gradients meet or where a gradient meets level ground. To provide comfortable ride on a horizontal curve, the level of the outer rail is raised above the level of the inner rail. This is known as superelevation.

13.1 CIRCULAR CURVES

This section describes the defining parameters, elements, and methods of setting out circular curves.

Radius or degree of a curve

A curve is defined either by its radius or by its degree. The degree of a curve (D) is the angle subtended at its centre by a 30.5 m or 100 ft arc.

The value of the degree of the curve can be determined as indicated below.

Circumference of a circle = $2\pi R$

Angle subtended at the centre by a circle with this circumference = 360°

Angle subtended at the centre by a 30.5 m arc, or degree of curve

$$= \frac{360°}{2\pi R} \times 30.5$$

= 1750/R (approx., R is in metres)

In cases where the radius is very large, the arc of a circle is almost equal to the chord connecting the two ends of the arc. The degree of the curve is thus given by the following formulae:

$$D = 1750/R \text{ (when } R \text{ is in metres)}$$

$$D = 5730/R \text{ (when } R \text{ is in feet)}$$

A 2° curve, therefore, has a radius of 1750/2 = 875 m.

Relationship between radius and versine of a curve

Versine is the perpendicular distance of the midpoint of a chord from the arc of a circle. The relationship between the radius and versine of a curve can be established as shown in Fig. 13.1. Let R be the radius of the curve, C be the length of the chord, and V be the versine of a chord of length C.

AC and DE being two chords meeting perpendicularly at a common point B, simple geometry can prove that

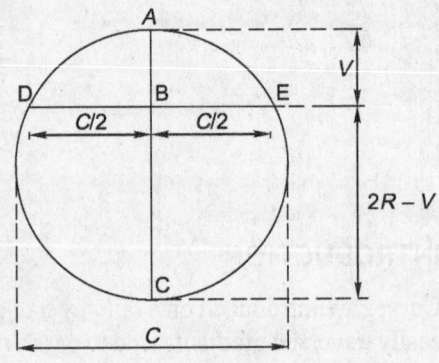

$$AB \times BC = DB \times BE$$

or

$$V(2R - V) = (C/2) \times (C/2)$$

or

$$2RV - V^2 = C^2/4$$

Fig. 13.1 Relation between radius and versine of a curve

V being very small, V^2 can be neglected. Therefore,

$$2RV = \frac{C^2}{4}$$

or

$$V = \frac{C^2}{8R} \tag{13.1}$$

In Eqn (13.1), V, C, and R are in the same unit, say, metres or centimetres. This general equation can be used to determine versines if the chord and the radius of a curve are known.

Case I: Values in metric units Equation (13.1) can also be written as

$$\frac{V}{100} = \frac{C^2}{8R}$$

Here, R is the radius of the curve, C is the chord length in metres, and V is the versine in centimetres, or

$$V = \frac{C^2 \times 100}{8R}$$

$$\frac{12.5C^2}{R} \text{cm} \quad \text{or} \quad \frac{125C^2}{R} \text{mm} \tag{13.2}$$

Case II: Values in fps units When R_1 is the radius in feet, C_1 is the chord length in feet, and V_1 is the versine in inches, Eqn (13.1) can be written as:

$$\frac{V_1}{12} = \frac{C_1^2}{8R_1}$$

or

$$V_1 = \frac{1.5C_1^2}{R_1} \tag{13.3}$$

Using Eqs (13.2) and (13.3), the radius of the curve can be calculated once the versine and chord length are known.

Determination of degree of a curve in field

For determining the degree of the curve in the field, a chord length of either 11.8 m or 62 ft is adopted. The relationship between the degree and versine of a curve is very simple for these chord lengths as follows:

Versine on a 11.8 m chord

$$V = \frac{12.5C^2}{R} \text{cm} \quad [\text{from Eqn (13.2)}]$$

$$D = \frac{1750}{R} \quad (\text{as specified before})$$

From the two equations given above, the degree of the curve for a 11.8 m chord can be determined as follows.

Substituting the value of $R = 12.5C^2/V$,

$$D = \frac{1750V}{12.5C^2} = \frac{1750V}{12.5 \times (11.8)^2}$$

$$= V \text{ approx. (cm)}$$

Versine on a 62 ft chord

$$V_1 = \frac{1.5C^2}{R_1} \text{in} \quad [\text{from Eqn (13.3)}]$$

$$D = \frac{5730}{R_1} \quad (\text{as specified before})$$

The degree of the curve for a 62 ft chord can be determined as follows:

Substituting the value of

$$R_1 = \frac{5730V_1}{1.5C_1^2} = \frac{5730V_1}{1.5 \times (62)^2}$$

$$= V_1 \text{ approx. (in.)}$$

This important relationship is helpful in determining the degree of the curve at any point by measuring the versine either in centimetres on a 11.8 m chord or in inches on a 62 ft chord. The curve can be of as many degrees as there are centimetres or inches of the versine for the chord lengths given above.

Maximum degree of a curve

The maximum permissible degree of a curve on a track depends on various factors such as gauge, wheel base of the vehicle, maximum permissible superelevation, and other such allied factors. The maximum degree or the minimum radius of the curve permitted on Indian Railways for various gauges is given in Table 13.1.

Table 13.1 Maximum permissible degree of curves

Gauge	On plain track		On turnouts	
	Max. degree	*Min. radius (m)*	*Max. degree*	*Min. radius (m)*
BG	10	175	8	218
MG	16	109	15	116
NG	40	44	17	103

Elements of a circular curve

In Fig. 13.2, AO and BO are two tangents of a circular curve which meet or intersect at a point O, called the *point of intersection* or *apex*. T_1 and T_2 are the points where the curve touches the tangents, called *tangent points* (TP). OT_1 and OT_2 are the tangent lengths of the curve and are equal in the case of a simple curve. T_1T_2 is the chord and EF is the versine of the same. The angle AOB formed between the tangents AO and OB is called the *angle of intersection* ($\angle 1$) and the angle BOO_1 is the *angle of deflection* ($\angle \phi$). The following are some of the important relations between these elements:

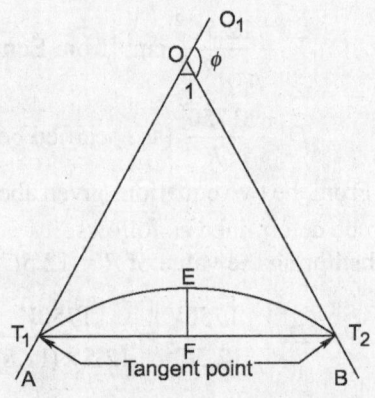

Fig. 13.2 Elements of a circular curve

$$\angle 1 + \angle \phi = 180°$$

Tangent length $OT_1 = OT_2 = R \tan \dfrac{\phi}{2}$

T_1T_2 = length of the long chord = $2\,R \sin \dfrac{\phi}{2}$

Length of the curve = $\dfrac{2\pi R}{360} \times \phi = \dfrac{\pi R \phi}{180}$

13.1.1 Setting Out a Circular Curve

A circular curve is generally set out by any one of the following methods.

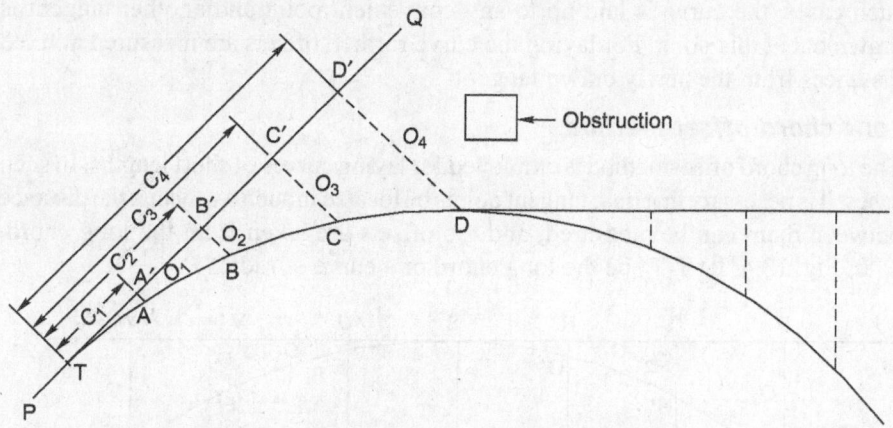

Fig. 13.3 Curve setting by tangential offset method

Tangential offset method

The tangential offset method is employed for setting out a short curve of a length of about 100 m (300 ft). It is generally used for laying turnout curves.

In Fig. 13.3, let PQ be the straight alignment and T be the tangent point for a curve of a known radius. Let AA′, BB′, CC′, etc., be perpendicular offsets from the tangent. It can be proved that

$$\text{Value of offset } O_1 = \frac{C_1^2}{2R}$$

where C_1 is the length of the chord along the tangent. Similarly,

$$O_2 = \frac{C_2^2}{2R}$$

$$O_3 = \frac{C_3^2}{2R}$$

$$O_n = \frac{C_n^2}{2R}$$

The various steps involved in the laying out of a curve using this method are as follows:

(a) Extend the straight alignment PT to TQ with the help of a ranging rod. TQ is now the tangential direction.

(b) Measure lengths C_1, C_2, C_3, etc., along the tangential direction and calculate the offsets O_1, O_2, O_3, etc., for these lengths as per the formulae given above. For simplicity, the values of C_1, C_2, C_3, etc., may be taken in multiples of three or so.

(c) Measure the perpendicular offsets O_1, O_2, O_3, etc., from the points A, B, C, etc., and locate the points A′, B′, C′, etc., on the curve.

In practice, sometimes it becomes difficult to extend the tangent length beyond a certain point due to the presence of some obstruction or because the offsets become too large to measure accurately as the length of the curve increases. In such cases, the curve is laid up to any convenient point and another tangent is drawn out at this point. For laying the curve further, offsets are measured at fixed distances from the newly drawn tangent.

Long chord offset method

The long chord offset method is employed for laying curves of short lengths. In such cases, it is necessary that both tangent points be located in such a way that the distance between them can be measured, and the offsets are taken from the long chord.

In Fig. 13.4, let T_1T_2 be the long chord of a curve of radius R.

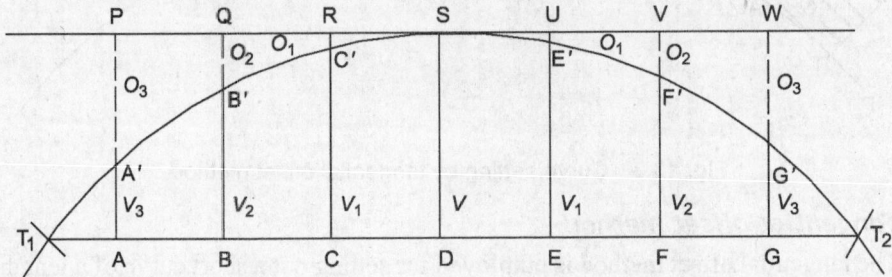

Fig. 13.4   Long chord offset method

Let the length of the long chord be C and let it be divided into eight equal parts T_1 A, AB, BC, CD, etc., where each part has a length $x = C/8$. Let PW be a line parallel to the long chord and let O_1, O_2, and O_3 be the offsets taken from points R, Q, and P.

Versine V from the long chord C is calculated by using the equation

$$V = \frac{C^2}{8R}$$

$$= DS = \frac{(T_1T_2)^2}{8R} \tag{13.4}$$

Offset O_1 from the line PW is calculated by using the equation

$$O_1 = \frac{x^2}{2R} = \frac{(RS^2)}{2R} = \frac{C^2}{128R} \qquad \text{(since Rs} = \frac{C}{8})$$

or

$$RC' = \frac{C^2}{128R} \tag{13.5}$$

Using Eqns (13.4) and (13.5), the values of the perpendicular offsets V_1, V_2, V_3, etc. can be calculated as follows:

$$V_1 = \frac{C^2}{8R} - \frac{C^2}{128R} = \frac{15}{16} \times \frac{C^2}{8R} = \frac{15}{16}V$$

$$V_2 = \frac{C^2}{8R} - \frac{(2C)^2}{128R} = \frac{12}{16} \times \frac{C^2}{8R} = \frac{12}{16}V$$

$$V_3 = \frac{C^2}{8R} - \frac{(3C)^2}{128R} = \frac{7}{16} \times \frac{C^2}{8R} = \frac{7}{16}V$$

During fieldwork, first the long chord is marked on the ground and its length measured. Then points A, B, C, etc., are marked by dividing this long chord into eight equal parts. The values of the perpendicular offsets V_1, V_2, V_3, etc., are then calculated and the points A', B', C', etc., identified on the curve.

Quartering of versine method

The quartering of versine method (Fig. 13.5) is also used for laying curves of short lengths, of about 100 m (300 ft). In this method, first the location of the two tangent points (T_1 and T_2) is determined and then the distance between them is measured. The versine (V) is then calculated using Eqn (13.2)

$$V_3 = \frac{125C^2}{R} \text{ (mm)}$$

V is measured in the perpendicular direction at the central point O of the long chord. The tangent points T_1 and T_2 are joined and the distance AT_1 measured. As AT_1 is almost half the length of chord T_1T_2 and as versines are proportional to the square of the chord, the versine of chord AT_1 is $V/4$.

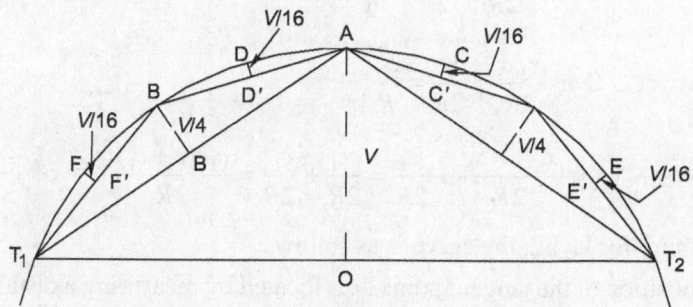

Fig. 13.5 Quartering of versine method

For laying the curve in the field, the versine $V/4$ is measured at the central point B on chord AT_1 and the position of point B is thus fixed. Similarly, a point is also fixed on the second half of the curve. AB is further taken as a sub-chord and the versine on this sub-chord is measured as $V/16$. In this way the points D and F are also fixed. The curve can thus be laid by marking half-chords and quartering the versines on these half-chords.

Chord deflection method

The chord deflection method of laying curves is one of the most popular methods on Indian Railways. The method is particularly suited to confined locations, as most of the work is done in the immediate proximity of the curves. In Fig. 13.6, let T_1 be the tangent point and A, B, C, D, etc., be the successive points on the curve. Let X_1, X_2, X_3, and X_4 be the lengths of chords T_1A, AB, BC, and CD. In practice, all the chords are of equal length. Let the value of these chords be c. The last chord may be of a different length. Let its value be c_1. It can be proved that

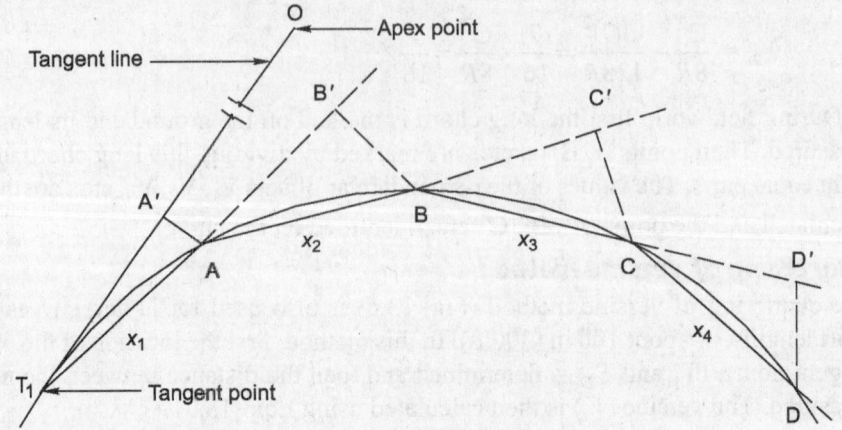

Fig. 13.6 Chord deflection method

First offset $A'A = \dfrac{x_1^2}{2R} = \dfrac{c^2}{2R}$

Second offset $B'B = \dfrac{x_1 x_2}{2R} + \dfrac{x_2^2}{2R} = \dfrac{c^2}{R}$

Third offset $C'C = \dfrac{x_2 x_3}{2R} + \dfrac{x_3^2}{2R} = \dfrac{c^2}{R}$

List offset $N'N = \dfrac{x_{n-1} \times x_n}{2R} + \dfrac{x_n^2}{2R} = \dfrac{cc_1}{2R} + \dfrac{c_1^2}{2R} = \dfrac{c_1 c(c + c_1)}{2R}$

The procedure for laying the curve is as follows:

1. The position of the tangent point T_1 is located by measuring a distance equal to the tangent length $R \tan \phi/2$ from the apex point O. In this case, ϕ is the deflection angle.
2. A length equal to the first chord (c) is measured along the tangent line T_1O and the point A′ is marked.
3. The zero end of the tape is placed at the tangent point T_1. It is then swung and the arc A_1A marked. Then the first offset on the arc is measured. The value of the offset is $c^2/2R$. The position of point A is thus fixed.
4. The chord T_1A is extended to point B and AB′ is marked as the second chord length equal to c.
5. The position of point B is then fixed on the curve since the value of the second offset is known and is equal to c^2/R.
6. Similarly, the positions of other points C, D, etc., are also located.
7. The last point on the curve is located by taking the value of the offset as $c_1(c + c_1)/R$, where c_1 is the length of the last chord.

The various points on the curve should be set with great precision because if any point is fixed inaccurately, its error is carried forward to all subsequent points.

Theodolite method

The theodolite method for setting out curves is also a very popular method on Indian Railways, particularly when accuracy is required. This method is also known as Rankine's method of tangential angles. In this method, the curve is set out using tangential angles with the help of a theodolite and a chain or a tape.

In Fig. 13.7, let A, B, C, D, etc., be successive points on a curve with lengths $T_1A = x_1$, $AB = x_2$, $BC = x_3$, $CD = x_4$, etc. Let $\delta_1, \delta_2, \delta_3, \delta_4$, be the tangential angles OT_1A, AT_1B, BT_1C, and CT_1D made by the successive chords amongst themselves.

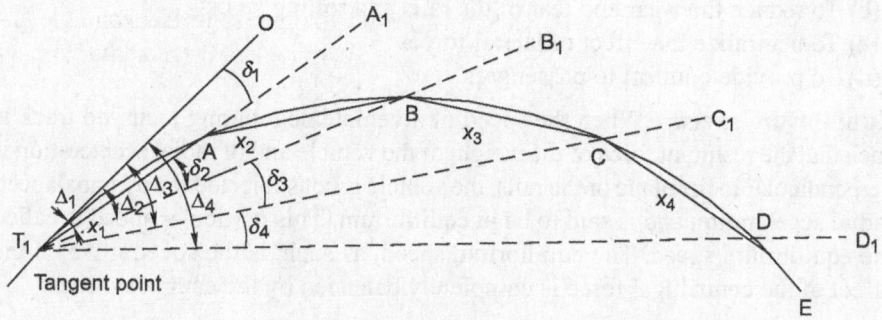

Fig. 13.7 Theodolite method

Let $\Delta_1, \Delta_2, \Delta_3$, and Δ_4 be the deflection angles of the chord from the deflection line.

Angle subtended at centre by a 100-ft chord $= D°$

Tangential angle for a 100-ft chord $= D/2$

Tangential angle for an x-ft chord, $\delta = (D/2) \times (1/100)x$ degree
$$= (5730/2R) \times (1/100) \times 60x \text{ minutes}$$
$$= 1719(x/R)$$

where δ is the deflection angle in minutes, x is the chord length in feet, and R is the radius in feet. It is seen that:

$$\Delta_1 = \delta_1$$
$$\Delta_2 = \delta_1 + \delta_2 = \Delta_1 + \delta_2$$
$$\Delta_3 = \delta_1 + \delta_2 + \delta_3 = \Delta_1 + \delta_2$$

The procedure followed for setting the curve is as follows:

(a) The theodolite is set on the tangent point T_1 in the direction of T_1O.

(b) The theodolite is rotated by an angle δ_1, which is already calculated, and the line T_1A_1 is set.

(c) The distance x_1 is measured on the line T_1A_1 in order to locate the point A.

(d) Now the theodolite is rotated by a deflection angle δ_2 to set it in the direction of T_1B_1 and point B is located by measuring AB as the chord length x_2.

(e) Similarly, the other points C, D, E, etc., are located on the curve by rotating the theodolite to the required deflection angles till the last point on the curve is reached.

(f) If higher precision is required, the curve can also be set by using two theodolites.

13.2 SUPERELEVATION

The following terms are frequently used in the design of horizontal curves.

Superelevation or cant (C_a)   It is the difference in height between the outer and the inner rail on a curve. It is provided by gradually raising the outer rail above the level of the inner rail. The inner rail, also known as the gradient rail, is taken as the reference rail and is normally maintained at its original level. The main functions of superelevation are the following:

(a) To ensure a better distribution of load on both rails
(b) To reduce the wear and tear of the rails and rolling stock
(c) To neutralize the effect of lateral forces
(d) To provide comfort to passengers

Equilibrium speed   When the speed of a vehicle negotiating a curved track is such that the resultant force of the weight of the vehicle and of radial acceleration is perpendicular to the plane of the rails, the vehicle is not subjected to any unbalanced radial acceleration and is said to be in equilibrium. This particular speed is called the equilibrium speed. The equilibrium speed, as such, is the speed at which the effect of the centrifugal force is completely balanced by the cant provided.

Maximum permissible speed   This is the highest speed permitted to a train on a curve taking into consideration the radius of curvature, actual cant, cant deficiency, cant excess, and the length of transition. On curves where the maximum permissible speed is less than the maximum sectional speed of the section of the line, permanent speed restriction becomes necessary.

Cant deficiency (C_d)   It occurs when a train travels around a curve at a speed higher than the equilibrium speed. It is the difference between the theoretical cant required for such high speeds and the actual cant provided.

Cant excess (C_e)   It occurs when a train travels around a curve at a speed lower than the equilibrium speed. It is the difference between the actual cant provided and the theoretical cant required for such a low speed.

Cant gradient and cant deficiency gradient   These indicate the increase or decrease in the cant or the deficiency of cant in a given length of transition. A gradient of 1 in 1000 means that a cant or a deficiency of cant of 1 mm is attained or lost in every 1000 mm of transition length.

Rate of change of cant or cant deficiency   This is the rate at which cant deficiency increases while passing over the transition curve, e.g., a rate of 35 mm per second means that a vehicle will experience a change in cant or a cant deficiency of 35 mm in each second of travel over the transition when travelling at the maximum permissible speed.

13.2.1 Centrifugal Force on a Curved Track

A vehicle has a tendency to travel in a straight direction, which is tangential to the curve, even when it moves on a circular curve. As a result, the vehicle is subjected to a constant radial acceleration.

Radial acceleration $= a = V^2/R$

where V is the velocity (metres per second) and R is the radius of curve (metres). This radial acceleration produces a centrifugal force which acts in a radial direction away from the centre. The value of the centrifugal force is given by the formula:

Force = mass × acceleration

$$F = m \times (V^2/R)$$

$$= (W/g) \times (V^2/R)$$

where F is the centrifugal force (tonnes), W is the weight of the vehicle (tonnes), V is the speed (m/s), g is the acceleration due to gravity (m/s^2), and R is the radius of the curve in metres.

To counteract the effect of the centrifugal force, the outer rail of the curve is elevated with respect to the inner rail by an amount equal to the *superelevation*. A state of equilibrium is reached when both the wheels exert equal pressure on the rails and the superelevation is enough to bring the resultant of the centrifugal force and the force exerted by the weight of the vehicle at right angles to the plane of the top surface of the rails. In this state of equilibrium, the difference in the heights of the outer and inner rails of the curve is known as *equilibrium superelevation*.

13.2.2 Equilibrium Superelevation

In Fig. 13.8, if θ is the angle that the inclined plane makes with the horizontal line, then

$$\tan \theta = \frac{\text{superelevation}}{\text{gauge}} = \frac{e}{G}$$

Also,

$$\tan \theta = \frac{\text{centrifugal force}}{\text{weight}} = \frac{F}{W}$$

From these equations

$$\frac{e}{G} = \frac{F}{W}$$

or

$$e = F \times \frac{G}{W}$$

$$e = \frac{W}{g} \times \frac{V^2}{R} \times \frac{G}{W} = \frac{GV^2}{gR}$$

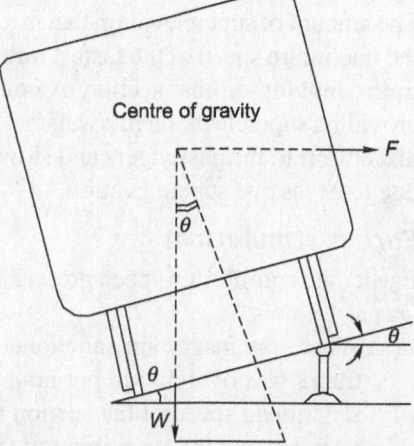

Fig. 13.8 Equilibrium superelevation

Here, e is the equilibrium superelevation, G is the gauge, V is the velocity, g is the acceleration due to gravity, and R is the radius of the curve. In the metric system equilibrium superelevation is given by the formula:

$$e = \frac{GV^2}{127R} \tag{13.6}$$

where e is the superelevation in millimetres, V is the speed in km per hour, R is the radius of the curve in metres, and G is the dynamic gauge in millimetres, which is equal to the sum of the gauge and the width of the rail head in millimetres. This is equal to 1750 mm for BG tracks and 1058 mm for MG tracks.

13.2.3 Thumb Rules for Calculating Superelevation in the Field

A field engineer can adopt the following thumb rules for determining the superelevation of any curve.

(a) Superelevation for BG in cm

$$= \left(\frac{\text{speed in km ph}}{10}\right)^2 \times \frac{\text{degree of curve}}{13}$$

(b) For MG tracks the value of superelevation is taken as three-fifths of the value calculated using the above formula. The equilibrium speed is used in this formula.

For example, if the maximum sanctioned speed (MSS) of the section is 100 km per hour, the equilibrium speed may be taken as 75 per cent of the MSS, i.e., 75 km per hour. The superelevation for a 1° curve as calculated by the thumb rule is as follows:

$$\text{SE} = \left(\frac{75}{10}\right)^2 \times \frac{1}{13} = 4.32 \text{ cm} = 43.2 \text{ mm}$$

Note that presuming that the MSS is 100 km per hour, the thumb rule is that for every 1° of curve, the cant is approximately 43 mm for BG tracks and 25 mm for MG tracks.

13.2.4 Equilibrium Speed for Providing Superelevation

The amount of superelevation that is to be provided on a curve depends not only on the maximum speed of the fastest train, but also on the average speed of the goods traffic moving on that section. A compromise, therefore, has to be achieved by providing superelevation in a way that fast trains run smoothly without causing any discomfort to the passengers and slow trains run safely without fear of derailment due to excessive superelevation.

Earlier stipulations

Earlier the equilibrium speed prescribed on a level track under average conditions was as follows:

(a) Where the maximum sanctioned speed of the section on both BG and MG tracks was over 50 km per hour (30 mile/h), three-fourths of the maximum sanctioned speed of the section was taken as the equilibrium speed, subject to a choice between minimum speed of 50 km per hour (30 mile/h) and the safe speed of the curve, whichever was less.

(b) Where the maximum sanctioned speed of the section on both BG and MG tracks was 50 km per hour (30 mile/h) or less, the maximum sanctioned speed of the section or the safe speed of the curve, whichever was less, was taken as the equilibrium speed.

Revised standards

The standards for deriving the equilibrium speed stated in the preceding section have been revised by Indian Railways recently. As per the revised standards, the chief engineer (CE) should decide the equilibrium speed that would be required for the determination of the cant to be provided on a curve after careful deliberation and taking into consideration the following factors:

(a) The maximum permissible speed which can actually be achieved both by fast trains and by goods trains

(b) Permanent and temporary speed restrictions

(c) Number of stoppages

(d) Gradients

(e) Composition of both slow and fast trains

After deciding the equilibrium speed as described, the amount of superelevation to be provided is calculated using the following formula:

$$e = \frac{GV^2}{127R} = \frac{13.76V^2}{R} \text{ (for BG)}$$

$$= \frac{8.33V^2}{R} \text{ (for MG)}$$

Here, e is the superelevation in mm, V is the speed in km per hour, G is the dynamic gauge (1750 mm for BG and 1058 mm for MG tracks), and R is the radius of the curve in metres.

13.2.5 Maximum Value of Superelevation

The maximum value of superelevation has been laid down based on experiments carried out in Europe on a standard gauge for the overturning velocity, taking into consideration the track maintenance standards. The maximum value of superelevation generally adopted on on many railways around the world is one-tenth to one-twelth of the gauge. The values of maximum superelevation prescribed on Indian Railways are given in Table 13.2.

Table 13.2 Maximum value of superelvation

Gauge	Group	Limiting value of cant (mm)	
		Under normal conditions	*With special permission of CE*
BG	A	165	185
BG	B and C	165	–
BG	D and E	140	–
MG	All routes	90	100
NG	–	65	75

According to Table 13.2, a cant of 185 mm may be provided for the purpose of setting up permanent structures, etc., besides curves that have been laid on new construction sites and doublings on group A routes, which have the potential for allowing an increase in speed in the future. The transition length should also be provided on the basis of this cant of 185 mm for the purpose of planning and laying curves.

13.2.6 Cant Deficiency and Cant Excess

Cant deficiency is the difference between the equilibrium cant that is necessary for the maximum permissible speed on a curve and the actual cant provided. Cant deficiency is limited due to two considerations:

1. Higher cant deficiency causes greater discomfort to passengers

2. Higher cant deficiency leads to greater unbalanced centrifugal force, which in turn leads to the requirement of stonger tracks and fastenings to withstand the resultant greater lateral forces. The maximum values of cant deficiency prescribed on Indian Railways are given in Table 13.3.

Table 13.3 Allowable cant deficiency

Gauge	Group	Normal cant deficiency (mm)	Remarks
BG	A and B	75	For BG group
BG	C, D, and E	75	For A and B routes; 100 mm cant deficiency permitted only for nominated rolling stock and routes with the approval of the CE
MG	All routs	50	
NG	–	40	

The limiting values of cant excess have also been prescribed. Cant excess should not be more than 75 mm on BG and 65 mm on MG for all types of rolling stock. Cant excess should be worked out taking into consideration the booked speed of the trains running on a particular section. In the case of a section that carries predominantly goods traffic, cant excess should be kept low to minimize wear on the inner rail. Table 13.4 lists the limiting values of the various parameters that concern a curve.

Table 13.4 Limiting values of various parameters concerning curves

Parameter	Limiting values	
	BG	MG and NG
Maximum degree	10°	16° for MG and 40° for NG
Maximum cant	Groups A, B, and C—165 mm Groups D and E—140 mm	90 mm (100 mm with special permission of chief engineer)
Maximum cant deficiency	In normal cases—75 mm (in special cases, 100 mm for group A and B routes with a nominated rolling stock and with permission of the chief engineer)	50 mm for MG 40 mm for NG
Cant excess	75 mm	65 mm
Maximum cant gradient	1 in 720 (in exceptional cases, 1 in 360 with permission of CE)	1 in 720
Rate of change of cant or cant deficiency	Desirable—35 mm/sec Maximum—55 mm/sec	Desirable—35 mm/sec Maximum—35 mm/sec
Minimum cant deficiency in turnout	75 mm	50 mm

13.2.7 Negative Superelevation

When the main line lies on a curve and has a turnout of contrary flexure leading to a branch line, the superelevation necessary for the average speed of trains running over the main line curve cannot be provided. In Fig. 13.9, AB, which is the outer rail of the main line curve, must be higher than CD. For the branch line, however,

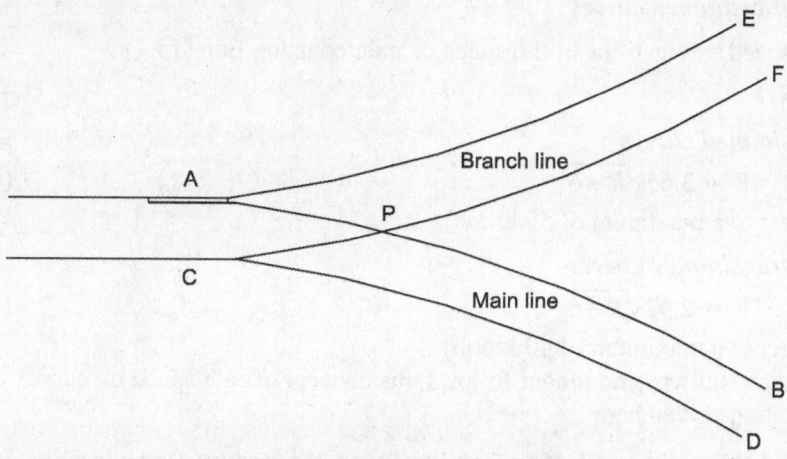

Fig. 13.9 Negative superelevation

CF should be higher than AE or point C should be higher than point A. These two contradictory conditions cannot be met within one layout. In such cases, the branch line curve has a negative superelevation and, therefore, speeds on both tracks must be restricted, particularly on the branch line.

The provision of negative superelevation for the branch line and the reduction in speed over the main line can be calculated as follows:

(i) The equilibrium superelevation for the branch line curve is first calculated using the formula

$$e = \frac{GV^2}{127R}$$

(ii) The equilibrium superelevation e is reduced by the permissible cant deficiency C_d and the resultant superelevation to be provided is

$$x = e - C_d$$

where x is the superelevation, e is the equilibrium superelevation, and C_d is 75 mm for BG and 50 mm for MG. The value of C_d is generally higher than that of e, and, therefore, x is normally negative. The branch line thus has a negative superelevation of x.

(iii) The maximum permissible speed on the main line, which has a superelevation of x, is then calculated by adding the allowable cant deficiency $(x + C_d)$. The safe speed is also calculated and the smaller of the two values is taken as the maximum permissible speed on the main line curve.

13.3 SAFE SPEED ON CURVES

For all practical purposes safe speed refers to a speed which protects a carriage from the danger of overturning and derailment and provides a certain margin of safety. Earlier it was calculated empirically by applying Martin's formula:

For BG and MG
Transitioned curves

$$V = 3.65\sqrt{R - 6} \tag{13.7}$$

where V is the speed in km per hour and R is the radius in metres.

Non-transitioned curves

Safe speed = four-fifths of the speed calculated using Eqn (13.7).

For NG

Transitioned curves

$$V = 3.65\sqrt{R-6} \tag{13.8}$$

(subject to a maximum of 50 kmph).

Non-transitioned curves

$$V = 2.92\sqrt{R-6} \tag{13.9}$$

(subject to a maximum of 40 kmph).

Indian Railways no longer follows this concept of safe speed on curves or the stipulations given here.

13.3.1 New Formula for Determining Maximum Permissible Speed on Transitioned Curves

Earlier, Martin's formula was used to work out the maximum permissible speed or safe speed on curves. This empirical formula has been changed by applying a formula based on theoretical considerations as per the recommendations of the committee of directors, chief engineers, and the ACRS. The maximum speed for transitioned curves is now determined as per the revised formulae given below:

For BG

$$V = \sqrt{\frac{(C_a + C_d) \times R}{13.76}} = 0.27\sqrt{(C_a + C_d) \times R} \tag{13.10}$$

where V is the maximum speed in km per hour, C_a is the actual cant in millimetres, C_d is the permitted cant deficiency in millimetres, and R is the radius in millimetres. This equation is derived from Eqn (13.6) for equilibrium superelevation and is based on the assumption that $G = 1750$ mm, which is the centre-to-centre distance between the rail heads of a BG track with 52 g rails.

For MG

$$V = 0.347\sqrt{(C_a + C_d)R} \tag{13.11}$$

This is based on the assumption that the centre-to-centre (c/c) distance between the rail heads of an MG track is 1058 mm.

For NG (762 mm.)

$$V = 3.65\sqrt{R-6} \text{ (subject to a maximum of 50 kmph)} \tag{13.12}$$

13.3.2 New Criteria for Determining Maximum Speed on Curves Without Transition

As per the procedure being followed at present, the determination of the maximum permissible speed on curves without transitions involves the concept of *virtual transitions*. The linear velocity of a train moving with uniform velocity on a straight track begins to change into angular velocity as soon as the first bogie reaches the tangent point. This change continues till the rear bogie reaches the tangent point, at which moment the train acquires full angular velocity. The change in the motion of the train from a straight line to a curve takes place over the shortest distance between the bogie centres and is considered a *virtual transition*. Normally, this distance is 14.6 m on BG, 13.7 m on MG, and 10.3 m on NG, commencing on a straight line at half the distance before the tangent point and terminating on the curve at half the distance beyond the tangent point. The deficiency of cant is considered as being gained over the length of the virtual transition and the cant has to be gained in a similar manner. The cant gradient must not be steeper than 1 in 360 on BG and 1 in 720 on MG and NG under any circumstance.

The safe speed should be worked out on the basis of the the the cant that can be practically provided based on these criteria, and increased by the permissible amount of cant deficiency. In the case of non-transitioned curves, where no cant is provided, the safe speed for the curve can be worked out by calculating the permissible cant deficiency after taking into consideration the rate at which the cant deficincy is gained or lost over the virtual transition.

13.3.3 Maximum Permissible Speed on a Curve

The maximum permissible speed on a curve is the minimum value of the speed that is calculated after determining the four different speed limits listed in Table 13.5.

Table 13.5 Calculation of permissible speed on curves

Type of curve	*Procedure for calculating maximum permissible speed or safe speed*
Fully transitioned curve	For BG $V = 0.27\sqrt{R(C_a + C_d)}$
	For MG $V = 0.347\sqrt{R(C_a + C_d)}$
	For NG $V = 3.65\sqrt{R - 6}$ (subject to a maximum of 50 kmph)
Non-transitioned curve with cant on virtual transition	Cant to be gained over virtual transition is 14.6 m on BG, 13.7 m on MG, and 10.3 m on NG, and the cant gradient is to be calculated accordingly.
	The cant gradient is not to exceed 1 in 360 (2.8 mm/m) on BG and 1 in 720 (1.4 mm/m) on MG and NG.
Non-transitioned curves with no cant	Calculate permissible cant deficiency that is to be gained or lost over the virtual transition.
	The desirable value of rate of change of cant deficiency is 35 mm/sec for BG and 55 mm/sec for MG.
Curves with inadequate transition	Calculate the actual cant or cant deficiency which can be provided taking into consideration its limiting value.
	The cant or cant deficiency has to be run over the transition length. The rate of change of cant or cant deficiency should not exceed its limiting value. For BG, the desirable value is 35 mm/sec and the maximum permissible value is 55 mm/sec.

The first three speed limits are taken into account for the calculation of maximum permissible speed, particularly if the length of the transition curve can be increased. For high-speed routes, however, the fourth speed limit is also very important, as cases may arise when the length of the transition curve cannot be altered easily.

(i) Maximum sanctioned speed of the section This is the maximum permissible speed authorized by the commissioner of railway safety. This is determined after an analysis of the condition of the track, the standard of interlocking, the type of locomotive and rolling stock used, and other such factors.

(ii) Maximum speed of the section based on cant deficiency This is the speed calculated using the formula given in Table 13.5. First, the equilibrium speed is decided after taking various factors into consideration and the equilibrium super-elevation (C_a) calculated. The cant deficiency (C_d) is then added to the equilibrium superelevation and the maximum speed is calculated as per this increased superelevalion ($C_a + C_d$).

(iii) Maximum speed taking into consideration speed of goods train and cant excess Cant (C_a) is calculated based on the speed of slow moving traffic, i.e., goods train. This speed is decided for each section after taking various factors into account, but generally its value is 65 km per hour for BG and 50 km per hour for MG.

The maximum value of cant excess (C_e) is added to this cant and it should be ensured that the cant for the maximum speed does not exceed the value of the sum of the actual cant + and the cant excess ($C_a + C_e$).

(iv) Speed corresponding to the length of the transition curves This is the least value of speed calculated after considering the various lengths of transition curves given by the formulae listed in Table 13.6.

Table 13.6 Various lengths of transition curves to be considered when calculating speed

Criteria for length of transition curve	Desirable length of transition curve*	Minimum length of transition curve
When the rate of change of cant is taken as 35 mm/sec for normal cases and 55 mm/sec for exceptional cases	$\dfrac{C_a \cdot V_m}{125}$ $(0.008\, C_a V_m)$	$\dfrac{C_a \cdot V_m}{198}$
When the rate of change of cant deficiency is taken as 35 mm/sec for normal cases and 55 mm/sec for exceptional cases	$\dfrac{C_d \cdot V_m}{125}$ $(0.008\, C_d V_m)$	$\dfrac{C_d \cdot V_m}{198}$
Taking the cant gradient into account	Cant gradient not to exceed 1 in 720	Cant gradient not to exceed 1 in 360 for BG and 1 in 720 for MG and NG

* Notation used in the table: C_a is the value of actual cant in mm, V_m is the maximum permissible speed in kmph, and C_d is the cant deficiency in mm.

The following points may be noted when calculating the maximum permissible speed on a curve.

(a) Criterion (iv) is to be used only in cases where the length of the transition curve cannot be increased due to site restrictions. The rate of change of cant or cant deficiency has been permitted at a rate of 55 mm/sec purely as an interim measure for the existing curves on BG tracks.

(b) For high-speed BG routes, when the speed is restricted as a result of the rate of change of cant deficiency exceeding 55 mm/sec, it is necessary to limit the cant deficiency to a value lower than 100 mm in such a way that optimum results are obtained. In this situation, the maximum permissible speed is determined for a cant deficiency less than 100 mm, but gives a higher value of the maximum permissible speed. This concept is further explained with the help of the following solved problems.

Example 13.1 Calculate the superelevation and maximum permissible speed for a 2° BG transitioned curve on a high-speed route with a maximum sanctioned speed of 110 kmph. The speed for calculating the equilibrium superelevation as decided by the chief engineer is 80 kmph and the booked speed of goods trains is 50 kmph.

Solution

(i) $R = \dfrac{1750}{D} = \dfrac{1750}{2} = 875$ m

(ii) Superelevation for equilibrium speed $= \dfrac{GV^2}{127R}$

where $G = 1750$ mm (c/c distance of 52-kg rail), $V = 80$ kmph and $R = 875$ m

$$SE = \frac{1750 \times 80^2}{127 \times 875} = 100.8 \text{ mm}$$

(iii) Superelevation for maximum sanctioned speed (110 kmph):

$$\frac{GV^2}{127R} = \frac{1750 \times 110^2}{127 \times 875} = 190.6 \text{ mm}$$

Cant deficiency $= 190.6 - 100.8 = 89.8$ mm
(which is less than 100 mm and hence permissible).

(iv) Superelevation for goods trains with a booked speed of (50 kmph)

$$\frac{GV^2}{127R} = \frac{1750 \times 50^2}{127 \times 875} = 39.4 \text{ mm}$$

Cant excess $= 100.8 - 39.4 = 61.4$ mm (which is less than 75 mm and hence permissible).

(v) Maximum speed potential or safe speed of the curve as per theoretical considerations, being a high-speed route:

$$V = \sqrt{\frac{(C_a + C_d) \times R}{13.76}} = 0.27 \sqrt{(C_a + C_d) \times R}$$

where $C_a = 100.8$ mm, C_d 89.8 mm, and $R = 875$ m.

$$V = \sqrt{\frac{(100.8 + 89.8) \times 875}{13.76}} = 110.1 \text{ kmph}$$

(vi) The maximum permissible speed on the curve is the least of the following:
- Maximum sanctioned speed, i.e., 110 kmph
- Maximum or safe speed over the curve based on theoretical considerations, i.e., 110.1 kmph
- There is no speed constraint due to the transition length of the curve

Therefore, the maximum permissible speed over the curve is 110 kmph and the superelevation to be provided is 100.8 mm or approx. 100 mm.

Simplified method of calculating permissible cant and speed

Often a simplified method is used for calculating the permissible cant and the maximum permissible speed in the field. This simplified method is applicable to most cases except those involving very flat curves.

Step 1 Calculate the cant for the maximum sanctioned speed of the section, say, 110 km/h using the standard formula $C = GV^2/127R$. This is C_{110}.

Step 2 Calculate the cant using the same standard formula as for the slowest traffic, i.e., for a goods train which may be running at, say, 50 kmph. This is C_{50}. To this add cant excess. This becomes $C_{50} + C_e$.

Step 3 Calculate the cant for equilibrium speed (if decided) using the same standard formula. Let it be 80 kmph. This value is C_{80}.

Step 4 Adopt the lowest of the three values obtained from the preceding steps and that becomes the permissible cant (C_a). The three values are C_{110}, $C_{50} + C_e$, and C_{80}.

Step 5 Taking this cant value (C_a), add the cant deficiency and find the maximum permissible speed using the Eqn (13.10).

Solution to Example 13.1

Step 1

$$C_{110} = \frac{GV^2}{127R} = \frac{1750 \times 110 \times 110}{127 \times 875} = 190.6 \text{ mm} \tag{i}$$

Step 2

$$C_{50} = \frac{GV^2}{127R} = \frac{1750 \times 50 \times 50}{127 \times 875} = 39.4 \text{ mm}$$

On adding cant excess,

$$C_a + C_e = 39.4 + 75 = 114.4 \text{ mm} \tag{ii}$$

Step 3

$$C_{80} = \frac{GV^2}{127R} = \frac{1750 \times 80 \times 80}{127 \times 875} = 100.8 \text{ mm} \tag{iii}$$

Step 4 The lowest of the three values calculated in the preceding steps is 100.8 mm. Therefore, 100 mm is adopted as the actual cant.

Step 5 Cant to be provided = 100 mm, cant deficiency = 75 mm

$$V = 0.27\sqrt{(C_a + C_d) \times R} = 0.27\sqrt{(100 + 75) \times 875}$$

$$= 110.1 = 110 \text{ kmph approx.}$$

Therefore, the maximum cant to be provided is 100 mm and the maximum permissible speed is 110 kmph.

Example 13.2 Calculate the superelevation, maximum permissible speed, and transition length for a 3° curve on a high-speed BG section with a maximum sanctioned speed of 110 kmph. Assume the equilibrium speed to be 80 kmph and the booked speed of the goods train to be 50 kmph.

Solution

(i) Radius of curve = $\dfrac{1750}{D} = \dfrac{1750}{3} = 583.3$ m

(ii) Equilibrium superelevation for 80 km/h $= \dfrac{GV^2}{127R} = \dfrac{1750 \times 80^2}{127 \times 583.3} = 151.2$ mm

(iii) Equilibrium superelevation for maximum sanctioned speed (110 kmph)

$$= \frac{1750 \times 110^2}{127 \times 583.3} = 285.5 \text{ mm}$$

(iv) Cant deficiency = 285.8 mm – 151.2 mm = 134.6 mm

This value of cant deficiency is more than 100 mm (the permitted value of C_d), therefore, take C_d as 100 mm. Now,

Actual cant = 285.8 – 100 = 185.8 mm

However, actual cant is to be limited to 165 mm, and, therefore, this value will be adopted.

(v) Equilibrium superelevation for a goods train with a speed of 50 kmph

$$= \frac{1750 \times 50^2}{127 \times 583.3} = 59 \text{ mm}$$

(vi) Cant excess = actual cant – 59 mm

$$= 165 - 59 = 106 \text{ mm}$$

which is in excess of 75 mm—the permitted value. With 75 mm taken as cant excess, the actual cant to be provided now is 75 + 59 mm = 134 mm. Therefore, a cant of 135 mm should be provided (rounding off to the higher multiple of 5).

(vii) Safe speed or speed potential (for high-speed route)

$$= 0.27\sqrt{(C_a + C_d) \times R}$$

$$= 0.27\sqrt{(135 + 100) \times 583.3}$$

$$= 99.6 \text{ kmph}$$

(or approx. 100 kmph)

(viii) Maximum permissible speed on the curve is the least of the following:
- Maximum permissible speed of the section, i.e., 110 kmph
- Safe speed on the curve, i.e., 100 kmph

The maximum permissible speed on the curve is, therefore, 100 km/h.

(ix) The length of transition is the maximum value from among the following:
- When taking the rate of change of cant into consideration (35 mm/sec),
$$L = 0.008 \, (C_a \times V_m) = 0.008 \times 135 \times 100 \text{ m} = 108 \text{ m}$$

- When taking the rate of change of cant deficiency into consideration (35 mm/sec),
$$L = 0.008 \, (C_d \times V_m)$$
$$= 0.008 \times 100 \times 100 \text{ m}$$
$$= 80 \text{ m}$$

- When taking the cant gradient into consideration (1 in 720),
$$L = 0.72 \times e = 0.72 \times 135 \text{ m} = 97.2 \text{ m}$$

Therefore, the superelevation to be provided is 135 mm, the maximum permissible speed over the curve is 100 kmph, and the length of transition curve is 108 m.

Example 13.3 Calculate the maximum permissible speed on a curve of a high-speed BG group A route having the following particulars: degree of the curve = 1°, superelevation = 80 mm, length of transition curve = 120 m, maximum speed likely to be sanctioned for the section =160 kmph.

Solution

(i) Radius of curve = 1750/D = 1750/1 = 1750 m

(ii) Safe speed over the curve as per theoretical considerations, this being a high-speed route,

$$V = 0.27 \, \sqrt{(C_a + C_d) \times R}$$

where C_d = 100 mm (assumed) C_a = 80 mm, R = 1750 m

$$V = 0.27 \sqrt{(80 + 100) \times 1750} = 151.3 \text{ kmph}$$

(iii) Speed based on transition length:

(a) Rate of gain of cant (not to exceed 55 mm/sec)

$$V_m = \frac{198L}{Ca} = \frac{198 \times 120}{80} = 297.0 \text{ kmph}$$

(b) Rate of gain of cant deficiency (not to exceed 55 mm/sec)

$$V_m = \frac{198L}{C_d} = \frac{198 \times 120}{100} = 237.6 \text{ kmph}$$

(c) Cant gradient:

$$= 1 \text{ in } \frac{120 \times 1000 \text{ mm}}{80 \text{ mm}} = 1 \text{ in } 1500$$

(which is not steeper than 1 in 720).

(iv) Maximum permissible speed is the least of the following:
- Maximum sanctioned speed of the section, i.e., 160 km per hour
- Safe speed based on theoretical considerations, i.e., 151.3 km per hour
- Speed based on the transition length, i.e., 237.6 km per hour

Therefore, the maximum permissible speed over the curve is 151.3 km per hour or about 150 km per hour. As the controlling factor in this case is the safe speed based on theoretical considerations (and not the rate of change of C_d), hence no further analysis is necessary.

Example 13.4 Calculate the maximum permissible speed on a 1° curve on a Rajdhani route with a maximum sanctioned speed of 130 km per hour. The superelevation provided is 50 mm and the transition length is 60 m. The transition length of the curve cannot be increased due to the proximity of the yard.

Solution
(i) Radius of the curve = $1750/D = 1750/1 = 1750$ m
(ii) Safe speed on the curve as per thoretical considerations,

$$V = 0.27 \sqrt{(C_a + C_d) \ R}$$

where

$$C_a = 50 \text{ mm } C_d = 100 \text{ mm}, R = 1750 \text{ m}$$

$$V = 0.27 \sqrt{(50 + 100) \times 1750} = 138.3 \text{ kmph}$$

(iii) Speed based on transition length:
(a) Rate of change of cant (not to exceed 55 mm/sec)

$$V_m = \frac{198L}{C_a} = \frac{198 \times 60}{50} = 237.6 \text{ kmph}$$

(b) Rate of change of cant deficiency (not to exceed 55 mm/sec)

$$V_m = \frac{198L}{C_d} = \frac{198 \times 60}{100} = 118.8 \text{ kmph}$$

(c) Cant gradient = $\dfrac{60 \times 1000 \text{ mm}}{50 \text{ mm}} = 1$ in 1200

(which is not steeper than 1 in 720).

(iv) Maximum sanctioned speed on the curve is the least of the following:
- Maximum speed sanctioned for the section, i.e., 130 kmph
- Safe speed based on theoretical considerations, i.e., 138.3 kmph
- Speed based on transition length, i.e., 118.3 kmph

In this case, the speed has to be restricted to 118.8 km per hour, because of the constraint of transition length. A cant deficiency of 100 mm has been assumed, which is its maximum possible value. On the field, the cant deficiency may be somewhat lower, giving a lower rate of change of C_d for the given transition length and a higher permissible speed. The optimum value of this maximum permissible speed can be found from the following equation:

Equilibrium superelevation = actual cant + cant deficiency for maximum permissible speed for a given transition length

or

$$\frac{GV^2}{127R} = \text{actual cant} + \frac{198L}{V}$$

where $G = 1750$ mm, $R = 1750$ m, $L = 60$ m, and V is the maximum permissible speed. Therefore,

$$\frac{1750V^2}{127 \times 1750} = 50 + \frac{198 \times 60}{V}$$

Solving this equation, $V = 133$ kmph. This value however, cannot be more than the MSS of the section, i.e., 130 kmph. Therefore, the maximum permissible speed over the curve is 130 kmph.

With $V = 130$ kmph, $C_d = \dfrac{198 \times L}{V} = \dfrac{198 \times 60}{130}$

$$= 91.4 \text{ mm}$$

which is less than 100 mm. Therefore, the maximum permissible speed over the circular curve is 130 kmph and that over the transition curve is 118 kmph.

Example 13.5 A BG branch line track takes off as a contrary flexure through a 1-in-12 turnout from a main line track of a 3° curvature. Due to the turnout, the maximum permissible speed on the branch line is 30 kmph. Calculate the negative superelevation to be provided on the branch line track and the maximum permissible speed on the main line track (when it takes off from a straight track).

Solution
 (i) For a branch line track, the degree of the curve is $4 - 3 = 1°$.

Radius $= 1750/D = 1750/1 = 1750$ m

$$e = \frac{GV^2}{127R} = \frac{1676 \times 30^2}{1270 \times 1750} = 6.8 \text{ mm}$$

After rounding it off to a higher multiple of 5, it is taken as 10 mm.
 (ii) The value of negative superelevation for a branch line track,
    $x = e - C_d = 10 \text{ mm} - 75 \text{ mm} = 65 \text{ mm (negative)}$
 (iii) The superelevation to be provided on the main line track is 65 mm, which is the same as the superelevation of the branch line track, but in the opposite direction.
 (iv) The maximum permissible speed is calculated by taking the actual superelevation of the main line track (65 mm) and adding it to the cant deficiency (75 mm), and then using this value of superelevation, i.e., 140 mm (65 + 75) in the formula for equilibrium speed. The main line track has a 3° curve, i.e., 1750/3 = 583.3 m radius.
    Therefore, the maximum permissible speed on the main line track,

$$e = \frac{GV^2}{127R} = \frac{1676 \times V^2}{127 \times 583.3} = 140 \text{ mm}$$

or

$$V = \sqrt{\frac{127 \times 583.3 \times 140}{1676}} = 78.7 \text{ kmph}$$

Alternatively, the maximum permissible speed can also be calculated as follows:

$$V = 0.27 \sqrt{(C_a + C_d) \times R}$$

$$= 0.27 \sqrt{(65 + 75) \times 583.3}$$

$$= 77.16 \text{ kmph}$$

Therefore, the maximum permissible speed on the main line track is 77.16 kmph. After rounding it off to a lower multiple of 5, it becomes 75 kmph.

13.4 TRANSITION CURVE

As soon as a train commences motion on a circular curve from a straight line track, it is subjected to a sudden centrifugal force, which not only causes discomfort to the passengers, but also distorts the track alignment and affects the stability of the rolling stock. In order to smoothen the shift from the straight line to the curve, transition curves are provided on either side of the circular curve so that the centrifugal force is built up gradually as the superelevation slowly runs out at a uniform rate (Fig. 13.10). A transition curve is, therefore, the cure for an uncomfortable ride, in which the degree of the curvature and the gain of superelevation are uniform throughout its length, starting from zero at the tangent point to the specified value at the circular curve. The following are the objectives of a transition curve.

(a) To decrease the radius of the curvature gradually in a planned way from infinity at the straight line to the specified value of the radius of a circular curve in order to help the vehicle negotiate the curve smoothly.

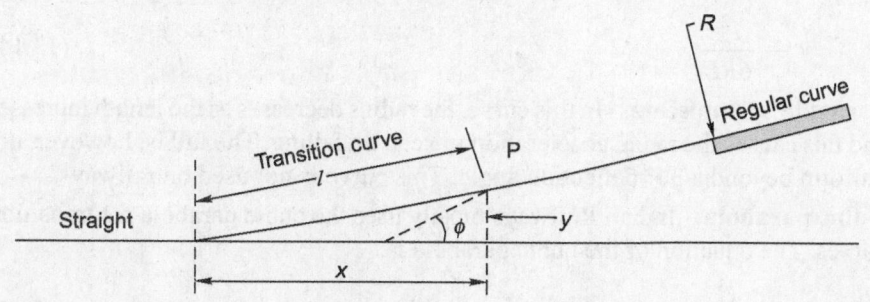

Fig. 13.10 Transition curve

(b) To provide a gradual increase of the superelevation starting from zero at the straight line to the desired superelevation at the circular curve.

(c) To ensure a gradual increase or decrease of centrifugal forces so as to enable the vehicles to negotiate a curve smoothly.

13.4.1 Requirements of an Ideal Transition Curve

The transition curve should satisfy the following conditions.

(a) It should be tangential to the straight line of the track, i.e., it should start from the straight part of the track with a zero curvature.

(b) It should join the circular curve tangentially, i.e., it should finally have the same curvature as that of the circular curve.

(c) Its curvature should increase at the same rate as the superelevation.

(d) The length of the transition curve should be adequate to attain the final superelevation, which increases gradually at a specified rate.

13.4.2 Types of Transition Curves

The types of transition curves that can be theoretically provided are described here. The shapes of these curves are illustrated in Fig. 13.11.

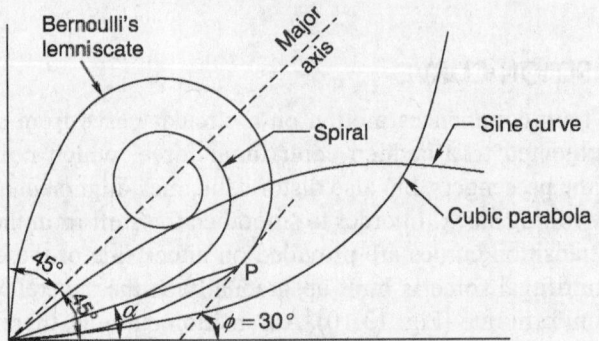

Fig. 13.11 Different types of transition curves

Euler's spiral This is an ideal transition curve, but is not preferred due to mathematical complications. The equation for Euler's spiral is:

$$\phi = \frac{l^2}{2RL} \tag{13.13}$$

Cubical spiral This is also a good transition curve, but quite difficult to set on the field.

$$y = \frac{l^2}{6RL} \tag{13.14}$$

Bernoulli's lemniscate In this curve, the radius decreases as the length increases and this causes the radial acceleration to keep on falling. The fall is, however, not uniform beyond a 30° deflection angle. This curve is not used on railways.

Cubic parabola Indian Railways mostly uses the cubic parabola for transition curves. The equation of the cubic parabola is:

$$y = \frac{x^3}{6RL} \tag{13.15}$$

In this curve, both the curvature and the cant increase at a linear rate. The cant of the transition curve from the straight to the curved track is so arranged that the inner rail continues to be at the same level while the outer rail is raised in the linear form throughout the length of the curve. A straight line ramp is provided for such transition curves.

The notations used in Eqs (13.13) to (13.15) are as follows: ϕ is the angle between the straight line track and the tangent to the transition curve, l is the distance of any point on the transition curve from the take-off point, L is the length of the transition curve, x is the horizontal coordinate on the transition curve, y is the vertical coordinate on the transition curve, and R is the radius of the circular curve.

S-shaped transition curve In an S-shaped transition curve, the curvature and superelevation assume the shape of two quadratic parabolas. Instead of a straight line ramp, an S-type parabola ramp is provided with this transition curve. The special feature of this curve is that the shift required ('shift' is explained in the following section) in this case is only half of the normal shift provided for a straight line ramp. The value of shift is:

$$S = \frac{L^2}{48R} \tag{13.16}$$

Further, the gradient is at the centre and is twice steeper than in the case of a straight line ramp. This curve is desirable in special conditions—when the shift is restricted due to site conditions.

The Railway Board has decided that on Indian Railways, transition curves will normally be laid in the shape of a cubic parabola.

13.4.3 Shift

For the main circular curve to fit in the transition curve, which is laid in the shape of a cubic parabola, it is required to be moved inward by a measure known as the 'shift' (Fig. 13.12). The value of shift can be calculated using the formula

$$S = \frac{L^2}{24R} \tag{13.17}$$

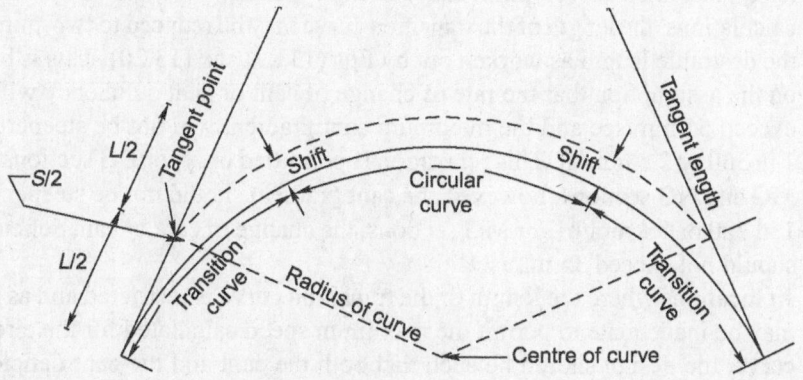

Fig. 13.12 Shift

where S is the shift in metres, L is the length of the transition curve in metres, and R is the radius in metres.

The offset (in centimetres) from the straight line to any point on the transition curve is calculated using the equation:

$$y = 16.7 \ \frac{x^3}{LR} \tag{13.18}$$

where y is the offset from the staight line in centimetres, x is the distance from the commencement of the curve in metres, L is the length of transition in metres, and R is the radius of curve in metres.

13.4.4 Length of Transition Curve

The length of the transition curve prescribed on Indian Railways is the maximum of the following three values:

$$L = 0.008 C_a \times V_m = \frac{C_a \times V_m}{125} \tag{13.19}$$

$$L = 0.008 C_d \times V_m = \frac{C_d \times V_m}{125} \tag{13.20}$$

$$L = 0.72 C_a \tag{13.21}$$

where L is the length of the curve in metres, Ca is the actual cant or superelevation in millimetres, and C_d is the cant deficiency in millimetres.

Equations (13.19) and (13.20) are based on a rate of change of a cant or cant deficiency of 35 mm/sec. Equation (13.21) is based on a maximum cant gradient of 1 in a 720 or 1.4 mm/m.

Other provisions made to meet the requirements of special situations are as follows:

(a) When deciding the length of transition curves, particularly on high-speed routes, future speeds expected to be implemented on those tracks, such as 160 km per hour for group A routes and 130 km per hour for group B routes, may be taken into account.

(b) In exceptional cases, when there is no space available for providing full length transition curves, particularly on high-speed routes as per the preceding calculations, the length of the transition curve may be reduced to two-thirds of the desirable length as worked out by Eqs (13.19) and (13.20). This is based on the assumption that the rate of change of cant or cant deficiency will not exceed 55 mm/sec and the maximum cant gradient will not be steeper than 1 in 360 or 2.8 mm/m. This relaxation is permitted only for BG sections. For MG and NG sections, however, the cant gradient should not be steeper than 1 in 720 or 1.4 mm/h. For MG sections, the change of cant or cant deficiency should not exceed 35 mm/sec.

(c) At locations where the length of the transition curve is restricted and as such may be inadequate to permit the maximum speed calculated for the circular curve, the design should be such that both the cant and the cant deficiency are lowered, which will reduce the maximum speed on the transition curve to permit the highest speed on the curve as a whole.

Example 13.6 A curve of 600 m radius on a BG section has a limited transition of 40 m length. Calculate the maximum permissible speed and superelevation for the same. The maximum sectional speed (MSS) is 100 kmph.

Solution In a normal situation, a curve of 600 m radius will have quite a long transition curve for an MSS of 100 kmph. However, as the transition curve has been restricted to 40 m, the cant should be so selected that the speed on the main circular curve is equal to the speed on the transition curve as a whole.

(i) For the circular curve, the maximum speed is calculated from Eqn (13.10):

$$V = 0.27 \sqrt{R(C_a + C_d)}$$

The most favourable value of speed is obtained when $C_a = C_d$.

(ii) For the transition curve, the maximum change of cant is taken as 55 mm/sec and the maximum speed is then calculated:

$$L = \frac{C_a \times V_m}{198} \text{ or } V = \frac{198L}{C_a}$$

Therefore,

$$0.27\sqrt{R(C_a + C_d)} = \frac{198L}{C_a}$$

or

$$0.27\sqrt{600 \times 2C_a} = \frac{198 \times 40}{C_a}$$

On solving this equation, $C_a = 89.50$ mm $\cong 90$ mm.

(iii) On limiting the value of C_d to 75 mm,

$$\text{Maximum speed} = 0.27 \sqrt{R(C_a + C_d)}$$

$$= 0.27 \sqrt{600(90 + 75)}$$

$$= 84.95 \text{ or approx. } 85 \text{ kmph}$$

(iv) Cant gradient $= \dfrac{90 \text{ mm}}{40 \times 1000}$

$$= 1 \text{ in } 444$$

This is within the permissible limits of 1:360.

Therefore, the maximum permissible speed is 85 kmph and the superelevation to be provided is 90 mm.

13.4.5 Laying a Transition Curve

A transition curve is laid in the following steps (Fig. 13.13).

1. The length of the transition curve is calculated by the formulae given in Eqs (13.19) to (13.21).
2. This transition length is divided into an even number of equal parts, usually eight.
3. The equations for a cubic parabola and the shift [Eqs (13.15) and (13.17)], reproduced here, are used for calculations.

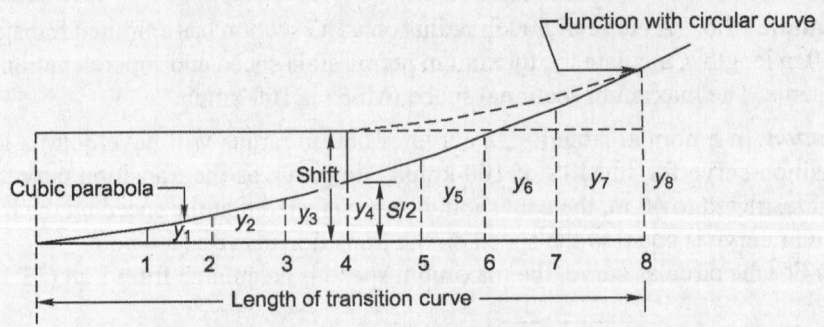

Fig. 13.13 Laying a transition curve

$$y = \frac{x^3}{6RL} = 0.167 \frac{x^3}{RL} \qquad\qquad (i)$$

$$S = \frac{L^2}{24R} = 0.042 \frac{L^2}{R} \qquad\qquad (ii)$$

(all measurements are in the same units).

4. The shift is calculated using Formula (ii).

5. The ordinates are then calculated at points 1, 2, 3, etc., using Formula (i).

$$y_1 = \frac{(L/8)^3}{6RL} = \frac{S}{128}$$

$$y_2 = \frac{(L/4)^3}{6RL} = \frac{S}{16}$$

$$y_3 = \frac{(3L/8)^3}{6RL} = 0.211S$$

$$y_4 = \frac{(L/2)^3}{6RL} = 0.500S$$

$$y_5 = \frac{(5L/8)^3}{6RL} = 0.976S$$

$$y_6 = \frac{(6L/8)^3}{6RL} = 1.688S$$

$$y_7 = \frac{(7L/8)^3}{6RL} = 2.680S$$

$$y_8 = \frac{L^3}{6RL} = 4S$$

6. The point at which the transition curve starts is then determined approximately by shifting the existing tangent point backwards by distance equal to half the length of the transition curve.

7. The offsets y_1, y_2, y_3, etc., are measured perpendicular to the tangent to get the profile of the transition curve.

13.5 COMPOUND CURVE

A compound curve (Fig. 13.14) is formed by the combination of two circular curves of different radii curving in the same direction. A common transition curve may be provided between the two circular curves of a compound curve. Assuming that such a connecting curve is to be traversed at a uniform speed, the length of the transition curve connecting the two circular curves can be obtained from the formula

$$L = 0.008 \, (C_{a1} - C_{a2}) \times V$$

or

$$L = 0.008 \, (C_{d1} - C_{d2}) \times V_m, \text{ whichever is greater}$$

where C_{a1} and C_{d1} are the cant and cant deficiency for curve 1 and C_{a2} and C_{d2} are the cant and cant deficiency for curve 2 in millimetres. L is the length of the transition curve in metres, and V_m is the maximum permissible speed in km per hour.

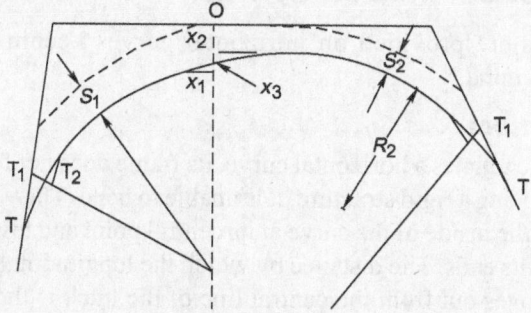

Fig. 13.14 Compound curve

13.6 REVERSE CURVE

A reverse curve (Fig. 13.15) is formed by the combination of two circular curves with opposite curvatures. A common transition curve may be provided between the two circular curves of a reverse curve. The total length of the transition curve, from the common circular curve to the individual circular curve, may be obtained in the same manner as explained for a compound curve in Section 13.5.

It has been stipulated that for high-speed group A and B routes, a minimum straight length of 50 m should be kept between the two curves constituting a reverse curve. In the case of a high-speed MG route, the distance to be kept should be 30 m. Straight lines between the circular curves measuring less than 50 m on BG sections of group A and B routes and less than 30 m on high-speed MG routes should be eliminated by suitably extending the transition lengths. When doing so, it should be ensured that the rate of change of cant and versine along the two transition lengths being extended is kept the same. When such straight lines between reverse curves cannot be eliminated and their lengths cannot be increased to over 50 m in the case of BG routes and 30 m in the case of MG routes, speeds in excess of 130 km per hour on BG routes and 100 km per hour on MG routes should not be permitted.

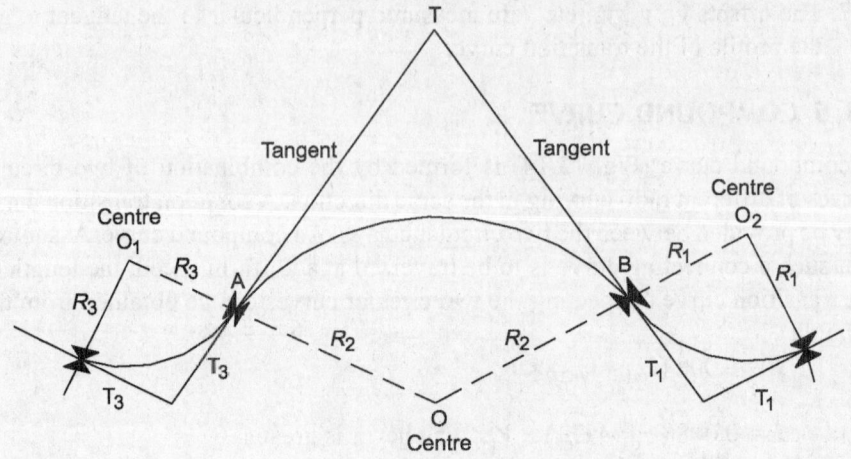

Fig. 13.15 Reverse curve

13.7 EXTRA CLEARANCE ON CURVES

Extra clearances are provided on horizontal curves keeping the following considerations in mind.

Effect of curvature

When a vehicle negotiates a horizontal curve, its frame does not follow the path of the curve, since, being a rigid structure, it is unable to bend. The vehicle, therefore, projects towards the inside of the curve at its central point and towards the outside of the curve near its ends. The distance by which the longitudinal axis of the body of the vehicle moves out from the central line of the track is the extra clearance required (Fig. 13.16).

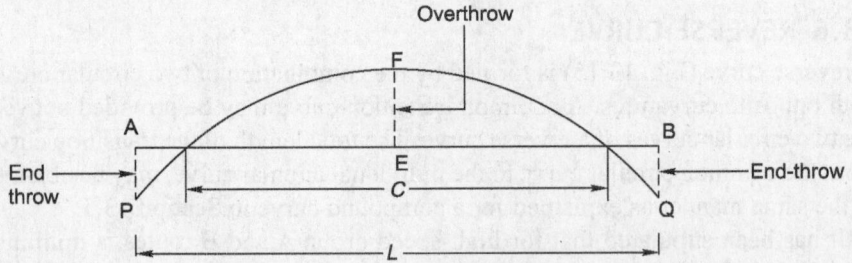

Fig. 13.16 Effect of curvature on long vehicle

(i) The extra clearance required at the centre of the vehicle, which projects towards the inside of the curve, is called overthrow and is given by the equation:

$$EF = \frac{C^2}{8R} \tag{13.22}$$

(ii) The extra clearance required at the ends of the vehicle, which projects towards the outside of the curve, is called end-throw and is given by the equation

$$\text{AP or BQ} = \frac{L^2 - C^2}{8R} \tag{13.23}$$

where L is the length of the vehicle, C is the centre-to-centre distance between the bogies, and R is the radius of the curve.

Effect of leaning due to superelevation

On account of the superelevation provided on a curve, the vehicle leans towards the inside of the curve, thereby requiring extra clearance as shown in Fig. 13.17. The extra clearance required for leaning is as follows:

Lean $= he/G$

where h is the height of the vehicle, e is the superelevation, and G is the gauge.

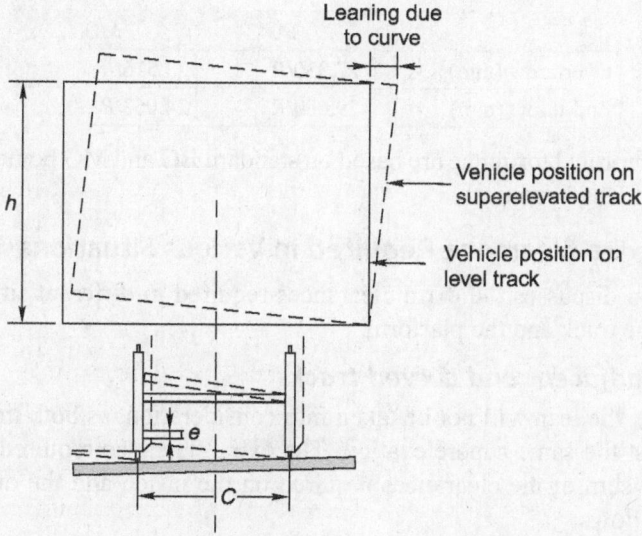

Fig. 13.17 Effect of lean due to superelevation

In case the superelevation is not known, it is suggested that its value be assumed to be 70 mm up to a 1° curve and 115 mm for curves above 1°. No extra clearance, however, is required for leaning on the outside of the curve.

Effect of sway of vehicles

On account of unbalanced centrifugal forces caused due to cant deficiency or cant excess, the vehicles tend to experience an additional sway. The extra clearance required on the inside of the curve due to the sway is taken as one-fourth of the clearance necessary due to leaning.

On summarizing, the total extra clearance (in mm) required on curves is as follows:

(i) Extra clearance inside the curve = overthrow + lean + sway

$$\text{Ec}_1 = \frac{C^2}{8R} + \frac{eh}{G} + \frac{1}{4} \times \frac{eh}{G} \tag{13.24}$$

(ii) Extra clearance outside the curve = end-throw

$$Ec_2 = \frac{L^2 - C^2}{8R} \tag{13.25}$$

Here, C is the centre-to-centre distance between bogies, which is 14,785 mm for BG routes and 13,715 mm for MG routes, R is the radius of the curve in mm, L is the length of a bogie, which is 21,340 mm for BG routes and 19,510 mm for MG routes, e is the superelevation in mm, h is the height of the vehicle, which is 3350 mm for BG tracks and 3200 mm for MG tracks, and G is the gauge, which is 1676 mm for BG tracks and 1000 mm for MG tracks.

The empirical formulae normally adopted in the field for determining the extra clearance due to the curvature effect are shown in Table 13.7.

Table 13.7 Extra clearance on borizontal curves

	BG	MG
Overthrow (mm)	27,330/R	23,516/R
End-throw (mm)	29,600/R	24,063/R

These empirical formulae are based on standard BG and MG bogie lengths and the value of R is in metres.

13.7.1 Extra Clearance Required in Various Situations

This section discusses the extra clearances required in different situations with regard to the track and the platform.

Between adjacent and curved tracks

In this case, the lean will not be taken into consideration, as both the tracks will have almost the same superelevation. The extra clearance required in this case will be the sum of the clearances required on the inside and the outside of the curve as follows:

$$Ec = (Ec_1 - lean) + Ec_2$$

$$= Overthrow + sway + end\text{-}throw$$

$$\frac{C^2}{8R} + \frac{1}{4} \times \frac{eh}{G} + \frac{L^2 - C^2}{8R} \tag{13.26}$$

where e is the superelevation in millimetres, h is the height of the vehicle (3.35 m for BG and 3.2 m for MG), and G is the gauge.

For adjacent tracks with structures in between

When there is a structure between two adjacent tracks, each track is treated independently and extra clearances are provided by considering each track with respect to the structure.

For platforms

In the case of platforms, it has been observed that the provision of extra clearance on curves as discussed may lead to excessive gap between the footboard and the platform. It is, therefore, stipulated that next to platforms this extra clearance be reduced by 51 mm (2 in.) on the inside of the curve and 25 mm (1 in.) on the outside of the curve.

Example 13.7   Two high-level platforms are to be provided on the inside as well as the outside of a 2° curve on a BG track with a superelevation of 100 mm. What should be the required extra clearances for these platforms, both on the inside and the outside of the curve, (Length of bogie = 21,340 mm, c/c bogie distance = 14,785 mm, height of platform = 840 mm.)

Solution

Radius of the curve = 1750/*D* = 1750/2° = 875 m.

(i) Extra clearance required on the inside of the curve,

Ec_1 overthrow + lean + sway – 51 mm

$$\frac{C^2}{8R} + \frac{eh}{G} + \frac{1}{4} \times \frac{eh}{G} - 51 \text{ mm}$$

In this example, C = 14,785 mm, R = 875,000 mm, e = 100 mm, G = 176 mm, and h = 840 mm. Therefore,

$$Ec_1 = \frac{14,785^2}{8 \times 875,000} + \frac{100 \times 840}{1676} + \frac{1}{4} \times \frac{100 \times 840}{1676} - 51 \text{ mm}$$

$$= 42.88 \text{ mm} = 45 \text{ mm approx.}$$

(ii) Extra clearance required on the outside of the curve,

$$Ec_2 = \text{End-throw} - 25 \text{ mm} = \frac{L^2 - C^2}{8R} - 25$$

where L = 21,340 mm, C = 14,875 mm, and R = 875,000 mm. Therefore,

$$Ec_2 = \frac{21,340^2 - 14785^2}{8 \times 875,000} - 25$$

$$= 33.83 - 25 \text{ mm} = 8.83 \text{ mm or approx. 10 mm}$$

Therefore, an extra clearance of 45 mm should be provided for the outside platform on the inner side of the curve and of 10 mm for the inside platform on the outer side of the curve.

13.8 WIDENING OF GAUGE ON CURVES

A vehicle normally assumes the central position on a straight track and the flanges of the wheels stay clear of the rails. The situation, however, changes on a curved track. As soon as the vehicle moves onto a curve, the flange of the outside wheel of the leading axle continues to travel in a straight line till it rubs against the rail. Due to the coning of wheels, the outside wheel travels a longer distance compared to the inner wheel. This, however, becomes impossible for the vehicle as a whole, since the rigidity of the wheel base causes the trailing axle to occupy a different position. In an effort to make up for the difference in the distance travelled by the outer wheel and the inner wheel, the inside wheels slip backward and the outer wheels skid forward. A close study of the running of vehicles on curves indicates that the wear of flanges eases the passage of the vehicle around curves, as it has the effect of increasing the gauge. The widening of the gauge on a curve has, in fact, the same effect and tends to decrease the wear and tear on both the wheel and the track.

The stipulations laid down with regard to the gauge on straight tracks and curves on Indian Railways are given in Table 13.8.

Table 13.8 Gauge standard for curves

Type of track	Gauge tolerances for BG	Gauge tolerances for MG and NG
Straight track including curves of 350 m for BG, 290 m for MG, and 400 m and more for NG	−5 mm to + 3 mm	MG: − 2 mm to + 3 mm NG: − 3 mm to + 3 mm
For curves of radius less than 350 m for BG, 290 m for MG, and 400 m to 100 m for NG	Up to + 10 mm	Up to +10 mm
For curves with radius less than 100 m for NG	−	Up to + 15 mm

The widening of the gauge on curves can be calculated using the formula:

$$\text{Extra width on curves } (w) = \frac{13(B+L)^2}{R} \qquad (13.27)$$

where B is the wheel base of the vehicle in metres, R is the radius of the curve in metres, $L = 0.02(h^2 + Dh)^{1/2}$ is the lap of the flange in metres, h is the depth of flange below top of the rail, and D is the diameter of the wheel of the vehicle.

Example 13.8   The wheel base of a vehicle moving on a BG track is 6 m. The diameter of the wheels is 1524 mm and the flanges project 32 mm below the top of the rail. Determine the extra width of the gauge required if the radius of the curve is 168 m. Also indicate the extra width of gauge actually provided as per Indian Railways standards.

Solution

(i) Lap of flange $L = 0.02 \sqrt{h^2 + Dh}$

where $h = 3.2$ cm is the depth of the flange below the top of the rail and $D = 152.4$ cm is the diameter of the wheel. Therefore,

$$L = 0.02 \sqrt{h^2 + Dh}$$

$$= 0.02 \sqrt{3.2^2 + (152.4 \times 3.2)} = 0.446 \text{ m}$$

(ii) Extra width of gauge $(w) = \dfrac{13(B + L)^2}{R}$

$$= \frac{13(6 + 0.446)^2}{168} = 3.21 \text{ cm} = 32.1 \text{ mm}$$

(iii) As per Indian Railways standards, an extra width of 10 mm is provided for curves with a radius less than 350 m in actual practice.

13.9 VERTICAL CURVES

An angle is formed at the point where two different gradients meet, forming a summit or a sag as explained in Fig. 13.18. The angle formed at the point of contact of the gradients is smoothened by providing a curve called the vertical curve in the vertical plane. In the absence of a vertical curve, vehicles are likely to have a rough run on the track. Besides this, a change in the gradient may also cause bunching of vehicles in the sags and a variation in the tension of couplings in the summits, resulting in train parting and an uncomfortable ride. To avoid these ill effects, the change in gradient is smoothened by providing a vertical curve. A rising gradient is normally considered positive and a failing gradient is considered negative.

A vertical curve is normally designed as a circular curve. The circular profile ensures a uniform rate of change of gradient, which controls the rotational acceleration.

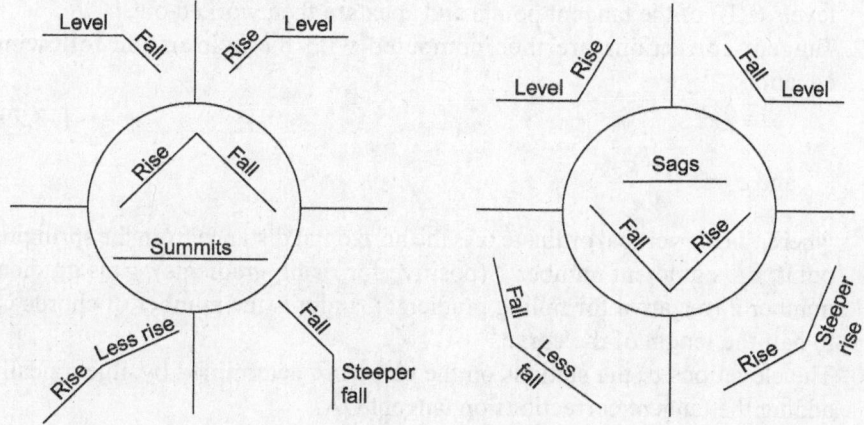

Fig. 13.18 Summits and sags in vertical curves

13.9.1 Calculating the Length of a Vertical Curve (Old Method)

The length of a vertical curve depends upon the algebraic difference between the gradients and the type of curve formed (summit or sag). The rate of change of gradient in the case of summits should not exceed 0.1 per cent between successive 30.5 m (100 ft) chords, whereas the corresponding figure for sags is 0.05 per cent per 30.5 m (100 ft) chord. The required length of a vertical curve for achieving the maximum permissible speed is given by the formula

$$L = (a/r) \times 30.5 \text{ m} \tag{13.28}$$

where L is the length of the vertical curve in m, a is the per cent algebraic difference between successive gradients, and r is the rate of change of the gradient, which is 0.1 per cent for summit curves and 0.05 per cent for sag curves.

13.9.2 Existing Provisions on Indian Railways

As per the existing provisions, vertical curves are provided only at the junction of gradients, when the algebraic difference between the gradients is equal to or more than 0.4 per cent. The minimum radii for vertical curves are given in Table 13.9.

Table 13.9 Minimum radii for vertical curves

Broad gauge (BG)		Metre gauge (MG)	
Group	*Min. radius (m)*	*Group*	*Min. radius (m)*
A	4000	All routes	2500
B	3000		
C, D, and E	2500		

13.9.3 Setting a Vertical Curve

A vertical curve can be set by various methods, such as the tangent correction method and the chord deflection method. The tangent correction method, which is considered simpler than the other methods and is more convenient for the field staff, is described here (Fig. 13.19). It involves the following steps.

1. The length of the vertical curve is first calculated. The chainages and reduced levels (RL) of the tangent points and apex are then worked out.
2. Tangent corrections are then computed with the help of the following equation:

$$y = cx^2 \tag{13.29}$$

$$\text{and } c = \frac{g_1 - g_2}{4.n}$$

where y is the vertical ordinate, x is the horizontal distance from the springing point, g_1 is gradient number 1 (positive for rising gradients), g_2 is gradient number 2 (negative for falling gradients), and n is the number of chords up to half the length of the curve.

3. The elevations of the stations on the curve are determined by algebraically adding the tangent corrections on tangent OA.

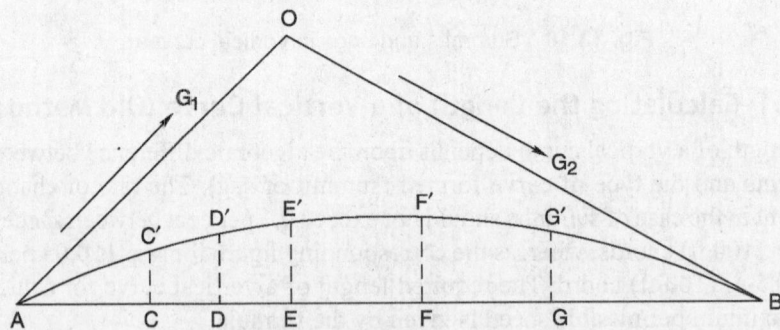

Fig. 13.19 Setting out a vertical curve

Example 13.9 Calculate the length of the vertical curve between two gradients meeting in a summit, one rising at a rate of 1 in 100 and the other falling at a rate of 1 in 200.

Solution

Gradient of the rising track (1 in 100) = 1% (+)
Gradient of the falling track (1 in 200) = 0.5% (−)

Change of gradient $(a) = 1 - (-0.5) = 1 + 0.5\% = + 1.5\%$
Rate of change of gradient (r) for summit curve $= 0.1\%$

$$\text{Length of vertical curve} = \frac{1.5}{0.1}\frac{a}{r} \times 30.5 \text{ m} = \frac{1.5}{0.1} \times 30.5$$

$$= 457.5 \text{ m}$$

13.9.4 New Method of Calculating Length of Vertical Curve

According to the new method, the length of a vertical curve is calculated as follows:

$$L = RQ \qquad\qquad (13.30)$$

where L is the length of the vertical curve, R is the radius of the vertical curve as per the existing provisions given in Table 13.9, and Q is the difference in the percentage of gradients (expressed in radians).

It is seen that the length of the vertical curve calculated as per the new practice is relatively small compared to the length calculated using the old method. The length of the vertical curve according to the new practice is considered very reasonable for the purpose of laying the curve in the field, as can be seen from the next solved example.

Note that when the change in gradient (a) is positive it forms a summit and when it is negative it forms a sag.

Example 13.10 A rising gradient of 1 in 100 meets a falling gradient of 1 in 200 on a group A route. The intersection point has a chainage of 1000 m and its RL is 100 m. Calculate the length of the vertical curve, and the RL and the chainage of the various points in order to set a vertical curve at this location.

Solution

First gradient $= 1$ in 100 (rising) $= + 1\%$
Second gradient $= 1$ in 200 (falling) $= -0.5\%$
Difference in gradient $= (+ 1) - (-0.5) = 1.5\%$
Length of vertical curve $= L = RQ$
$\qquad\qquad\qquad = 4000 \times (1.5/100) = 60 \text{ m}$
Chainage of point A (refer Fig. 13.19) $= 1000 - 30 = 970 \text{ m}$
Chainage of point B $= 1000 + 30 = 1030 \text{ m}$
RL of point A $= 100 - (30/100) = 99.70 \text{ m}$
RL of point B $= 100 - (30/200) = 99.85 \text{ m}$
Increase in RL for 60 m $= 99.85 - 99.70 = 0.15 \text{ m}$

$$\text{Increase in RL for 10 m} = \frac{0.15 \times 10}{60} = 0.025 \text{ m}$$

$$\text{First offset on vertical curve} = \frac{x(L - x)}{2R}$$

where $L = 60$ m, $R = 4000$ m, and $x = 10$ m.

The calculations for the RL of different points on the curve are shown in Table 13.10.

Table 13.10 Setting a vertical curve

Chain-age	Point	RL of point	Offset = $[x(L-x)]/2R$	Points on vertical curve	RL of points on curve
970	A	99.700	0.0000	A	99.7000
980	C	$99.7000 + 0.025$ $= 99.725$	$\dfrac{10 \times 50}{2 \times 4000} = 0.0625$	C′	$99.725 + 0.0625$ $= 99.7875$
990	D	$99.725 + 0.025$ $= 99.750$	$\dfrac{20 \times 40}{2 \times 4000} = 0.1000$	D′	$99.750 + 0.1000$ $= 99.850$
1000	E	$99.750 + 0.025$ $= 99.775$	$\dfrac{30 \times 30}{2 \times 4000} = 0.1125$	E′	$99.775 + 0.1125$ $= 99.8875$
1010	F	$99.775 + 0.025$ $= 99.800$	$\dfrac{40 \times 20}{2 \times 4000} = 0.1000$	F′	$99.800 + 0.100$ $= 99.900$
1020	G	$99.800 + 0.025$ $= 99.825$	$\dfrac{50 \times 10}{2 \times 4000} = 0.0625$	G′	$99.825 + 0.0625$ $= 99.8875$
1030	B	$99.825 + 0.025$ $= 99.850$	0.0000	B	99.850

13.10 REALIGNMENT OF CURVES

A rail curve is likely to get distorted from its original alignment with the passage of time due to the following reasons:

(a) Unbalanced loading on both the inner and outer rails due to cant excess at slower speeds or cant deficiency at higher speeds instead of the equilibrium speed for which the cant has been provided.

(b) Effect of large horizontal forces exerted on the rails by passing trains. These forces tend to make a curve flatter at certain locations and sharper at others and the radius of the curve thus varies from place to place. These result in a rough ride on the curve due to the change in the radial acceleration from place to place. Realignment of the curve, therefore, becomes necessary to restore the smooth running of vehicles on these curves.

13.10.1 Criteria for Realignment of Curves

The *Indian Railway Way Manual* and the *Indian Railway Works Manual* had earlier prescribed that a curve should be realigned when, during an inspection, the running on a curve is found to be unsatisfactory. No hard and fast rule was laid as to when a curve should be realigned. Subsequently, the Railway Board prescribed the following criteria for the realignment of a curve.

Cumulative frequency diagram

For group A and B routes, the need for curve realignment should be determined by drawing a cumulative frequency diagram showing the variation of the field versine over the theoretical versine. The versine variations measured on a 20-m chord should be limited to 4 mm and 5 mm for group A and B lines, respectively.

Realignment should be taken up when the cumulative percentage of the versines lying within these limits is less than 80.

Station-to-station versine difference

The type of ride over a curve depends not only on the difference between the actual and the proposed versine, but also on the station-to-station variation of the actual versine values. The station-to-station variation of versine determines the rate of change of radial acceleration, on which the comfort of the ride would depend. The following stipulations have been made regarding the different gauges adopted on the Indian Railways.

Broad gauge   On curves where speeds in excess of 100 km per hour are permitted, the station-to-station variation of versines at stations 10 m apart should not exceed 15 mm, and for speeds of 100 km per hour and less, these variations should not exceed 20 mm or 20 per cent of the average versine of the circular portion, whichever is more.

Metre gauge   On curves which permit speeds in excess of 75 km per hour, the station-to-station variation of the versine at stations 10 m apart should not exceed 15 mm. For speeds of 75 km per hour and less, such variations should not exceed 20 mm or 20 per cent of the average versine of the circular portion, whichever is more.

The decision to completely realign a curve should be taken after ascertaining the type of ride the curve provides, on the basis of the cumulative frequency diagram or the distribution of the variation of versines between stations as described here.

Curve realignment can also be taken up under the following circumstances.

Unsatisfactory running of track

For other routes, curve realignment should be taken up when a curve is found to be unsatisfactory as a result of inspection done by trolley, from the footplate of the locomotive, by rear carriage, or as a result of various track tests that may have been carried out.

Local adjustment

When there is an abrupt variation of versines between adjacent stations, local adjustments should be done to achieve a versine variation, which is within reasonable limits. Such corrections should be carried out before complete curve realignment is taken up.

13.10.2 String Lining Method of Realignment of Curves

The realignment of existing curves using a theodolite is difficult and laborious work. Therefore, curves are realigned by measuring the versines with the help of a string and then correcting these versines. This method is known as the string lining method on Indian Railways. It is based on the following basic principles:

 (a) The sum of all versines taken on equal chords of any two curves between the same tangents are equal. It follows that the final value of the sum of the differences between the existing and proposed versine must be zero.
 (b) The throw at any station is equal to twice the second summation of the differences of the proposed and existing versine up to the previous station.

Procedure

Realigning a curve using the string lining method consists of the following three operations:

(i) Survey of the existing curve for measurement of versines

(ii) Computation of slews, including provision of proper transition and superelevation for the revised alignment

(iii) Slewing of the curve to the revised alignment

Survey of existing curves

Existing curves are surveyed as follows

(i) Versine readings are taken on the gauge face of the outer rail of the curve at 10-m intervals, using 20 m chords.

(ii) Versine readings are taken with the help of a nylon fishing cord. The cord is kept tight and at a preferred distance of 20 mm away from the gauge face side of the outer rail, with the help of a special gadget.

(iii) Versine readings should be taken for at least six stations beyond the apparent tangent point.

Computation of slews

Slews are computed as follows:

1. The length of the transition curve is determined based on the permissible speed and degree of curvature as per standard practice. The versine gradient, i.e., the rate of change of versine per unit length, is then calculated once the length of the transition curve and the theoretical versine proposed to be adopted are known. After calculating the versine gradient, the versines proposed to be adopted for the transition length can be easily computed.

2. The theoretical ideal versine proposed to be adopted for the transition length is calculated either exactly by detailed mathematical calculations or approximately by geometrical methods. In the geometrical method, the versines are plotted on a graph with respect to the number of stations and an average figure of versines is estimated by drawing a mean line in between the peaks and depressions of the graph as shown in Fig. 13.20.

3. A tabular statement is then prepared as shown in Table 13.11. In this table, the station numbers and existing versines (VE) are given in columns 1 and 2.

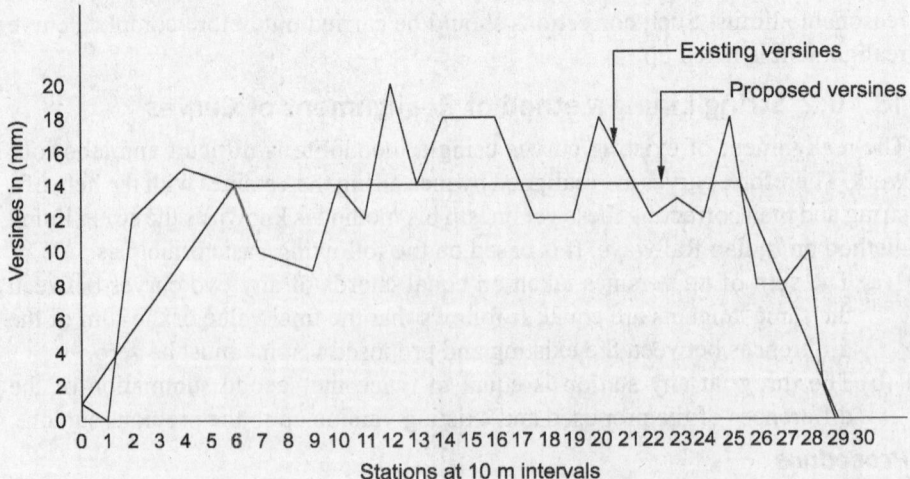

Fig. 13.20 Graphical solution to find average versines

Table 13.11 Realignment of curve based on string lining method

Station	VE	VP	VP – VE	S (VP – VE)	SS (VP – VE)	CC	S (CC)	SS (CC)	Half slew [SS (VP – VE) + SS (CC)]	Full slew (2 × col. 10)	Resultant slew (VP + CC)	SE
0	1	1	0	0			0	0				0
1	0	3	+3	+3	0		0	0	0	0	3	4
2	6	6	0	+3	+3		0	0	+3	+6	6	8
3	15	9	−6	−3	+6		0	0	+6	+12	9	12
4	14	12	−2	−5	+3		0	0	+3	+6	12	16
5	15	15	0	−5	−2		0	0	−2	−4	15	20
6	9	15	+6	+1	−7		0	0	−7	−14	15	20
7	9	15	+6	+7	−6		0	0	−6	−12	15	20
8	17	15	−2	+5	+1	+1	+1	0	+1	+2	16	20
9	13	15	+2	+7	+6	+1	+2	+1	+7	+14	16	20
10	21	15	−6	+1	+13		+2	+3	+16	+32	15	20
11	18	15	−3	−2	+14		+2	+5	+19	+38	15	20
12	15	15	0	−2	+14		+2	+9	+19	+38	15	20
13	18	15	−3	−5	+10		+2	+9	+19	+38	15	20
14	21	15	−6	−11	+5		+2	+11	+16	+32	15	20
15	9	15	+6	−5	−6		+2	+13	+7	+14	15	20
16	13	15	+2	−3	−11		+2	+15	+4	+8	15	20
17	13	15	+2	−1	−14		+2	+17	+3	+6	15	20

Contd.

Table 13.11 (Contd.)

Station	VE	VP	VP – VE	S (VP – VE)	SS (VP – VE)	Correcting couple (CC)			Half slew [SS (VP – VE) + SS (CC)]	Full slew (2 × col. 10)	Resultant slew (VP + CC)	SE
						CC	S (CC)	SS (CC)				
18	13	15	+2	+1	−15		+2	+19	+4	+8	15	20
19	16	15	−1	0	−14		+2	+21	+7	+14	15	20
20	21	15	−6	−6	−14		+2	+23	+9	+18	15	20
21	14	15	+1	−5	−20		+2	+25	+5	+10	15	20
22	11	15	+4	−1	−25	−1	+1	+27	+2	+4	14	20
23	14	15	+1	0	−26	−1	0	+28	+2	+4	14	20
24	12	15	+3	+3	−26		0	+28	+2	+4	15	20
25	21	15	−6	−3	−23		0	+28	+5	15	15	20
26	9	12	+3	0	−26		0	+28	+2	+4	12	16
27	9	12	+3	0	−26		0	+28	+2	4	9	12
28	7	6	−1	−1	−26		0	+28	+2	+4	6	8
29	3	3	0	−1	−27		0	+28	+1	+2	3	4
30	0	1	+1	0	−28		0	+28	0	0	1	6

All dimensions are in millimetres. Positive slews indicate inward slews and negative slews indicate outward slews.

The versines proposed to be adopted (VP) for the circular curve as well as for the transition length, as calculated before, are given in column three. It should be ensured that the sums of the proposed and existing versines are equal.

4. Versine differences between the existing and proposed versines (VP – VE) are then calculated for each station and given in column 6, shifting it by half the station.

5. The first summation of the versine difference is S(VP – VE) written in column 5 and in between, shifting is done by half the station, i.e., nth row of column $5 + (n + 1)$th row of column $4 = (n + 1)$th row of column 5—this must be zero in the end.

6. The second summation of versines SS(VP – VE) is calculated by adding to it the first summation of versine difference. This is written in column 6, shifting it by half the station again. It should be ensured that the second summation of the difference of versines, which is also equal to half the slew, is zero at the first and last stations and at obligatory points, if any. If this condition is not satisfied, correcting couples are applied as described next.

7. The correcting couples (CC) are applied (column 7) by changing the proposed versines in such a way that this brings down the second summation of the difference of versines at one place and makes the second summation negative at another place, keeping a proper distance, i.e., the correct number of stations, in between.

   The first and second summations of the correcting couples, S(CC) and SS(CC), are done in columns 8 and 9, respectively. The correcting couple is applied depending upon the value of the second summation of versine difference derived against the last station, so that the final value, after adding the effect of the couple, becomes zero. Similarly, correcting couples are applied to control the slews at obligatory points. Otherwise too, slews must be limited to the minimum possible values in the entire curve.

8. The resultant half slews (column 6 + column 9) and full slews (2 × columns 10) are shown in columns 10 and 11. The final versines (column 3 + column 7) to be adopted are written in column 12.

9. The value of cant to be provided, rate of introduction of cant, and points of zero and maximum cant are also calculated. These are shown in column 13.

Slewing curve to the new alignment

The following points should be kept in mind when a curve is slewed to the new alignment.

(a) A positive slew indicates an inward slew and a negative slew indicates an outward slew.

(b) The curve should be slewed to an accuracy of ±2 mm. After the realignment, the versine of the new curve should be measured for uniformity.

(c) The necessary superelevation, which is already determined, should be given to the curve. The superelevation is zero at the tangent point.

(d) Curve indication posts should be fixed at important locations for checks to be carried out during maintenance, if required.

Use of computers for calculation of slew

As the normal method for calculating slew is slow and tedious, computerizing the same would be well appreciated. Eastern Railway has already developed two programs on the IBM 1401 computer, one for calculating slews for simple curves and the other for calculating the same for compound curves with obligatory points so as to obtain solutions for curve realignment. The time taken in computing a realignment solutions for a curve with 150 stations is about 1.5 minutes. Besides saving time, computerized calculations provide much better precision.

13.10.3 Curve Correctors Cum Recorders

Indian Railways has procured about 50 curve correctors, which continuously record the versines of curves and help in improving their alignments. The utility of this equipment increases if it can also be used for measuring gauge variations and unevenness, and this can be done easily by making suitable modifications to it. Accordingly, RDSO has developed special attachments to be used with existing curve correctors for recording the unevenness and gauge parameters, thus converting the existing curve correctors into track recorder cum curve correctors. This new equipment measures the following:

(a) Alignment over a 10 m chord
(b) Unevenness over a 10 m chord
(c) Sleeper-to-sleeper gauge variation

13.10.4 Realignment of Curves on Double or Multiple Lines

On double or multiple tracks, each curve should be string lined independently. No attempt should be made to realign a curve by slewing it to a uniform centre-to-centre distance from another realigned curve due to the following reasons:

(a) The existing track centres may not be uniform, and a relatively small throw on one track may entail a much larger (even prohibitively larger) throw on the adjacent track.
(b) It is nearly impossible to measure the centre-to-centre distance of curved tracks along the true radial line, and a small error in the angular direction of measurement would mean an appreciable error in the true radial distance.
(c) The transition lengths at the entry and exit may measure differently, which make it impracticable to maintain uniform centres on them, even though the degree of the circular curves is nearly the same.

13.11 CUTTING RAILS ON CURVES

Rails on curves are usually laid with square joints. The inner rail gradually gains the lead over the outer rail on a curved track. The excess length D, which the inner rail gains over the outer rail for a length L of the circular curve is calculated as follows (Fig. 13.21).

(i) Difference in circumference of outer rail and inner rail (i.e., gain):
$$2\pi R - 2\pi R (R - G) = 2\pi G$$

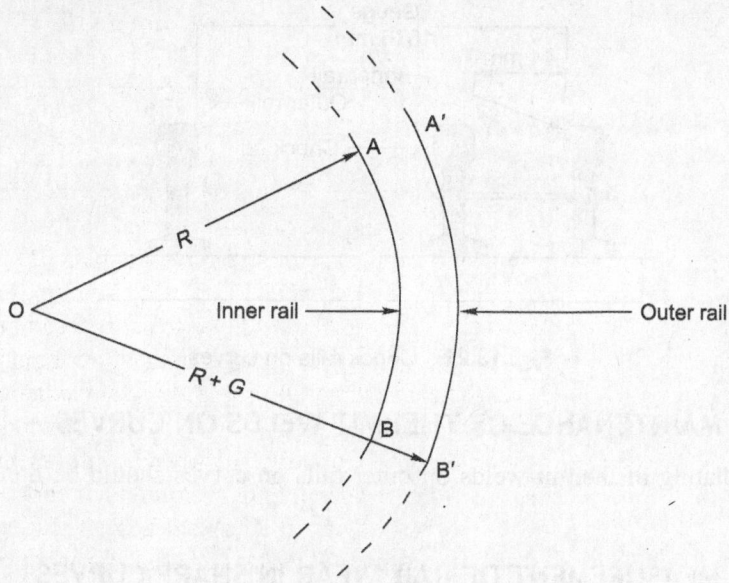

Fig. 13.21 Cutting rails on curves

(ii) Gain for length $2\pi R = 2\pi G$

$$\text{Gain for length L} \quad = \frac{2\pi G}{2\pi R} \times L = \frac{GL}{R}$$

$$\frac{DL}{1000} \text{ for BG and } \frac{DL}{1654} \text{ for MG}$$

Here, L is the length of outer rail of the circular curve, R is the radius of outer rail of the circular curve, D is the degree of curvature of outer rail, and G is the dynamic gauge (i.e., gauge + width of rail head), which is 1750 mm for BG and 1058 mm for MG.

Normally, when the inner rail of the curve leads over the outer rail by an amount equal to half the pitch of the fish bolt holes, the inner rail is cut by an amount equal to one full pitch and another hole is drilled for fastening the joint with a fish plate. The number of rails to be cut for a particular curve is worked out based upon the degree and length of the curve and the pitch of the bolt holes.

13.12 CHECK RAILS ON CURVES

Check rails (Fig. 13.22) are provided parallel to the inner rail on sharp curves to reduce the lateral wear on the outer rail. They also prevent the outer wheel flange from mounting the outer rail and thus decrease the chances of derailment of vehicles. Check rails wear out quite fast but since, normally, these are worn-out rails, further wear is not considered objectionable.

According to the stipulations presently laid down on Indian Railways, check rails are provided on the gauge face side of the inner rails on curves sharper than 8° on BG, 10° on MG, and 14° on NG routes. The minimum clearance prescribed for check rails is 44 mm for BG and MG routes and 41 mm for NG routes.

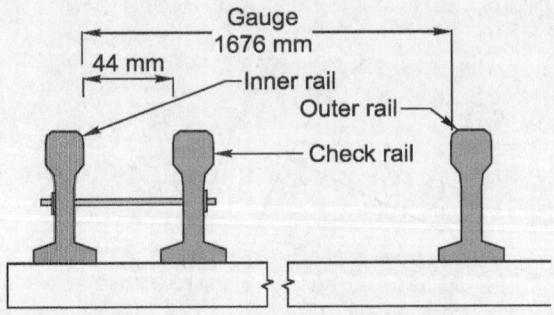

Fig. 13.22   Check rails on curves

13.13 MAINTENANCE OF THERMIT WELDS ON CURVES

Joggle plating of thermit welds on outer rails on curves should be done using clamps.

13.14 MEASUREMENT OF RAIL WEAR IN SHARP CURVES

The wear of rails of curves having radius of 600 m or less on BG and 300 m or less on MG shall be periodically recorded. (Periodicity to be prescribed by Railways.) The lateral, vertical, and total loss of section are to be recorded. Proper record of measurements should be maintained.

Curves for high-speed trains

Flat curves are generally adopted on high-speed tracks. Flat curves become necessary in view of restriction on maximum values of cant deficiency and cant excess along with maximum speed of operation. The minimum radius of curvature for the high-speed lines generally varies from 4000 m to 7000 m for standard gauge.

Note: Full details of curves for high-speed trains can be seen in Chapter 32.

SUMMARY

The design of the horizontal and vertical curves is extremely important for railway tracks. Both the speed and safety of the vehicles greatly depend on the design of the horizontal curves. Important stipulations laid down on Indian Railways with respect to the design of horizontal curves have been highlighted in this chapter. The methods of calculating superelevation, lengths of transition curves, and extra clearance on horizontal curves have also been discussed.

REVIEW QUESTIONS

1. What is superelevation? Why is it necessary to provide superelevation on the curves of a railway track?
2. A 8° curve track diverges from a main curve of 5° in the opposite direction in the layout of a BG yard, calculate the superelevation and the speed on the branch line when the maximum speed permitted on the main line is 45 kmph.

3. Explain the objective of providing transition curves on either side of a circular curve.

4. Why is the widening of gauge required on sharp curves? Determine the extent of gauge widening required for a BG track on a 5° curve, given the following data: $B = 6$ cm is the rigid wheel base, $D = 1.5$ m is the diameter of wheel, and $h = 3.2$ cm is the depth of flange below rail top.

5. For a main line and a branch line on 5° curves, calculate the superelevation and the speed on the branch line if the maximum speed permitted on the main line is 48 kmph.

6. Define the terms equilibrium cant and cant deficiency on a railway track. Calculate cant deficiency for a 4° curve on a BG track.

7. A 5° curve diverges from a 3° main curve in the reverse direction in the layout of a BG yard. If the speed on the branch line is restricted to 35 kmph, determine the restricted speed on the main line.

8. Calculate the maximum permissible speed on a curve on a Rajdhani route with a maximum sanctioned speed of 130 kmph. The superelevation provided is 50 mm and the transition length is 60 m. Also, the transition length of the curve cannot be increased due to proximity of the yard.

9. Calculate the superelevation to be provided for a 1.5°. transitioned curve on a high-speed route. The maximum sanctioned speed in the section is 120 kmph. The speed for calculating the superelevation is set at 88 kmph. What is the length of the transition to be provided?

10. Explain the following terms and state the circumstances under which they occur:
    (a) Negative superelevation
    (b) Grade compensation on curves
    (c) Realignment of curves

11. Clearly explain the string lining method to realign the original curve.

12. Explain 'degree of a simple rail track curve' and derive a simple expression for the radius of a curve in terms of its degree.

13. A transition curve is to be laid to join the ends of a 3° circular curve with straight track. The length of the transition curve is 100 m. Work out the shift and the offset at every 20 m interval for setting the transition curve. Describe how the transition curve can be set.

14. Define equilibrium speed and cant deficiency. A-1 in-8.5 BG turnout takes off from the outside of a 2° curve. Allowing for a speed of 20 mph (32 kmph) on the turnout and 3 inch. (7.5 cm) cant deficiency, both for the line and the turnout, find the maximum permissible speed for the main line.

15. Work out the complete data needed for setting out both the circular and transition portions of a 3° curve for a BG track. The deflection angle is 20°–30°. Cant gradient should be taken as 1 in 720. Maximum speed is 100 kmph. Describe the process of laying out the curve on ground with the data that has been worked out.

16. What would be the equilibrium cant on a curved MG track with a 7° curve for an average speed of 50 kmph? Also calculate the maximum permissible

speed after allowing a maximum cant deficiency of 5 cm. The formula given by Railway Board is

$$V = 4.35 \sqrt{R - 67}$$

17. The design speed on a BG track for a circular curve with transitions is 96 kmph. Determine the length of transition. Assume suitable values for the other data required to solve the problem.

18. What are the elements to be considered for the geometric design of a railway track? A BG branch line tracks off as a contrary flexure through a 1-in-12 turnout from the main line of a 2° curve. The maximum permissible speed on the branch line is 40 kmph. Calculate the negative superelevation to be provided on the branch line and the maximum permissible speed on the main line track.

19. Why are vertical curves provided? Calculate the length of vertical curve between two gradients, one rising at a rate of 1 in 200 and the other falling at a rate of 1 in 400.

20. What are the extra clearances provided on a curved track? Describe the extra clearances provided between two adjacent curved tracks.

21. Establish a relationship between the radius and versine of a curve. How is the degree of a curve determined in the field?

22. Enumerate the various methods of setting out a circular curve. Describe the tangential offset method for laying a circular curve.

23. What do you understand by equilibrium superelevation? Describe the thumb rules for calculating superelevation in the field.

Choose the correct answer from the choices given.

24. Versine in cm on a chord of length ……. is equal to degree of the curve:
    - (a) 9.2 m
    - (b) 10.5 m
    - (c) 11.8 m
    - (d) 12.7 m

25. Versine in inches on a chord of length …….. is equal to degree of the curve.
    - (a) 35 ft
    - (b) 47 ft.
    - (c) 55 ft
    - (d) 62 ft

26. Maximum degree of curve permitted for BC plain track and on turnout side is:
    - (a) 8° and 6°
    - (b) 10° and 8°
    - (c) 10° and 7°
    - (d) 10° and 10°

27. The cant requirement for 2° BG curve with an equilibrium speed of 90 kmph will be:
    - (a) 65 mm
    - (b) 95 mm
    - (c) 120 mm
    - (d) 127 mm

28. Degree of a curve of radius 875 m is:
    - (a) 1°
    - (b) 1.5°
    - (c) 2°
    - (d) 3°

29. The maximum cant deficiency prescribed on BG is:
    - (a) 40 mm
    - (b) 59 mm
    - (c) 75 mm
    - (d) 100 mm

30. The maximum allowable superelevation on BG in India is:
    (a) 76 mm   (b) 102 mm
    (c) 124 mm   (d) 165 mm

31. The radius of curvature of a 1° curve equals to about:
    (a) 1600 m   (b) 1000 m
    (c) 1750 m   (d) 850 m

32. The degree of curvature allowed is more in:
    (a) BG   (b) NG
    (c) MG   (d) equal in all above

33. The degree of curvature, normally, does not exceed 10° in:
    (a) NG   (b) MG
    (c) BG   (d) All of the above

34. Length of vertical curve on 'A' route of BG and algebraic difference G 1.5% between the two grades:
    (a) 30 m   (b) 45 m
    (c) 60 m   (d) 75 m

14

Points and Crossings

INTRODUCTION

Points and crossings are provided to help transfer railway vehicles from one track to another. The tracks may be parallel to, diverging from, or converging with each other. Points and crossings are necessary because the wheels of railway vehicles are provided with inside flanges and, therefore, they require this special arrangement in order to navigate their way on the rails. The points or switches aid in diverting the vehicles and the crossings provide gaps in the rails so as to help the flanged wheels to roll over them. A complete set of points and crossings, along with lead rails, is called a *turnout*.

14.1 IMPORTANT TERMS

The following terms are often used in the design of points and crossings.

Turnout   It is an arrangement of points and crossings with lead rails by means of which the rolling stock may be diverted from one track to another. Figure 14.1(a) shows the various constituents of a turnout. The details of these constituents are given in Table 14.1.

Table 14.1   Parts of a turnout

Name of the main assembly	Various constituents of the assembly
Set of switches (Figs 14.1 and 14.2)	A pair of stock rails, a pair of tongue rails, a pair of heel blocks, several slide chairs, two or more stretcher bars, and a gauge tie plate
Crossing	A nose consisting of a point rail and splice rails, two wing rails, and two check rails
Lead rails (Fig. 14.1)	Four sets of lead rails

Direction of a turnout   A turnout is designated as a right-hand or a left-hand turnout depending on whether it diverts the traffic to the right or to the left. In Fig. 14.1(a), the turnout is a right-hand turnout because it diverts the traffic towards the right side. Figure 14.1(b) shows a left-hand turnout. The direction of a point (or turnout) is known as the *facing direction* if a vehicle approaching the turnout or a point has to first face the thin end of the switch. The direction is *trailing direction* if the vehicle has to negotiate a switch in the trailing direction, that is, the vehicle first negotiates the crossing and then finally traverses on the switch from its thick end to its thin end. Therefore, when standing at the toe of a switch, if one looks in the direction of the crossing, it is called the *facing direction* and the opposite direction is called the *trailing direction*.

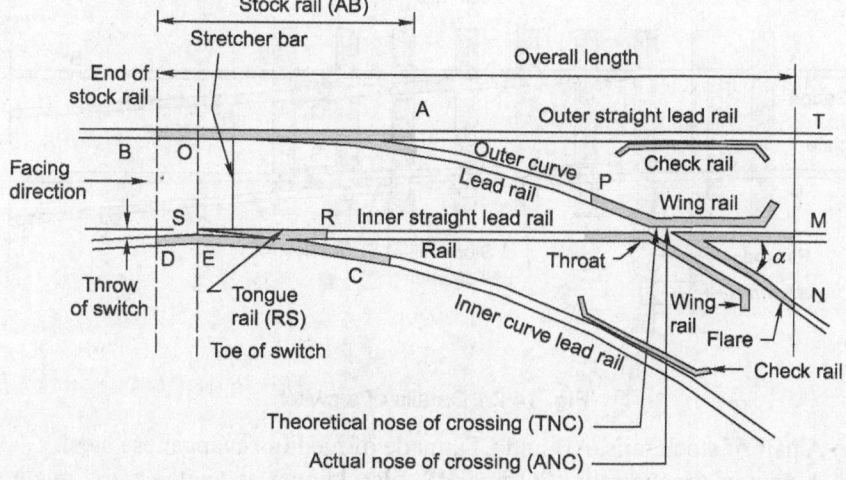

Fig. 14.1 (a)   Constituents of a turnout

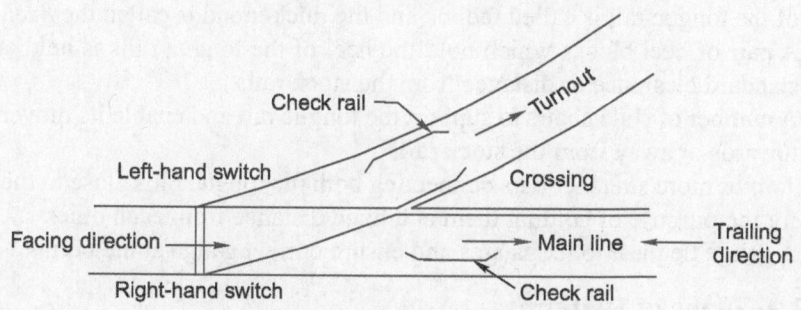

Fig. 14.1 (b)   Left-hand turnout

Tongue rail   It is a tapered movable rail, made of high-carbon or -manganese steel to withstand wear. At its thicker end, it is attached to a running rail. A tongue rail is also called a *switch rail*.

Stock rail   It is the running rail against which a tongue rail operates.

Points or switch   A pair of tongue and stock rails with the necessary connections and fittings forms a switch.

Crossing It is a device introduced at the junction where two rails cross each other to permit the wheel flange of a railway vehicle to pass from one track to another.

14.2 SWITCHES

A set of points or switches consists of the following main constituents (Fig. 14.2).

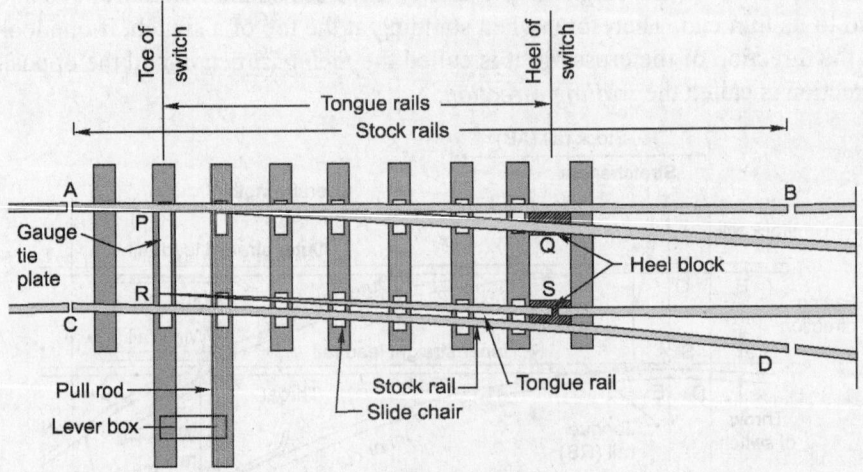

Fig. 14.2 Details of a switch

(a) A pair of stock rails, AB and CD, made of medium-manganese steel.

(b) A pair of tongue rails, PQ and RS, also known as switch rails, made of medium-manganese steel to withstand wear. The tongue rails are machined to a very thin section to obtain a snug fit with the stock rail. The tapered end of the tongue rail is called the *toe* and the thicker end is called the *heel*.

(c) A pair of heel blocks which hold the heel of the tongue rails is held at the standard clearance or distance from the stock rails.

(d) A number of slide chairs to support the tongue rail and enable its movement towards or away from the stock rail.

(e) Two or more stretcher bars connecting both the tongue rails close to the toe, for the purpose of holding them at a fixed distance from each other.

(f) A gauge tie plate to fix gauges and ensure correct gauge at the points.

14.2.1 Types of Switches

Switches are of two types, namely *stud switch* and *split switch*. In a stud type of switch, no separate tongue rail is provided and some portion of the track is moved from one side to the other side. Stud switches are no more in use on Indian Railways. They have been replaced by split switches. These consist of a pair of stock rails and a pair of tongue rails. Split switches may also be of two types—loose heel type and fixed heel type. These are discussed below.

Loose heel type In this type of split switch, the switch or tongue rail finishes at the heel of the switch to enable movement of the free end of the tongue rail. The fish plates holding the tongue rail may be straight or slightly bent. The tongue rail is fastened to the stock rail with the help of a fishing fit block and four bolts. All the fish bolts in the lead rail are tightened while those in the tongue rail are kept loose or snug to allow free movement of the tongue. As the discontinuity of the track at the heel is a weakness in the structure, the use of these switches is not preferred.

Fixed heel type   In this type of split switch, the tongue rail does not end at the heel of the switch, but extends further and is rigidly connected. The movement at the toe of the switch is made possible on account of the flexibility of the tongue rail.

Toe of switches

The toe of the switches may be of the following types.

Undercut switch   In this switch the foot of the stock rail is planed to accommodate the tongue rail (Fig. 14.3).

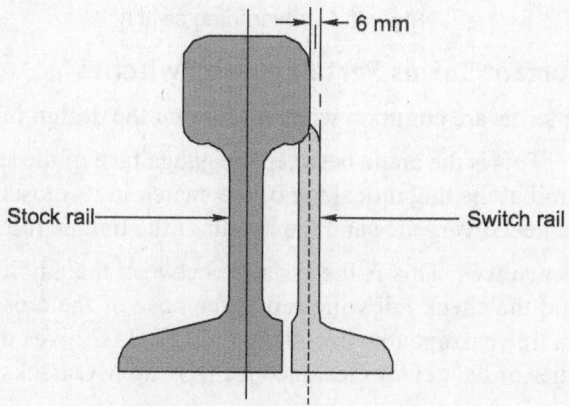

Fig. 14.3   Undercut switch

Overriding switch   In this case, the stock rail occupies the full section and the tongue rail is planed to a 6 mm (0.25")-thick edge, which overrides the foot of the stock rail (Fig. 14.4). The switch rail is kept 6 mm (0.25") higher than the stock rail from the heel to the point towards the toe where the planing starts. This is done to eliminate the possibility of splitting caused by any false flange moving in the trailing direction. This design is considered to be an economical and superior design due to the following reasons:

(a) Since the stock rail is uncut, it is much stronger.
(b) Manufacturing work is confined only to the tongue rail, which is very economical.
(c) Although the tongue rail has a thin edge of only 6 mm (0.25"), it is supported by the stock rail for the entire weakened portion of its length. As such, the combined strength of the rails between the sleepers is greater than that of the tongue rail alone in the undercut switch.

Overriding switches have been standardized on Indian Railways.

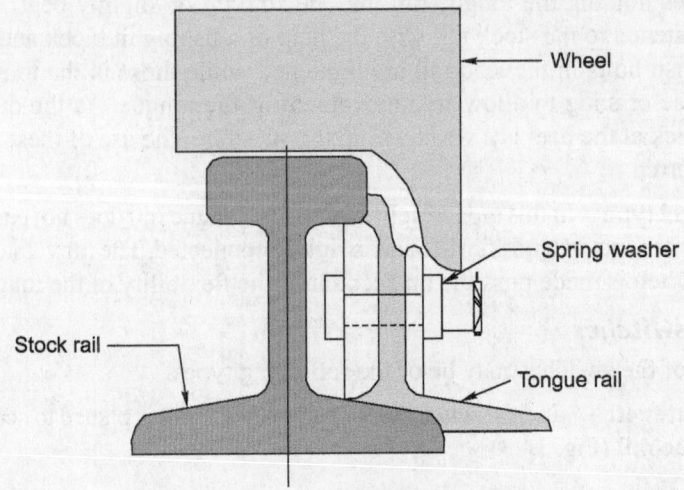

Fig. 14.4 Overriding switch

14.2.2 Important Terms Pertaining to Switches

The following terms are common when discussing the design of switches.

Switch angle This is the angle between the gauge face of the stock rail and that of the tongue rail at the theoretical toe of the switch in its closed position. It is a function of the heel divergence and the length of the tongue rail.

Flangeway clearance This is the distance between the adjoining faces of the running rail and the check rail/wing rail at the nose of the crossing. It is meant for providing a free passage to wheel flanges. Table 14.2 gives the minimum and maximum values of flangeway clearance for BG and MG tracks.

Table 14.2 Flangeway clearance

Gauge	Flangeway clearance	
	Maximum value (mm)	Minimum value (mm)
BG	48	44
MG	44	41

Heel divergence This is the distance between the gauge faces of the stock rail and the tongue rail at the heel of the switch. It is made up of the flangeway clearance and the width of the tongue rail head that lies at the heel.

Throw of the switch This is the distance through which the tongue rail moves laterally at the toe of the switch to allow movement of the trains. Its limiting values are 95–115 mm for BG routes and 89–100 mm for MG routes.

14.3 DESIGN OF TONGUE RAILS

A tongue rail may be either straight or curved. Straight tongue rails have the advantage that they are easily manufactured and can be used for right-hand as well as left-hand turnouts. However, trains get jolted while negotiating with straight tongue rail turnouts because of the abrupt change in the alignment. Straight tongue rails are normally used for 1-in-8.5 and 1-in-12 turnouts on Indian Railways.

Curved tongue rails are shaped according to the curvature of the turnout from the toe to the heel of the switch. Curved tongue rails allow for smooth turning of trains, but can only be used for the specific curvature for which they are designed. Curved switches are normally used for 1-in-16 and 1-in-20 IRS (Indian Railway Standard) turnouts on Indian Railways. Recently Indian Railways has also started laying 1-in-8.5 and 1-in-12 turnouts with curved switches on important lines.

14.3.1 Length of Tongue Rails

The length of a tongue rail from heel to toe varies with the gauge and angle of the switch. The longer the length of the tongue rail, the smoother the entry to the switch because of the smaller angle the switch rail would make with the fixed heel divergence. The longer length of the tongue rail, however, occupies too much layout space in station yards where a number of turnouts have to be laid in limited space. The length of the tongue rail should be more than the rigid wheel base of a four-wheeled wagon to preclude the possibility of derailment in case the points move from their position when a train is running on the switch. Table 14.3 gives the standard lengths of switches (tongue rails) for BG and MG tracks.

Table 14.3 Length of tongue rail

Gauge and type	Length of tongue rail (mm)				
	1-in-8.5 straight	1-in-12 straight	1-in-12 curved	1-in-16 curved	1-in-20 curved
BG (90 R)	4725	6400	7730	9750	1,1150
MG (75 R)	4116*	5485*	6700		

* These dimensions hold good for NG tracks also.

14.4 CROSSING

A crossing or *frog* is a device introduced at the point where two gauge faces cross each other to permit the flanges of a railway vehicle to pass from one track to another (Fig. 14.5). To achieve this objective, a gap is provided from the throw to the nose of the crossing, over which the flanged wheel glides or jumps. In order to ensure that this flanged wheel negotiates the gap properly and does not strike the nose, the other wheel is guided with the help of check rails. A crossing consists of the following components, shown in Fig. 14.6.

(a) Two rails, *point rail* and *splice rail*, which are machined to form a nose. The point rail ends at the nose, whereas the splice rail joins it a little behind the nose. Theoretically, the point rail should end in a point and be made as thin as possible, but such a knife edge of the point rail would break off under the movement of traffic. The point rail, therefore, has its fine end slightly cut off to form a blunt nose, with a thickness of 6 mm (1/4"). The toe of the blunt nose is called the *actual nose of crossing* (ANC) and the theoretical point

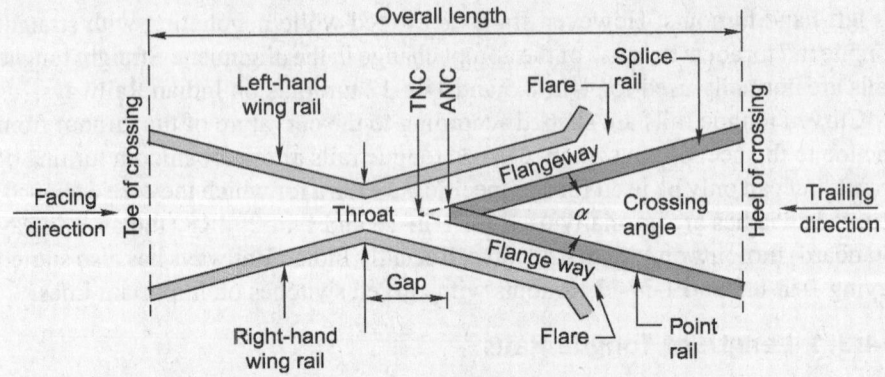

Fig. 14.5 Details of a crossing

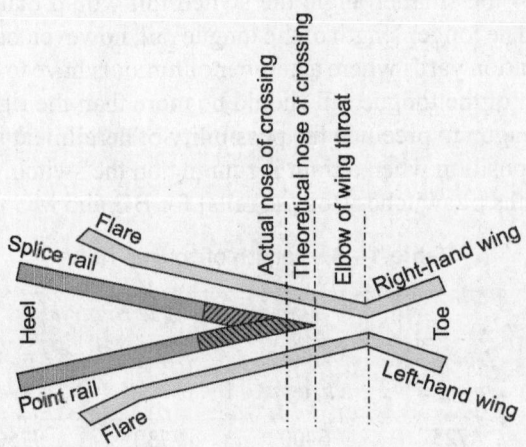

Fig. 14.6 Point rail and splice rail

where the gauge faces from both sides intersect is called the *theoretical nose of crossing* (TNC). The 'V' rail is planed to a depth of 6 mm (1/4") at the nose and runs out in 89 mm to stop a wheel running in the facing direction from hitting the nose.

(b) Two wing rails consisting of a right-hand and a left-hand wing rail that converge to form a throat and diverge again on either side of the nose. Wing rails are flared at the ends to facilitate the entry and exit of the flanged wheel in the gap.

(c) A pair of check rails to guide the wheel flanges and provide a path for them, thereby preventing them from moving sideways, which would otherwise may result in the wheel hitting the nose of the crossing as it moves in the facing direction.

14.4.1 Types of Crossings

A crossing may be of the following types.

(a) An *acute angle crossing* or 'V' crossing in which the intersection of the two gauge faces forms an acute angle. For example, when a right rail crosses a left rail, it makes an acute crossing. Thus, unlike rail crossings form an acute crossing.

(b) An *obtuse* or *diamond crossing* in which the two gauge faces meet at an obtuse angle. When a right or left rail crosses a similar rail, it makes an obtuse crossing.

(c) A *square crossing* in which two tracks cross at right angles. Such crossings are rarely used in actual practice (Fig. 14.7).

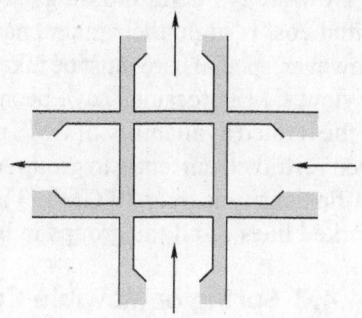

Fig. 14.7 Square crossing

For manufacturing purposes, crossings can also be classified as follows.

Built-up crossing In a built-up crossing, two wing rails and a V section consisting of splice and point rails are assembled together by means of bolts and distance blocks to form a crossing. This type of crossing is commonly used on Indian Railways. The advantage of such crossings is that their initial cost is low and that repairs can be carried out simply by welding or replacing each constituent separately. A crossing becomes unserviceable when wear is more than 10 mm (3/8"). A built-up crossing, however, lacks rigidity. The bolts require frequent checking and sometimes break under fast and heavy traffic.

Cast steel crossing This is a one-piece crossing with no bolts and, therefore, requiring very little maintenance. Comparatively, it is a more rigid crossing since it consists of one complete mass. The initial cost of such a crossing is, however, quite high and its repair and maintenance pose a number of problems. Recently, cast manganese steel (CMS) crossings, which have longer life, have also been adopted.

Combined rail and cast crossing This is a combination of a built-up and cast steel crossing and consists of a cast steel nose finished to ordinary rail faces to form the two legs of the crossing. Though it allows the welding of worn-out wing rails, the nose is still liable to fracture suddenly.

14.4.2 CMS Crossing

Due to increase in traffic and the use of heavier axle loads, the ordinary built-up crossings manufactured from medium-manganese rails are subjected to very heavy wear and tear, especially in fast lines and suburban sections with electric traction. Past experience has shown that the life of such crossings varies from six months to two years, depending on their location and the service conditions. CMS crossings possess higher strength, offer more resistance to wear, and consequently have a longer life. The following are the main advantages of CMS crossings:

(a) Less wear and tear.
(b) Longer life: The average life of a CMS crossing is about four times more than that of an ordinary built-up crossing.
(c) CMS crossings are free from bolts as well as other components that normally tend to get loose as a result of the movement of traffic.

Now-a-days CMS crossings are preferred on Indian Railways. Though their initial cost is high, their maintenance cost is relatively less and they last longer. However, special care must be taken in their laying and maintenance. Keeping this in view, CMS crossings have been standardized on Indian Railways. On account of the limited availability of CMS crossings in the country, their use has, however, been restricted currently to group A routes and those lines of other routes on which traffic density is over 20 GMT. These should also be reserved for use on heavily worked lines of all the groups in busy yards.

14.4.3 Spring or Movable Crossing

In a spring crossing, one wing rail is movable and is held against the V of the crossing with a strong helical spring while the other wing rail is kept fixed. When a vehicle passes on the main track, the movable wing rail is snug with the crossing and the vehicle does not need to negotiate any gap at the crossing. In case the vehicle has to pass over a turnout track, the movable wing is forced out by the wheel flanges and the vehicle has to negotiate a gap as in a normal turnout.

This type of crossing is useful when there is high-speed traffic on the main track and slow-speed traffic on the turnout track.

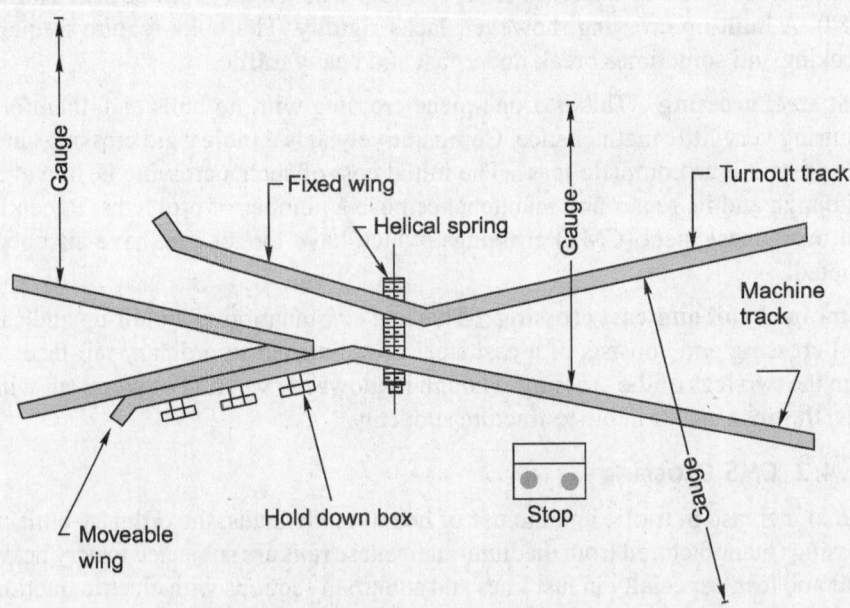

Fig. 14.8 Spring or movable crossing

14.4.4 Raised Check Rails for Obtuse Crossings

In order to provide a guided pathway in the throat portion of a 1-in-8.5 BG obtuse diamond crossing, the check rails are raised by welding a 25 mm thick MS plate. This arrangement is considered satisfactory for BG as well as MG routes.

14.4.5 Position of Sleepers at Points and Crossings

Sleepers are normally perpendicular to the track. At points and crossings, a situation arises where the sleepers have to cater to the main line as well as to the turnout portion of the track. For this purpose, longer sleepers are used for some length of the track as shown in Fig. 14.9.

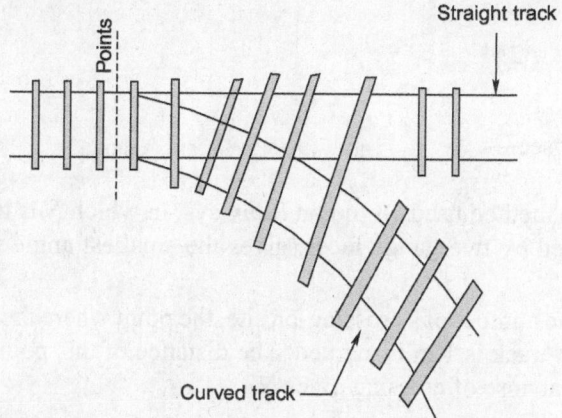

Fig. 14.9 Sleepers for points and crossings

14.5 NUMBER AND ANGLE OF CROSSING

A crossing is designated either by the angle the gauge faces make with each other or, more commonly, by the number of the crossing, represented by N. There are three methods of measuring the number of a crossing, and the value of N also depends upon the method adopted. All these methods are illustrated in Fig. 14.10.

Centre line method

This method is used in Britain and the US. In this method, N is measured along the centre line of the crossing.

$$\cot \frac{\alpha}{2} = N \div \frac{1}{2}$$

or

$$N = \frac{1}{2} \cot \frac{\alpha}{2}$$

Right angle method

This method is used on Indian Railways. In this method, N is measured along the base of a right-angled triangle. This method is also called *Coles method*.

$$\cot \alpha = \frac{N}{1}$$

or

$$N = \cot \alpha$$

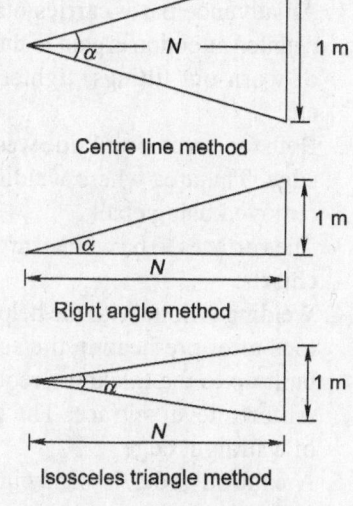

Centre line method

Right angle method

Isosceles triangle method

Fig. 14.10 Different methods of measuring number (N) and angle of crossing

Isosceles triangle method

In this method, N is taken as one of the equal sides of an isosceles triangle.

$$\sin\frac{\alpha}{2} = \frac{1/2}{N} = \frac{1}{2N}$$

or

$$\operatorname{cosec}\frac{\alpha}{2} = 2N$$

$$N = \frac{1}{2}\operatorname{cosec}\frac{\alpha}{2}$$

The right angle method used on Indian Railways, in which N is the cotangent of the angle formed by two gauge faces, gives the smallest angle for the same value of N.

To determine the number of a crossing-on site, the point where the offset gauge face of the turnout track is 1 m is marked. The distance of this point (in metres) from the theoretical nose of crossing gives N.

14.6 RECONDITIONING OF WORN-OUT CROSSINGS

Generally, noses of crossings and wing rails undergo the maximum amount of wear in a turnout. The limiting wear for a crossing is 10 mm, after which it is required to be replaced. A worn-out crossing is generally reconditioned at the stage when the wear is only 6 mm (1/4"). In the case of tongue rails, the limit of vertical wear for 52 kg and 90 R rails is 6 mm (1/4") and that of lateral wear is 8 mm. Similarly, the limit of vertical wear for 60 R and 75 R rails is 6 mm and that of lateral wear is 5 mm. Normally, gas welding is adopted to recondition crossings at the site itself. The sequence of operation is as follows:

1. An advance party carries out the preliminary work in which complete and detailed attention is paid to the turnout including through packing, replacement of worn-out fittings, tightening of fittings, squaring, spacing of sleepers, etc.
2. Both the vertical and side wear are measured with the help of an 1.8 m straight edge. The area where welding is to be done is cleaned, and burns, etc., are removed using chalk.
3. The surfaces to be welded are also cleaned, and burns, etc., are removed using chisels.
4. Welding is done with the help of an oxyacetylene flame using suitable welding rods after pre-heating the surface for about 5 minutes. When the section is built up to the thickness required, the deposit metal is hammered to make a uniform level surface. The prepared surface is then checked with the help of a straight edge.
5. A caution order is sent out while the work is in progress and no speed restriction is necessary.
6. One welding party consisting of one permanent way *mistry* (craftsman), two welders, and six *khalasis* (labourers) including lookout men can weld one

crossing or two pairs of switches every working day. The consumable items required for reconditioning work are listed in Table 14.4.

Table 14.4 Consumables required for reconditioning of crossings

Component	Requirement of oxygen (m^3)	Requirement of acetylene (m^3)	Requirement of welding rods (kg)
One crossing	5.7	6.5	1.60
One pair of switches	2.3	3.0	0.75

14.7 TURNOUTS

The simplest arrangement of points and crossing can be found on a turnout taking off from a straight track. There are two standard methods prevalent for designing a turnout. These are the (a) Coles method and (b) IRS method.

These methods are described in detail in the following sections.

The important terms used in describing the design of turnouts are defined as follows:

Curve lead (CL)   This is the distance from the tangent point (T) to the theoretical nose of crossing (TNC) measured along the length of the main track.

Switch lead (SL)   This is the distance from the tangent point (T) to the heel of the switch (TL) measured along the length of the main track.

Lead of crossing (L)   This is the distance measured along the length of the main track as follows:

Lead of crossing (L) = curve lead (CL) – switch lead (SL)

Gauge (G)   This is the gauge of the track.

Heel divergence (d)   This is the distance between the main line and the turnout side at the heel.

Angle of crossing (α)   This is the angle between the main line and the tangent of the turnout line.

Radius of turnout (R)   This is the radius of the turnout. It may be clarified that the radius of the turnout is equal to the radius of the centre line of the turnout (R_1) plus half the gauge width.

$$R = R_1 + 0.5G$$

As the radius of a curve is quite large, for practical purposes, R may be taken to be equal to R_1.

Special fittings with turnouts

Some of the special fittings required for use with turnouts are enumerated as follows:

Distance blocks   Special types of distance blocks with fishing fit surfaces are provided at the nose of the crossing to prevent any vertical movement between the wing rail and the nose of the crossing.

Flat bearing plates   As turnouts do not have any cant, flat bearing plates are provided under the sleepers.

Spherical washers These are special types of washers and consist of two pieces with a spherical point of contact between them. This permits the two surfaces to lie at any angle to each other. These washers are used for connecting two surfaces that are not parallel to one another. Normally, tapered washers are necessary for connecting such surfaces. Spherical washers can adjust to the uneven bearings of the head or nut of a bolt and so are used on all bolts in the heel and the distance blocks behind the heel on the left-hand side of the track.

Slide chairs These are provided under tongue rails to allow them to move laterally. These are different for ordinary switches and overriding switches.

Grade off chairs These are special chairs provided behind the heel of the switches to give a suitable ramp to the tongue rail, which is raised by 6 mm at the heel.

Gauge tie plates These are provided over the sleepers directly under the toe of the switches, and under the nose of the crossing to ensure proper gauge at these locations.

Stretcher bars These are provided to maintain the two tongue rails at an exact distance.

Coles method

This is a method used for designing a turnout taking off from a straight track (Fig. 14.11). The curvature begins from a point on the straight main track ahead of the toe of the switch at the theoretical toe of switch (TTS) and ends at the theoretical nose of crossing (TNC). The heel of the switch is located at the point where the offset of the curve is equal to the heel divergence. Theoretically, there would be no kinks in this layout, had the tongue rail been curved as also the wing rail up to the TNC. Since tongue rails and wing rails are not curved generally, there are the following three kinks in this layout.

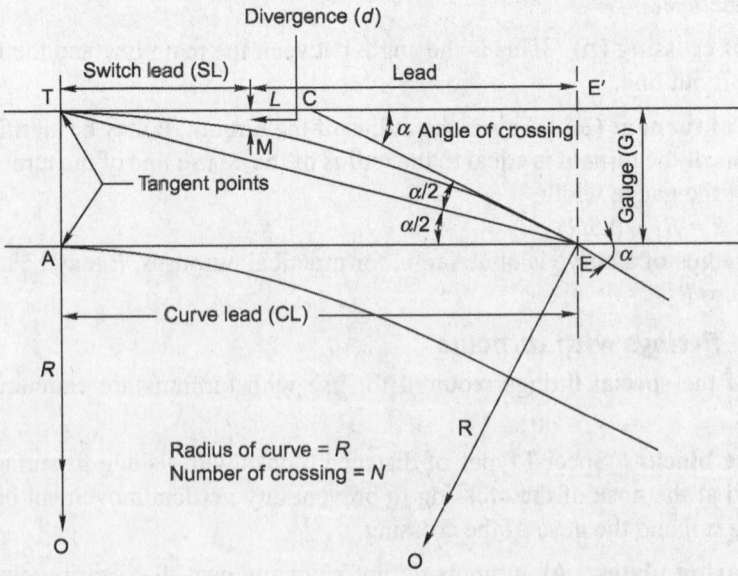

Fig. 14.11 Turnout from a straight track (Coles method)

(a) The first kink is formed at the actual toe of the switch.

(b) The second kink is formed at the heel of the switch.

(c) The third kink is formed at the first distance block of the crossing.

The notations used in Fig. 14.11 are the following.

Curve lead (CL) $= AE = TE'$

Switch lead (SL) $= TL$

Lead of crossing $(L) = LE'$

Gauge of track $(G) = AT = EE'$

Angle of the crossing $(\alpha) = \angle CEA = \angle ECE'$

Heel of divergence $(d) = LM$

Number of the crossing $(N) = \cot \alpha$

Radius of outer rail of turnout curve $(R) = OE = OT$

(O is the centre of the turnout curve)

Calculations

Curve lead (CL) In $\triangle ATE$,

$$AT = G \text{ and } \angle AET = \frac{\alpha}{2}$$

$$\tan\frac{\alpha}{2} = \frac{AT}{AE} = \frac{G}{\text{curve lead}}$$

or

$$\text{Curve lead} = G \cot\frac{\alpha}{2}$$

Also,

$$\text{Curve lead} = E'C + CT$$

$$= E'C + CE \text{ (as } CT = CE, \text{ tangents from a comman point)}$$

$$= G \cot \alpha + G \operatorname{cosec} \alpha$$

$$= GN + G\sqrt{1 + N^2} \text{ (as } \cot \alpha = N)$$

or

$$= 2GN \text{ (approximately)}$$

Switch lead (SL) TL is the length of the tangent with an offset $LM = d =$ heel divergence.

From the properties of triangles,

$$SL \times SL = d(2R - d)$$

or

$$\text{Switch lead} = \sqrt{2Rd - d^2}$$

Lead of crossing (L)

$$L = \text{curve lead} - \text{switch lead}$$

$$= G \cot\frac{\alpha}{2} - \sqrt{2Rd - d^2}$$

Radius of curve (R) In $\triangle AOE$,

$$OE = OT = R, OA = R - G$$

$$OE^2 = OA^2 + AE^2$$
$$OE^2 = (R - G)^2 + (\text{curve lead})^2$$

or

$$R^2 = (R - G)^2 + (GN + G\sqrt{1 + N^2})$$

$$= R^2 - 2RG + G^2 + G^2N + G^2(1 + N^2) + 2\,G^2N\sqrt{1 + N^2}$$

$$2RG = 2G^2(1 + N^2) + 2G^2N\sqrt{1 + N^2}$$

or

$$R = G(1 + N^2) + GN\sqrt{1 + N^2}$$

$$= 1.5G + 2GN^2 \ (\text{approximately})$$

Summarizing the formulae derived,

$$\text{Curve lead (CL)} \quad = G \cot \frac{\alpha}{2} \text{ or } 2GN \text{ approx.} \tag{14.1}$$

$$\text{Switch lead (SL)} \quad = \sqrt{2Rd - d^2} \tag{14.2}$$

$$\text{Lead of crossing } (L) = G \cot \frac{\alpha}{2} - \sqrt{2Rd - d^2}$$

$$= 2GN - \sqrt{2Rd - d^2} \tag{14.3}$$

$$\text{Radius of curve } (R) = 1.5G + 2GN^2 \tag{14.4}$$

$$\text{Heel divergence } (d) = \frac{(SL)^2}{2\left(R + \dfrac{G}{2}\right)} \tag{14.5}$$

Example 14.1 Calculate the lead and radius of a 1-in-8.5 BG turnout for 90 R rails using Coles method.

Solution

$$G = 1.676 \text{ m} \qquad d = 120 \text{ mm}$$
$$\alpha = 6° 42' 35'' \qquad N = 8.5$$

(i) Curve lead (CL) $= GN + G\sqrt{1 + N^2}$

$$= 1.676 \times 8.5 + 1.676 \sqrt{1 + 8.5^2}$$

$$= 28.6 \text{ m}$$

(ii) Radius of turnout curve $(R) = 1.5G + 2GN^2$

$$= 1.5 \times 1.676 + 2 \times 1.676 \times (8.5)^2$$

$$= 245 \text{ m}$$

(iii) Switch lead (SL) $= \sqrt{2Rd - d^2}$

$$= \sqrt{2 \times 245 \times 0.12 \ 0.12^2}$$

$$= 7.67 \text{ m}$$

(iv) Lead $= CL - SL = 28.6 - 7.7 = 20.9 \text{ m}$

IRS method

In this layout (Fig. 14.12), the curve begins from the heel of the switch and ends at the toe of the crossing, which is at the centre of the first distance block. The crossing is straight and no kink is experienced at this point. The only kink occurs at the toe of the switch. This is the standard layout used on Indian Railways. The calculations involved in this method are somewhat complicated and hence this method is used only when precision is required.

Lead of crossing (L)   In ΔBMH,

$$BM = MH \text{ (as both are tangents)}$$

$$\angle MHB = \angle MHB = \frac{\alpha - \beta}{2}$$

Therefore $\angle BHC = \dfrac{\alpha + \beta}{2}$

$$BC = AD - (AB + CD) = G - (d + h \sin\alpha)$$

Therefore, crossing lead

$$L = (G - d - h \sin \alpha) \cot \frac{\alpha + \beta}{2} + h \cos \alpha \tag{14.6}$$

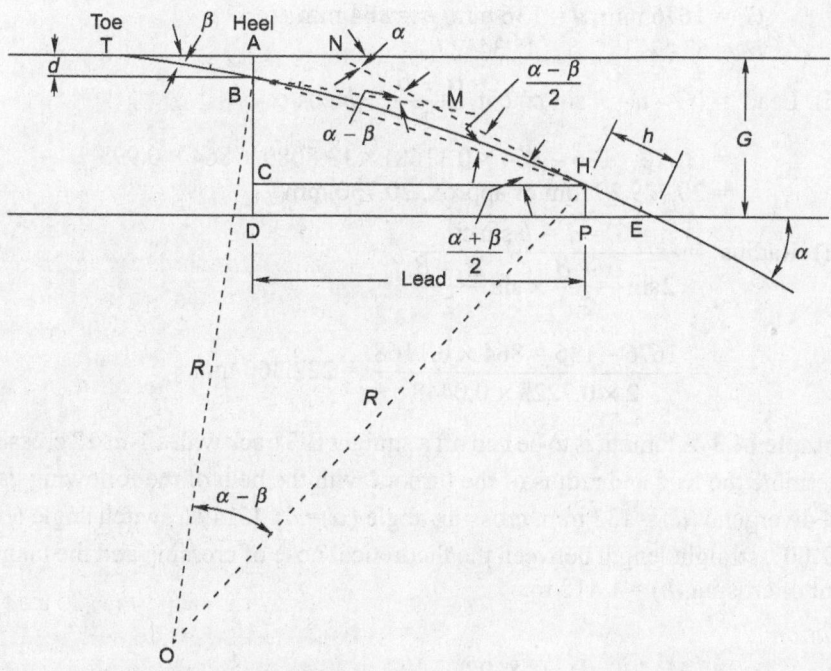

Fig. 14.12   Turnout from a straight track (IRS method)

Radius of curve (R) ΔOBH,

$$\angle BOH = \alpha - \beta$$

$$BH = 2 R \sin \frac{\alpha - \beta}{2} \tag{14.7}$$

In $\triangle$BHC,

$$BH = \frac{BC}{\sin\dfrac{\alpha + \beta}{2}} = \frac{G - d - h\sin\alpha}{\sin\dfrac{\alpha + \beta}{2}} \qquad (14.8)$$

Equating Eqs (14.7) and (14.8)

$$2R\sin\frac{\alpha - \beta}{2} = \frac{G - d - h\sin\alpha}{\sin\dfrac{\alpha + \beta}{2}}$$

or

$$R = \frac{G - d - h\sin\alpha}{2\sin\dfrac{\alpha + \beta}{2} \times \sin\dfrac{\alpha - \beta}{2}}$$

$$= \frac{G - d - h\sin\alpha}{\cos\beta - \cos\alpha} \qquad (14.9)$$

Example 14.2 Calculate the lead and radius of a 1-in-8.5 BG turnout with straight switches. Use the IRS method.

Solution

$$G = 1676 \text{ mm}, d = 136 \text{ mm}, h = 864 \text{ mm}$$
$$\alpha = 6° 42' 35'', \beta = 1° 34' 27''$$

(i) Lead $= (G - d - h\sin\alpha)\cot\dfrac{\alpha + \beta}{2} + h\cos\alpha$

$$= (1676 - 136 - 864 \times 0.1168) \times 13.8089 + 864 \times 0.993$$
$$= 20{,}729.89 \text{ mm or approx. } 20{,}730 \text{ mm}$$

(ii) Radius $= \dfrac{G - d - h\sin\alpha}{2\sin\dfrac{\alpha + \beta}{2} \times \sin\dfrac{\alpha - \beta}{2}}$

$$= \frac{1676 - 136 - 864 \times 0.1168}{2 \times 0.7223 \times 0.0448} = 222{,}360 \text{ mm}$$

Example 14.3 A turnout is to be laid off a straight BG track with a 1-in-12 crossing. Determine the lead and radius of the turnout with the help of the following data: heel divergenc $(d) = 133$ mm, crossing angle $(\alpha) = 4° 45' 49''$, switch angle $(\beta) = 1° 8' 00''$, straight length between the theoretical nose of crossing and the tangent point of crosing $(h) = 1.418$ m.

Solution

$$\alpha = 4° 45' 49'', \beta = 1° 8' 0''$$
$$G = 1.676 \text{ m}, d = 0.133 \text{ m}$$
$$N = 12, h = 1.418 \text{ m}$$

(i) Radius $R = \dfrac{G - d - h\sin\alpha}{\cos\beta - \cos\alpha}$

$$= \frac{1.676 - 0.133 - 1.418\sin 4° 45' 49''}{\cos 1° 8' 0'' - \cos 4° 45' 49''}$$

$$= 437.38 \text{ m}$$

(ii) Crossing lead $(L) = h \cos \alpha + (G - d - h \sin \alpha) \cot \dfrac{\alpha + \beta}{2}$

$$= 1.418 \cos 4° 45' 49'' + (1.676 - 0.133 - 1.418 \sin 4° 45' 49'')$$
$$\times \cot 2° 56' 54''$$
$$= 1.418 \times 0.9965 + 1.425 \times 19.415$$
$$= 29.084 \text{ m}$$

Standard turnouts and permissible speeds

On Indian Railways, normally 1-in-8.5 turnouts are used for goods trains while 1-in-12 and 1-in-16 turnouts are used for passenger trains. Recently 1-in-20 and 1-in-24 turnouts have also been designed by the RDSO, to be used to permit higher speeds for fast trains on the turnout side. The maximum speeds permitted on these turnouts are given in Table 14.5.

Table 14.5 Permissible speeds on turnouts

Gauge	Type of turnout	Switch angle	Permissible speed (kmph)
BG	1 in 8.5	1° 34' 27"	10* for straight switch and 15 for curved switch for 52/60 kg rails on PSC sleepers
BG[†]	1-in-8.5	Symmetrical split (SS) 0° 27' 35"	30 for curved switch as well as SS with 52/60 kg on PSC sleepers; 15* for curved switch for 52/60 kg on PSC sleepers[‡]
BG	1-in-16	1° 8' 0" 0° 24' 27"	50 or 60[¶]
MG	1-in-8.5	1° 35' 30" 0° 29' 14"	10 for straight as well as curved switch
MG[†]	1-in-12	1° 09' 38" 0° 24' 27"	15 for straight switch and 15 for partly curved switch
MG	1-in-16	0° 24' 27"	30

Source: Indian Railway Permanent Way Manual (IRPWM)—Correction slip no. 94 dated 1 June 2004.

* As per the Indian Railway way and works (IRWW) manual, a speed of 15 kmph was originally permitted on 1-in-8.5 turnouts. However, due to a subsequent number of derailments of passenger trains on turn-in curves, the speeds on these turnouts have now been reduced to 10 km/h only.

[†] The figures in the second row correspond to curved switches.

[‡] A speed of 30 kmph is also permitted on 1-in-12 turnouts on those interlocked sections where all turnouts over which a running train may pass are 1-in-12 throughout the section and the locomotives are fitted with speedometers. In all other cases, speed is restricted to 15 kmph only.

[¶] 60 kmph permitted only for high-speed turnouts to Drg. No. RDST/T-403.

14.8 TURNOUT WITH CURVED SWITCHES

The following formulae are used for the calculation of turnouts with curved switches.

$$R = \frac{G - t - h \sin \alpha}{2 \sin \dfrac{\alpha + \beta}{2} \times \sin \dfrac{\alpha - \beta}{2}} = \frac{G - t - h \sin \alpha}{\cos \beta - \cos \alpha} \qquad (14.10)$$

$$I = R \sin \alpha - (G - t - h \sin \alpha) \cot \frac{\alpha + \beta}{2} \qquad (14.11)$$

$$V = G - \{h \sin \alpha + R(1 - \cos \alpha)\} \qquad (14.12)$$

$$\text{Switch lead} = \sqrt{2R(d - y) - (d - y)^2 - 1} \qquad (14.13)$$

$$\text{Lead} = (G - t - h \sin \alpha) \cot \frac{\alpha + \beta}{2} - \text{SL} - h \cos \alpha \qquad (14.14)$$

Here, R is the radius of the outer lead rail, G is the gauge, h is the lead of the straight leg of the crossing ahead of TNC up to the TP of the lead curve, t is the thickness of the switch at the toe, I is the distance from the toe of the switch to the point where the tangent drawn to the extended lead curve is parallel to the main line gauge face, V is the distance between the main line gauge face and the tangent drawn to the lead curve from a distance l from the toe, y is the vertical ordinate along the Y-axis, α is the crossing angle, and β is the switch angle.

14.9 LAYOUT OF TURNOUT

To lay out a turnout in the field, the values of offsets from the gauge face of the straight track to the gauge face of the turnout may be adopted from Table 14.6.

Table 14.6 Laying turnouts in the field*

Distance from heel	Offset from gauge face of straight track to gauge face of turnout					
	1-in-8.5 4725-mm straight (ST) switch	1-in-12 6400-mm ST switch	1-in-12 7730-mm curved switch	1-in-8.5 4115-m ST switch	1-in-12 5485-mm ST switch	1-in-12 6700-mm curved switch
3000	–	–	–	241	–	–
4500	–	–	–	330	–	–
6000	382	293	322	437	313	362
7500	469	365	381	564	386	445
9000	565	403	445	709	468	537
10,500	672	466	515	–	559	638
12,000	790	512	589	–	659	747
13,500	917	606	668	–	769	–
15,000	1055	685	752	–	–	–
16,500	1202	767	841	–	–	–
18,000	–	856	935	–	–	–
19,500	–	950	1033	–	–	–
21,000	–	1047	1137	–	–	–
22,500	–	1151	1246	–	–	–
24,000	–	1260	1359	–	–	–
Total number of sleepers	51	70	70	37	47	47
Switch angle	1°34′27″	1°8′0″	0°27′35″	1°35′30″	1°9′38″	0°24′27″

* All dimensions are in millimetres.

(Contd.)

Table 14.6 *(Contd.)*

Distance from heel	Offset from gauge face of straight track to gauge face of turnout					
	1-in-8.5 4725-mm straight (ST) switch	*1-in-12 6400-mm ST switch*	*1-in-12 7730-mm curved switch*	*1-in-8.5 4115-m ST switch*	*1-in-12 5485-mm ST switch*	*1-in-12 6700-mm curved switch*
Crossing angle	6°42'35"	4°45'49"	4°45'49"	6°42'49"	4°45'49"	4°45'49"
Drg. no.	TA 20,104 and 20,804	TA 5268(M) and 20,801	TA 20,171 and 21,831	TA 20,171 and 21,004	TA 20,401 and 20,001	TA 20,484

14.10 TRENDS IN TURNOUT DESIGN ON INDIAN RAILWAYS

The main factors responsible for low speeds over turnouts on Indian Railways are as follows:

(a) A sudden change in the direction of the running edge upon entry onto the switch from a straight track
(b) Absence of a transition between the curved lead and the straight crossing
(c) Non-transitioned entry from the curved lead to the straight crossing
(d) Absence of superelevation over the turnout curve
(e) Gaps in the gauge face and the running table at the crossing
(f) Variation in cross level caused by raised switch rails

In order to achieve higher speeds on turnouts, it is necessary that all the limitations of the design of a turnout are overcome as far as possible. In European countries, the design of turnouts has been greatly improved and speeds of more than 100 km per hour are permitted on turnout curves. The main features of the design of these turnouts are the following:

(a) Long curved switches are provided to avoid the abrupt change in the direction of the vehicle at the entry to the switch.
(b) Switches and crossings are curved to the same radius as the lead curve or, alternatively, a transition curve is provided between the toe of the switch and the nose of the crossing. This provides a smooth passage to the trains on the turnout curve.
(c) Higher cant deficiency is permitted so that the disadvantage of not providing superelevation on the turnout curve is duly compensated.

In keeping with the trend in the railways of the world to permit higher speeds on turnouts, Indian Railways is considering standardization of high-speed turnouts for the following conditions of the track.

(a) For goods yards for a maximum permissible speed of 25 km per hour and for passenger yards for maximum permissible speed of 50 km per hour
(b) In peripheries of big yards for bypass lines for a maximum permissible speed of 75 km per hour
(c) At junction joints of single-line and double-line sections for a maximum permissible speed of 100 km per hour

A design of 1-in-12 turnouts for passenger yards with thick web tongue rails and CMS crossings (RDSO Drg. no. T-2733) has already been finalized for enabling a maximum permissible speed of 50 km per hour. Similarly, a new design of 1-in-24 turn outs for BG routes with curved switches and thick web tongue rails with a speed potential of 160 km per hour is being finalized by Indian Railways.

14.10.1 New High-speed Turnouts for Passenger Yard (1 in 12) for 50 kmph speed

The main features of 1-in-12 turnout with a speed potential of 50 km per hour are as follows (Fig. 14.13):

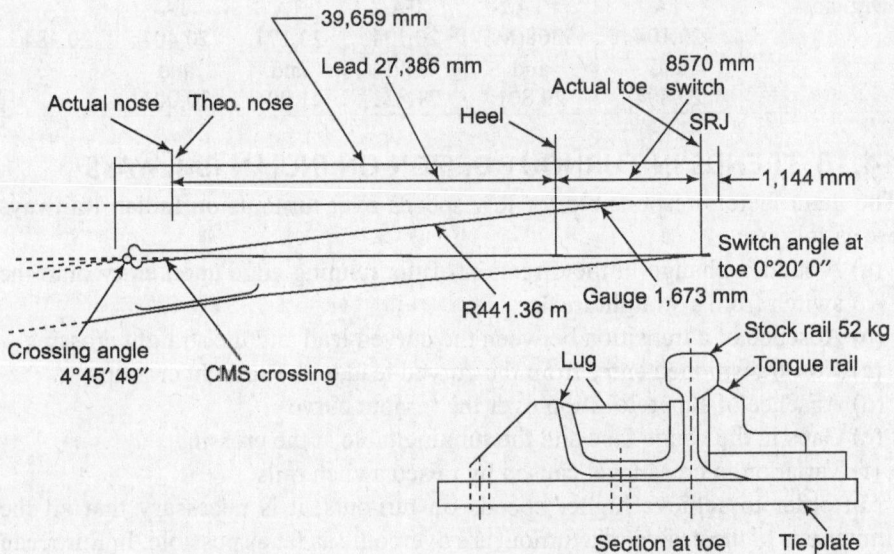

Fig. 14.13 New passenger turnout 1 in 12

(i) Its overall length is almost equal to the present standard 1:12 turnout, to facilitate easy replacement.

(ii) Thick web tongue rail has been used to provide high lateral rigidity and longer life to turnout.

(iii) The stock rail head prevents non-lateral rotation by using a special spring leaf clip to fasten the inner foot of the stock rail. Additionally, sturdy fittings are used on the outside of stock rail to minimize dynamic gauge widening.

Basic data about 1-in-12 turnout

- Gauge = 1673 mm
- Lead = 27, 386 mm
- Crossing angle = 4°45′49″
- Total length = 39659 mm
- Radius (R) = 141.36 m
- Switch angle at toe = 0°20′0″

Technical details of new 1-in-12 turnout for passenger yards having potential of 50 kmph

Track It consists of 52 kg in rail section having wooden sleepers 250 mm × 150 mm having 550 to 600 mm spacing the gauge is 1676 mm.

Switch It consists of non-overriding thick web switch (ZU-2-49 type) having an entry angle of 0°20′0″.

Crossing The crossing is of CMS type having an angle of 4°45′49″.

Crossing rail clearance
(a) At throat–44 mm
(b) Opposite nose–44 mm
(c) At end of flare–64 mm
To provide better guidance keep flare slope 1 in 31.

Check rails   These are provided at level with main rail.

Check rail clearance
(a) Opposite nose: 41 mm
(b) At end of flare slope: 63 mm; (To provide better guidance keep flare slope 1 in 68).

14.11 INSPECTION AND MAINTENANCE OF POINTS AND CROSSINGS

Points and crossings should be inspected in detail, as the quality of a train ride greatly depends on their maintenance. The following important points should be checked.

Condition of tongue rails and stock rails   There should be no wear on the top as well as the gauge face side of the tongue rail. Badly worn-out rails should be replaced. It should be ensured that the turnout side stock rail is provided with the requisite bend ahead of the toe of the switch; otherwise the alignment at this spot is bound to be kinky.

Condition of fittings of tongue and stock rails   The fittings should be tight and the spherical washers must be placed at their correct locations. The slide chairs should be cleaned and greased with graphite for smooth operation of the points. The fish plates should be provided with the correct amount of bend at the loose heel joint. A gauge tie plate should be added if provisions for the same have not been made.

Gauge and cross level at switch assembly   The gauge and cross levels should be checked for correctness at the following locations: (i) the stock joint, (ii) 150 mm (6") behind the toe of the switch, (iii) the mid-switch for the straight track and for the turnout side, and (iv) the heel of the switch for the straight track and for the turnout side.

Clearance between stock and tongue rails at the heel of the switch   The correct divergence to be provided at heel of the switch should be as follows:

1-in-16 or 1-in-12	1-in-8.5
BG—133 mm (5.25")	120 mm (4.25")
MG—117 mm (4.65")	120 mm (4.75")

Throw of the switch   The throw of the switch should be as follows:

	Recommended	Minimum
BG	115 mm (4.5")	95 mm (3.25")
MG	100 mm (4")	89 mm (3.5")

Condition of crossing and tongue rail   The condition of the crossings and of the fittings should be checked. The maximum vertical wear permitted on a point or wing rail is 10 mm and these should be reconditioned when the wear is 6 mm.

The burn burrs should also be removed and the fittings should be tightened. The maximum vertical wear permitted on a tongue rail is 6 mm, whereas the permitted lateral wear is 8 mm for 90 R and 52 kg rails and 5 mm for 60 R and 75 R rails. The tongue rail should be replaced or reconditioned before this value is reached. The Railway Board has now decided that the maximum vertical wear on wing rails and the nose of the crossings should be limited to 4 mm on the Rajdhani and Shatabdi routes and 6 mm on all other routes. The wear limits for CMS crossings are, however, 5.5 mm for Rajdhani and Shatabdi routes and 7.5 mm for all other routes.

Gauge and cross level of crossing assembly   The gauge and cross level should be checked at the following locations and should always be correct: (i) 1 m ahead of the nose on straight tracks and on turnouts, (ii) 150 mm (6") behind the ANC on straight tracks and on turnouts, and (iii) 1 m behind the ANC on straight tracks and on turnouts.

Check rails   The condition of check rails should be ascertained. Check rail clearances should be as follows:

	Maximum	Minimum
BG	48 mm	44 mm
MG	44 mm	41 mm

Lead curvature   The curvature should be checked either by the offset method or by the versine method. The curvature should be correct and uniform.

Cross levels on straight tracks and turnouts   The cross levels on straight tracks and turnouts should be checked to see that they are correct at all places.

Sleepers   The condition of the sleepers and their fittings should be checked and unserviceable sleepers should be replaced. The squaring and spacing of sleepers should be proper and they should be well packed.

Ballast and drainage   Enough quantity of ballast should be available so as to provide an adequate cushion. The drainage should be proper.

Any other defects   If there are any other defects in the layout, these should be checked and corrected.

Special attention is required to ensure that the sleepers are well packed, all fittings are tightened, gauge and cross levels are properly maintained, and the wear on the tongue rail as well as on the crossing is within permissible limits. It should also be ensured that proper distance blocks are provided at correct locations. The schedule of inspection followed on Indian Railways is given in Table 14.7.

Table 14.7   Schedule of inspection

PWI (permanent way inspector) III and PWI (in charge)	Once in 3 months in rotation for passenger lines and once in 6 months in rotation for other lines. The interval between two inspections for passenger lines of the same turnout should not exceed 4 months.
AEN (assistant engineer)	Once in 12 months for all passenger lines and test checking of 10 per cent of other points and crossings.
DEN (divisional engineer)	Test checking of certain number of points and crossings, particularly in running lines and those recommended for renewals.

14.12 MODERN TURNOUT DEVELOPED BY DELHI METRO (DMRC)

Delhi Metro has developed a modern turnout with special features having a speed potential of 50 km per hour for 1-in-12 layout and 25 km per hour for 1-in-8.5 layout. Some of the important technical features of turnout, are given below in Table 14.8.

Table 14.8 Important technical features of turnout

Item	*Detail of Item*
General design	• Turnout is designed having gauge of 1673 mm; UIC 60 kg canted rails with an inward slope of 1 in 20; The design will take LWR through turnout also.
Sleepers	• Concrete sleepers of fan-shaped design have been used for turnout. • 1-in-20 cant on rails has been achieved by providing a canted HD polythene pad between the rail pad and the PSC sleepers.
Details of switches	• The switch entry angle is 0°10′42″ for 1-in-12 turnout and 0°36′7″ for 1-in-8.5 turnout. • All switches are of 1080 grade head hardened rails suitable of being welded by alumino-thermic welding. • The minimum flange way clearance in switch portion is not less than 60 mm and opening at toe of switch is kept as 160 mm.
	• The fastening used is vossloh SKL-12 with a toe load of 1100 kg per clip.
Detail of crossing	• CMS crossings are provided with welded leg extensions of UIC 60,1080 grade head hardened rails. • The crossings are of CMS type and the welded layouts continue through LWR, which gives better riding quality.
	• The length of crossing without weldable lays used by Delhi Metro for 1-in-12 is 4.49 m and for 1-in-8.5 is 3.88 m, which is longer than the normal size of crossing used on Indian Railways.
Check rails	• All the check rails are higher by 25 mm above the running rails. • Each check rail is end flared by machining.

14.13 IMPROVEMENTS MADE IN POINTS AND CROSSINGS IN FOREIGN COUNTRIES

Utilizing modern technologies, a number of improvements have been carried out abroad in geometry, structure, design and manufacturing process of the points and crossing. Some of these developments are as follows:

Improvement in geometry by way of points and crossings Flatter entry angle, thereby reducing the angle of attack. Adoption of tangential layouts for high speeds, use of curved crossings, use of higher throw of the switches and provision of spiring setting device to get adequate clearance of JOH.

Improvements of switch portion Introduction of the thick web switch; use of the higher UTS rails, use of non-greasing bearing plates.

Improvements in lead portion Sturdier elastic fastenings throughout the turnout and jointless track.

Improvements in crossing area Use of the CMS crossing; use of welded forged crossings.

Intelligent turnouts Voest Alpine Company has developed electronic remote turnout monitoring system. It has contact loss inductive and strain guage sensors.

These sensors, measure, monitor, and report the condition of the important components of turnouts.

Plug in turnouts Normal turnouts need time and space for assembling the same, VAE has developed iplug in and play turnouts named 'Hydrostar'. Such devices provided in hollow sleepers make taping of these turnouts easy.

Astro rollers To facilitate movement of tongue rails, rollers are provided below the tongue rails. These are helpful in flat long switches. It is expected that with these new concepts and developments the speed potential, maintainability and the overall asset life of the turnouts will be greatly enhanced.

SUMMARY

A turnout is an integral part of a railway track. It is a combination of lead rails and points and crossings. These are provided when two tracks are to be connected or when a branch line is to be introduced. The various features and designs of a turnout have been discussed in this chapter. Design examples have also been included. It is possible to have various types of track junctions with different combinations of points and crossings. These are described in Chapter 15.

REVIEW QUESTIONS

1. Design a turnout for a BG track if the number of the crossing is 12 and the heel divergence is 114 mm. Assume a simple circular curve from the toe of the switch to the TNC.
2. Draw a neat sketch of a right-hand turnout taking off from a straight broad gauge track and name thereon the various component parts and important terms connected with the layout. Show the disposition of the sleepers.
3. A turnout is to be laid off a straight BG track with a 1-in-12 crossing. Determine the lead and radius for the turnout given the following data–heel divergence $d = 133$ mm, the straight length between the TNC and the tangent point of the crossing curve, $h = 1.418$ m, crossing angle $\alpha = 4°45'49''$, and switch angle $\beta = 1°9'00''$.
4. Draw a neat sketch of a left-hand turnout and name its various components. Describe any one method of designing a turnout and give the detailed procedure for calculating the (a) lead, (b) radius, and (c) heel divergence.
5. Calculate the elements required to set out a 1-in-8.5 turnout taking off from a straight BG track, with its curve starting from the heel of the switch and ending at a distance of 864 mm from the TNC, given that the heel divergence is 136 mm and the switch angle is $1°34'27''$. Make a freehand sketch showing the values of the calculated elements.
6. Explain with the help of neat sketches the points and crossings used in railways, indicating the precautions to be taken while laying the same.
7. On a straight BG track a turnout takes off at an angle of $6°42'35''$. Design the turnout when it is given that the switch angle is equal to $1°34'27''$ and the length of the switch rails is 4.73 m. The heel divergence is 11.43 cm. The straight arm is 0.85 m long.
8. Draw a neat sketch of a crossover between two parallel straight MG tracks spaced at 5 m centre-to-centre distance. Show the position of the switches

as they would be when trains would be diverted from one track to the other. Also, show the following particulars on the sketch of the crossover: (a) overall length of turnout, (b) overall length of the crossover, (c) any two wing rails, (d) any two check rails, (e) any two stock rails, (f) any two switch rails, (g) two heel blocks, (h) one pull rod.

9. Calculate all the necessary elements for a 1-in-12 turnout taking off from a straight BG track, with its curve starting from the toe of the switch, i.e., tangential to the gauge face of the outer main rail and passing through the TNC, given that the heel divergence is 11.4 cm.
   (a) Find the crossing angle of the 1-in-12 crossing using the right angle method.
   (b) Draw a labelled section for the points at the toe of the switch.

10. A crossover is to be laid connecting two BG parallel tracks spaced 4.5 m apart. Assuming that 1-in-8.5 crossings are to be used, work out the various details required for setting out the crossover.

11. Calculate the principal dimensions required to connect a parallel siding to a main line with the help of a 1-in-8.5 turnout, the spacing between these BG parallel tracks being 6 m c/c. For a 1-in-8.5 crossing the detailed dimensions are as follows: theoretical length of switch rail = 4950 mm, actual length of switch rail = 4725 mm, heel divergence = 136 mm, distance between the ends of the switch and stock rails = 840 mm, distance between TNC and the first distance block (tangent point) at the toe of the crossing = 864 mm. Draw a detailed dimensioned sketch for this turnout connecting two parallel tracks and label therein the various parts of the turnout.

12. Find the crossing angle for a 1-in-12 crossing using the centre line method.

13. What are the standard turnouts prescribed on Indian Railways? What are the special fittings that are provided with turnouts and what are their functions?

14. When is it necessary to recondition worn-out crossings? Describe in detail one of the methods of reconditioning a crossing.

15. What are the various points required to be checked during the inspection of points and crossings? Give the schedule laid down by Indian Railways for the inspection of these points and crossings.

16. Describe the main constituents of a crossing. Draw neat sketches to show a point rail and a splice rail.

17. What is a CMS crossing? Describe its advantages.

18. Differentiate between the following.
    (a) Stud switch and split switch
    (b) Stock rail and tongue rail
    (c) Flangeway clearance and heel divergence
    (d) Flat bearing plate and anticreep bearing plate
    (e) Slide chairs and grade-off chairs

Choose the correct answer from the choices given.

19. For determining the number or angle of crossing, the method used on Indian Railways is:
    (a) Centre line method   (b) Right angle method
    (c) Isosceles triangle method   (d) any of these

20. The maximum throw of an existing switch for MG is:
    - (a) 89 mm
    - (b) 95 mm
    - (c) 100 mm
    - (d) 115 mm
21. Vertical wear limit of 52 kg tongue rail is:
    - (a) 8 mm
    - (b) 5 mm
    - (c) 3 mm
    - (d) none of these
22. The minimum throw of an existing switch for BG is:
    - (a) 89 mm
    - (b) 95 mm
    - (c) 100 mm
    - (d) 115 mm
23. Lubrication of outer rail of gauge face curve is done:
    - (a) once a week
    - (b) twice a week
    - (c) once in fortnight
    - (d) none of these
24. The maximum value of throw of switch for a BG track is:
    - (a) 89 mm
    - (b) 95 mm
    - (c) 100 mm
    - (d) 115 mm
25. On Indian Railways, the number of a crossing is defined as:
    - (a) sine of angle of crossing
    - (b) cosine of angle of crossing
    - (c) tangent of angle of crossing
    - (d) cotangent of angle of crossing
26. In a diamond crossing, the number of noses are:
    - (a) 2
    - (b) 3
    - (c) 4
    - (d) 6
27. The overall length of a turnout is the distance between the end of stock rail and:
    - (a) heel of crossing
    - (b) actual nose of crossing
    - (c) throat of crossing
    - (d) toe of crossing
28. The heel divergence for BG turnout for 1-in-12 straight switch is:
    - (a) 117 mm
    - (b) 120 mm
    - (c) 133 mm
    - (d) 136 mm
29. The switch angle for a BG 1-in-12 straight switch:
    - (a) 1° 34′27″
    - (b) 1° 8′0″
    - (c) 0° 47′27″
    - (d) 0° 27′35″
30. A turnout is designated as right-hand turnout depending upon whether:
    - (a) the turnout is situated on right side of railway station
    - (b) the traffic is diverted on right side
    - (c) the turnout is assembled on right side of cabin
    - (d) none of these
31. The maximum check rail clearance for BG on point and crossing is:
    - (a) 41 mm
    - (b) 44 m
    - (c) 48 mm
    - (d) 51 m

Track Junctions and Simple Track Layouts

INTRODUCTION

Track junctions are formed by the combination of points and crossings. Their main objective is to transfer rail vehicles from one track to another or to enable them to cross from one track to another. Depending upon the requirements of traffic, there can be several types of track junctions with simple track layouts. The most commonly used layouts are discussed in the following sections.

15.1 TURNOUT OF SIMILAR FLEXURE

A turnout of similar flexure (Fig. 15.1) continues to run in the same direction as the main line curve even after branching off from it. The degree of the turnout curve will be higher than that of the main line curve. The degree and radius of the turnout curve are given by the formulae:

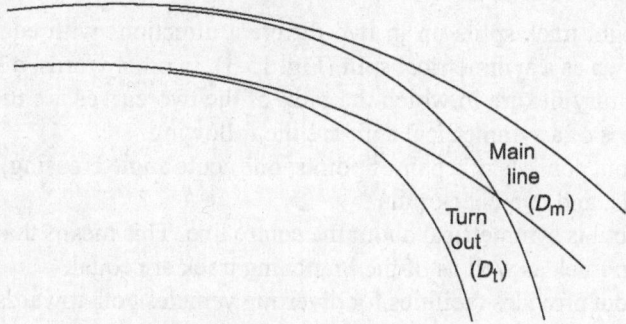

Main
line
(D_m)

Turn
out
(D_t)

Fig. 15.1 Turnout of similar flexure

$$D_t = D_s + D_m \tag{15.1}$$

$$R_t = \frac{R_m R_s}{R_m + R_s} \tag{15.2}$$

where D_s is the degree of the outer rail of the turnout curve from the straight track, D_m is the degree of the rail of the main track on which the crossing lies, i.e., the inner rail in Fig. 15.1, D_t is the degree of the rail of the turnout curve on which the crossing lies, i.e., the outer rail, R_s is the radius of the outer rail of the turnout curve from the straight track, and R_t is the radius of the rail of the turnout curve on which the crossing lies, i.e., the outer rail.

15.2 TURNOUT OF CONTRARY FLEXURE

A turnout of contrary flexure (Fig. 15.2) takes off towards the direction opposite to that of the main line curve. In this case, the degree and radius of the turnout curve are given by the following formulae:

$$D_t = D_s - D_m \tag{15.3}$$

$$R_t = \frac{R_s R_m}{R_m - R_s} \tag{15.4}$$

Here, D_m is the degree of the rail of the main track on which the crossing lies, i.e., the outer rail in Fig. 15.2.

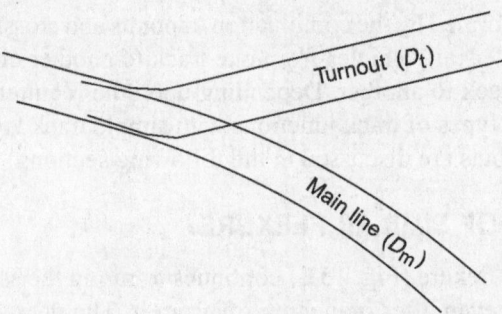

Turnout (D_t)

Main line (D_m)

Fig. 15.2 Turnout of contrary flexure

15.3 SYMMETRICAL SPLIT

When a straight track splits up in two different directions with equal radii, the layout is known as a symmetrical split (Fig. 15.3). In other words, a symmetrical split is a contrary flexure in which the radii of the two curves are the same. The salient features of a symmetrical split are the following:

(a) The layout consists of a pair of points, one acute angle crossing, four curved lead rails, and two check rails.

(b) The layout is symmetrical about the centre line. This means that the radii of the main track as well as of the branching track are equal.

(c) The layout provides facilities for diverting vehicles both towards the left and the right.

(d) It is suitable for locations with space constraints, as it occupies comparatively much less space than a turnout from the straight track.

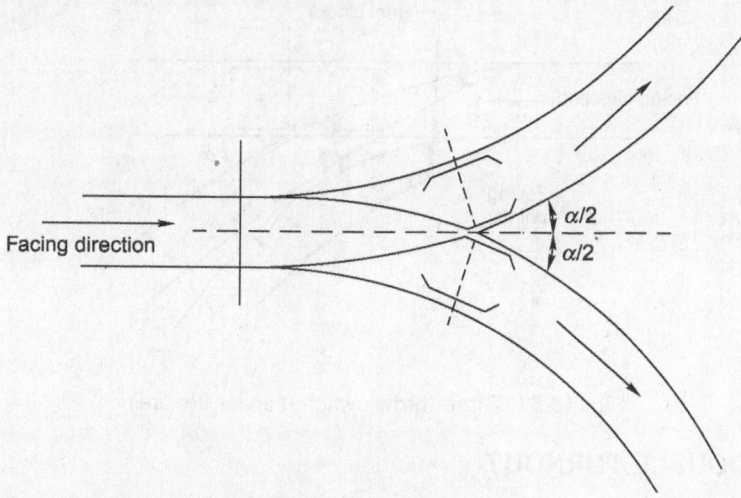

Fig. 15.3 Symmetrical.split

15.4 THREE-THROW SWITCH

In a three-throw arrangement, two turnouts take off from the same point of a main line track. A three-throw switch can have contrary flexure or similar flexure, as shown in Figs 15.4 and 15.5, respectively. Three-throw switches are used in congested goods yards and at entry points to locomotive yards, where there is much limitation of space.

A three-throw switch has two switches and each switch has two tongue rails placed side by side. There is a combined heel block for both the tongue rails of the switch. The switches can be operated in such a way that movement is possible in three different directions, that is, straight, to the right, and to the left. Three-throw switches are obsolete now as they may prove to be hazardous, particularly at higher speeds, because the use of double switches may lead to derailments.

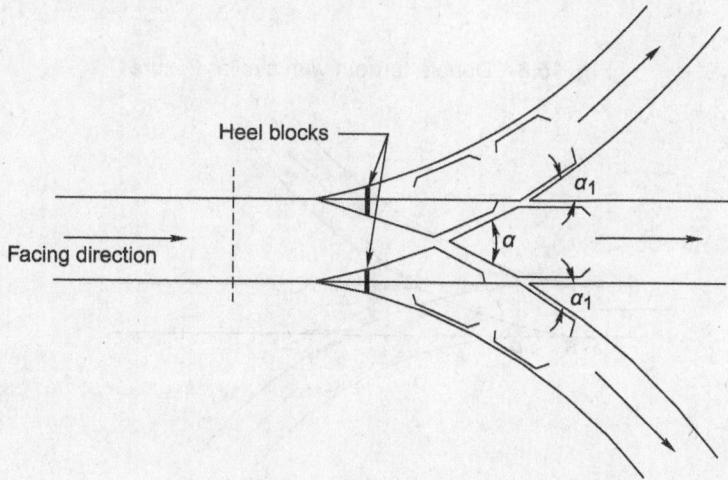

Fig. 15.4 Three-throw switch (contrary flexure)

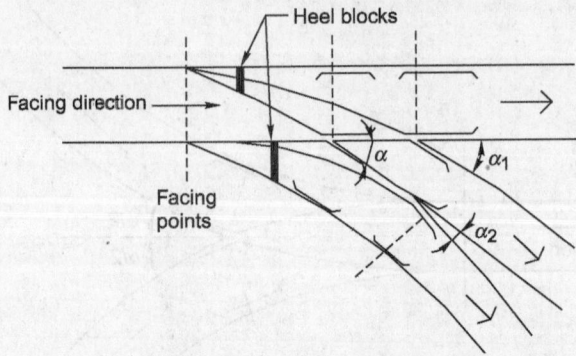

Fig. 15.5   Three-throw switch (similar flexure)

15.5 DOUBLE TURNOUT

A double turnout or *tandem* is an improvement over a three-throw switch. In a double turnout, turnouts are staggered and take off from the main line at two different places. This eliminates the defects of a three-throw switch, as the heels of the two switches are kept at a certain distance from each other. The distance between the two sets of switches should be adequate to allow room for the usual throw of the point.

Double turnouts can be of similar flexure, when the two turnouts take off on the same side of track (Fig. 15.6) or of contrary flexure, when the two turnouts take off in two different directions (Fig. 15.7).

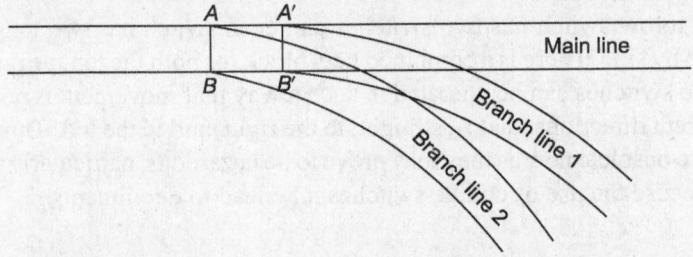

Fig. 15.6   Double turnout with similar flexure

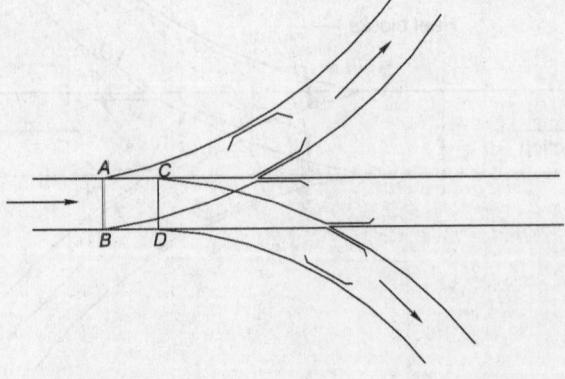

Fig. 15.7   Double turnout with contrary flexure

Double turnouts are mostly used in congested areas, particularly where traffic is heavy, so as to economize on space.

15.6 CROSSOVER BETWEEN TWO PARALLEL TRACKS WITH AN INTERMEDIATE STRAIGHT LENGTH

The crossover between two parallel tracks with an intermediate straight length can be designed by applying any one of the following two methods:

15.6.1 Coles Design

Coles design is a simple layout. In this design, two parallel tracks at a distance D from each other are connected by a crossover with a small length of the straight portion of the track lying between the two theoretical noses of the crossing. The straight portion of the track (ST) can be calculated using the formula:

$$\text{Straight track (ST)} = (D - G)\,N - G\,\sqrt{1 + N^2} \qquad (15.5)$$

where G is the gauge of the track and N is the number of the crossing. The overall length (OL) of the crossover from the tangent point of one track to the tangent point of the other track is found by adding the lengths of the curve leads of the two turnouts and the length of the straight portion in between the two TNC (Fig. 15.8).

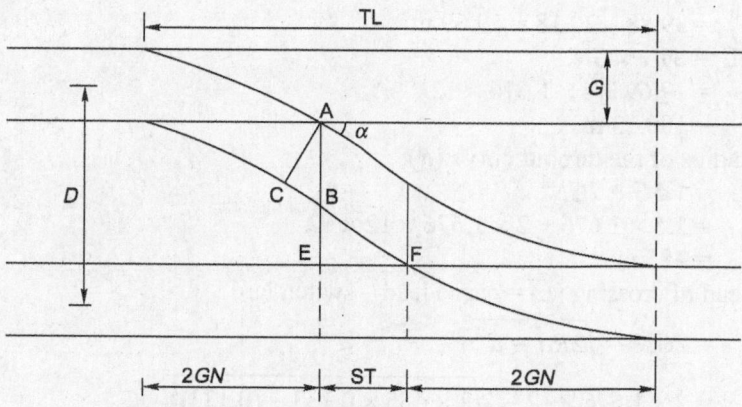

Fig. 15.8 Crossover between two parallel tracks

$$\text{Overall length} = \text{OL of one turnout} + \text{ST} + \text{OL of other turnout}$$

$$= 2GN + (D - G)\,N - G\,\sqrt{1 + N^2} + 2GN$$

$$= (D - G)\,N + G\,(4N - \sqrt{1 + N^2}) \qquad (15.6)$$

Since the value of N^2 is very large as compared to 1, the value $\sqrt{1 + N^2}$ can be taken approximately as N. Simplifying Eqn (15.6),

$$\text{Total length (TL)} = (D + 2G)\,N$$

$$= 2GN + \text{ST} + 2GN$$

$$= 4GN + \text{ST} \qquad (15.7)$$

Example 15.1 A 1-in-8.5 crossover exists between two BG parallel tracks with their centres 5 m apart. Find the length of the straight track and the overall length of the crossover. Use Coles method. Given $D = 5$ m, $N = 8.5$, $G = 1.676$ m.

Solution

    (i) ST $= (D - G) N - G \sqrt{1 + N^2}$

             $= (5 - 1.676)8.5 - 1.676 \sqrt{1 + N^2}$

             $= 13.91$ m

   (ii) OL $=$ ST $+ 4GN$

             $= 13.91 + 4 \times 1.676 \times 8.5 = 70.89$ m

Example 15.2 A crossover is laid between two BG straight tracks placed at a distance of 5 m c/c. Calculate the (i) overall length, (ii) radius of the curved lead, (iii) lead distance. Heel divergence of 1-in-12 crossing $= 133$ mm.

Solution The crossing number is equal to 12 and the intermediate portion is straight.

    (i) ST $= (D - G) N - G \sqrt{1 + N^2}$

      where $D = 5$ m, $G = 1.676$ m, and $N = 12$

      ST $= (5 - 1.676)$ 12 $- 1.676 \sqrt{1 + 144}$

          $= 39.88 - 20.18 = 19.69$ m

   (ii) OL $=$ ST $+ 4GN$

          $= 19.69 + 4 \times 1.676 \times 12$

          $= 100.13$ m

  (iii) Radius of the turnout curve (R):

      R $= 1.5G + 2GN^2$

         $= 1.5 \times 1.676 + 2 \times 1.676 \times 12 \times 12$

         $= 485$ m

  (iv) Lead of crossing (L) $=$ curve lead $-$ switch lard

        $= 2GN - \sqrt{2Rd - d^2}$

        $= 2 \times 1.676 \times 12 - \sqrt{2 \times 485 \times 0.133 - (0.133)^2}$

        $= 40.2 - 11.4 = 28.8$ m

15.6.2 IRS Design

In IRS design, the distance from the TNC measured along the straight track is given by the formula:

$$ST = (D - G - G \sec \alpha) \cot \alpha \tag{15.8}$$

On simplification

$$ST = D \cot \alpha - G \cot \alpha / 2 \tag{15.9}$$

where ST is the distance from TNC to TNC along the straight track, D is the distance from centre to centre of two tracks, G is the gauge, and α is the angle of crossing.

Similarly, the distance from TNC to TNC along the crossover is given by the formula (Fig. 15.8)

$$CF = (D - G - G \sec \alpha) \csc \alpha + G \tan \alpha \qquad (15.10)$$

where CF is the distance from TNC to TNC along the crossover, D is the distance from centre to centre of two tracks, G is the gauge, and α is the angle of crossing.

Example 15.3 A 1-in-12 crossover of IRS type is laid between two BG parallel tracks with their centres 5 m apart. Calculate ST and the distance from TNC to TNC along the crossover.

Solution

$$G = 1.676 \text{ m}, N = 12, D = 5.0 \text{ m}, \alpha = 4° 45' 49$$

(i) $ST = D \cot \alpha - G \cot \alpha / 2$

$$= 5 \times 12 - 1.676 \times 24.04$$

$$= 19.7 \text{ m}$$

(ii) $CF = (D - G - G \sec \alpha) \csc \alpha + G \tan \alpha$

$$= (5.0 - 1.676 - 1.682) \times 12.04 + 1.676 \times (1/12)$$

$$= 19.91 \text{ m}$$

15.7 DIAMOND CROSSING

A diamond crossing is provided when two tracks of either the same gauge or of different gauges cross each other. It consists of two acute crossings (A and C) and two obtuse crossings (B and D). A typical diamond crossing consisting of two tracks of the same gauge crossing each other, is shown in Fig. 15.9.

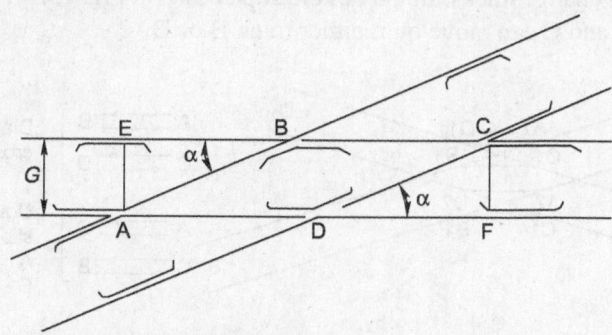

Fig. 15.9 Diamond crossing

In the layout, ABCD is a rhombus with four equal sides. The length of the various constituents may be calculated as follows:

$$EB = DF = AE \cdot \cot \alpha = GN$$

$$AB = BC = G \csc \alpha$$

Diagonal $AC = G \csc \alpha / 2$

Diagonal $BD = G \sec \alpha / 2$

It can be seen from the layout that the length of the gap at points B and D increases as the angle of crossing decreases. Longer gaps increase the chances of the wheels, particularly of a small diameter, being deflected to the wrong side of the nose. On Indian Railways, the flattest diamond crossing permitted for BG and MG routes is 1 in 8.5.

Along with diamond crossings, single or double slips may also be provided to allow the vehicles to pass from one track to another.

15.7.1 Single Slip and Double Slip

In a diamond crossing, the tracks cross each other, but the trains from either track cannot change track. Slips are provided to allow vehicles to change track.

The slip arrangement can be either single slip or double slip. In single slips, there are two sets of joints, the vehicle from only one direction can change tracks. In the single slip shown in Fig. 15.10, the train on track A can change to track D, whereas the train on track C remains on the same track, continuing onto track D.

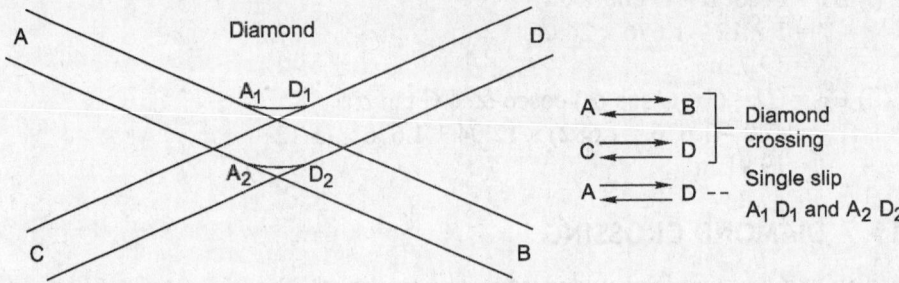

Fig. 15.10 Single slip

In the case of double slips, there are four sets of points, and trains from both directions can change tracks. In the double slip shown in Fig. 15.11, the trains on both tracks A and C can move onto either track B or D.

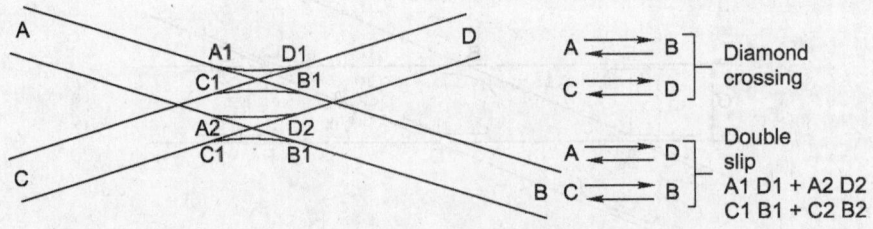

Fig. 15.11 Double slip

15.7.2 Improvements in the Design of Diamond Crossings

In order to smoothen the ride over a diamond crossing, the following improvements are generally made.
(a) Provision of 25 mm higher check rails
(b) Reduction in the check rail clearance by 3 mm in the case of obtuse crossings.

Example 15.4 Two BG tracks cross each other at an angle of 1 in 10. Calculate the important dimensions of the diamond crossing.

Solution The data given is as follows (Fig. 15.9):
(i) Number of crossing the $(N) = 10$
Gauge $(G) = 1.676$ m

(ii) $N = \cot \alpha$ or $10 = \cot \alpha$

Therefore, $\alpha = 5°42'38''$

(iii) $EB = DF = AE \cot \alpha = GN$

$= 1.676 \times 10 = 16.76$ m

(iv) $AB = BC = G \cosec \alpha$

$= 1.676 \times 10.05 = 16.85$ m

(v) $AC = G \cosec \alpha/2$

$= 1.676 \times \cosec \dfrac{5°42'38''}{2}$

$= 1.676 \times 20.10 = 33.70$ m

(vi) $BD = G \sec \alpha/2$

$= 1.676 \sec \dfrac{5°42'38''}{2}$

$= 1.676 \times 1.014 = 1.70$ m

15.8 SCISSORS CROSSOVER

A scissors crossover (Fig. 15.12) enables transferring a vehicle from one track to another track and vice versa. It is provided where lack of space does not permit the provision of two separate crossovers. It consists of four pairs of switches, six acute crossings, two obtuse crossings, check rails, etc.

Fig. 15.12 Scissors crossover

The scissors crossovers commonly used are of three types depending on the distance between the two parallel tracks they join. A brief description of these crossovers is as follows.

(a) In the first type, the acute crossing of the diamond falls within the lead of the main line turnout. In this case, the lead of the main line turnout is considerably reduced and hence this is not a satisfactory arrangement.

(b) In the second type, the acute crossing of the diamond falls opposite the crossing of the main line turnout. Here, both the crossings lie opposite each other, resulting in a simultaneous drop of the wheel and this results in jolting. This is also not a desirable type of layout.

(c) In the third type of scissors crossover, the acute crossing falls outside the lead of the main crossing. Thus, the acute crossing of the diamond is far away from the crossing of the main line track. This is the most satisfactory arrangement out of these three layouts.

15.9 GAUNTLETTED TRACK

Gauntletted track is a temporary diversion provided on a double-line track to allow one of the tracks to shift and pass through the other track. Both the tracks run together on the same sleepers. It proves to be a useful connection when one side of a bridge on a double-line section is required to be blocked for major repairs or rebuilding. The speciality of this layout is that there are two crossings at the ends and no switches [Fig. 15.13 (a)].

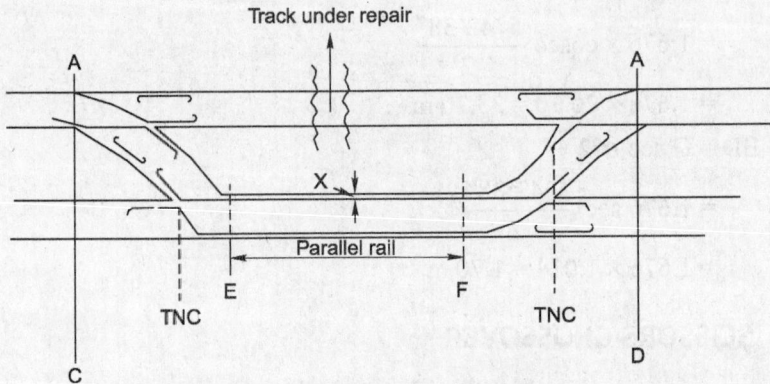

Fig. 15.13 (a)   Gauntletted track

Gauntletted tracks are also used on sections where trains have to operate on mixed gauges, say, both BG and MG, for short stretches. In such cases, both the tracks are laid on the same set of wooden sleepers [Fig. 15.13(b)].

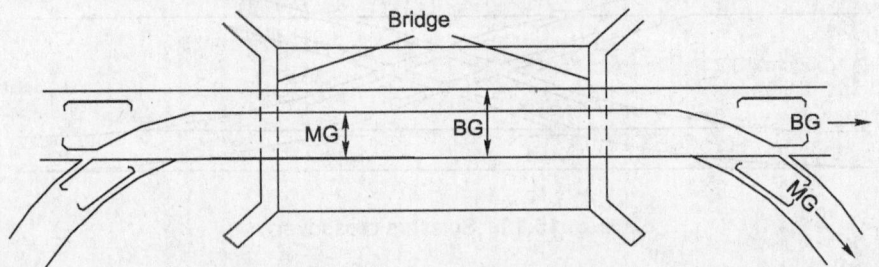

Fig. 15.13 (b)   Gauntletted track for mixed gauge

The salient features of the gauntletted track are as follows:
(a) Two tracks are laid on the same sleepers with two sets of crossings without any switches.
(b) Gauntletted tracks can be economically used for mixed gauge, that is, say, for tracks with both BG and MG.
(c) This layout is used when part of a double-line bridge is under repair. It is also used to economize the cost of a double-line bridge.

15.10 GATHERING LINE

A gathering line (also called a *ladder track*) is a track where a number of parallel tracks gather or merge. Alternatively, a number of parallel tracks also branch off from a gathering line. A gathering line is defined by the turnout angles and the angle of inclination of the ladder track to the parallel tracks (Fig. 15.14).

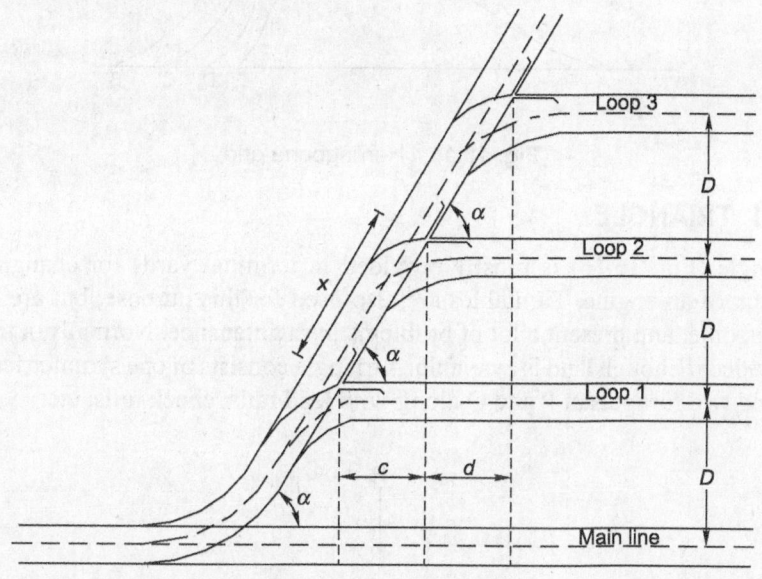

Fig. 15.14 Gathering line

Gathering line at crossing angle

When the angle of inclination of the gathering line is the same as that of the turnout, it is said to be laid at the angle of crossing. In this situation, there is some gap between the back leg of the crossing of the turnout and the stock joint of the next turnout and a closure rail has to be used. The angle of the ladder track being equal to the angle of crossing, the two tracks intersect at the theoretical nose of crossing and no curve is introduced at the turnout crossing to connect the parallel tracks.

Gathering line at limiting angle

In this case the angle of the gathering line is greater than the crossing angle and a curve follows the back leg of the crossing. The back leg of the crossing is followed by the stock joint of the next turnout and no space is wasted. The limiting angle of the gathering line is given by following formula:

$$\text{Sine of limiting angle} = \frac{\text{Space between two adjacent parallel tracks}}{\text{Overall length of turnout}} \quad (15.11)$$

$$= \frac{D}{x}$$

Gathering lines can also be laid at 2α or 3α, that is, at twice or thrice the crossing angle. Such gathering lines are generally found in marshalling yards and are known as *balloon layouts*. This layout of a marshalling yard based on the *Herringbone grid*, is used when the various sidings of the marshalling yard are almost of equal length. This is not a very popular design (Fig. 15.15).

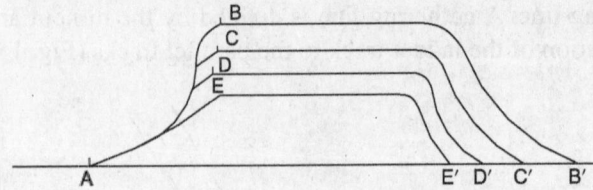

Fig. 15.15 Herringbone grid

15.11 TRIANGLE

A triangle (Fig. 15.16) is mostly provided in terminal yards for changing the direction of an engine. Turntables are also used for this purpose, but are costly, cumbersome, and present a lot of problems in maintenance. Normally, a triangle is provided if enough land is available. A triangle consists of one symmetrical split at R and two turnouts at P and Q along with lead rails, check rails, etc.

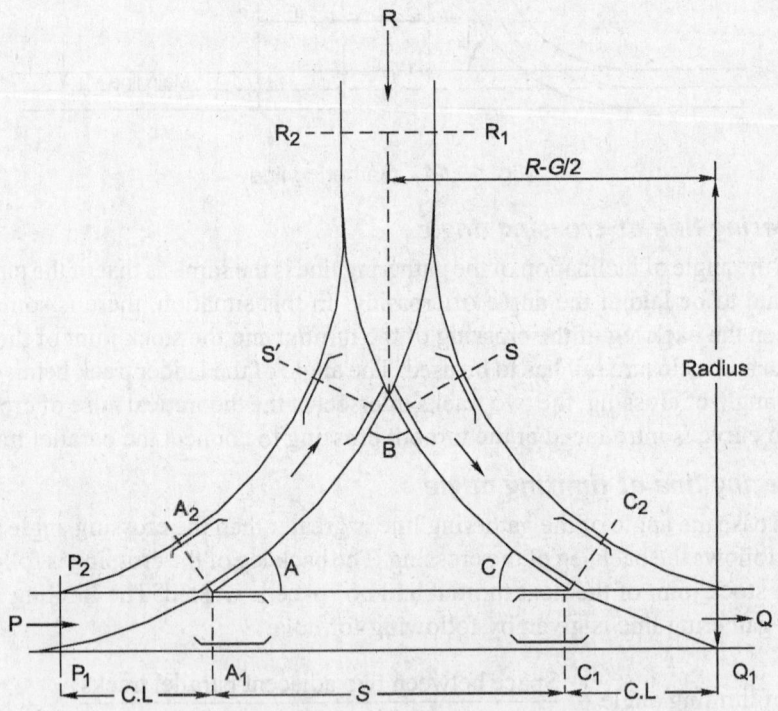

Fig. 15.16 Triangle

To change the direction of an engine standing at P, it is first taken to R, then to Q, and then back to P. By following these movements, the direction of the engine gets changed. The concept of change of direction of the engine was more relevant in the case of steam locomotives and is not applicable to electric and diesel locomotives, which can be operated conveniently from both sides. With the phasing out of steam locomotives from Indian Railways, the triangle is almost redundant.

15.12 DOUBLE JUNCTIONS

A double junction (Fig. 15.17) is required when two or more main line tracks are running and other tracks are branching off from these main line tracks in the same direction. The layout of a double junction consists of ordinary turnouts with one or more diamond crossings depending upon the number of parallel tracks.

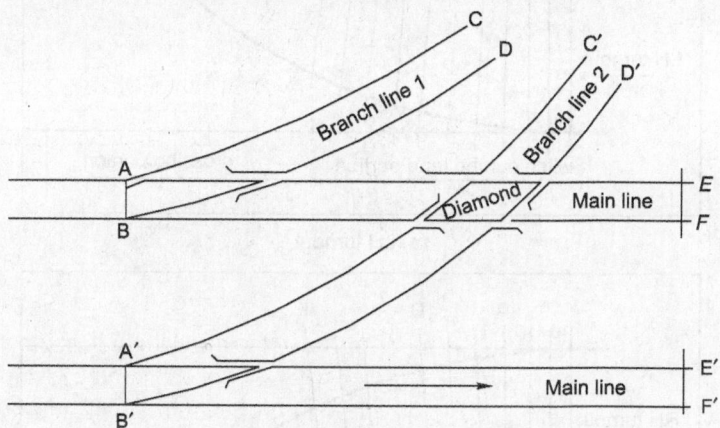

Fig. 15.17 Double junction

Double junctions may occur either on straight or curved main lines and the branch lines may also be either single or double lines. These types of junctions are quite common in congested yards.

15.13 FAN SHAPED LAYOUT WITH CONCRETE SLEEPERS

Fan shaped layout was designed in 1990 especially for concrete sleepers so as to use the same set of sleepers for both RH and LH turnouts.

Earlier, wooden and ST sleepers had holes and the same sleepers were being used for LH as well as RH turnouts without any problem. However, it was difficult to use the same concrete sleepers for both LH and RH turnouts because the contact between inserts and rail is a line contact along the insert face and this is not possible in case of concrete sleepers as the insert have the same length.

In order to solve this problem the fan shaped layout with the following system of laying of concrete sleepers was resorted to. Some important points about the fan shaped layout are given below:

(i) In a fan shaped layout the sleepers in the switch and the crossing portion have the same orientation as in the design, that is, perpendicular to the main line in the switch and at right angles to the bisector of the crossing in the crossing portion.

(ii) In the lead portion the sleepers are laid at the bisector of the angle made between the perpendicular drawn from the main line and the tangent drawn at the various points on the lead curve, that is, $\phi/2$.

(iii) The sleepers spacing as measured along the straight track will be slightly more than on the lead curve (opposite end).

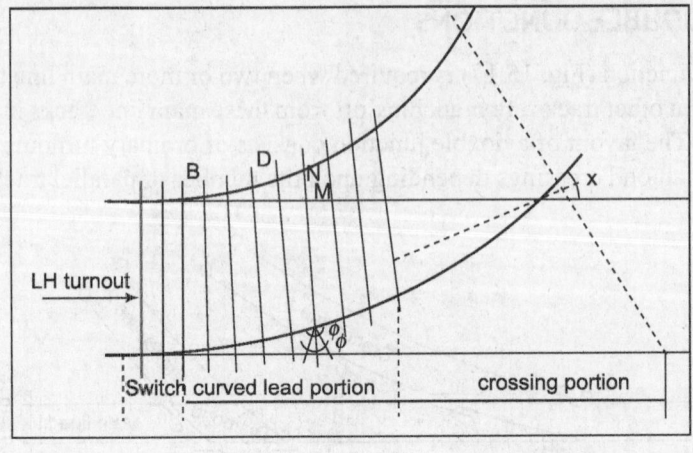

(a) LH turnout

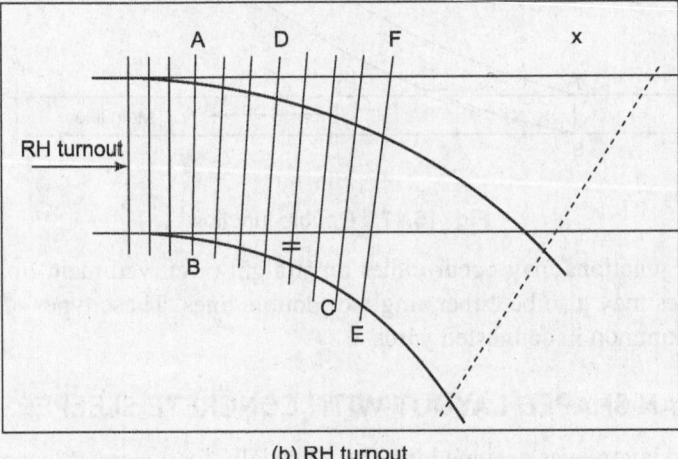

(b) RH turnout

Fig. 15.18 Fan shaped layout for concrete sleepers

Note: It may be seen that because of slightly less spacing on the inner rail and wider spacing on the outer rail and tilting of sleepers in the lead portion, the layout gives the shape of a 'fan' and thus this is called 'fan' shaped layout.

15.13.1 Special Features of Fan Shaped Layouts

The following are the special features of fan shaped layouts:

(i) The fan type will enable the same sleepers to be used for LH or RH turnouts with slight lateral shifting.

(ii) To minimize the number of the design, all PSC turnout sleepers for BG are designed for 60 kg rail section, and for 90R rails section on MG. A smaller rail section turnout can be used with these sleepers by using thicker liners.

(iii) To accommodate different rail sections in the switch portion, the slide chair for different rail sections have common hole spacings so that any slide can be fixed to the sleepers.

(iv) Since 90 R rails will not be used in BG running lines in future, the PSC turnout designs for BG have been evolved for 1/12 60 kg and 52 kg turnouts only.

(v) Only CMS crossings will be used on PSC sleepers layouts.

(vi) The gauge of the concrete sleepers for 1-in-12 BG layout has been fixed at 1673 mm.

15.13.2 Technical Details of Fan Shaped Layout with Concrete Sleepers

The number of concrete sleepers for 1-in-8.5 and 1-in-12 fan-shaped turnout are as given in Table 15.1.

Table 15.1 Number of sleepers and the spacing of sleepers

Type of turnout	Number of main sleepers	Number of special sleepers on approaches	Number of sleepers on exit side	Total number of sleepers
1 in 8.5	54	5	4 on each side, total 8	54 + 5 + 8 = 67
1 in 12	83	5	4 on each side, total 8	83 + 5 + 8 = 96

Notes: (i) **Approach sleepers** *First approach sleeper 60S (2750 mm) is designed for housing a right angled crank. The remaining four approach sleepers each of 2750 mm length are designed to runout cant from 1 in 20 to level.*

(ii) **Exit sleepers** *The exit sleepers 1 and 2 have a length of 2550 mm which is reduced as compared to approach sleepers to avoid interfacing with the sleepers in the adjacent track. These sleepers do not have any notches/dowels.*
The other two exit sleepers (3 and 4) have a length of 3750 mm to accommodate 3 piece gauge tie plates and point machine.

SUMMARY

Track junctions are required to transfer railway vehicles from one track to another or enable a train to cross a track. Several types of junctions can be laid using different combinations of points and crossings. The commonly used layouts have been discussed in this chapter with the help of sketches and solved examples.

REVIEW QUESTIONS

1. A BG turnout takes off a straight track, with a 1-in-12 crossing. Given: heel divergence (d) is 13.3 cm, distance between TNC and tangent point of crossing (h) is 1.346 m, angle of crossing (α) and switch angle (β) are 4° 45′ 49″ and 1°9′0″, respectively. Calculate the lead of the turnout and the radius of the curve. What would be the lead if no straight length is provided at the crossing?

2. A BG branch line takes off in contrary flexure through a 1-in-12 turnout from the main line in a symmetrical split. The maximum permissible speed on the branch line is 40 kmph. Calculate the negative superelevation to be provided on the branch line and the maximum permissible speed on the main line (the degree of turnout curvature for the 1-in-12 crossing is about 4°).

3. Write short notes on the following:
   (a) Gathering line
   (b) Scissors crossover

4. What is a crossover? A crossover with a 1-in-12 turnout and a straight intermediate portion is to be laid between two parallel straight BG tracks. Determine the various design elements. (Assume $D = 5.0$ m)

5. Draw a neat sketch of an RH crossover between two parallel straight railway tracks and list the principal components and terms associated with its layout.

6. Draw a neat sketch of a symmetrical split of a railway track.

7. Draw a neat sketch of a diamond crossing and list its important features. Explain why a square diamond crossing is not desirable.

8. Two parallel railway lines are to be connected by a reverse curve, each section having the same radius. The centre lines of the two tracks are 8 m apart and the maximum distance between the tangent points is 32 m. Find the maximum allowable radius that can be used.

9. Sketch a typical diamond crossing and label all its components. Design a diamond crossing between two BG tracks crossing each other at an angle of 1 in 10.

10. A turnout takes off a 4° curved BG track in contrary flexure. What would be the degree of curvature for the turnout, if a 1-in-8.5 crossing is used?

11. Draw a scissors crossover showing the different rail pieces and gaps distinctly.

12. Calculate the (a) lead distance, (b) radius of the curved lead rail, and (c) overall length of the crossover installed between two BG straight parallel tracks spaced at 5 m from centre to centre. Two 1-in-12 crossings are connected by an intermediate straight track in this layout. Also make a dimensioned sketch of the crossover in question, properly labelling the wing rail, switch rails, and crossing angle.

Choose the correct answer from the choices given.

13. In case of fan shaped layout, the number of sleepers to be laid perpendicular to main line under the switch portion for a 1-in-12 turnout are:
    (a) 1 to 3   (b) 1 to 5
    (c) 1 to 20   (d) 1 to 25

14. For a fan shaped 1-in-12 layout, the length of the longest sleeper is:
    (a) 2750 mm   (b) 3500 mm
    (c) 4200 mm   (d) 4900 mm

15. For a fan shaped 1-in-8.5 layout, the total number of sleepers are:
    (a) 60   (b) 67
    (c) 80   (d) 96

16. In case of a turnout of a similar flexure the degree of turnout curve is:
    (a) higher than the main line curve
    (b) equal to the main line curve
    (c) lower than the main line curve
    (d) depends upon the portion of the yard

17. A three-throw switch can have the following type of flexure:
    (a) contrary flexure   (b) similar flexure
    (c) contrary or similar flexure   (d) none of the these

18. The demand crossing permitted on Indian Railways is:
    (a) 1 in 6   (b) 1 in 8.5
    (c) 1 in 10   (d) 1 in 12
19. A triangle which is meant for changing the direction of an engine is useful for:
    (a) steam locomotive only   (b) diesel locomotive only
    (c) electric locomotive only   (d) all of the above

Rail Joints and Welding of Rails

INTRODUCTION

Although a rail joint has always been an integral part of the railway track, it is looked upon as a necessary evil because of the various problems that it presents. Earlier, rails were rolled in short lengths due to difficulties in rolling and the problem of transportation. With increase in temperature, rails expand and this expansion needs to be considered at the joints. It was, therefore, felt that the longer the rail, the larger the required expansion gap, and this too limited the length of the rail. A rail joint is thus an inevitable feature of railway tracks, even though it presents a lot of problems in the maintenance of the permanent way. This chapter discusses the various types of rail joints and their suitability on a railway track.

16.1 ILL EFFECTS OF A RAIL JOINT

A rail joint is the weakest link in the track. At a joint, there is a break in the continuity of the rail in both the horizontal and the vertical planes because of the presence of the expansion gap and imperfection in the levels of rail heads. A severe jolt is also experienced at the rail joint when the wheels of vehicles negotiate the expansion gap. This jolt loosens the ballast under the sleeper bed, making the maintenance of the joint difficult. The fittings at the joint also become loose, causing heavy wear and tear of the track material. Some of the problems associated with the rail joint are as follows.

Maintenance effort

Due to the impact of moving loads on the joint, the packing under the sleeper loosens and the geometry of the track gets distorted very quickly because of which the joint requires frequent attention. It is generally seen that about 30 per cent extra labour is required for maintenance of a joint.

Bonded main line 6-bolt rail joint on a segment of 76.9 kg/m rail. Note how bolts are oppositely oriented to prevent complete separation of the joint in the event of being struck by a wheel during a derailment.(Courtesy: Sturmovik This file is licensed under the Creative Commons Attribution-Share Alike 3.0 Unported license; http://creativecommons.org/licenses/by-sa/3.0/deed.en)

Lifespan

The life of rails, sleepers, and fastenings gets adversely affected due to the extra stresses created by the impact of moving loads on the rail joint. The rail ends particularly get battered and hogged and chances of rail fracture at joints are considerably high due to fatigue stresses in the rail ends.

Noise effect

A lot of noise pollution is created due to rail joints, making rail travel uncomfortable.

Sabotage chances

Wherever there is a rail joint, there is a potential danger of the removal of fish plates and rails by miscreants and greater susceptibility to sabotage.

Impact on quality

The quality of the track suffers because of excessive wear and tear of track components and rolling stock caused by rail joints.

Fuel consumption

The presence of rail joints results in increased fuel consumption because of the extra effort required by the locomotive to haul the train over these joints.

16.2 REQUIREMENTS OF AN IDEAL RAIL JOINT

An ideal rail joint provides the same strength and stiffness as the parent rail. The characteristics of an ideal rail joint are briefly summarized here.

Holding the rail ends An ideal rail joint should hold both the rail ends in their precise location in the horizontal as well as the vertical planes to provide as much continuity in the track as possible. This helps in avoiding wheel jumping or the deviation of the wheel from its normal path of movement.

Strength An ideal rail joint should have the same strength and stiffness as the parent rails it joins.

Expansion gap The joint should provide an adequate expansion gap for the free expansion and contraction of rails caused by changes in temperature

Flexibility It should provide flexibility for the easy replacement of rails, whenever required.

Provision for wear It should provide for the wear of the rail ends, which is likely to occur under normal operating conditions.

Elasticity It should provide adequate elasticity as well as resistance to longitudinal forces so as to ensure a trouble-free track.

Cost The initial as well as maintenance costs of an ideal rail joint should be minimal.

16.3 TYPES OF RAIL JOINTS

The nomenclature of rail joints depends upon the position of the sleepers or the joints.

16.3.1 Classification According to Position of Sleepers

Three types of rail joints come under this category.

Supported joint
In this type of joint, the ends of the rails are supported directly on the sleeper. It was expected that supporting the joint would reduce the wear and tear of the rails, as there would be no cantilever action. In practice, however, the support tends to slightly raise the height of the rail ends. As such, the run on a supported joint is normally hard. There is also wear and tear of the sleeper supporting the joint and its maintenance presents quite a problem. The duplex sleeper is an example of a supported joint (Fig. 16.1).

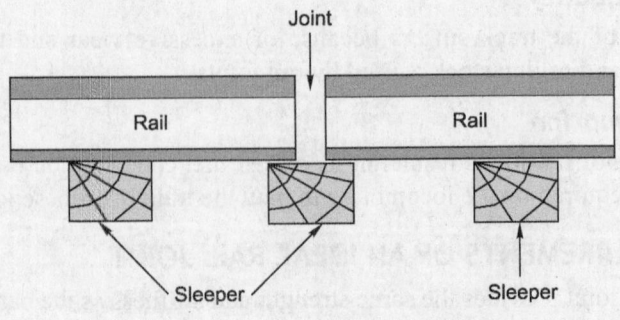

Fig. 16.1 Supported rail joint

Suspended joint

In this type of joint, the ends of the rails are suspended between two sleepers and some portion of the rail is cantilevered at the joint. As a result of cantilever action, the packing under the sleepers of the joint becomes loose particularly due to the hammering action of the moving train loads. Suspended joints are the most common type of joints adopted by railway systems worldwide, including India (Fig. 16.2).

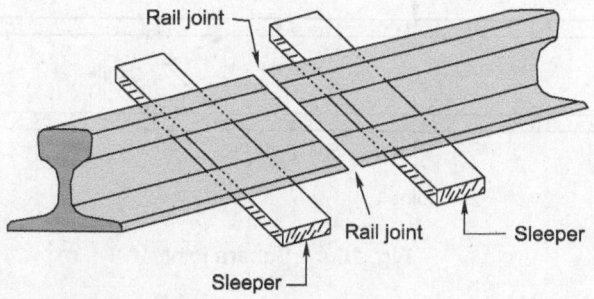

Fig. 16.2 Suspended joint

Bridge joints

The bridge joint is similar to the suspended joint except that the two sleepers on either side of a bridge joint are connected by means of a metal flat [Fig. 16.3(a)] or a corrugated plate known as a bridge plate [Fig. 16.3(b)]. This type of joint is generally not used on Indian Railways.

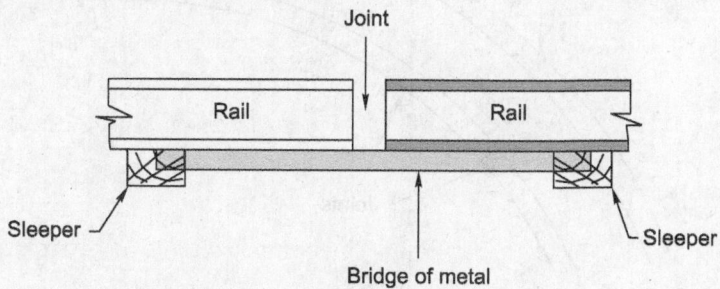

Fig. 16.3 (a)   Bridge joint with metal flat

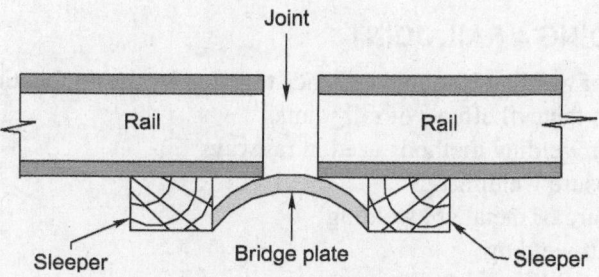

Fig. 16.3 (b)   Bridge joint with bridge plate

16.3.2 Classification Based on the Position of the Joint

Two types of rail joints fall in this category.

Square joint   In this case, the joints in one rail are exactly opposite to the joints in the other rail. This type of joint is most common on Indian Railways (Fig. 16.4).

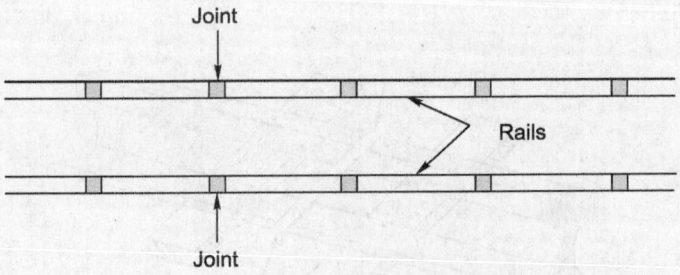

Fig. 16.4   Square joint

Staggered joint   In this case, the joints in one rail are somewhat staggered and are not opposite the joints in the other rail. Staggered joints are normally preferred on curved tracks because they hinder the centrifugal force that pushes the track outward (Fig. 16.5).

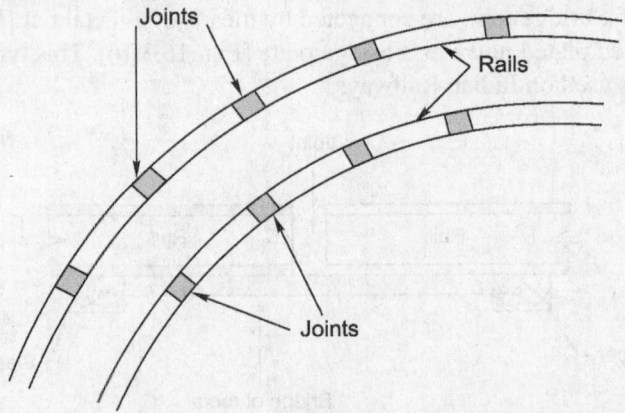

Fig. 16.5   Staggered joint

16.4 WELDING A RAIL JOINT

The purpose of welding is to join rail ends together by the application of heat and thus eliminate the evil effects of rail joints.

There are four welding methods used in railways.

  (a) Gas pressure welding
  (b) Electric arc or metal arc welding
  (c) Flash butt welding
  (d) Thermit welding

The detailed description of these methods is given below.

Welded rail joint (Courtesy: LosHawlos, Wikipedia)

16.5 GAS PRESSURE WELDING

In gas pressure welding, the necessary heat is produced by the combination of oxygen and acetylene gases. The rail ends to be welded are brought together and heat is applied through a burner connected to oxygen and acetylene cylinders by means of regulators and tubes. A temperature of about 1200°C is achieved. At this temperature, the metal of the rail ends melts, resulting in the fusion and welding together of the ends.

The rails to be welded are clamped at the wall by applying a pressure of 40 tonne pressure, heated to a temperature of about 1200°C to 1400°C, and butted with an upset pressure of about 20 tonnes. Then the joint is again heated to a temperature of 850°C and allowed to cool naturally. It has been seen that this method of welding is cheaper as compared to flash butt welding. The quality of this welding joint is also claimed to be quite good. There are both stationary and mobile units available for gas pressure welding.

The process, though simple, has not yet been adopted on a large scale on Indian Railways. The main reason is its limited output and the difficult and irregular availability of gas. India has only one plant that offers gas pressure welding, which is located at Bandel on the ER (Eastern Railway) and the progress in this plant has been nominal.

16.6 ELECTRIC OR METAL ARC WELDING

In electric or metal arc welding, heat is generated by passing an electric current across a gap between two conductors. A metal electrode is energized by a voltage source and then brought close to another metal object, thereby producing an arc of electric current between the two objects. A lot of heat is generated by this electric arc, causing the two rail ends to fuse or weld.

This type of welding can be done using any of the following methods.
  (a) Insert plate technique
  (b) Scheron process
  (c) Enclosed space technique
  Indian Railways has recently started welding rail joints using the metal arc process on a trial basis and the performance so far has been satisfactory.

16.7 FLASH BUTT WELDING

In flash butt welding, heat is generated by the electric resistance method. The ends of the two rails to be welded are firmly clamped into the jaws of a welding machine. One of the jaws is stationary, while the other is moveable and as such the gap between the two rail ends can be adjusted. It is not necessary to specially prepare the rail ends, though these can be preheated with an oxy-acetylene torch, if necessary. The rail ends are brought so close together that they almost touch each other. An electric current of 35 kA is passed between the interfaces of the two rails, developing a voltage of 5 V. The rails are subjected to a predetermined number of preheats (15 for 52 kg rails and 13 for 90 R rails) before they are welded. A lot of flashing (sparking) occurs and considerable heat is generated by the passage of electrical current between the rail ends. The rail ends are automatically moved to and fro by the machine till the temperature rises to a fusion limit in the range of 1000°C to 1500°C. At this juncture, the rail ends are pressed together with an upset pressure of about 37 tonnes and final flashing takes place joining the two rail ends together. The process is so well regulated that any steel that might have been oxidized during the preheating phases gets completely eliminated. The total time taken for welding a joint is 150–200 seconds and the loss in rail length is about 20 mm for each weld. In the case of 90 R rails, the total welding time is 161 seconds, which includes a burn-off period of 20 seconds, on-preheat time of 65 seconds (13 × 5 sec), off-preheat time of 36 seconds (12 × 3 sec), and final flashing time of 40 seconds.

  High-quality welded joints are produced by the flash butt welding method. The cost of a welded joint using this method is also quite low compared to other methods of welding. The method, however, can be adopted most economically and efficiently only in a workshop, for which capital investment is required.

  Flash butt welding is the standard method of welding of rails on Indian Railways. Most railways have one or more flash butt welding plants where rails are welded together. The existing plants on Indian Railways are listed in Table 16.1.

Table 16.1 List of welding plants on Indian Railways

Name of the railway	Site of flash butt welding plant
Central Railway	Chalisgaon and Kalyan
Eastern Railway	Bandel and Mughalsarai
Northern Railway	Meerut
Southern Railway	Arakonam
South Eastern Railway	Jharsuguda
South Central Railway	Moulali
Western Railway	Sabarmati
North Central Railway	Gonda

The stepwise procedure for the flash butt welding of rails is as follows.

Pre-straightening of rails   The rails are straightened before they are welded in order to ensure that the welded rail has a good alignment.

End cleaning   The ends of the rails are cleaned for a length of 150–225 mm using electric or pneumatic grinders.

Adjustment of rail ends   The rail ends are then brought together in the flash butt welding machine and aligned longitudinally and vertically by suitably adjusting the machine.

Welding   The rail ends are then welded in the flash butt welding machine. Most machines on Indian Railways are manufactured by AI. Welders, Inverness, Scotland. The important characteristics of a typical machine manufactured by AI. Welders are presented in Table 16.2.

Table 16.2   Characteristics of a welding machine

Characteristics	52 kg rail	90 R rail
Maximum clamping force	60 tonnes	60 tonnes
Maximum butting force	37 tonnes	37 tonnes
Temperature achieved	1500°C	1500°C
Number of preheats	15	13
Preheat time	5.5 sec	5.0 sec
Total time of welding	190 sec	161 sec
Flashing stroke	13 mm	13 mm
Butting stroke	10 mm	10 mm

Stripping   As soon as the rails are welded, they are made to pass through a stripping machine, where all the extra metal, called upset metal, is chipped off.

Hot chipping   In case there is no stripping machine available, the extra material on the rail head is chipped off manually using pneumatic chisels while the metal is still hot.

Spray cooling   After the hot metal is chipped off, the rails are cooled by spray cooling.

Profiling   The rails are then correctly profiled.

Post straightening   The rails are straightened in the post straightening machine, which removes both horizontal and vertical kinks, if any, so as to ensure perfect alignment in both directions.

Ultrasonic inspection   The rails as well as the welds are examined to ensure that there are no flaws in them. This is particularly important for second-hand rails.

Examination and inspection   The rail ends are finally examined and inspected with regard to specified tolerances so that the welded surface has a good finish.

16.7.1 Output and Cost

The average time taken for welding a joint is about 6 minutes for 52 kg rails and 5.5 minutes for 90 R rails, and about 70 to 90 joints can be welded per eight hour shift. The flash butt welding plant at Meerut (Northern Railway) welds about 120 joints per day by working in two shifts. The approximate cost comes to about ₹ 1000 per weld including overheads and depreciation charges.

16.7.2 Welding Recorder

The quality of welding can be checked using a 'welding recorder', which automatically records all the parameters that control the quality of a weld. The following parameters are recorded by this device.

(a) Primary amperage
(b) Voltage
(c) Butting pressure
(d) Loss of length

A graphical study of the records of these parameters helps in judging the quality of the welding, after which the desired action can be taken if any one of these parameters is found to be improperly regulated. A few of these welding recorders have recently been purchased by Indian Railways and are being used in flash butt welding plants.

16.7.3 Automatic Flash Butt Welding Machine

Indian Railways has procured a few of the latest superior quality Mark IV type of flash butt welding machines (APHF-60). The new design of this machine permits the welding of rails of sections up to 60 kg/m or above made up of medium manganese and of the wear resistant type. Most of the operations in this machine are automatic. These machines are capable of aligning and de-twisting rail ends to facilitate the formation of high-quality welded joints. The technical characteristics of this machine are given in Table 16.3.

Table 16.3 Characteristics of APHF-60

Characteristics	*Value*
Maximum clamping force	122 tonnes
Maximum forging force	61 tonnes
Horizontal and vertical adjustment	6 mm
Maximum starting welding gap	190 mm
Maximum finished welding gap	100 mm
Maximum force of rail alignment and anti-twist unit	15.5 tonnes
Average out put of the machine	20 welds/h

The new automatic flash butt welding machine has many supplementary machines such as the grinding machine, pre and post straightening machines, the short blasting or brushing machine for end cleaning, and generators. The cost of the ensemble of supplementary machines is about ₹ 90 million whereas the cost of the main welding machine is about ₹ 40 million. The new welding machine is able to perform most of the operations automatically and the time taken for welding a 52 kg rail is approximately 70 seconds. The average output of the machine is 20 welds per hour.

16.7.4 Manual for Flash Butt Welding of Rails

The code of practice for the flash butt welding of rails has been standardized on Indian Railways from time to time. The latest instructions in this regard are contained in the *Manual for Flash Butt Welding of Rails 1994*. The manual describes the type and suitability of the rails to be welded and the general procedure to be followed, and enlists the tolerances for the finished joints as well as the acceptance tests the joints must undergo to ensure quality control.

Tolerances for flash butt welded joints

Each completed flash butt welded joint should be checked for its straightness, alignment, and finish using 1m- and 10 cm-long straight edges. The permissible tolerances are given in Table 16.4. These tolerances also apply to thermit welded joints barring the web zone, where the tolerance specified is +10 mm and −0.0 mm.

Table 16.4 Limits of tolerances

Item	Tolerances for welds with new rails	Tolerances for welds with old rails
Vertical misalignment	+0.3 mm and −0.0 mm at the centre of a 1m straight edge	±0.5 mm at the centre of a 1 m straight edge
Lateral misalignment	±0.3 mm at the centre of a 1-m straight edge	±0.5 mm at the centre of a 1 m straight edge
Head finishing (one side)	±0.25 mm on the gauge side at the centre of a 10-cm straight edge	±0.3 mm on the gauge side at the centre of a 10 cm straight edge
Head finishing (on top table surface)	+0.2 mm and −0.0 mm at the centre of a 10-cm straight edge	±0.2 mm at the centre of a 10 cm straight edge
Web zone (underside of head, web, top of base, and both fillets on each side)	+0.3 mm and −0.0 mm of the parent contour	+3.0 mm and −0.0 mm of the parent contour

Testing of rail joints

A rail joint should be tested for its strength and hardness before it is considered acceptable for use in railways. The following tests are prescribed on Indian Railways.

Transverse test   One joint should be tested using the transverse test daily before work starts in all flash butt welding depots where there is no provision of welding recorders. In depots where recorders have been provided, one in every 1000 joints should be tested using the transverse test.

In the transverse test, a 1.5m-long test piece with a weld in the centre is taken and placed on two cylindrical supports that have a diameter of 30 to 50 mm and are placed 1 m apart. When pressure is applied in the form of a load at the centre of the test piece, it should show the minimum recommended deflection without any sign of cracking (Fig. 16.6).

Metallurgical test   A macro graphic examination of the flash butt weld is done after every 5000 welds. This test checks the presence of any porosity due to cracks, slag inclusion, or other welding defects.

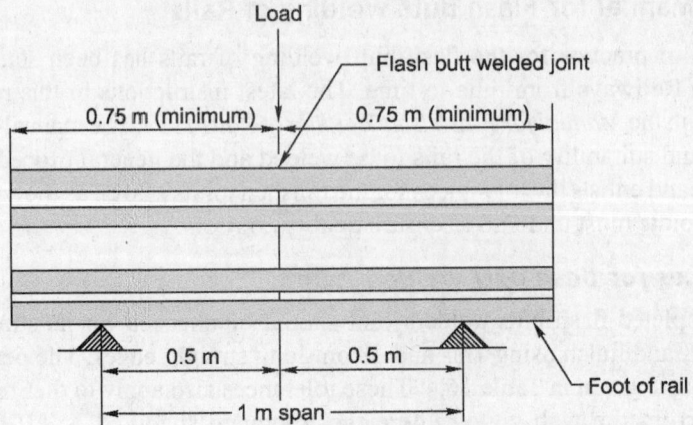

Fig. 16.6 Transverse test

Hardness test A hardness test may also be carried out for the welded head affected zone. The Brinell hardness number (BHN) should be between 210 and 250, presuming that the BHN of the parent rail is 230.

Ultrasonic flaw detection Every joint is USFD (ultrasonic flaw detection) tested using normal 45°/37°, 70°, and 80° probes to cover the head, web, and foot.

Fatigue test Testing is done for two stress ranges with a 20 per cent reversal.
$$+27.5 \text{ kg/mm}^2 \text{ to } -5.5 \text{ kg/mm}^2 \text{ (range 33 kg/mm}^2)$$
$$+25 \text{ kg/mm}^2 \text{ to } -5 \text{ kg/mm}^2 \text{ (range 30 kg/mm}^2)$$

Welding of second-hand rails

Second-hand rails can be welded conveniently in flash butt welding depots after being cropped for use on branch lines. European countries implement the welding of second-hand rails on a large scale in order to economize. The aspects that require particular attention in the welding of second-hand rails are as follows:

(a) Checking of the dimensions of old rails as per specifications
(b) Matching the old rails
(c) Sawing the rail ends
(d) Planing the rail head
(e) Permissible wear of the rails to be welded
(f) Marking the gauge side
(g) Ultrasonic inspection of the rails

16.7.5 Operations involved

Various operations involved in welding of second-hand rails by the flash butt welding technique are shown in the flow chart shown (Fig. 16.7).

16.8 THERMIT WELDING OF RAILS

Thermit welding of rails is the only form of site welding that is adopted universally. The method was first developed by Gold Schmidt of Germany towards the end of the nineteenth century. A code of practice for welding rail joints using the alumino-

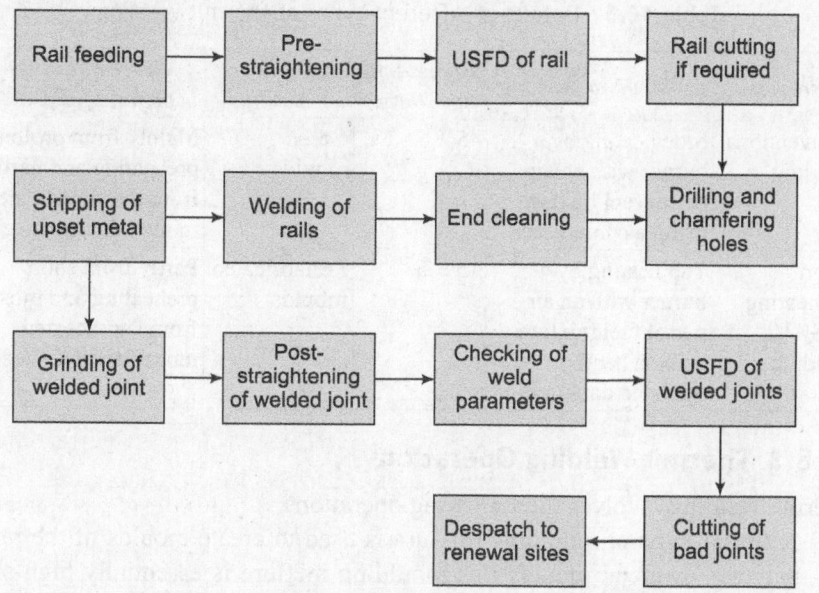

Fig. 16.7 Operations involved in flash butt welding

thermic process has been developed on Indian Railways. The code defines the method of welding and the precautions and steps to be taken before, during, and after welding for the production of satisfactory weld joints.

16.8.1 General Principles

The principle behind this process is that when a mixture of finely divided aluminium and iron oxide, called *thermit mixture*, is ignited, a chemical reaction takes place which results in the evolution of heat and the production of iron and aluminium oxide:

$$Fe_2O_3 + 2Al = 2Al_2O_3 + 2Fe + heat$$

In this reaction, 159 g of iron oxide combines with 54 g of aluminium to give 102 g of aluminium oxide, 112 g of iron, and 182 kcal of heat. The reaction is exothermic and it takes about 15–25 seconds to achieve a temperature of about 2450°C. The released iron is in the molten state and welds the rail ends, which are kept enveloped in molten boxes. The aluminium oxide, being lighter however, floats on top and forms the slag.

16.8.2 Different Types of Thermit Welding

There are two types of alumino-thermic welding processes sanctioned for the welding of rails on Indian Railways. These are conventional welding and SKV welding. SKV is the short form of the German phrase 'Schweiss-Verfahran mit Kurz vorwarmung' meaning the short preheat welding method. The technique is therefore also termed SPW (short preheat welding). The Railway Board, as a matter of general policy, has decided that the SKV welding technique should be introduced as soon as possible on Indian Railways. Table 16.5 gives the details of these two types of thermit welding.

Table 16.5 Details of different types of thermit welding

Process	*Mode of preheating*	*Preheating time (min.)*	*Type of mould*	*Source of energy to achieve full fusion*
Conventional welding	Side heating by a burner with an air–petrol fuel mixture using a compressor	45 ± 5	Green moulds	Mainly from prolonged preheating and partly from the superheated molten thermit steel
Short preheating or SKV welding	Top heating by a burner with an air–petrol fuel mixture using a hand-operated compressor	15 ± 5	Prefabricated moulds	Partly from short preheating and mostly from superheated molten thermit steel

16.8.3 Thermit Welding Operations

Thermit welding involves the following operations.

(a) A special type of moulding mixture is used to create moulds of the rail in halves. For green moulds this moulding mixture is essentially high silica sand mixed with bentonite sieved to the required gradation so that it is coarse enough to permit ventilation. The sand should neither be too dry nor too wet. It is mixed with dextrin (a form of molasses) to make it as pliable as desired. The moulds are clamped at the rail joint in such a way that there is adequate peripheral clearance around the rail profile. Normally, green sand moulds are used for conventional thermit welding and prefabricated carbon dioxide sand moulds are used for SKV welding.

(b) After fixing and luting the moulds, the rail ends are heated with a blue flame so as to attain a temperature of 950°C to 1000°C for the conventional process and 600°C for the short preheating process. In case of conventional welding, heating should be continued till the rail ends have turned yellowish red or orange, which can be checked visually through a coloured glass.

   An opening is provided in the mould through which heat is supplied by the means of burners that use any one of the following fuels:

   (i) Air and petrol
   (ii) Oxygen and cooking gas (LPG—liquefied petroleum gas)
   (iii) Oxygen and propane
   The time taken for preheating is about 30–45 minutes for conventional welding and 10–12 minutes for short preheating (SKV) welding.

(c) A special type of crucible lined with magnetite is fixed near the rail joint in such a way that, when required, it can be swung a round and brought exactly over the joint. A hole is provided in the bottom of the crucible which is plugged with a closing pin and has asbestos wool sprinkled over it to protect it from the molten steel. Powdered slag is then strewn over the asbestos wool so that it lies undisturbed. (Fig. 16.8).

(d) The thermit mixture is then placed inside the crucible. About 4–7 kg of the mixture is required for conventional thermit welding and about 9.0–15 kg of it is required for SKV welding.

(e) As soon as preheating is completed, the thermit mixture is ignited using special igniters made of barium peroxide and aluminium. A violent reaction takes place in the crucible that leads to the evolution of heat, and the thermit mixture turns into a molten bath. The slag, being lighter, floats to the top and the molten iron remains at the bottom. The reaction takes place for about 15–25 seconds and an extra margin of about 5 seconds is kept for the separation of the slag. A temperature of about 2540°C is reached during the process.

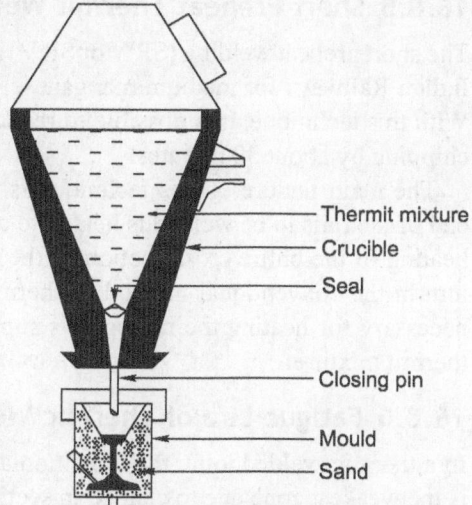

Thermit mixture
Crucible
Seal
Closing pin
Mould
Sand

Fig. 16.8 Thermit welding equipment

(f) The crucible is then swung around and the closing pin is taped up. Molten iron flows down and fills the peripheral area around the mould. The crucible is then swung further and the slag flows out.

(g) The molten thermit steel fuses around the preheated surface of the rail ends and a homogeneous weld is made.

(h) The moulds are removed after about 5 minutes. When demoulding, only the head, and not the foot and the web, should be exposed. The excess metal is chipped off from top of the rail and the gauge face while it is still red hot.

16.8.4 Post-welding Operations

The following operations are carried out after the welding of the rail ends is complete.

(a) The rail ends are cooled for 3–4 minutes; controlled cooling is required for alloy steel rail joints.

(b) In order to ascertain that the rail profile is correct, the finish of the welded joint is achieved either by the use of hand files or portable grinders.

(c) The welded joint is now ready. The sleepers are shifted to their original positions and properly packed. At least 30 minutes should have elapsed since the pouring of the metal before the first train is allowed to pass on the welded joint.

(d) The USFD testing of new welds made by thermit welding should be completed within 30 days of executing the welds.

The thermit process is a very convenient form of welding when work needs to be carried out at the site. No extra power is required and there is enough heat generated during the chemical reaction. The welded joint, however, is found to be weak in strength as compared to the flash butt welded joint. The conventional process has generally been abandoned on Indian Railways to pave way for the short preheat process.

16.8.5 Short Preheat Thermit Welding Technique

The short preheat welding (SPW or SKV) method has recently been developed on Indian Railways for medium manganese, wear resistant, and special alloy rails. With this technique, it is possible to reduce the total time taken for welding and chipping by about 30 minutes.

The main feature of this technique is that only a length of 3–5 mm at each end of the rails to be welded is heated to a temperature of 6000°C, as against the heating of the entire cross section of the rail to 1000°C over a length of 10–15 mm in the conventional method of thermit welding. The large quantity of heat necessary for heating the rail ends is supplied by the use of a large quantity of thermit mixture.

16.8.6 Fatigue Life of Thermic Welded Joint

In a thermic welded joint, the transition area zone from collar to the rail section is the weakest zone due to change in sectional area and change in the metallurgy on account of casting and heat effect. Any improvement on this area by way of reducing the thickness of the collar and reduction in the heat affected zone will improve the fatigue life. With short preheating using gases, the heat effected zone gets minimized to 5 mm as against 15 mm with SKV welding process compared to the conventional welding process.

Some welding experts feel that if it is possible to clean the root and carry out the polishing of this area, then the fatigue life of joints can be improved by almost 30 per cent.

The short preheat process is best suited for in-situ welding on busy routes, as with this process it is possible to do welding in 30 minutes as against the minimum 60 minutes required for conventional thermite welding.

16.8.7 Conventional Welding Versus SKV Welding

The salient features of conventional thermit welding in the case of 52 kg medium manganese rails versus those of short preheat welding are listed in Table 16.6.

Table 16.6 Important features of conventional and SKV welding

Item/characteristic	Conventional alumino-thermic welding	Short preheat thermit welding (SKV)
Weight of the thermit mixture	7 kg	10.5–13 kg*
Preheating time with petrol and air	45 min.	10–12 min.
Total time taken from mould assembly to final chipping	60–65 min.	20–30 min.
Initial gap between rail ends	12–14 mm	23–25 mm
Preheating technique	Side heating	Top heating
Type of moulds	Green sand mould	Prefabricated mould
Petrol consumption	3.4 l	1.0 l

* Lately a new SKV welding technique with thin collar has been designed for use with lesser amounts of thermit mixture.

16.8.8 Precautions During Thermit Welding

In order to ensure the quality of thermit welded joints, the following precautions should be taken:

Follow prescribed procedure   Thermit welding of joints should be carried out strictly as per the prescribed norms. The horizontal and vertical alignment of two rail ends require special attention at the joint. In particular, care should be taken to see that the rail ends are square and that their alignment is perfect.

Equipment in good order   All the relating equipment and gadgets should be in working order and be available at the site. The important welding equipment and gadgets are rail thermometer, rail tensor, stop watch, 10 cm straight edge, feeler gauge, leather glove, blue goggles, wire brush, slag container, spatula, and first aid box.

Qualified welder   Thermit welding should be done only by a qualified welder who holds a valid competency certificate.

Effective supervision   Thermit welding should be done only under the supervision of a qualified PWI/PWM (permanent way inspector/permanent way mistry) with a valid competency certificate.

End cropping   Second-hand rails should not be welded before their ends have been cropped. The rail ends should be cropped vertically and thoroughly cleaned with kerosene oil with the help of a brush.

Proper gaps   In order to get good results, proper gaps should be ensured between the two rails to be welded. The standard gaps recommended are the following:

Conventional welding	11 ± 1 mm
SPW or SKV welding	24 ± 1 mm
50 mm welding	50 ± 2 mm

Adequate block   When the conventional method is used for the thermit welding of rails on a running line, the work should normally not be completed in a time block of less than 75 minutes. In the case of SKV/SPW welds, the same work should not be done in a block of less than 50 minutes.

Use of rail tensor   A rail tensor must be used for maintaining the correct gap when thermit welding rails in a decreasing range of temperature and also when repair welding on LWR/CWR (long welded rail/continuous welded rail) tracks. In the case of repair welding, 100 m on either side of the weld should be destressed in order to get good results.

Work to be done on cess   In the field, thermit welding should be done on the cess as far as possible to ensure the quality of the welded joints. Luting should be done after ensuring that the moisture content is minimum so as to improve the quality of the weld. In the case of cess welding, rails should be supported by about ten wooden blocks under each rail seat.

Adequate pressure   Welding should normally be done at a pressure of 7.0–7.6 kg/cm^2 (100–110 psi). The time taken for preheating should normally be about 10–12 minutes.

Use of wooden planks   The portion of the rail to be welded should be kept on wooden planks to ensure that moisture does not enter these portions.

Finishing of joint   After welding, the joint should be given a proper finish on both the gauge as well as the non-gauge side and any extra collar should be removed in order to enable SFD testing.

Joggled fish plate   After thermit welding of LWR, the joint should be joggle fish plated and supported on wooden blocks till it has cleared the USFD test.

16.8.9 Testing of Thermit Welded Joints

A rail joint should be tested for its strength and hardness before it can be accepted for use by the railways. To this end, the following tests are prescribed on Indian Railways.

Reaction test   The characteristic reaction of the thermit mixture when it is placed in a standard crucible is scrutinized to ensure that it conforms to the specified standards. The alumino-thermit steel is extracted out of the melted metal and its chemical composition is determined. The aluminium content should be between 0.3 per cent and 0.7 per cent. The reaction test should be carried out on the mixture for every 250 portions or part thereof.

Hardness test   The Brinell hardness test is carried out in welded zones, in heat-affected zones, on the parent metal of the rail, and at the top and sides of the head of the test weld using a 3000 kg load and 10 mm diameter ball for 10 seconds. The average Brinell hardness number (BHN) for welded and heat-affected zones as well as for the parent metal of different rail sections should be as given in Table 16.7.

Table 16.7   Value of BHN for different rail sections

Description of metal	Medium manganese IRS T-12	90 kg/mm² UTS and UIC rails	UIC chrome manganese alloy steel	UIC chrome vanadium alloy steel
Parent metal hardness	230	280	310	310
Welded zone hardness	230–250	280–300	310–330	310–330
Heat-affected zone hardness	210–250	260–300	290–330	290–330

Transverse breaking load test   The test weld is positioned on cylindrical or semi-cylindrical supports of diameter 30–50 mm at a distance of 1 m from centre to centre, with the weld placed at the centre of the span and loaded in such a manner that the foot of the rail is in tension. The load is gradually increased till a rupture occurs in the weld. The test weld should withstand the minimum deflection that has been specified for all the different sections and types of rails.

One out of every 100 welded joints should be picked up at random and be subjected to both the hardness and transverse tests. For 90 UTS rails weighing 50–60 kg/m, the minimum breaking load is 80 tonnes with a minimum deflection of 15 mm at the centre. The tolerances for the various dimensions of thermit-welded joints are the same as specified for flash butt welded joints.

The important features of flash butt welding and thermit welding are compared in Table 16.8.

Table 16.8 Comparison of flash butt and thermit welding

Description of item	Flash butt welding	Thermit welding
Principles of welding	By passing a 35,000 A electric current between two rail ends	By initiating an exothermic chemical reaction between iron oxide and aluminium
Quality of welding	Excellent	Good
Strength of welding	Good in fatigue	Weak in fatigue
Time required for welding	About 3–6 min.	10–12 min. for SKV and 30–45 min. for conventional
Place of welding	Normally in workshop	On site
Cost of welding	₹ 400–600 per weld	₹ 700–1200 per weld
Tolerance	Very high	Normal
Control on the quality of welding	Quality can be controlled with the help of a welding recorder	Quality control is possible only by working diligently and no monitoring is possible

16.9 RECENT DEVELOPMENTS IN WELDING TECHNIQUES

The welding methods described above have been further improved in the recent past by employing new techniques and equipment. These techniques and equipment are described below in detail.

Automatic welding recorder   The invention of a welding recorder for controlling the quality of welding in the flash butt welding method is a recent development. The recorder is able to not only identify the defects in the flash butt welded joint, but also indicate the reasons for the same. A very sophisticated recording system has recently been developed on the German Railways with the help of Siemens. This system gives visible and audible indications whenever any parameter controlling the quality of the weld transgresses the predetermined limits. The recorder helps considerably in exercising proper control on the quality of welding.

Thermit welding without preheating   A new method of thermit welding rails that does not involve preheating is currently being tested on German Railways. The necessary heat is produced by much larger quantities of a specially-manufactured alumino-thermit mixture. This method also involves the use of special prefabricated moulds made from pure quartz sand. This method of welding requires considerably less amount of time due to the prior solidification of the material in the mould. A similar method is also being tried on Indian Railways on an experimental basis.

Flame cutting and welding technique   A new technique of 'Flame cutting and welding' has been developed recently on Indian Railways. This technique is proposed to be adopted to get desirable gap in a welded track. In this technique time in flame cutting is only 2 minutes as against 15 to 20 minutes for cutting by Hacksaw. Similarly, the new technique of welding with wide gaps (50 mm or 75 mm) has recently been standardized on Indian Railways. In the new technique a gap of 50 mm or 70 mm gap is created by cutting the defective joint. The gap is held in position with the help of a tensor. This gap is fitted with wider mould and heated for 20 minutes. Welding portion, which is in bigger quantity, is ignited and after the reaction is completed, the same is poured into the moulds. Demoulding is done after about 5 minutes and the joint is finished to perfect tolerance.

Wide gap (50 mm or 75 mm) welding technique   At present, the fractured rail welds have to be replaced by a long rail closure. This rail piece is inserted in the track after creating a gap of about the same length and welded with the existing rail on both the ends. Thus, in order to remove one defective/fractured alumino thermit (AT) weld, two AT welds are created. The whole process of replacing the fractured/defective weld with closure rail piece involves considerable amount of manpower and block time as well as wastage of rails. With the development of 75-mm wide gap welding technique, the defective/fractured weld can be replaced with a single manpower and block time required for the execution of AT welds also reduces considerely.

The major procedural difference between the standard 25 mm gap welding and 50/75 mm wide gap welding arise mainly due to larger quantity of thermic steel the latter in due to increased volume and solidifiation. Hence, this results in temperature dependent this results in post welding activities. The time required for the executing of wide gap welding joint is slightly longer than a standard 25 mm weld, but still lower than the time required for two 25 mm standard welds.

Project for acquiring state of the art AT welding technology   With the objective of improvement in the quality of AT weld, a work to acquire state of the art technology to improve the quality of alumino thermic welds has been sanctioned by the Railways Board.

The objective of the project is to acquire the state of the art technique of aluminothermit (AT) welding as a complete package encompassing the entire gamut of activities involved in the welding. The package would involve improvement in the consumables and equipment, training of welders and supervisors and the latest technological advances in this field from global leaders. It is expected that all aspects of alumino-thermit welding would get upgraded by this project.

This is expected though one of the major hurdles in improving the quality of AT welding is the element of human error. The same can be reduced by (a) use of one-shot crucible (b) use of auto thimble (c) acquiring skills.

Mobile flash butt welding machines   In order to produce good quality welded joints, Indian Railways has recently procured a few K335 type mobile flash welding machines manufactured by Messers Plasser and Theureri as shown in Fig. 16.9.

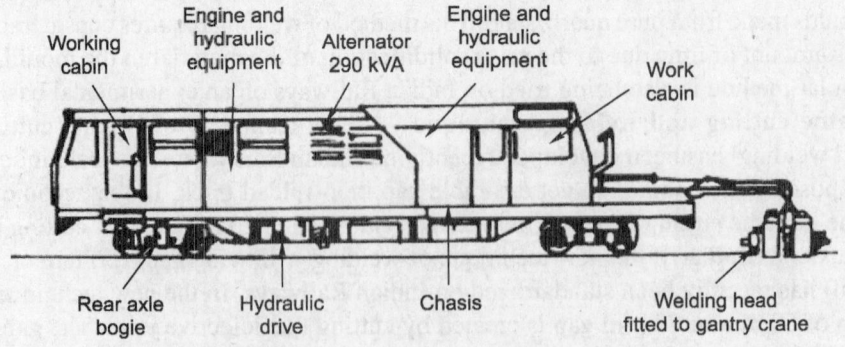

Fig. 16.9   Mobile flash butt welding equipment

The new machine is self-propelled with a separate electric generator and can weld about 60 joints per day under field conditions. The approximate cost of welding comes to about ₹ 500 per joint using this machine.

SUMMARY

Rail joints are used to join rails in series and are the weakest links in the track. Fish-plated joints are particularly weak. The continuous hammering at the joints results in the battering and wear of rail ends. Welded rail joints are the most suitable as they satisfy almost all the requirements of an ideal rail joint. There are several methods of welding rails but the flash butt and thermit welding methods are normally preferred. Flash butt welding is cheaper than thermit welding and also gives better results. Recent developments in welding techniques have made it possible to produce high-quality welded joints.

REVIEW QUESTIONS

1. List the requirements of an ideal rail joint. What are the problems that arise due to rail joints? Enumerate the different types of rail joints.
2. Why is welding of rails necessary? Obtain an expression for the force required to prevent the expansion of rails due to temperature variations.
3. Name the different methods of welding rails. Describe any one method.
4. What are the different methods of welding rails? What are their advantages and disadvantages? Elaborate on the tests conducted to check the quality of the weld.
5. How are rails welded using the thermit welding method? What are its relative advantages/disadvantages vis-à-vis the flash butt welding method?
6. What are the specifications on Indian Railways for flash butt welding of rails? What are the tolerances prescribed for flash butt welded joints? Briefly describe the various tests required to be conducted before a welded joint is accepted for use on a railway track.
7. What are the recent developments in welding techniques on Indian Railways?
8. Describe in detail the various steps involved in flash butt welding. What are the various tests for determining whether the welds are of the right quality?
9. What is the principle of thermit welding? Describe the various steps involved in the thermit welding of a rail joint.
10. Write short notes on (a) SKV welding, (b) suspended joint, (c) welding of second-hand rails, (d) automatic welding recorder.

Choose the correct answer from the choices given.

11. Duplex sleeper is an example of:
    (a) supported rail joint   (b) suspended rail joint
    (c) bridge joint   (d) none of these
12. The rail joint normally used on Indian Railways is:
    (a) supported rail joint   (b) suspended rail joint
    (c) bridge joint   (d) none of these
13. Thermit welding or alumino thermic of rails was originally started by:

(a) John Thermit of UK   (b) Gold Schmidt of Germany

(c) K. Alumino and W. Thermit   (d) none of them

14. In thermit welding, the powder used for ignition is composed of:

(a) aluminium   (b) iron oxide

(c) both (a) and (b)   (d) aluminium oxide

15. In case of 90 UTS rails prior to flange cutting, the rail end shall be preheated to the extent of:

(a) 300–350°C   (b) 250–300°C

(c) 200–250°C   (d) none of these

16. After pouring metal in the welding technique by AT welding the first train can be allowed after:

(a) 10 minutes   (b) 20 minutes

(c) 30 minutes   (d) 45 minutes

17. In flash butt welding, the maximum clamping force applied is:

(a) 100   (b) 122

(c) 150   (d) 200

18. A welded rail joint is:

(a) generally supported on a sleeper

(b) rails supported on a metal plate

(c) left suspended

(d) supported on ballast

19. Preheating time for AT welding for air–petrol is:

(a) 10 to 12 minutes   (b) 8 to 10 minutes

(c) 6 to 8 minutes   (d) 6 minutes

Modern Welded Railway Track

INTRODUCTION

As discussed in an earlier chapter, a rail joint is the weakest link in the railway track. The ill effects of the rail joint are well known, including the fact that a track with joints requires about 30 per cent extra maintenance work as compared to a plain track. The best remedy for problems caused by rail joints lies in welding the rails and reducing the number of joints to the extent possible. The modern welded railway track incorporates systematic welding of rails, which provides it the potential to carry trains at faster speeds, provide better riding conditions, and reduce maintenance costs.

The following terms are commonly used with respect to welded tracks.

Long-welded rail (LWR)   It is a welded rail in which the central portion does not undergo any longitudinal contraction or expansion due to temperature variations (thermal expansion). Normally, a rail with a length greater than 250 m on BG and 300 m on MG functions as an LWR (Fig. 17.1). In the Indian setting, the maximum length of the LWR is normally restricted to one block section.

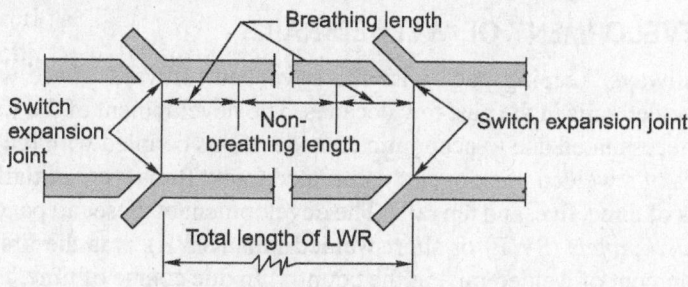

Fig. 17.1   Long-welded rail

Continuous welded rail (CWR)   It is a type of LWR that continues through station yards, including points and crossings.

Short-welded rail (SWR)   It is a welded rail that contracts and expands throughout its length.

Breathing length   It is the length at each end of an LWR that is subjected to expansion or contraction on account of variations in temperature.

Anchor length (l_a)   It is the length of the track that is required to resist the pull exerted by the rail tensor on the rails.

Switch expansion joint (SEJ)   It is an expansion joint installed at each end of an LWR to permit the expansion or contraction of the adjoining breathing lengths due to temperature variations.

Buffer rails   These are a set of rails provided at the ends of an LWR to allow the expansion or contraction of the breathing lengths due to temperature variations.

Destressing   It is an operation undertaken with or without the use of rail tensors to attain a stress-free LWR at a specified rail temperature.

Rail temperature   It is the temperature of the rail as recorded by an approved rail thermometer at the site. This is different from the ambient temperature, which is the atmospheric temperature as reported by the meteorological department.

Mean rail temperature (T_m)   The mean rail temperature for a section of rail is the average of the maximum and minimum rail temperatures recorded for the section.

Installation temperature (T_i)   This is the average rail temperature achieved when the rails are being fastened to the sleepers at the time of installation of LWRs.

Standard installation temperature (T_s)   This is the installation temperature at which a standard gap of 6 mm is provided for fish-plated joints.

Prevailing rail temperature (T_p)   This is the prevailing temperature of the rail at the time of any operation connected with destressing is being carried out.

Stress free temperature (T_o)   This is the rail temperature at which the rail is free of thermal stresses.

Destressing temperature (T_d)   This is the average rail temperature at the time of fastening of rails to sleepers after destressing an LWR without the use of rail tensors.

17.1 DEVELOPMENT OF WELDED RAILS

Indian Railways, keeping pace with the advanced railways of the world, has adopted welded rails in the past few decades. The development of the welded rail has been necessitated due to economic considerations coupled with the technical advantages of a welded track over a fish-plated track. Rails were initially welded into panels of three, five, and ten rails. The development of these rail panels, called *short-welded panels* (SWP) or short-welded rails (SWR), was the first stage in the development of welded rails in the country. In due course of time, there were considerable advancements in track technology, and a new concept of 'locking up of stresses' in the rail, which thereby resisted any changes in the length of the rail, was developed. Accordingly, long-welded rails of varying lengths of up to 1 km were laid to eliminate rail joints and make the most of all the advantages that welding offers.

The concept of continuous welded rails is now developing rapidly and welded rails are being laid continuously so as to get the full advantage of a jointless track.

17.2 THEORY OF LONG-WELDED RAILS

It is well known that metals expand and contract with increase or decrease in temperature, i.e., undergo thermal expansion. Thus, a rail expands and contracts depending upon the variations in temperature. The expansion of a rail is a function of the coefficient of the linear expansion of the rail material, the length of the rail, and the variations in the rail temperature. Normally, a free rail would undergo alterations in its length corresponding to the variations in rail temperature, but as rails are fastened to sleepers, which in turn are embedded in the ballast, their expansion and contraction due to temperature changes are restricted. The restraint put on the thermal expansion of rail gives rise to locked-up internal stresses in the rail metal. The resulting force, known as the thermal force, is given by the following equation:

$$P = EA\alpha T \tag{17.1}$$

where P is the force in tonnes, E is the modulus of elasticity of rail steel = 2.15×10^6 kg/cm^2 or 2150 tonnes/cm^2, A is the cross-sectional area of steel in cm^2 and depends upon the individual rail section (for a 52 kg rail it is 66.15 m^2), α is the coefficient of linear expansion = 0.00001152 per °C, and T is the temperature variation in °C. Substituting the values of E, A, α, and T, the force for every 1° rise of temperature for a 52 kg rail can be derived as follows:

$$P = (2.15 \times 10^6) \times 66.15 \times 0.00001152 \times 1 \times 10^{-3}$$
$$= 1.638 \text{ tonnes per } °C$$

The values of E and α are fixed for each type of rail steel. The value of the cross-sectional area depends upon the sectional weight of the rail. Substituting the value of sectional weight in kg/m in Eqn (17.1), the force P can also be given by the formula

$$P = 31.5AT$$

where P is the force in kilograms, A is the sectional weight in kg/m, and T is the temperature variation in °C. For a 52 kg rail

$$P = 31.5 \times 52 \text{ kg per unit } °C$$
$$= 1638 \text{ kg} = 1.638 \text{ tonnes per } °C$$

17.2.1 Longitudinal Thermal Expansion of LWR and Breathing Length

In the case of long-welded rail, the thermal expansion of the rail takes place at the rail ends because of temperature variations and the inability of the resisting force offered by the rail and the ballast to overcome the same. An LWR continues to expand at its ends up to that particular length at which an adequate resisting force is developed towards the centre. A stage is finally reached at a particular length of the rail from its ends when the resistance offered by the track structure becomes equal to the thermal forces created as a result of temperature variations. There is no alternation in the rail length beyond this point. The cumulative value of the expansion or contraction of these end portions of the rail (breathing lengths) is given by the formula

$$\delta\ell = \frac{\ell\alpha T}{2} \tag{17.2}$$

where, $\delta\ell$ is the amount of expansion or contraction of the rail, ℓ is the breathing length of the rail, α is the coefficient of thermal expansion of the rail, and T is the variation in temperature. This value of expansion or contraction of the rail is half the value that would have been attained if the rail had been free to expand on rollers without any ballast resistance. This alteration in length is confined to only a certain portion at the ends of the LWR. The central portion of the LWR, where the force is constant, is immobile and does not undergo any change in its length.

The portion at the end of the LWR which undergoes thermal expansion is called the *breathing length*. On Indian Railways this length is equal about 100 m at either end of the rail in the case of BG tracks.

Example 17.1 Calculate the minimum theoretical length of LWR beyond which the central portion of rail would not be subjected to any thermal expansion, given the following data: cross-sectional area of a 52 kg rail section = 66.15 cm^2, coefficient of thermal expansion of rail steel = 11.5×10^{-6} per °C, temperature variation = 30° C, modulus of elasticity of rail steel = 2×10^6 kg/cm^2, sleeper spacing = 65 cm, and average restraining force per sleeper per rail = 300 kg.

Solution
(a) Using Eqn (17.1) and the given values, the thermal force P is given by:
$$P = (2 \times 10^6) \times 66.15 \times 11.5 \times 10^{-6} \times 30$$
$$= 45.6 \text{ tonnes}$$
(b) Resistance offered per sleeper = 300 kg = 0.3 tonnes
(c) Number of sleepers required to restrain a force of 45.6 tonnes:

$$\frac{45.6}{0.3} = 152 \text{ sleepers at each end}$$

(d) Breathing length at each end when sleeper spacing is 0.65 m:
$$= 152 \times 0.65 = 98.8 \text{ m}$$
$$= 100 \text{ m approx. at either end}$$
(e) Total breathing length on both sides presuming the zero portion in centre:
$$= 100 \times 2 \text{ m} = 200 \text{ m}$$
Therefore, the minimum theoretical length of the LWR is 200 m.

17.3 PROHIBITED LOCATIONS FOR LWR

Due to technical problems arising on account of the use of LWRs, the Indian Railways have specified locations where the laying of LWRs is prohibited. These locations are listed below.
(a) New constructions and doublings of lines, where the formation and the track are not fully stabilized. However, if mechanical compaction of the earthwork is done, LWRs may be laid on the new lines at the initial stage itself with the approval of the chief engineer.
(b) Locations where rails are subjected to heavy wear, corrugation, or corrosion, or require frequent renewal.

(c) Locations where the formation is weak and track deformations are excessive, which may lead to buckling.

(d) Locations where the formation soil is susceptible to pumping and the ballast is likely to get heavily contaminated, thereby necessitating the frequent opening of the track and the screening of the ballast.

(e) Locations where frequent breaches, flooding, and subsidence may occur.

17.3.1 Alignment and Curvature of LWR

(i) LWR/CWR shall not be laid on curves sharper than 440 m radius both for BG and MG. However, in Temperature Zone-I LWR/CWR may be laid on curves up to 360 m radius (5° curve) on BG with the following additional precautions.

(ii) Shoulder ballast for curves sharper than 440 m radius should be increased to 600 mm on the outside of curve and should be provided for 100 m beyond the tangent point.

(iii) Reference marks should be provided at every 50 m interval to record creep, if any.

(iv) Each curve of length greater than 250 m should preferably be provided with switch expansion joint (SEJ) on either side. SEJ should be located in straight track at 100 m away from the tangent point.

(v) LWR/CWR may be continued through reverse curves not sharper than 875 m radius. For reverse curves sharper than 1500 m radius, shoulder ballast of 600 mm over a length of 100 m on either side of the common point should be provided.

The minimum radius of vertical curves for LWR, as recommended on Indian Railway is given in Table 17.1.

Table 17.1 Minimum radius of the vertical curve

BG	Minimum Radius (m)	MG	Minimum Radius (m)
Group A	4000		
Group B	3000	All routes of MG	2500
Group CD and E	2500		

Longitudinal resistance of sleepers The longitudinal ballast resistance in kg per metre rail for different types of sleeper in BG and MG is generally as given in Table 17.1(a):

Table 17.1(a) Longitudinal resistance of sleepers for BG and MG

Type of sleeper	Longitudinal ballast resistance in kg/m^2			
	BG			MG
	1310 sleepers/km	1540 sleepers/km	1660 sleeper/km	1540 sleepers/km
PRC	12.93	13.28	13.74	—
ST	11.48	12.14	12.68	3.86
CST-9	9.92	10.65	11.04	3.47
Wooden	7.85	7.97	8.06	5.0

Note: The values given above are indicative and can vary as per site conditions.

17.4 TRACK STRUCTURE FOR LWR

The minimum length required for a rail to function as an LWR depends upon the range of temperature variation, the section of the rail, the resistance offered by the ballast to the thermal expansion of the sleepers, and the resistance offered by the rail and sleeper assembly to any thermal expansion of the rails. Normally, a rail length of about 100 m on BG and 150 m on MG is subjected to thermal expansion at each rail end. Thus a length of more than 200 m on BG and 300 m on MG is generally necessary for a welded rail panel to function as an LWR.

The minimum track structure should be 52 kg rail on PSC sleeper, M + 7 Sleeper density with 300 mm clean ballast cushion.

17.4.1 Formation

The formation should not pose any problems in the laying of LWRs. In stretches where the formation is bad, it should be stabilized before the LWRs are laid. A cross slope of 1 in 40 should be provided at the time of screening and laying of LWRs. In the case of concrete sleeper tracks, an extra cess width is provided to the extent of 90 cm for embankments and 60 cm for cuttings.

17.4.2 Ballast

A clean ballast cushion of a minimum depth of 250 mm (10") should be provided below the bottom of the sleeper for LWRs. In order to increase resistance, the shoulder width of the ballast should measure 350 mm in the case of straight tracks and the inside of curves and 500 mm in the case of the outer ends of curves. The ballast should also be humped to a height of 150 mm on both shoulders.

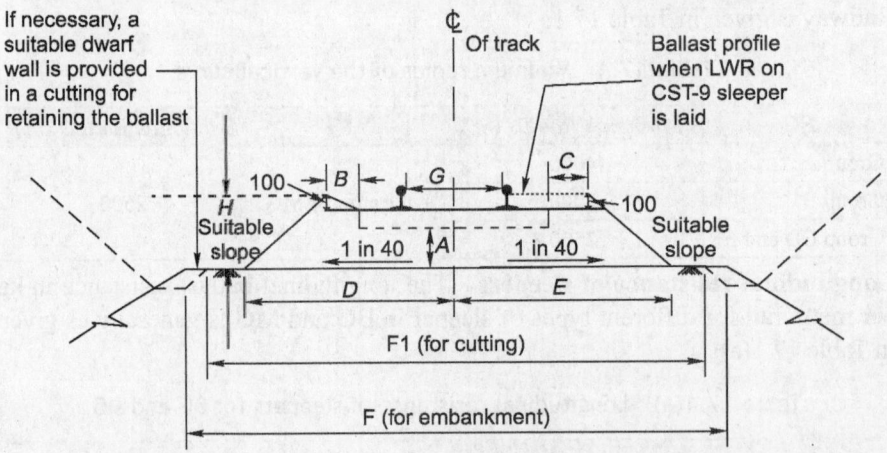

Fig. 17.2 Ballast profile of single-line LWR track (all dimensions in mm)

Table 17.2 Ballast profile of single-line LWR track

Gauge	Sleeper	A	B	C	D	E	F	F₁	H
	Wooden	250	350	500	2270	2420	6850	6250	540
BG	Steel	250	350	500	2280	2430	6850	6250	590

(Contd.)

Table 17.2 Contd.

Gauge	Sleeper	A	B	C	D	E	F	F₁	H
	Concrete	250	350	500	2525	2675	6850	6250	640
	Wooden	250	350	500	1760	1930	5850	5250	510
MG	ST	250	350	500	1790	1940	5850	5250	520
	CST/9	250	350	500	1730	1880	5850	5250	510
	Concrete	250	350	500	2025	2175	5850	5250	590

All dimensions are in millimetres.

The typical ballast profiles of LWRs for single-line and double-line BG concrete sleeper tracks with a 250-mm (10") ballast cushion are given in Figs 17.2 and 17.3, respectively and are also, respectively, summarized in Tables 17.2 and 17.3.

Table 17.3 Ballast profile of double-line LWR track

Gauge	Sleeper	A	B	C	D	E	F	F₁	H	J
	Wooden	250	350	500	2300	2340	12,155	10,210	570	5250
BG	Steel	250	350	500	2310	2350	12,155	10,210	580	5250
	Concrete	250	350	500	2525	2460	12,155	10,980	700	5250
	Wooden	250	350	500	1890	1810	9810	9210	535	3960
MG	ST	250	350	500	1890	1810	9810	9210	535	3960
	CST/9	250	350	500	1750	1810	9810	9210	535	3960
	Concrete	250	350	500	2025	1970	9810	9210	595	3960

All dimensions are in millimetres.

If followed by a non-LWR track, the ballast section and depth of the LWR track should be continued over the switch expansion joint or buffer rail assembly and also up to the rail lengths beyond it.

17.4.3 Sleepers Prescribed for LWR and CWR

The sleepers prescribed for LWR and CWR tracks are listed below.

Broad gauge (BG)

The features of BG sleepers are as follows:
 (i) Concrete sleepers with elastic fastenings
 (ii) Untreated or hard (U category) wooden sleepers with fastenings for speeds not exceeding 160 km per hour
 (iii) Steel trough sleepers with elastic fastenings for speeds not exceeding 130 m per hour (as an interim measure for speeds up to 160 m per hour)

 However, as an interim measure, steel trough sleepers with loose jaws and keys, untreated wooden sleepers, and CST-9 sleepers are also permitted.

Metre gauge (MG)

The prescribed sleepers for MG tracks are concrete sleepers, steel trough sleepers, durable wooden sleepers, and CST-9 sleepers.

Sleeper density for MG

On MG tracks, the sleeper density should be 1540 sleepers per km per hour for speeds up to 100 km per hour and 1660 sleepers per km for speeds above 100 km

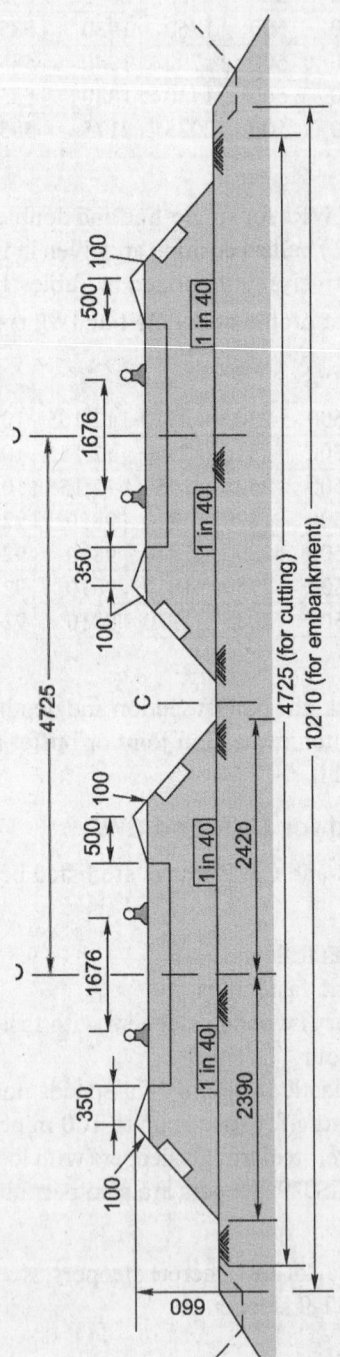

Fig. 17.3 Ballast profile in mm for double-line LWR track

per hour. On the existing LWR/CWR tracks, the decision to retain or change the existing sleeper density is taken by the chief engineer.

Table 17.4 Sleeper density for LWR/CWR for different temperature zones

Type of sleeper	Temperature zones	Sleeper density (sleeper per km)
PRC sleeper	I & II	1310
PRC sleeper	III & IV	1540
Other sleepers	I, II, III & IV	1540

17.4.4 Rails

LWRs should be laid with 60 kg, 52 kg, or 90 R rails on BG and 90 R or 75 R rails on MG barring the portions already laid with 60 R rails. Generally, new rails with fish bolt holes should not be used for LWRs. Further, level crossings should not fall within the breathing lengths of LWRs and switch expansion joints or buffer rails should be located at the end of the LWRs.

17.5 RAIL TEMPERATURE AND ITS MEASUREMENT

Rail temperature is the temperature of the rail as recorded by a rail thermometer at the site. In northern India, the maximum rail temperature in summer is about 20°C above the maximum ambient temperature and the minimum rail temperature is about 2°C to 3°C below the minimum ambient temperature.

The track on Indian Railways has been geographically divided into four rail temperature zones as per the following criteria:

Zone I	40°C to 50°C
Zone II	51°C to 60°C
Zone III	61°C to 70°C
Zone IV	71°C to 76°C

The mean annual rail temperature (T_m) is the average of the maximum and minimum rail temperatures during the year. The value of T_m, therefore, depends upon the range of temperature. The maximum and minimum rail temperatures for one year should be recorded in order to determine the mean annual rail temperature, although a record of five years is preferable.

Rail thermometers are used for measuring the rail temperature. The different types of thermometers available on Indian Railways are the following.

(a) Embedded type

(b) Dial type

(c) Continuous recording-type

The most commonly used amongst these is the embedded-type thermometer (Fig. 17.4). This type of thermometer consists of an ordinary thermometer inserted into a cavity in a piece of rail head. The cavity is filled with mercury and sealed. The

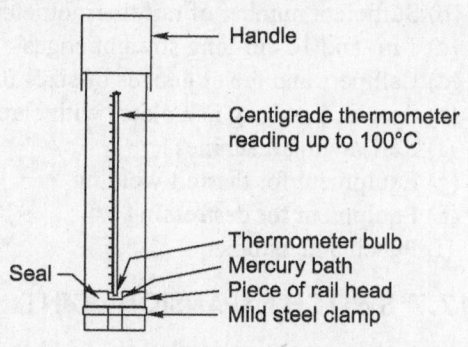

Fig. 17.4 Embedded-type rail thermometer

rail piece containing the thermometer is exposed to the same conditions as the rails in the track. The temperature recorded by the thermometer at that time is the rail temperature. It takes about 25–30 minutes to record the rail temperature.

17.6 MAINTENANCE OF LWR

Some important points which generally help in ensuring the safety and effective maintenance of LWR tracks are as follows:

(a) A well-compacted ballast bed should be available below the sleepers at all times to give adequate lateral and longitudinal resistance to the track in order to prevent buckling and excessive alterations in the sleeper lengths of the LWR track.

(b) The track should be left undisturbed as far as possible and only essential track maintenance work should be carried out at temperatures close to the destressing temperature (T_d).

(c) Regular maintenance should be completed well before the onset of summer and confined to the hours when the rail temperature is below $T_d + 10°C$.

(d) The track should preferably be maintained mechanically with 'on-track' tie tamping machines.

(e) At no time during manual maintenance should more than 30 sleeper spaces be opened in a continuous stretch between two stretches of 30 sleeper lengths that are fully boxed.

(f) Packing or renewal of a single isolated sleeper may be done by a gang mate, provided that at least 50 sleeper spaces on either side are left untouched for a minimum of 24 hours after such work is done.

17.6.1 Special Equipment for Maintenance of LWR Tracks

The staff should be allocated additional equipment as detailed below for the systematic maintenance of LWRs and for emergencies.

Additional equipment with gangs

(a) A pair of joggled fish plates with special clamps
(b) 30 mm-long rail closure pieces
(c) A rail thermometer with markings for temperature ranges
(d) Special 1 m-long fish plates with clamps

Additional equipment with PWIs

(a) Sufficient number of joggled fish plates and clamps
(b) Sufficient number of rail thermometers
(c) 1 m- and 10 cm-long straight edges
(d) Callipers and feeler gauges of sizes 0.1 mm to 2 mm
(e) Special 1 m-long fish plates with clamps
(f) Rail sawing machines
(g) Equipment for thermit welding
(h) Equipment for destressing
(i) Punch and hammer

17.7 SWITCH EXPANSION JOINT

The switch expansion joint (Fig. 17.5) is a device installed at the end of LWRs to allow the thermal expansion of the breathing length. Presuming a breathing length of 100 m for a BG track and a maximum temperature variation of ±35°C from the mean temperature in India, the breathing length would shift or move by approximately ±20 mm from its mean position. SEJs, however, are designed to accommodate a total length modification of ±60 mm. The extra allowance has been made to cater to the likely occurrence of fractures in the breathing length, creep, and the wrong positioning of the SEJ. Switch expansion joints are situated at the ends of LWRs, i.e., after a length of 1 km or so. They are positioned in the facing direction in the case of double lines and in the direction of heavier traffic in the case of single lines.

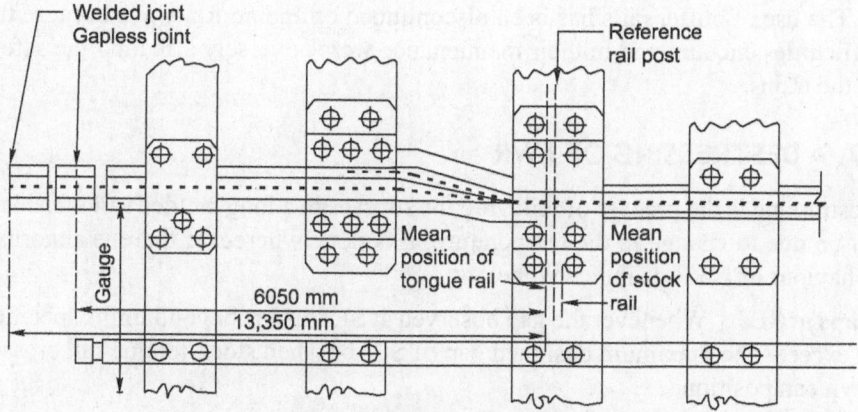

Fig. 17.5   Switch expansion joint

17.8 BUFFER RAILS

Buffer rails can also be used at the end of LWRs instead of switch expansion joints. Buffer rails are ordinary rails but of a much higher standard with respect to proper alignment at the fish-plated joints and proper gaps between the rail ends. A gap of 7.5 mm is left at the end of each buffer rail; the ends of these buffer rails should butt evenly against each other with thermal expansion. To permit free expansion and contraction, buffer rails are laid on wooden sleepers with MS canted bearing plates and 22 mm-diameter rail screws. The portion of the LWR that bears the buffer rail should be maintained by measured shovel packing only. The buffer rails should be used at the end of an LWR only when an SEJ is not readily available.

The use of buffer rails has been discontinued on Indian Railways because the difficulties encountered in their maintenance were adversely affecting the safety of the trains.

17.9 DESTRESSING OF LWR

Destressing is the process of relieving the stresses in a long welded rail which are set up due to change in the temperature. It is done whereever there is abnormal behaviour of LWR as indicated below.

Gaps at SEJ   Whenever the gap observed at SEJ differs beyond limits specified or exceeds the maximum designed gap of SEJ or when stock/tongue rail crosses the mean position.

After special maintenance operations   After special maintenance operations, viz., deep screening, lifting or lowering of track, major realignment of curves, sleeper renewals, other casual renewals and rehabilitation of bridges and formation causing disturbance to track.

Unusual occurrences   After unusual occurrences, viz. rail fractures or replacement of defective rail/glued joint, damage to SEJ/buffer rails, buckling and breaches, etc.

Temporary repairs   If number of locations where temporary repairs have been done exceeds 3 per km.

17.9.1 Procedure for Destressing without Use of Rail Tensors

(i) A traffic block of adequate duration, say about 3 hours, is taken at such a time when the rail temperature is between $T_m + 5°c$ and $T_m + 10°c$, for 60 kg/ 52 kg rail or Time to $T_m + °c$ for lighter rail sections. Before the block is taken, a speed restriction of 30 kmph is imposed and fastenings on alternate sleepers loosened.

(ii) During the block closure, rails are disconnected and the SEJs are adjusted to be in the mean position.

(iii) The sleeper fastenings on both the rails are loosened, starting from the SEJ to the centre of LWR.

(iv) The rails are lifted and placed on the rollers at about every 15th sleeper. To permit free expansion, the rails are also struck with wooden hammers to help in destressing.

(v) The rollers are then removed and the fastenings tightened, starting from the centre of LWR towards SEJ. The tightening of the fastenings must be completed in the defined temperature range of destressings.

(vi) A cut rail is provided between the SEJ and LWR keeping adequate provision for thermit welding.

(vii) Thermit welding of joints is done subsequently to complete all the operations of destressing.

Destressing of the rails should be done simultaneously. While destressing on curves, the rails should be provided with the lateral supports at an interval of 10 sleepers in the inside of curves and at an interval of 30 sleepers on the outside of the curve. At the time of destressing, a gap of 7 to 8 mm should be provided at each end of the fish-plated joints of the buffer rail assembly.

Note: Side rollers shall also be used while undertaking destressing on curved track. Side support on the inside of curve should be spaced at every ninth sleeper. Outside supports shall be used in addition at the rate of one for every three inside supports.

The total number of side rollers (n) required on a curve is calculated using the following equation.

$$n = \frac{\text{Radius of curve}(R) \times \text{No. of sleepers per rail length}}{50 \times (T_o - T_p)} \tag{17.3}$$

17.9.2 Destressing with the Use of Rail-Tensors

The progress of destressing work is greatly handicapped because of the following two limitations:

(i) The destressing has to be done between specified range of temperatures only.

(ii) A long duration of about a 3 hours has to be arranged for destressing.

To overcome these difficulties, Indian Railways has procured a few rail tensors both of hydraulic type called a hydro-stressor, and of mechanical type called a mechanical rail tensor' (Fig. 17.6).

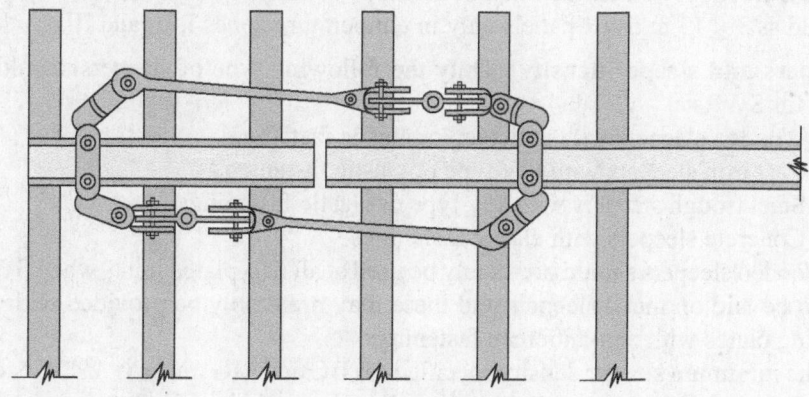

Fig. 17.6 Mechanical rail tensor

A hydrostressor consists essentially of a hydraulic pump, which transmits force through connecting vans to clamps that grip the rail. In case of a mechanical tensor, the force is exerted mechanically by longitudinal jacks. By means of this force the rails can be pushed or pulled to a desired length. In case of pushing, the force should not exceed 30 tonnes, otherwise the track is likely to buckle. The rails are however, normally pulled only during the destressing operations. The rail tensor is capable of destressing a rail panel at any time, when the prevailing rail temperature is less than the destressing temperature.

Efforts are being made to develop non-infringing type jacks cum hydrostressors that has the advantage of allowing traffic on the track with tensor applied.

17.10 SHORT-WELDED RAILS

A short-welded rail (SWR) is a welded rail that undergoes thermal expansion throughout its length. These rails are composed of three, five, or ten rails welded together.

It has been experienced on Indian Railways that short-welded panels of more than three rail lengths produce excessive thermal forces and it is felt that the extra effort required for their proper maintenance cannot be economically and technically justified. SWR of three rail lengths (3×13 m for BG and 3×12 m for MG) have, therefore, been standardized on Indian Railways.

17.10.1 Track Specifications for SWR

Indian Railways enumerates the following specifications for a track to be laid with SWRs.

Foundation and alignment An SWR should be laid on a stable and efficiently drained formation. It should not be laid on a track curved sharper than 3° on metal or concrete sleepers and 2° on wooden sleepers in the case of BG tracks. In the case of MG tracks it should not be laid on curves sharper than 5° on metal or concrete sleepers and 3° on wooden sleepers.

Rails Only new rails or good-quality second-hand rails must be used for SWRs. The minimum rail section specified for three rail panels is 90 R for BG tracks and 60 R for MG tracks for all temperature zones. However, 60 kg rails should, however, be laid as 3×13 m SWR panels only in temperature zones I, II, and III.

Sleepers and sleeper density Only the following type of sleepers should be used for SWRs:
  (a) Wooden sleepers with anticreep or elastic fastenings
  (b) Cast iron sleepers with key-type or elastic fastenings
  (c) Steel trough sleepers with key-type or elastic fastenings
  (d) Concrete sleepers with elastic fastenings

Wooden sleepers should preferably be used at all fish-plated joints when SWRs are to be laid on metal sleepers and these may preferably be provided with MS bearing plates with non-anticreep fastenings.

The minimum sleeper density specified for BG and MG tracks is $M + 4$, except in the case of 60 kg rails on BG tracks, where it should be $M + 7$.

Ballast Only stone ballast should be used with SWRs. The minimum ballast cushion below the sleeper should measure 200 mm both on BG and MG tracks. An extra 100 mm width of shoulder ballast over and above the standard ballast section on a straight track should be provided on the outer edges of the curves.

17.10.2 Regular Maintenance Operations of SWRs

Regular track maintenance work including all operations involving packing, lifting, aligning, local adjustments of curves, screening of ballast other than deep screening, and scattered renewal of sleepers may be carried out without restriction when the rail temperature is below $T_m + 25°C$.

All major lifting, major realignment of track, deep screening, and renewal of sleepers in continuous lengths should normally be done when the rail temperature is below $T_m + 15°C$ after suitable precautions have been taken. If it becomes necessary to undertake such work at rail temperatures exceeding $T_m + 15°C$, adequate speed restrictions should be imposed.

17.11 CONTINUOUS WELDED RAILS

On Indian Railways, the length of LWR has been limited to only 1 km, taking into consideration the convenience of laying, destressing, and maintenance. Welded panels longer than 1 km have also been tried and are known as *continuous welded rails*. Such rails have been laid from station to station, but the conventional insulation joints and turnouts have been left out and isolated by switch expansion joints. Trials for LWR/CWR passing through points and crossings are also in progress.

The theory behind continuous welded rails is the same as for LWR. Once the concept of locking up of longitudinal thermal forces is accepted, there is no reason why the length of LWR should be limited to only 1 km. However, the switch expansion joints or buffer joints that are provided after every 1 km of LWR have been a source of weakness in the track, requiring heavy maintenance. It was to avoid this effort that continuous welded rails were laid from station to station. In fact, on European Railways, particularly on German and British Railways, LWR or continuous welded rails have been laid for several miles together, without the inclusion of any SEJ, which pass through stations, yards, etc. The important features of SWR, LWR, and CWR are presented in Table 17.5.

Table 17.5 Comparison of SWR, LWR, and CWR

Item	SWR	LWR	CWR
Definition	A welded rail, which expands and contracts throughout its length	A welded rail, the central portion of which does not undergo any thermal expansion	A welded rail like the LWR, which has to be destressed in stages
Length	3–5 rail lengths	0.5–1 km	Above 1 km
Expansion arrangement	Expansion gap	Switch expansion joint	Switch expansion joint
Relevant theory and occurrence of stresses	Normal expansion/ contraction theory	Theory of locked-up stresses in the rail with no stresses	Same as LWR
Destressing of rails	Not required	Can be done in one stage	Has to be done in stages

(Contd.)

Table 17.5 Contd.

Item	SWR	LWR	CWR
Maintenance precautions	Can be maintained anytime in any way	Regulated maintenance in specified temperature ranges	Same as LWR
Cost of laying	Minimum	More than SWR	Almost the same as LWR
Comfort in travelling	Minimum due to a large number of joints	Better than SWR	Best
Type of maintenance	Manual	Both manual and mechanized	Same as LWR

17.12 BUCKLING OF TRACK

A rail track is liable to get distorted, particularly in hot weather when the compressive forces in the track exceed the lateral or longitudinal resistance of the track. The buckling of the track is a matter of grave concern as it may lead to derailments and even serious accidents.

17.12.1 Causes

A track can buckle due to the following reasons:
  (a) Inadequate resistance to track due to deficiencies in the ballast
  (b) Ineffective or missing fastenings
  (c) Laying, destressing, maintaining, or raising the track outside the specified rail temperature range, especially in hot weather
  (d) Failure to lubricate the SEJs in time
  (e) Excessive creep, jammed joints, sunken portions in a welded track

17.12.2 Symptoms

Buckling in a track becomes noticeable when the track displays the following symptoms.
  (a) Presence of kinks in the track
  (b) Absence of gaps in the SWR portion of the track in the morning hours of hot days
  (c) Expansion/contraction at SEJ is ±20 mm more than the theoretical range given in the LWR manual
  (d) High percentage of hollow sleepers

17.12.3 Precautions

Buckling can be avoided by taking the following precautions.
  (a) Proper expansion gaps as specified in the manual should be provided in the SWR portion of the track.
  (b) As buckling is likely to occur between the 11th and 17th hour of the day, rosters of key men should be so adjusted that there is proper patrolling of the LWR portions of the track when the temperature exceeds $T_m + 20°C$.

(c) No work of track maintenance including packing, laying, aligning, major/minor realignment of tracks, screening of ballast should be done outside the specified temperature.

(d) Wherever the track structure is weak and vulnerable to buckling, immediate action should be taken to strengthen the provision of extra shoulder ballast, increase in sleeper density, provision of adequate anticreep fastenings, replacement and tightening of missing and loose fastenings, etc.

17.12.4 Actions When Buckling Takes Place

As soon as a tendency towards buckling is detected in the track, traffic should be suspended and the track should be fully protected. The track should be stabilized by heaping the ballast on the shoulders up to the top of the web of the rail. When buckling takes place, traffic on the affected track should be suspended and remedial work should be carried out in the following stages under the personal supervision of a permanent way inspector (PWI).

(a) The temperature of the rail is brought down as far as possible by pouring water on the rails.

(b) Emergency or permanent repairs and destressing should be carried as specified in the LWR manual.

(c) In the case of fish-plated or SWR tracks, a gentle reverse curve may be provided in the rear of the buckled track to ease out the stress. The buckled rail should then be cut at two places that are more than 4 m apart. The track should then be slewed to correct the alignment and rails of the required lengths should be cut and inserted to close the gaps.

SUMMARY

Short-welded rails (SWRs) are rails that are welded in panels of two, three, or five rails. In an SWR, the entire length of the rail is subjected to expansion or contraction due to changes in temperature (thermal expansion). In the case of an LWR (long-welded rail), however, only the end portions of the rails are subject to thermal expansion. The central portion of an LWR remains clamped and does not undergo any change in length. Continuous welded rails (CWRs) are rails that are welded in lengths greater than 1 km. LWRs and CWRs are part of modern high-speed tracks and require very little maintenance as compared to fish-plated tracks.

REVIEW QUESTIONS

1. Welded rails have played an important role in the modernization of the railway system. Explain the benefits of using welded rail tracks. Do they have any associated disadvantages?

2. Given the following track data, calculate the minimum theoretical length of a LWR beyond which the central portion of the rail would not be subjected to thermal expansion.

Rail section adopted—52 kg of cross-sectional area = 65.15 cm^2

Coefficient of thermal expansion of rail steel = 11.5×10^{-6}/°C

Temperature variation in rail after laying of track = 30°C

Modulus of elasticity of rail steel $= 2 \times 10^6$ kg/cm^2

Sleeper spacing $= 65$ cm

Average restraining force per sleeper per rail $= 300$ kg

3. What are the advantages of welding rail joints? What initial precautions should be taken to prevent the possibility of buckling in LWR tracks?

4. What are the different steps involved in the maintenance of LWRs? Explain the function of switch expansion joints.

5. Discuss the concept of LWR. How can an existing SWR track be converted into an LWR track?

6. What are the recommendations regarding track structure with respect to formation, ballast, sleepers, and rails when LWRs are to be used? What are the permitted and prohibited locations for LWR on Indian Railways?

7. Discuss the theory regarding the maintenance of LWR tracks and how the same is put into practice.

8. What is the concept behind LWR? What is the measurement of the breathing length and the expansion in this portion? How is this expansion accommodated?

9. What do you understand by SWR? Briefly describe the standard specifications for SWR.

10. What is buckling of a track? What are its causes? Briefly describe the steps to be taken when buckling takes place.

11. Compare the salient features of SWR, LWR, and continuous welded rails.

Choose the correct answer from the choices given.

12. Indicate which of the statements about LWR is incorrect:
    (a) It is a welded rail, the central part of which does not undergo any longitudinal movement due to temperature variations.
    (b) Normally a length greater than 250 m on BG and 300 m on MG will function as LWR.
    (c) The maximum length of LWR is restricted to 3 km.
    (d) none of these

13. A short welded rail is formed when:
    (a) 3 rails are welded on BG lines.
    (b) 3 rails are welded on MG lines.
    (c) the welded rail contracts/expands throughout its length.
    (d) all of the above.

14. For a reverse curve, sharper than 1500 m radius, a shoulder ballast of .... mm over a length of 100 m on either side of common point should be laid for LWR:
    (a) 300 mm   (b) 450 mm
    (c) 600 mm   (d) 750 mm

15. The length at each end of long welded rail, which is subjected to expansion or contraction on account of temperature variations is called:
    (a) anchor length   (b) switch length
    (c) buffer length   (d) breathing length

16. The gradient on which LWR can be laid should not be steeper than:
    (a) 1 in 250   (b) 1 in 200
    (c) 1 in 150   (d) 1 in 100

17. The formula for thermal force in LWR/CWR is:
    (a) $L\alpha (T_p - T_o)$
    (b) $L\alpha (T_p - T_o)$
    (c) $EA\alpha$
    (d) $EA\alpha t$

18. LWR/CWR can be laid on vertical curve where algebaric difference between the two grades is:
    (a) 2 mm/m
    (b) 3 mm/m
    (c) 4 mm/m
    (d) 5 mm/m

19. The maximum radius of a vertical curve on 'A' route is:
    (a) 2000 m
    (b) 3000 m
    (c) 4000 m
    (d) 5000 m

20. The minimum shoulder ballast on the outside of a curve in LWR track:
    (a) 200 mm
    (b) 350 mm
    (c) 500 mm
    (d) 600 mm

21. 1-in-100 gradient is the limit prescribed for an LWR because:
    (a) there are difficulties in laying an LWR to steeper gradients
    (b) LWRs are not permitted in ghat sections
    (c) traction and braking forces increase with grade
    (d) the frictional forces increase

Track Maintenance

INTRODUCTION

Railway tracks can be maintained either conventionally by manual labour or by the application of modern methods of track maintenance, such as mechanical tamping or measured shovel packing. In India, maintaining tracks has traditionally been a manual activity and the 'calendar system of maintenance' has taken deep roots. In this system, a timetable or programme that outlines the track maintenance work to be done by the gangs in the course of a year is drawn out and generally followed. As per the timetable or calendar, the 12-month maintenance cycle consists of the following operations:

(a) Through packing
(b) Systematic overhauling
(c) Picking up slacks

This chapter discusses these operations together with the organizational structure of track maintenance incorporated by the Indian Railways.

18.1 NECESSITY AND ADVANTAGES OF TRACK MAINTENANCE

The railway track should be maintained properly in order to enable trains to run safely at the highest permissible speeds and to provide passengers a reasonable level of comfort during the ride. Track maintenance becomes a necessity due to the following reasons:

(a) Due to the constant movement of heavy and high-speed trains, the packing under the sleepers becomes loose and the track geometry gets disturbed. The gauge, alignment, and longitudinal as well as cross levels of the track thus get affected adversely and the safety of the track is jeopardized.
(b) Due to the vibrations and impact of high-speed trains, the fittings of the track become loose and there is heavy wear and tear of the track and its components.

(c) The track and its components get worn out as a result of the weathering effect of rain, sun, and sand.

A well-maintained track offers a safe and comfortable journey to passengers. If the track is not maintained properly, it will cause discomfort to the passengers and in extreme cases may even give rise to hazardous conditions that can lead to derailments and a consequential loss of life and property. Track maintenance ensures that such situations do not arise. The other advantages of track maintenance are as follows:

(a) If the track is suitably maintained, the life of the track as well as that of the rolling stock increases since there is lesser wear and tear of their components.

(b) Regular track maintenance helps in reducing operating costs and fuel consumption.

(c) Small maintenance jobs done at the appropriate time, such as tightening a bolt or key, hammering the dog spike, etc., help in avoiding loss of the concerned fitting and thus saving on the associated expenditure.

(d) When track maintenance is neglected for a long time, it may render the track beyond repair, calling for heavy track renewals that entail huge expenses.

18.2 ESSENTIALS OF TRACK MAINTENANCE

As mentioned earlier, a well-maintained track provides the base for a safe and comfortable journey. Therefore, for a track to serve its purpose well, the following characteristics are required:

(a) The gauge should be correct or within the specified limits.

(b) There should be no difference in cross levels except on curves, where cross levels vary in order to provide superelevation.

(c) Longitudinal levels should be uniform.

(d) The alignment should be straight and kink-free.

(e) The ballast should be adequate and the sleepers should be well packed.

(f) There should be no excessive wear and tear of the track and all its components and fittings should be complete.

(g) The track drainage should be good and the formation should be well maintained.

To achieve these standards, the major maintenance operations performed in a calendar year are described in the following sections.

18.2.1 Through Packing

Through packing is carried out in a systematic and sequential manner as described here.

Opening of road The ballast is dug out on either side of the rail seat for a depth of 50 mm (2") below the bottom of the sleeper with the help of a shovel with a wire claw. On the outside, the width of the opening should extend up to the end of the sleeper. On the inside, it should extend from the rail seat to a distance of 450 mm (18") in the case of BG, 350 mm (14") in the case of MG, and 250 mm (10") in the case of NG.

Examination of rails, sleepers, and fastenings The rails, sleepers, and fastenings to be used are thoroughly examined. Defective sleepers are removed and loose fastenings are tightened. Any kinks in the rails are removed using a Jim Crow.

Squaring of sleepers

The sleepers get out of square quite frequently resulting in gauge variations and kinks. To avoid this, one of the rails is taken as the sighting rail and the correct sleeper spacing is marked on it. The position of the sleeper is checked with reference to the second rail with the help of a T-square. The sleepers are attended to after their defects have been established, which may include their being out of square or at incorrect spacing.

Aligning the track

The alignment of the track is normally checked visually, wherein the rail is visually assessed from a distance of about four rail lengths or so. Small errors in the alignment are corrected by slewing the track after loosening the cores at the ends and drawing out sufficient ballast at the end of the sleepers. Slewing is carried out by about six people by planting crowbars deep into the ballast at an angle not exceeding 30° from the vertical.

Gauging

The gauge should be checked and an attempt should be made to provide a uniform gauge within permissible tolerance limits. Table 18.1 lists the tolerances prescribed for gauge variation, keeping in mind the side wear that occurs at the time of laying the tracks. This is done to ensure a comfortable ride for the passengers, provided that uniform gauge can be maintained over long lengths.

Table 18.1   Gauge tolerance for different tracks

Type of track	Gauge tolerance for BG	Gauge tolerance for MG and NG
Straight track	−6 mm to +6 mm	−3 mm to +6 mm
On curves with radius more than 400 m for BG, 290 m for MG, and 175 m for NG	−6 mm to +15 mm	−6 mm to +15 mm
On curves with radius less than 400 m for BG, 290 m for MG, and 175 m for NG	Up to + 20 mm	Up to + 20 mm

The gauge is adjusted in accordance with the type of sleeper under consideration as described in the following.

Wooden sleepers   In the case of wooden sleepers, gauge adjustment is possible only by removing the dog spikes and refitting them at a new location. Therefore, gauge adjustment should be avoided as far as possible unless the gauge is quiet irregular. When the gauge must be adjusted, all the spikes on the inside and half of those on the outside are removed while the remaining half are loosened. The old spike holes are plugged and new holes are bored in correct places. The gauge on each sleeper is adjusted and the spikes are re-driven.

Steel trough sleepers   In the case of steel trough sleepers, gauge adjustment is done with the help of keys. When the gauge is slack, the keys on the inside are loosened while those on the outside are driven. The procedure is reversed when the gauge is tight. The maximum possible adjustment of the gauge is − 2.5 mm to +4.0 mm.

CST-9 sleepers   In the case of CST-9 sleepers, gauge adjustment is done with the help of cotters. Normally a gauge is adjusted by ± 5 mm. The maximum extent to which a gauge can be adjusted is –3 mm to +10 mm. It has been noticed that adjusting the gauge may sometimes disturb the alignment, which is taken care of prior to gauging as per the standard practice. In such cases, the track has to be realigned once gauging is completed.

Packing of sleepers

The base rail is identified by the mate and the dip or low joints are lifted correctly to ensure that the longitudinal level of the rail is perfect. The sleepers are then packed by applying the *scissors packing method.* Four men tackle one sleeper simultaneously, two at each rail. The ballast under the sleeper bed is properly packed by the men who stand back to back and work their beaters diagonally by lifting them up to chest level. While the packing is being carried out, the second rail is brought to the correct cross levels thereby ensuring perfect surfacing of the track. In the case of wooden and steel trough sleepers, it should be ensured that the sleepers are not centre-bound and that as such the trough is made in the ballast section at the centre of the sleepers. After packing is completed, the alignment and top should be checked carefully and minor adjustments made as needed.

Repacking of joint sleepers

The joint sleepers are then packed once again and the cross levels checked.

Boxing ballast section and dressing

Afterwards the ballast section is properly boxed and dressed with the help of a special template. The cess should also be dressed or covered similarly and its level maintained in a way that proper drainage is ensured. Through packing follows a programme, which requires that it is undertaken after the monsoon such that it extends from one end of the section to the other. Through packing must be carried out at least once every year.

A gangman normally accomplishes about 11 m to 12 m of through packing on BG, 16 m to 17 m on MG, and 23 m to 24 m on NG tracks.

18.2.2 Systematic Overhauling

The track should be overhauled periodically with the object of ensuring that the best possible standards of track conditions are met and maintained. The systematic overhauling of the track should normally commence after the completion of one cycle of through packing. It involves the following operations in sequence.

(a) Shallow screening and making up of ballast section
(b) Replacing damaged or broken fittings
(c) Including all items in through packing
(d) Making up the cess

The frequency of overhauling depends upon a number of factors such as the type and age of track structure, the maximum permissible speed and volume of traffic, the mode of traffic, the mode of traction, the rate of track deterioration, and the amount of rainfall in the region. On the basis of these factors the chief engineer decides the length of the track to be overhauled but normally the plan is so drawn that the systematic overhauling of a section is completed in about three to four years. The stretch of the track to be tackled in a particular year should be in the continuation of the length overhauled during the previous year. If possible, gap adjustment, including joint survey and adjustment of creep, should be done prior to systematic overhauling.

Lubrication of rail joints

The lubrication of rail joints is an important part of the work done on the permanent way and is incidental to systematic track maintenance. Joints are lubricated for the following purposes:

(a) To allow for free expansion and contraction of rail

(b) To reduce wear and tear on the fishing planes of rails and the fish plates

All rail joints are lubricated once a year during the moderate season. This is also known as the *oiling and greasing of fish plates*. This work is not done during the rainy season. The lubricant used is a paste of workable consistency that consists of the following proportions of plumbago, kerosene oil, and black oil:

- Plumbago (dry graphite)–5 kg
- Kerosene oil (second quality)–3.5 l
- Black or reclaimed oil–2.75 l

The above quantity is sufficient for 100 joints of 52 kg/90 R or for 125 joints of 75 R/90 R rails. Sometimes only plumbago and kerosene oil are used in the ratio of 3:2 for lubricating fish plates. Black oil is, however, used for oiling fish bolts and nuts.

Rail joints should be lubricated only after ensuring that their surfaces are properly cleaned, preferably with the help of wire brushes and clean jute. Joints should not be lubricated in extreme temperatures or when the rails are in tension as a result of creep. Joints should be opened one at a time for lubrication. Even when opening a joint, only one fish plate should be tackled at a time and at no time during the operation should there be less than one fish plate and three fish bolts connecting the two rails.

18.2.3 Picking Up Slacks

Slacks are those points in the track where the running of trains is faulty or substandard. Slacks generally occur in the following cases:

(a) Stretches of yielding formation

(b) Poorly maintained sections that have loose packing, bad alignment, and improper longitudinal and cross levels

(c) Improperly aligned curves

(d) Approaches to level crossings, girder bridges, etc., particularly in sags

(e) Portions of track with poor drainage

(f) Sections with an inadequate or unclean ballast cushion

(g) Other miscellaneous reasons

In every working season, a certain number of days in each week (normally one or two days) are allotted to the picking up of slacks, depending upon the monsoon pattern and other local conditions. However, no through packing is done during the rainy season and slacks are only picked up in order to keep the track safe and in good running condition. In areas with less than 750 mm of rainfall, the alloted time may not be used only to attend to the slack but also to carry out through packing. Slack may be sometimes picked up by packing only the following segments of the track:

(a) Joint sleepers and the two other sleepers on either side of the joint, i.e., first shoulder and second shoulder sleepers

(b) A few sleepers in the approaches to level crossings and bridges

(c) Intermediate sleepers

(d) Stretches of track that adversely affect the running of trains as revealed by inspection notes

It may be noted here that points and crossings should be attended to throughout the year. In sections with no points and crossings, the alloted time may be utilized for creep adjustment and other such track maintenance work. Two separate charts, one for main line work and another for yard work, are maintained by each gang and kept in the personal custody of the gang mate.

A recommended annual programme is drafted for regular track maintenance, in which each major activity is specified a certain time slot as per a fixed time table (calendar). Table 18.2 presents the time table for regular track maintenance. Concrete sleeper tracks are maintained by on-track tie tamping machines as explained in Chapter 20.

Table 18.2 Timetable for regular track maintenance

Period	*Work to be done*
Post-monsoon: for about six months after the end of the monsoon	Run down lengths in the entire gang beat to be dealt with in order to restore the quality of the section. One cycle of systematic through packing/systematic directed track maintenance from one end of the gang length to the other, including overhauling of nominated sections. Normally 4–5 days per week should be allotted for through packing/overhauling and the remaining days should be set aside for picking up slacks and for attending to bridge approaches, level crossings, and points and crossings over the entire gang beat. Other essential maintenance work such as lubrication of rail joints, joint gap adjustment, and realignment of curves should also be done during this period.
Pre-monsoon: for about two months prior to the break of the monsoon	Normally 2–4 days in a week should be devoted to clearing side and catch water drains, earthwork repairs of the cess, clearing waterways, and picking up slacks. Normal systematic maintenance should be carried out during the remaining days.
During the monsoon: for about four months	Attention given to the track as required. This will consist primarily of picking up slacks and attention to side and catch water drains and waterways. During abnormally heavy rains, the line should be patrolled by the gangs in addition to regular monsoon patrolling.

18.3 MEASURING EQUIPMENT AND MAINTENANCE TOOLS FOR TRACKS

The measuring equipment and tools commonly used for track maintenance, together with their purpose, are presented in Table 18.3. Sketches of some of the tools used for the maintenance of tracks are given in Fig. 18.1.

Table 18.3 Measuring equipment and maintenance tools

Name of equipment/tool	Function
Measuring equipment	
Rail gauge	To check the gauge
Straight edge and spirit level	To check the alignment as well as the cross level
Gauge-cum-level	To check the gauge as well as the cross level
Cant board	To check the difference in cross levels or the superelevation
Mallet or wooden hammer	To check the packing of the sleepers
Canne-a-boule	A modern equipment used to assess the voids under the sleepers
T-square	To check the squareness of sleepers
Stepped feeler gauge	To measure wear or clearances
Maintenance tools	
Sleeper tong	To carry sleepers
Rail tong	To lift and carry rails
Beater	To pack ballast under the sleeper
Crowbar	To correct track alignment and to lift the track for surfacing. Clawed crowbars are used for taking out dog spikes.
Jim Crow	To bend or break the rails
Spiking hammer	To drive spikes
Keying hammer	To drive keys
Spanner	To tighten bolts
Wire claw or ballast rake	To draw or pull out ballast while screening, packing, etc.
Phowrah (shovel)	To cut earth or to pull out ballast
Auger	To bore holes in wooden sleepers
Box spanner	For driving rail screws or plate screws
Wire basket	For screening the ballast
Pan iron motor	For leading earth/ballast

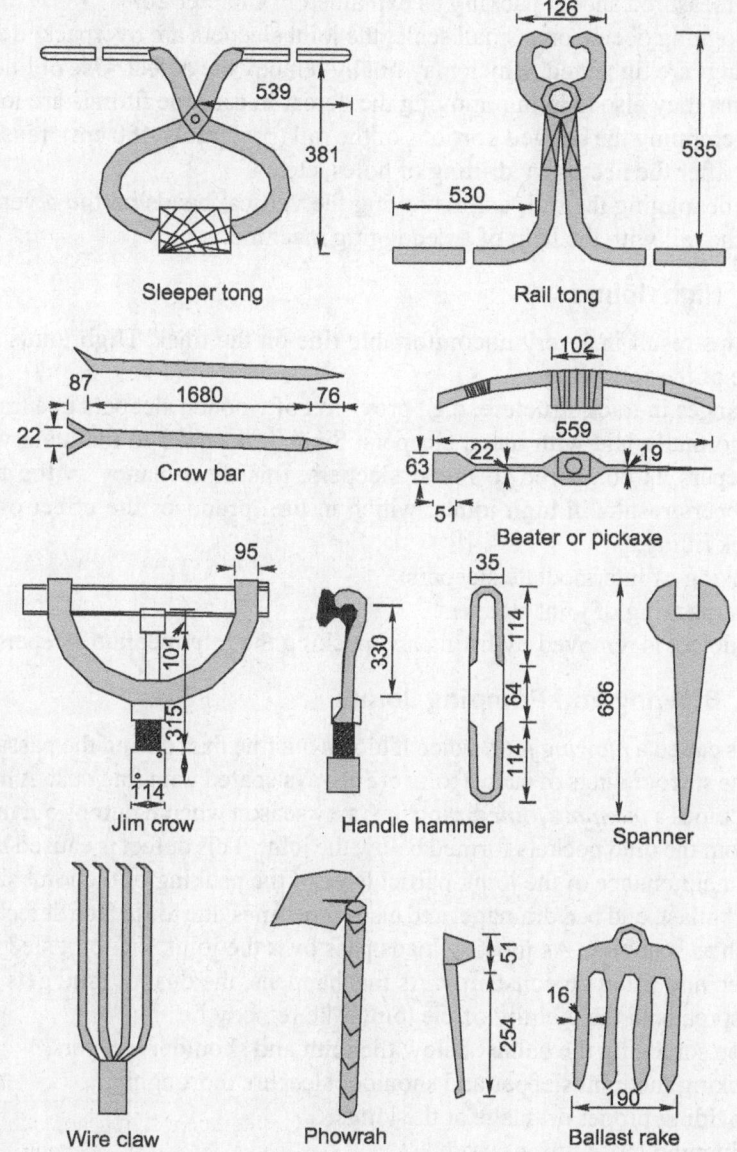

Fig. 18.1 Tools used for track maintenance

18.4 MAINTENANCE OF RAIL SURFACE

The surface of rails is susceptible to certain defects in the absence of proper care and maintenance. Each defect should be dealt with carefully at the proper time.

18.4.1 Hogged Joints

Hogging of a joint is a phenomenon in which the two rail ends at the joint get depressed on account of the poor maintenance of the rail joint, loose and faulty fastenings, and other such reasons. The hogging of joints brings about deterioration in the running quality of the track. The various techniques for removing this defect are as follows:

(a) By measured shovel packing as explained in Chapter 20.

(b) If hogging occurs on a small scale, the joint sleepers are overpacked and the fittings are tightened, which may finally remedy the defect. Use of liners and shims may also help in removing the defect in case the fittings are loose.

(c) By cropping the hogged portions of the rail (preferably 450 cm), reusing the rail after the necessary drilling of holes, etc.

(d) By dehogging the rail, i.e., removing the vertical bends on the reverse side of the rail with the help of a dehogging machine.

18.4.2 High Joints

High joints result in a very uncomfortable ride on the track. High joints are the outcome of the following:

(a) Changes in track structure, e.g., provision of wooden sleepers in a track that is normally laid with metal sleepers. Since it is easier to maintain wooden sleepers as compared to metal sleepers, this discrepancy in the type of sleepers results in high joints, which in turn produces the effect of camel back riding.

(b) Sinking of intermediate sleepers.

(c) Overpacking of joint sleepers.

This defect is removed by lifting and packing the intermediate sleepers.

18.4.3 Blowing and Pumping Joints

A joint is called a *blowing joint* when it blows out fine dust during the passage of a train. The surroundings of such a joint are always coated with fine dust. A blowing joint becomes a *pumping joint* during the rainy season when it pumps out mud and water from the mud pockets formed below the joint. This defect is caused because of poor maintenance of the joint, particularly of the packing of the joint sleepers, unclean ballast, and bad drainage, and also sometimes due to surface defects in the rail, such as scabbing. As moving loads pass over the joint, the joint sleepers get depressed and lifted up constantly. As this happens, the dust or mud gets sucked up and spreads in the vicinity of the joint. The remedy lies in:

(a) deep screening the ballast below the joint and shoulder sleepers,

(b) packing the joint sleeper and shoulder sleepers thoroughly,

(c) providing proper drainage at the joint,

(d) tightening loose fittings, and

(e) adjusting the creep, if any.

Lifting of track

Normally, lifting a track becomes necessary when the track undergoes regrading. This may be due to yard remodelling, construction of a bridge, etc., or in an effort to eliminate the sags that develop in the approaches to level crossings and bridges and at other locations made vulnerable due to defective maintenance or yielding formation. The points regarding the lifting of tracks that require special mention are as follows:

(a) A maximum of 75 mm (3″) of the track should be lifted at a time. Whenever heavy lifting is involved, it should be done in different stages, with each lift being not more than 75 mm.

(b) On single lines, lifting should commence from the downhill gradient and continue in the direction of the rising gradient. In the case of double lines, it should proceed in the direction opposite to that of the traffic, taking care not to exceed the easement grade.

(c) Lifting should be done under the supervision of a PWI after imposing suitable speed restrictions and setting up the obligatory engineering signals.

Lowering of track

The lowering of a track becomes necessary when the track is regraded for various reasons, such as yard remodelling, provision of level crossings, etc. It should be avoided until it becomes inevitable, as lowering the track makes it unstable and is quite a difficult, time-consuming, and costly proposal.

18.4.4 Longitudinal Sag in the Track

Normally a track between two rigid structures such as bridges, level crossings, etc., settles due to the passage of moving loads. The settlement of the track also takes place on yielding formation owing to the weakness of the formation and the puncturing of the ballast into the formation. Whereas a longitudinal unevenness in the shape of a vertical curve may not be noticeable, an irregular longitudinal sag may make the ride on the track uncomfortable. In such cases, a proper survey of the track should be carried out, pegs should be fixed at the correct longitudinal level with the help of a levelling instrument, and the track should then be lifted. The track should not be lifted more than 75 mm at a time. An adequate quantity of ballast should be collected in advance so that both packing and lifting can be done effectively. If excessive lifting is involved, the work should be done under speed restrictions.

18.4.5 Centre-bound Sleeper

The defect due to centre-bound sleeper is generally noticed in wooden and steel trough sleeper tracks. This defect occurs when, as a consequence of plying traffic, the sleeper starts to receive support at the centre instead of at the ends. If proper care is not taken during through packing and the middle portion of the sleeper is also packed, the defect can develop very early. Even under normal circumstances, the ballast under the sleeper ends, where the sleeper rests, gets more depressed compared to the ballast at the centre because of the impact of the moving loads and in the process the sleeper, instead of resting at the ends, starts to rest at the centre.

Centre binding of the sleepers leads to the rocking of the trains and is detrimental to the quality of the track. The defect can be removed by loosening the ballast at the centre of the sleeper. It is considered a good practice to make a small recess or depression in the ballast section at the centre of the sleeper.

18.5 DEEP SCREENING OF BALLAST

Deep screening of the ballast is done to ensure that a clean ballast cushion of the required depth is available below the lower half of the sleepers, which is necessary for providing proper drainage and elasticity to the track. In the absence of a clean ballast cushion of the desired depth, track geometry may get disturbed, affecting the performance of the track.

Deep screening is normally carried out under speed restrictions without obtaining a traffic block. It is desirable to proceed with deep screening in a direction opposite to that of the traffic. An adequate quantity of ballast, the necessary equipment and tools, and the required labour should all be arranged well in time. Work should not be carried out during monsoons. The procedure for deep screening is as follows:

(a) A group of four sleepers are tackled one at a time in sequence. As shown in Fig. 18.2, each of the sleepers (numbered 1, 2, 3, 4, etc.) is tackled individually.

(b) The ballast from spaces A and B is removed right down to the formation and wooden blocks are put in its place to support the track. Precautions are taken to avoid digging out the consolidated top of the formation.

(c) The ballast is removed from under sleeper 1, screened, placed under the same sleeper and packed.

(d) The wooden block is removed from space A.

(e) The ballast from space C is removed and placed after screening in space A. If additional ballast is required, it may be taken from the extra ballast trained out in advance along the side of the track.

(f) The wooden block that was removed from space A is placed in space C. Sleeper 2 is provided with wooden blocks on either side in order to support the track.

(g) The ballast under sleeper 2 is removed, a screened ballast is provided in its place, and the sleeper is packed.

(h) The ballast from space D is removed, screened, and placed in space B. Any extra ballast that may be needed is taken from the track.

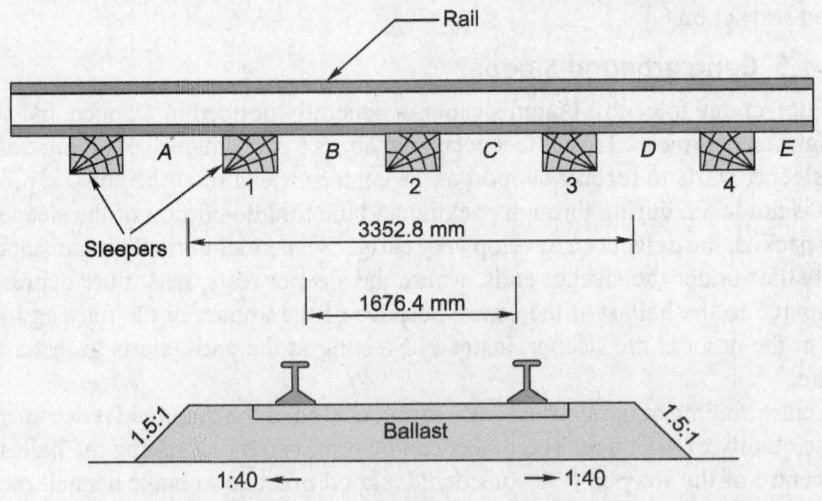

Fig. 18.2 Deep screening of ballast

(i) The wooden block from space B is removed and put in space D to support the rail.

(j) The procedure is repeated till the work is completed.

It can be seen from this procedure that work is done in such a way that when the ballast is being excavated from under one sleeper, there are at least four sleepers between this sleeper and the next sleeper that are being worked upon. The track is also lifted, if required. This should only be done at a rate of 50 mm (2") at a time. It should be ensured that the packing, cross level, and grading-off are satisfactory before closing the day's work.

Deep screening is done under a speed restriction of 15 km per hour. This speed can be released to normal speeds in the following ways.

Manual packing By manual packing four times in succession and by picking up slacks over a period of 21 days. (Fig. 18.3 (a))

Machine packing Packing with the help of tie machines, by tamping the track thrice followed by picking up slacks over a period of 10 days (Fig. 18.3 (b))

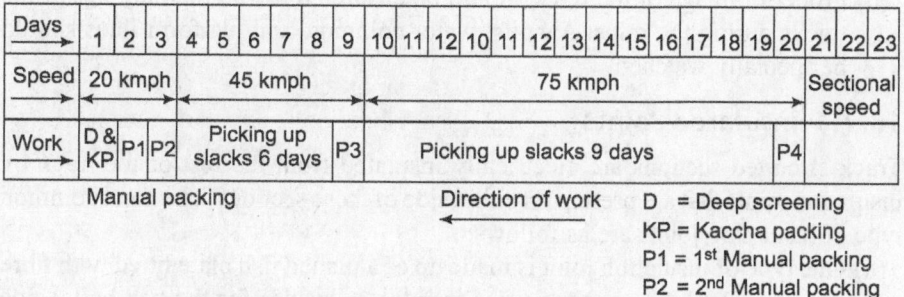

Fig. 18.3 (a) Speed restrictions with manual packing

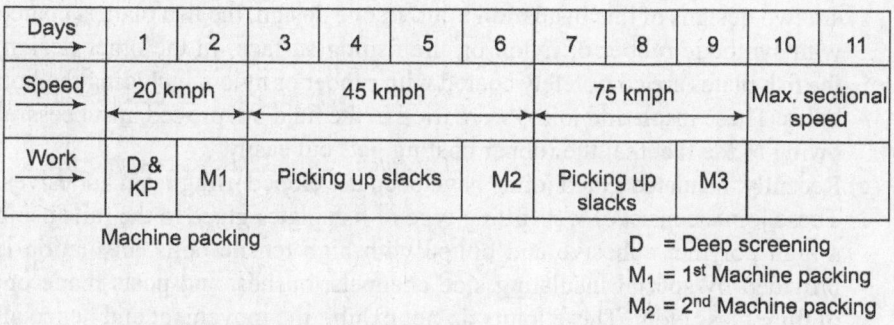

Fig. 18.3 (b) Speed restrictions with machine packing on BG

18.6 TRACK DRAINAGE

Track drainage can be defined as the interception, collection, and disposal of water from the track. It is accomplished by introducing a proper surface and sub-surface drainage system. This subject has been discussed in detail in Chapter 19.

18.7 MAINTENANCE OF TRACK IN TRACK-CIRCUITED LENGTHS

The length of track connected by an electric circuit to the signal cabin, block telegraph apparatus, etc., required for indication of light is called a track circuit. Thus, a track circuit functions to indicate whether the track is occupied or not.

Special precautions are necessary in the maintenance of track-circuited lengths as accidental short circuiting of a circuit in the track may cause serious delay in traffic. Some of the noteworthy points in this regard are as follows:

(a) The PWI should instruct the staff not to touch or place any tool or metal object across the two rails in the track, as this will cause short circuiting of the track circuit.

(b) All gauges, straight edges, and trolleys used in the track-circuited length should be insulated.

(c) The ballast must be clean throughout the track-circuited length and care should be taken to see that the ballast is kept clear of the rails and rail fastenings. The clearance from the foot of the rail must not be less than 40 mm.

(d) Proper drainage of the track should be ensured to avoid the flooding of the tracks during the rains. Ashpits, water columns, and platform lines should be specially watched.

18.7.1 Insulated Joints

Track-circuited sections are electrically insulated from the rest of the track by insulation joints that are present on either side of these sections. The most common type of insulation joints are as follows:

(a) One type of insulation joint is made up of a planed fish plate fitted with fibre or nylon insulation consisting of ferrules or bushes for the fish bolts, side channels between the fish plates and rails, and end posts between the rail ends.

(b) The other type of insulation joint consists of rubber-coated fish plates. There are two designs of this insulation joint. In one design, the fish plate is coated with synthetic rubber or nylon on the fishing surface. In the other design, the fish plates are completely coated with rubber or nylon, including the bolt holes. These insulation joints were tried in the field but proved unsuccessful owing to the fact that the rubber coating gets cut easily.

(c) Recently, insulated glued joints have been developed using resin adhesives. These joints consist of web-fitting type of fish plates glued to the rails using a high polymer adhesive and bolted with high tensile bolts. Insulation is provided by special insulating side channels, bushes, and posts made out of fibre glass cloth. These joints do not exhibit the movement and hence all maintenance problems are completely eliminated.

18.8 ORGANIZATION STRUCTURE FOR TRACK MAINTENANCE

Track maintenance on Indian Railways is a well-organized affair as described below.

18.8.1 Organization at the Headquarters

The primary duty of maintaining the track lies with the civil engineering branch headed by the principal chief engineer stationed at the headquarters of each zonal railway. He is assisted by a number of functional chief engineers, such as the chief track engineer (CTE), the chief bridge engineer (CBE), the chief engineer planning and design (CPDE), and the chief engineer general (CGE). Normally, one functional chief engineer is in charge of one division or more and supervises the work of that division in all civil engineering matters. Each chief engineer is assisted by deputy chief engineers, executive engineers, etc., who are responsible for planning, designing, and providing the necessary materials along with providing any other assistance that may be required by the field engineers.

18.8.2 Organization at the Divisions

The direct responsibility of the maintenance of civil engineering assets, including the track, lies with the division. In each division, there are one or more divisional superintending engineers (DSE) or senior divisional engineers (sr DEN) who

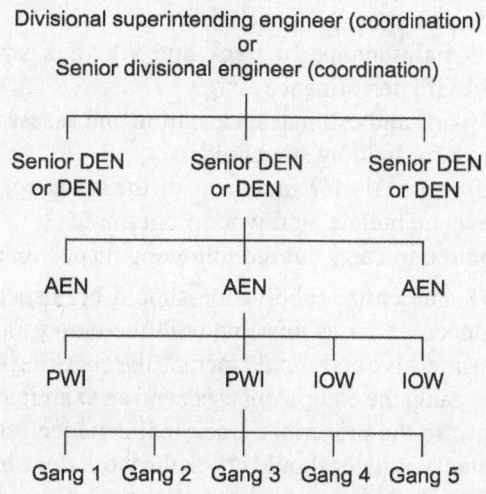

Fig. 18.4 Divisional organization for track maintenance

work under the administrative control of the divisional railway manager and the technical control of the chief engineer. Figure 18.4 gives the details of the divisional organization for track maintenance.

Each divisional superintending engineer (DSE) or senior divisional engineer has complete charge of a division. He or she is assisted by two to three divisional engineers (DENs), each with about 1000 integrated track kilometres under his or her charge. Every DEN is assisted by two to three assistant engineers (AENs), each in charge of about 400 kilometres of track. An AEN in turn is aided by two to three PWIs, who are directly responsible for the safety and maintenance of the track. Each AEN is also assisted by one or more inspector of works (IOW) who look after all the work. The track maintenance work is done by various gangs, each of which is headed by a mate who has about 10 to 20 gangmen working under him or her. Each gang has a keyman, who inspects the track daily to ensure its safety. The duties of AENs, PWIs, and other permanent way officials are explained briefly in the subsequent sections.

18.9 DUTIES OF PERMANENT WAY OFFICIALS

The permanent way officials in charge of maintaining the track have to ensure that the finest standards of track maintenance are followed as per the prescribed procedures at the minimum cost. The duties of all PW officials have, therefore, been specified so that they can work systematically to achieve the desired objectives. This chapter only outlines the essential duties of PW officials. In actual practice, however, they have to carry out many more miscellaneous functions as per the traditions and practices of Indian Railways.

18.9.1 Duties of Assistant Engineer

The assistant engineer is generally responsible for the maintenance and safety of all way and works under his or her charge, for the accuracy, quality, and progress of any new work that may be undertaken and for controlling all expenditure with respect to the budget allotment.

The essential duties of an AEN are as follows:

(a) Inspection and maintenance of track and all track structures to ensure satisfactory and safe performance
(b) Preparation of plans and estimates; execution and assessment of work
(c) Verification of stores held by stockholders
(d) Submission of proposals for inclusion in the track renewal programme, estimates of revenue budget, and work programme

An AEN is also required to carry out the following inspections:

Trolley inspection   The entire subdivision should be inspected once a month by an assistant engineer, as far as possible with the help of a push trolley. This inspection should be intensive and should include the checking of gang attendance, the work done by the gang, the equipment used, and an examination of gang charts/diaries with reference to the prescribed track maintenance schedule. During the inspection, the assistant engineer should check the work done by one or two gangs under each PWI and record his or her observations.

Fast train inspection   Either the foot plate of the engine or the last vehicle of a fast train should traverse the entire length of the subdivision once a month.

Inspection of level crossings   The assistant engineer should inspect all manned level crossings once every six months. He or she should examine the gatemen's knowledge of the rules and check the equipment, track, road approaches, and all other safety aspects of each crossing.

Checking of curves   The AEN should verify the versine and superelevation of at least one curve under the jurisdiction of every PWI every quarter.

Checking of points and crossings   Once a year the AEN should inspect all the points and crossings on passenger lines and 10 per cent of the points and crossings on other lines.

Monsoon patrolling   During monsoon patrolling, the assistant engineer should use either a train, push trolley, or motor trolley once every month to check the patrolman's work.

Scrutinizing of registers during inspection   The AEN should scrutinize the various registers maintained by the PWI, such as the creep register, curve register, and the points and crossings register.

Inspection of bridges   The AEN should inspect all the bridges of his or her subdivision once every year after the monsoon is over and record the inspection details in the bridge register. Tracks on girder bridges should also be inspected as part of the annual bridge inspection.

Inspection of office and stores   The AEN should inspect each office and store of all the PWIs and IOWs under his or her charge at least once a year. When checking stores, he or she should pay particular attention to the allocation of the imprest engineering indicators, protection equipment, and other important items in the stores.

An AEN also has the following additional duties:

(a) to ensure that all work is done as per the standard plans and specifications

(b) to witness the payments made under one inspector once every month

(c) to record the measurements of the ballast or to carry out a thorough check of its quality and quantity if the same has already been recorded by an inspector and to test check the measurements of other works

(d) to reach an accident site as early as possible and to take the necessary measures to restore traffic on the affected track

(e) to accompany any track recording or oscillograph car that runs in his or her section

(f) to exercise control on expenditure so as to contain it within the alloted budget

(g) to train probationers in their work

(h) to inspect water purification systems once every three months

(i) to inspect all steel structures once every six months

(j) to look after the welfare of the staff and to inculcate discipline in them

(k) to accompany the GM or other senior railway officials during inspection

(l) to coordinate with officials of other departments

18.9.2 Duties of a PWI in Overall Charge

The PWI is generally responsible for the following:

(a) Maintenance and inspection of the track to ensure satisfactory and safe performance

(b) Efficient execution of all works incidental to track maintenance, including track relaying work

(c) Accounts and periodical verification of the stores and tools in his or her charge

(d) Maintenance of land boundaries between stations and at important stations as may be specified by the administration

The PWI also carries out inspections of the following facets of a track:

Testing the track   He or she should run a test check on the foot plate of the engine of fast trains at least twice a month and in a rear brake van of a fast vehicle once a month, and make a note of sections where the quality of running is defective and get them rectified.

Inspection of track and gangs   The PWI should inspect the entire section with the help of a push trolley at least once a week or more often if necessary.

Level crossing inspection   This includes the following:

(a) He or she should check the equipment assigned to the gateman once a month.

(b) He or she should periodically examine knowledge of safety rules.

(c) He or she should ensure that all level crossings are safe.

Points and crossings inspection The PWI should inspect the points and crossings on passenger lines once in three months and those on other lines once in six months.

Curve inspection The PWI should check the versines and superelevation of each curve once in six months. Based on his or her observations, the PWI should take the appropriate action to correct the curve, if necessary.

Inspection diagram The PWI should maintain an inspection diagram of all inspections carried out during the month as per the schedule laid down in the pro forma and submit the same every month to the divisional engineer via the AEN, bringing out the reasons for failure in adhering to the schedules of inspections, if any.

Safety of track The PWI is directly responsible for the safety of the track. He or she should be vigilant so as to promptly locate faults in the permanent way and get them repaired without delay.

In addition to the inspections, a PWI also carries out the following duties:

(a) Accompanies high officials during their inspections along with the relevant records. The PWI should carry all the important measuring equipment, such as the gauge-cum-level, flange gauges, fishing chord, tape, and inspection hammer on these inspections

(b) Accompanies any track recording and oscillograph car that runs in his or her section

(c) Checks the proximity of trees that are likely to damage the track and get them removed

(d) Checks night patrolling at least once a month by train as well as by trolley

(e) Takes the necessary safety measures while executing maintenance work that affects the safety of the track

(f) Rushes to the site of an accident and take the necessary measures to safeguard the line and restore traffic

(g) Periodically inspects and supervises LWR tracks to ensure their safety

(h) Ensures the cleanliness of station yards

(i) Keeps proper records of the training out of ballast

(j) Witnessess the payments made out to the staff every month

(k) Looks after all establishment work, including the welfare of the staff working under his charge, and maintains their service records

(l) Ensures the safety of the track during the execution of work that affects the track

Based on the system of maintenance, the PWI in charge of the section should prepare detailed short-term plans covering a month's work (split into weekly programmes) at least a month in advance of the commencement of actual work. The PWI should ensure that adequate arrangements have been made for the requisite tools, materials, and manpower for the allotted task and that work is executed within the specified time. The following procedure of track maintenance is followed on Indian Railways:

(a) Each mate should be supplied with a gang chart and a gang register. The gang chart should have a record of the day-to-day track maintenance work to be done over the gang length, maintained by the permanent way inspector (PWI) according to specified instructions. The gang registers contain a record of the weekly programme of the work to be carried out, also maintained and entered by the PWI in charge of the section. At the end of the week, the PWI should qualitatively and quantitatively assess the completed work and record his/her observations in the gang register after a detailed inspection of the work done during the previous week.

(b) Gang charts or gang registers should be checked by the assistant engineer and divisional engineer during inspections. After inspecting the section by trolley, they should record their observations in the gang register.

(c) On withdrawal of old gang charts or gang registers and supply of fresh ones, the PWI should carefully analyse the work done and make a note of those stretches of the track that frequently gave trouble during the year, with a view to formulate such special measures as may be necessary.

18.9.3 Duties of Mates

Every mate should make sure that the length of line under his charge is kept safe for the passage of trains. Sections needing urgent attention should be taken care of without waiting for orders from PWIs.

18.9.4 Duties of Keymen

Once a day, the keyman should inspect both the track and the bridges on his/her beat. S/he should go along one rail on his outward journey and return along the opposite rail on his return journey.

18.9.5 Knowledge of Rules and Signals

Every mate, keyman, and gangman should have the correct knowledge of handling detonating signals and should be conversant with the following measures of track maintenance:

(a) Protecting the line in an emergency and during the execution of maintenance work
(b) Action to be taken when a train is noticed to have parted
(c) Knowledge of 'safety first' rules
(d) Action to be taken when sabotage is suspected
(e) Patrolling during emergencies

18.10 WORKING OF PERMANENT WAY GANGS

The track on Indian Railways is mostly maintained by permanent way gangs. Each permanent way gang has a strength of about 10 to 20 people and a beat of about 6 to 10 km. The gangs normally follow the annual programme set for regular track maintenance and complete at least one round of through packing in a year, depending on various circumstances. A gang works under the control of a mate who assigns track maintenance works to the permanent way gangmen.

18.10.1 Tools Used by a PW Gang

Each gang should have the following maintenance tools and equipment. Worn-out tools and equipment should be replaced every month:

(a) Gauge-cum-level
(b) One set of red hand-signal flags, two hand-signal lamps for the nights, and 12 detonators
(c) 30 cm-long steel scale, 1 m-long straight edge, square, hemp, cord, and marking chalk
(d) Wooden mallet or canne-a-boule, fish bolt spanner, keying and spiking hammer, and measured shovel packing (MSP) equipment if MSP is required
(e) A sufficient number of shovels, phowrahs, beaters, crowbars, ballast forks or rakes, and mortar pans or baskets

18.10.2 Knowledge of Safety Measures

The permanent way mistry (PWM) mate and all other gangmen should have complete knowledge of the following safety measures.

(a) Method of protecting the line in an emergency or during maintenance work that affects the running of trains, including methods of fixing detonators, banner flags, etc.
(b) Displaying the obligatory signals with or without a hand-signal flag during the day and with a hand-signal lamp during the night
(c) Action to be taken when a train is noticed to have parted
(d) Patrolling of lines on LWR tracks during heavy rains, storms, sandstorms, and during hot weather
(e) Awareness of 'safety first' rules

18.11 GANG STRENGTH FOR TRACK MAINTENANCE

A track is maintained manually by gangmen. Each gang has about 10 to 20 persons and a jurisdiction of about 4 to 6 km. In order to ensure uniformity of practice, Indian Railways has prescribed a standard formula known as the *special committee formula* for calculating the number of men to be included in each gang.

$$N = MKE = MKLU(1 + A + B + C)$$

where N is the number of men per km, M is the manpower factor, K is the correction factor due to modernization of track, standard of maintenance, etc., E is the number of equated track kilometres (ETKM) and is equal to $L \times U(1 + A + B + C)$, A, B, and C are variable factors, L is the length of a single track, and U is the traffic density that varies from 0.4 to 1.4 depending upon whether the section is a siding or a busy section with heavy gradients.

The manpower factor M is the weightage for the actual man-days required to carry out normal permanent way maintenance work in a year vis-à-vis the actual number of man-days available. The correction factor K is required as a result of the modernization of the track. It has been felt that the effort required to maintain SWR and LWR is much less as compared to a fish-plated track.

The variable factors, A, B, and C denote the following. A is the formation factor, which varies from 0 to 0.20 depending upon the type of soil, B is the alignment factor and is equal to 0 where the track is straight and 0.25 when the entire track is on a curve, and C is the rainfall factor and varies from 0.10 to 0.20 depending upon whether there is little or heavy rainfall.

18.11.1 Modifications of Existing Formula of Gang Strength

The Special Committee formula currently being adopted is as given below:

N = 0.95 MKE

The following modifications are recommended to the Special Committee formula to take care of the increased inputs required for the maintenance of track, increased GMT, higher speeds, heavier axle loads, and movement of BOXN type wagons, which were not envisaged by the Special Committee when they made their recommendations.

(i) For traffic density above 20 GMT per annum on BG Sections, the following values of factor (U) may be adopted

Traffic density	20–25 GMT	25–35 GMT	35–45 GMT	45–55 GMT
U	1.4	1.5	1.6	1.7

(ii) For BG sections with heavy haul operation or where five or more loaded trains consisting of BOXN or equivalent wagons operate per day the factor (U) may be further increased by the addition of 0.1.

(iii) To cater to satisfactory track maintenance for higher speeds, a further addition of 0.1 in factor (U) may be made both for BG and MG where the speeds are above 110 km per hour and 75 km per hour respectively.

18.12 YARDSTICK FOR WORKLOAD OF PERMANENT WAY OFFICIALS

A study of workload of PWIs and AENs of open line has been carried out by the Railway Board a few years back to assess the existing workload and lay down reasonable yardstick for their work for future. The workload on an open line for an AEN primarily consists of maintaining the track, bridges and buildings. There was earlier no common unit to quantify the total workload of an assistant engineer for purposes of comparison, etc. A new terminology called 'Integrated track kilometre' (ITKM) was, therefore, introduced which could quantity the work-load not only of track, but of buildings and bridges also for purposes of comparison and laying yardsticks.

18.12.1 Integrated Track Kilometres

Integrated track kilometre (ITKM) is a composite unit representing workload of an open line maintenance unit on account of track, buildings, and bridges.

ITKM is calculated as follows:

Equated track kilometre due to permanent way

This is calculated as follows:

(i) Running track kilometre is first found out depending upon route kilometre.

(ii) Weightage of non-running lines like sidings, etc., is taken as follows:

Siding	*BG*	*MG*	*NG*
Marshalling siding and busy yards	0.7	0.5	—
Other sidings	0.4	0.4	0.3

(i) Weightage for points and crossings is taken by equating 10 sets of points and crossings as equal to one track kilometre. The layouts equivalent to one set of points and crossings are as given below:

Turnout	1 set
Diamond	1 set
Cross over	2 set
Scissors	5 sets

(ii) Equated track kilometres due to permanent way is arrived at by adding all the above track kilometres.

Equated track kilometres due to buildings

Based on the principle of equivalent cost of maintenance it has been found that 1,500 m^2 of plinth area of building can be equated to one equated track kilometre. Knowing the total plinth area of the buildings, the equated track kilometre due to buildings, can be found out.

Equated track kilometres due to bridges

This is based on waterways of the bridges and is as follows:

100 m of minor bridges per track = 2 ETKM

100 m of major bridges per track = 3 ETKM

The ITKM is found out by summing the three ETKM as calculated above.

18.12.2 Yardstick for Workload of PWI

The following broad principles have been laid down for the workload of a PWI:

(i) The jurisdiction of a PWI should normally be about 60 to 70 route kilometre on single line, 50 route kilometre on double lines, and 40 route kilometre on multiple lines.

(ii) The equated track kilometre for a PWI is given in Table 18.4.

Table 18.4 Equated track kilometre for PWI

Gauge	*Single/double or multiple lines*	*Jurisdiction of a PWI*
BG (Trunk route and main line)	Single line	110–125 ETKM
—do—	Double line	150 ETKM
—do—	Multiple line	170 ETKM
MG (Trunk route and main line)	Single line	95–105 ETKM
MG (Trunk route and main line)	Double line	115–130 ETKM
BG and MG (Branch and tertiary line)	Single line	80–90 route km

18.12.3 Workload of a Sectional PWI

On trunk routes and main lines, a sectional PWI may have a route length of 30 to 40 km on a single line and an equated track kilometre of 40 for BG and 50 for

MG. On double-line section, a sectional PWI should normally have route length of 20 to 30 km and equated track km of 60 km for BG and 80 km for MG.

18.12.4 Work of AENs/DENs of Open Line

The following yardstick is laid for the workload of AENs.

Section	Workload of an AEN
BG main line section	425 ITKM units
BG suburban section	550 ITKM units
MG section	400 ITKM units
NG section	350 ITKM units
Purely building maintenance subdivision	425–450 ITKM units

AENs at divisional headquarters should have 60 ITKM less and those at the headquarters of zonal railway should have 100 ITKM less than the prescribed standards.

Separate yardsticks exist for field gazetted staff required for special works. An AEN/DEN is justified for works having 20/60 millions weighted outlay per year.

18.13 PROTECTION OF TRACK FOR ENGINEERING WORK

There are certain engineering works, such as track renewal work and bridge rehabilitation work that stretch over a few days and thus the track is required to be protected to ensure the safe journey of trains.

18.13.1 Engineering Indicators

When a track is under repair, trains are required to proceed with caution at restricted speeds and may even be required to stop. Temporary engineering indicators (Fig. 18.5) are set up at the affected portion of the track to alert the drivers to reduce the speed of (or even stop) the train and also to resume the normal speed once the affected portion has been treated. The following indicators are used.

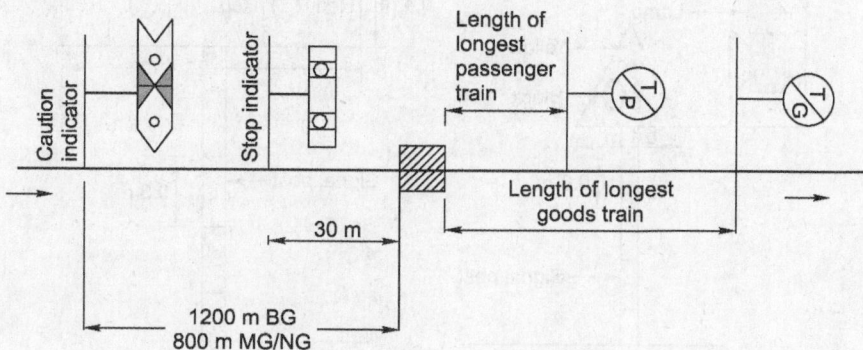

Fig. 18.5 Protection of track for works of long duration

Caution indicator This indicator cautions the driver to get ready to reduce the speed of the train (Fig. 18.6). It is placed at a distance from the stop indicator—1200 m away on BG tracks and 800 m away on MG and NG tracks.

Termination indicator This signal indicates that the driver can resume normal speed and that the speed restriction zone has ended (Fig. 18.7).

Speed indicator The driver has to reduce the speed of the train at the location bearing this indicator (Fig. 18.8).

Stop indicator The driver is required to stop at locations bearing this indicator (Fig. 18.9). It is normally placed 30 m away from the obstruction.

These indicators are also called *temporary fixed engineering signals* and are provided in the direction of the approaching train in the case of double-line tracks and in both directions in the case of single-line tracks. These signs or indicators should be luminous as per the latest policy of the Railway Board.

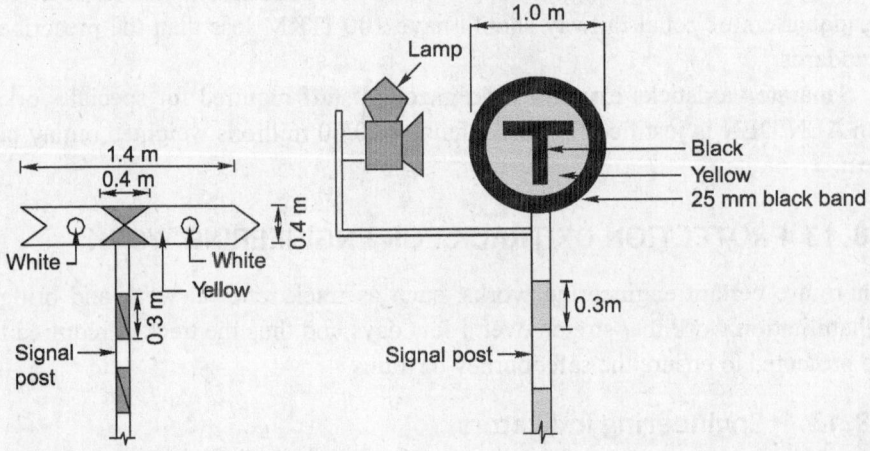

Fig. 18.6 Caution indicator   **Fig. 18.7** Termination indicator

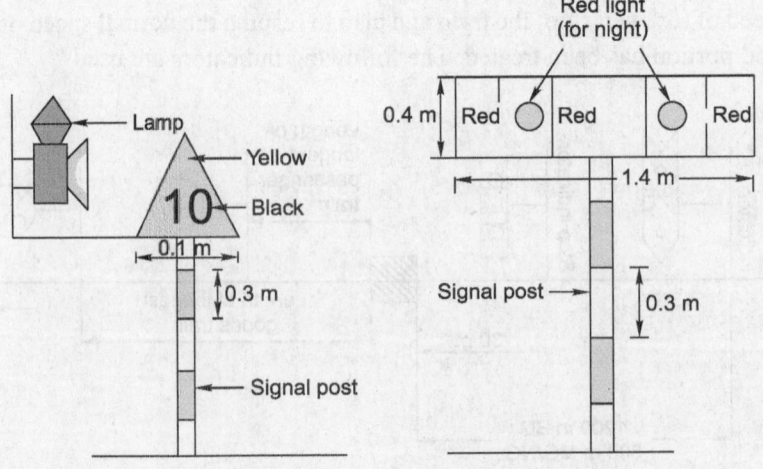

Fig. 18.8 Speed indicator   **Fig. 18.9** Stop indicator

18.14 PATROLLING OF RAILWAY TRACKS

Railway tracks are patrolled to ensure the safety of the track and of the traffic moving over it. Patrolling basically involves to and fro movement of the patrolman/ watchman along the track as per the specified programme in order to look out for any unusual occurrence that may threaten the safety of the track. The various types of patrolling that are in vogue on Indian Railways are presented below.

Keyman's daily patrol   The keyman inspects the track daily on foot. He or she normally inspects one rail as s/he moves forward and the other rail as s/he returns to the point from where s/he started. He or she tightens the keys and other fittings and ensures the safety of the track.

Gang patrol during abnormal rainfall or storm   Either at his or her own initiative or under instructions from the PWI, the mate organizes the patrolling of the track length affected by rains or storm. He or she, along with other gangmen, patrols the track and looks for any unusual occurrences that may be harmful to the track.

Night patrolling during monsoon   Night patrolling is done in a programmed way on specified sections of the railway track to detect damages such as breaches and settlements caused by floods to ensure safety.

18.15 TRACK TOLERANCES

The safety and comfort of travel depend primarily on track geometry and the standard at which it is maintained. In practice, it is not possible to obtain a flawless and perfect track; the parameters governing track geometry always show certain variations due to various reasons. Track tolerances may be defined as the limits of variability of various parameters pertaining to track geometry.

18.15.1 Track Parameters

Tolerances are generally laid down for the following track parameters:

Gauge variation   This is measured as the deviation from the nominal gauge, which is 1676 mm for BG and 1000 mm for MG. The gauge is measured as the minimum distance between the running faces of two rails.

Unevenness   This is measured in terms of the difference in longitudinal levels over a fixed base. Unevenness is generally measured over a base of length 3.5 m. It is measured separately for left and right rails.

Cross level difference   This is measured in terms of the relative differences in the level of two rail tops measured at the same point. Cross level difference also includes the variations in superelevation.

Twist   This is measured in terms of the change in cross levels per unit length of measurement. Twist is calculated after knowing the cross levels and the difference between two points over a fixed base of, say, 35 m and dividing the cross level difference by this base length. This is normally denoted as mm/m.

18.15.2 Safety Tolerances

Safety tolerances are the limits of variation beyond which the movement of traffic on the track becomes dangerous or unsafe. The kind of unsafe conditions that arise depend not only upon the condition of the track, but also on the type of vehicle, its riding characteristics, and its standard of maintenance. The factors that govern these unsafe conditions are so variable and indeterminate that almost none of the railway systems in the world have laid down any safety tolerances. It is generally believed that possibly the track would have to deteriorate to a great extent for it to reach these unsafe limits.

18.15.3 Track Tolerances for Good Riding Quality

The limits of track tolerance prescribed in order to guide engineering officials regarding the suitability of the standards of track maintenance on BG tracks with a sanctioned speed of above 100 km/hr and up to 140 km/hr are presented in Table 18.5.

There are no special tolerance limits for cross level defects. The track should be maintained at a standard that is generally superior to that of main line tracks where unrestricted speeds of up to 100 km/hr are permitted.

The provisions and tolerances mentioned here and elsewhere in the chapter have been provided with a view to maintain track geometry so as to ensure a comfortable

Table 18.5 Limits of track tolerances on Indian Railways

Parameter	*Value*	*Remarks*
Alignment defects (versine measured over 7.5 m under floating conditions)	5 mm for a straight track Curve ±5 mm Total change of versine from chord to chord <10 mm	Up to 10 mm at isolated locations Up to ±7 mm at isolated locations
Twist (measured on a base of 3.5 m)	2 mm/m for straight and curved track 1 mm/m for the transition of curves	Up to 3 mm/m at any isolated location Up to 2.1 mm/m for isolated locations
Gauge variation for BG (mm)	±6 on a straight track −6 to +15 mm on curves with radius of 350 m or more Up to +20 mm on curves with radius less than 350 m	−3 to +6 on a straight MG track −3 to +15 mm on curves with a radius 290 m or more Up to 20 mm on curves with radius less than 290 m
Unevenness of rail joint depressions	10 mm (measured on a chord of 3.5 m)	15 mm for isolated locations

ride and not from the point of view of stability or safety. (Source: ACS No. 96 of 2004 issued under Railway Board letter no. 2004/CE-II/CS-I dated 22 July).

18.16 TRACK RECORDING

Track recording consists of objective method of assessment of the quality of track with the help of sophisticated equipment. A continuous record of various track parameters is made with the help of track recording devices from which an assessment can be made about the running quality of the track. Track record is normally done by using any of the following types of equipment:

 (i) Hallade track recorder
 (ii) Oscillograph car
 (iii) Portable accelerometer and OMS-2000
 (iv) Track recording trolley
 (v) Amsler track recording car

Out of the various types of equipment used for track monitoring, 'Amser Track Recording Car' is the most sophisticated modern equipment. The following section elaborates on the same.

18.16.1 BG Amsler Car

Indian Railways have got track recording cars of the 'Amsler Type'. One track recording car was initially imported from Amsler and Company, Switzerland in 1964. The other track recording cars have, however, been copied out from this original model and developed indigenously in the country.

Amsler track recording car is a vehicle of the express passenger type, having two bogies. One of the bogies, which is a measuring bogie is a 3-axle bogie having a wheel base of 3.6 m. The central axle of the 3-axle bogie has flangeless cylindrical wheel, while the other two axis have have normal wheels having 1-in-20 taper.

Items of Recording in BG Amsler Car

Amsler track recording car gives a continuous record of the five track parameters viz. (i) unevenness, (ii) twist, (iii) gauge, (iv) superelevation (v) curvature or alignment.

Out of the above track parameters, only unevenness, twist and gauge are considered for assessing the quality of track. The recording of curvature or alignment, being unreliable with the present arrangements in the Amsler car, is not taken into consideration. Superelevaton is only recorded for theoretical interest and does not figure up anywhere in the results. Unevenness, twist, and superelevation are recorded under loaded conditions. The recording is done at a speed of 70 to 80 km per hour and the Amsler car covers about 350 km of track per day during daylight hours.

The important characteristics recorded by the Amsler car along with the scales adopted for BG are given in Table 18.6.

Table 18.6 Important characteristics recorded by Amsler car

Characteristics of records	*Scale*
Speed of train	1 cm = 20 kmph
Curvature of left rail (versine on 7.2 chord base)	1 mm = 1 mm
Curvature of right rail (versine on 7.2 chord base)	1 mm = 1 mm
Superelvation	1.5
Variation in track gauge	1.1
Twist of track on 3.6 m base	3.6 mm = 1 mm/m (1:1)
Vertical unevenness of left rail on 3.6 m base	1:1
Vertical unevenness of right rail on 3.6 m base	1:1

This item is not being recorded now in mechanical track recording car.

The parameters twist and unevenness are measured by the wheels of the Amsler car, whereas measurement, gauge, and curvature is done by its feeler. These parameters are subsequently transmitted mechanically to the recording table by a system of pulleys and cables.

18.16.2 Principles of Recording Track Parameters

Gauge The gauge is recorded at 14 mm below the rail surface by the spring loaded feelers pressed against the gauge face. The relative movements of the feelers are picked up by LVDTs and their signals are algebrically added to get the gauge.

Unevenness This is measured over two selected chord lengths of 3.6 m and 9.6 m. These midchord offsets are generated by running electronic chords, of 3.6 m and 9.6 m on inertial vertical profile, of left and right rail generated from the signals of coach accelerometers and LVDTs.

Twist TRC calculates the twist from dynamic cross level, i.e., difference in the vertical profile of the left and right rails on two different chords of 3.6 m and 4.8 m. Twist can also be recorded more precisely using gyroscope and inclinometer.

Alignment Lateral alignment of the left and right rails are measured over two selected chord lengths of 7.2 m and 9.6 m. These midchord offsets are generated by running electronic chords of 3.6 m and 9.6 m on inertial lateral profiles of the left and right rails generated from the signals of accelerometers and LVDTs.

SUMMARY

The track should be maintained properly to enable trains to run safely at maximum permissible speeds. Through packing, overhauling, and picking of slacks are the three main operations of maintenance in one calendar year. The money and labour to be spent on track maintenance should be optimally utilized to keep the track in good working condition. There are various railway officials who are assigned specific duties in regard to track maintenance. Also, there are limits assigned for the variability in the different parameters that affect track geometry, known as track tolerances.

REVIEW QUESTIONS

1. Categorize and briefly describe the various duties of a permanent way inspector.
2. Explain the system of annual maintenance of rail tracks on straight portions, including the several operations involved therein. Describe these operations in detail.
3. Describe the procedure involved in the annual through maintenance of a track commonly known as through packing.
4. What is the need for the proper maintenance of a track? Discuss the various methods that ensure that a track is well maintained.
5. What is meant by through packing? Describe the various steps involved in this procedure. What is the programme of annual track maintenance followed on Indian Railways?
6. Differentiate between the following.
   (a) Hogged joint and high joint
   (b) Lifting and lowering of track
   (c) Insulated joint and fish-plated joint
   (d) Defective sleeper and centre-bound sleeper

7. What do you understand by deep screening of ballast? Describe the procedure.
8. What are the duties of an assistant engineer on Indian Railways? What is the schedule of his or her inspection?
9. What is track recording? Give the description of working of the Amsler Car.
10. Write short notes on the following.
    (a) Gang strength
    (b) Blowing and pumping joint
    (c) Tools of a permanent way gang
    (d) Maintenance of track in track-circuited areas

Choose the correct answer from the choices given.

11. While doing through packing of BG track, ballast is opened out upto the end of the sleeper and on inside it is .... from the rail seat.
    (a) 250 mm   (b) 350 mm
    (c) 450 mm   (d) 500 mm
12. Gauge tolerances for BG on straight track are:
    (a) – 6 mm to + 6 mm   (b) – 3 mm to + 6 mm
    (c) – 6 mm to + 15 mm   (d) upto + 20 mm
13. The normal output of a gangman for through packing for MG is:
    (a) 23 to 24 m   (b) 18 to 20 m
    (c) 16 to 17 m   (d) 11 to 12 m
14. In deep screening of track, if track is packed manually, the normal speed is restored in:
    (a) 10 days   (b) 15 days
    (c) 21 days   (d) 30 days
15. In deep screening of track, if track is packed mechanically with the help of tampers, the normal speed is restored in:
    (a) 10 days   (b) 15 days
    (c) 21 days   (d) 30 days
16. Duration of pre-monsoon attention is:
    (a) 4 months   (b) 3 months
    (c) 2 months   (d) 1 month
17. The entire track must be deep screened at least once in:
    (a) 12 years   (b) 10 years
    (c) 8 years   (d) 5 years
18. BG electronic track recording car records alignment on the right and left rail at chord of:
    (a) 2.74 m   (b) 3.6 m
    (c) 7.2 m   (d) 9.6 m
19. BG electronic track records unevenness at a base of:
    (a) 2.74   (b) 3.6 m
    (c) 7.20 m   (d) 9.6 m

Track Drainage

INTRODUCTION

Track drainage can be defined as the interception, collection, and disposal of water from, upon, or under the track. It is accomplished by installing a proper surface and sub-surface drainage system.

19.1 NEED FOR PROPER TRACK DRAINAGE

Water is the greatest threat to a railway track, and the most prominent factor that adversely affects track maintenance is improper drainage. Excess water affects the stability of the embankment, and the bearing capacity of the soil and its resilience to shear gets considerably reduced. Railway engineers give maximum importance to the proper drainage of the track and the need to do so is felt because of the following concerns.

Settlement of embankment   Excess water may cause the embankment to settle. An unequal settlement may lead to variation in cross levels as well as longitudinal levels thereby affecting the safety and riding quality of the track.

Reduction in bearing capacity   The bearing capacity of the soil as well as its resistance to shear diminishes due to excess water in the soil. This in turn leads to numerous problems and finally affects the safety and stability of the track.

Failure of embankment   The percolation of water in the embankment increases the weight of the soil on the one hand and reduces its bearing capacity and shear resistance on the other. This makes the formation unstable. This is a common reason for the failure of embankments, which occurs in the form of slips.

Formation of ballast pockets   Excess moisture leads to punctures in the formation. The constant hammering action of running trains causes the sleepers to move up and down, resulting in ballast pockets, pumping sleepers, and other such problems. All these factors lead to poor riding quality.

Shrinkage and cracking of banks   The embankment soil cracks and shrinks once excess water dries up. The problem becomes acute in the case of poor soils. This in turn leads to many problems, such as the loss of ballast in the cracks and uneven settlement, which have an adverse effect on track maintenance.

Adverse effects of black cotton soil   There are certain soils, such as black cotton soil, which become plastic in nature due to the accumulation of excess water. The formation shrinks due to change in moisture content. The bearing capacity of the soil is greatly reduced due to the excess water. These problems imperil the safety of the track.

Formation of slush   Slush is formed due to the dynamic load of running trains. It is forced out, thus badly clogging the ballast. This makes the maintenance of the track very difficult.

19.2 SOURCES OF PERCOLATED WATER IN THE TRACK

Water can percolate to the formation of the track through any of the following methods:

By gravity   This includes water that collects due to rains, etc., moving into the subgrade because of the effect of gravity. This movement is resisted by the permeability of the soil. The effective ways of reducing the progress of water by this method are the following:
  (a) Provision for drainage on the top of the embankment in the form of cross fall, side drains, lowering of cess, etc.
  (b) Turfing of side slopes of the embankment.

By capillary action   At times water rises into the subgrade by way of capillary action. The capillary rise of water can be prevented by providing a pervious layer in the embankment, which serves as a capillary break.

From adjacent areas   In this case, water from some nearby source seeps into the subgrade. Water seepage can be reduced by taking the following steps:
  (a) Diverting the original source of water
  (b) Providing effective paved catch water drains
  (c) Providing inverted filters and underground drains

By hydroscopic action from atmosphere   The moisture present in the atmosphere is comparatively very small and has very little effect on the total moisture content of the soil.

19.3 REQUIREMENTS OF A GOOD TRACK DRAINAGE SYSTEM

A good drainage system should satisfy the following requirements:

Surface water should not percolate to track   One of the basic requirements of a good track drainage system is that surface water from rains and adjacent areas should not percolate and seep into the formation of the track.

Effective side drains   The size of the side drains should be adequate with a proper slope, so that they effectively carry all the surface water away.

Longitudinal drains to be saucer-shaped   The longitudinal drains provided between two tracks should preferably have a saucer-shaped cross section so that they can collect water from both sides.

Provision for clearing and inspection   The drains provided for drainage should be such that they can be inspected and cleared periodically.

Drain top to be below cess level   Normally, the drain top should not be above the cess level for effective drainage of the ballast bed.

No erosion of banks   The flow of water along the slope and across the track should not cause erosion of the banks or the slopes of the banks.

Formation to be of good soil   Ideally, the formation and subgrade should be made of a pervious, coarse-textured soil. Such soils are more permeable, retain less capillary water, and respond more favourably to a surface drainage system.

Proper sub-surface drainage   Arrangements should be made for a good sub-surface drainage system to drain off the water being retained is the track. This is more relevant in the case of defective formations.

Proper outfall   Longitudinal drains should be designed so as to provide a proper outfall, from where the water can eventually drain off.

Special arrangements for waterlogged areas and other difficult situations   A good track drainage system should have special arrangements for the drainage of waterlogged areas and for all other related perennial problems.

19.4 PRACTICAL TIPS FOR GOOD SURFACE DRAINAGE

The following measures will ensure that the track drainage system gives a good performance.

Maintain proper cess level   Tracks on embankments get drained as long as the proper cess levels are maintained and the ballast is clean. However, in cuttings and in yards, where water cannot recede freely and quickly from the track, a well-planned drainage system must be provided.

No vegetation   There should be no growth of vegetation in the track, as this indicates clogging of the ballast and a lack of adequate track drainage. Such stretches of tracks should be overhauled or deep screened. The ballast should be clean so that rainwater can easily flow out of the track.

Area below rail foot to be clear   About 25 to 50 mm of the area below the rail foot should be kept clear of any ballast, earth, or cinder on all lines inside and outside the yards. This would enable good surface flow as well as avoid corrosion and failure of track circuits.

Cleaning and repair of drains   All drains should be cleaned and repaired as a part of an annual through packing, not only on run-through lines but also on all other running lines in yards.

19.5 TRACK DRAINAGE SYSTEMS

As mentioned earlier, a good track drainage system should essentially ensure that no water percolates into the track at either the surface or the sub-surface levels, and the arrangements for the drainage of sub-surface water should be good and effective wherever required. Track drainage should be handled in two distinctive phases.

Surface drainage   Surface water due to rain or snow, or from adjacent areas should be drained off properly by designing well-planned and effective surface drains.

Sub-surface drainage   In case water percolates into the formation due to bad soil or such other reasons, the formation gets adversely affected and this has a bearing on the safety and stability of the track. Complete details of the same are given in Chapter 9.

In the following sections, surface drainage is dealt with in three different stages covering the entire length of the track.
 (a) Drainage in mid-sections between railway stations
 (b) Drainage in station yards
 (c) Drainage of station platforms

19.5.1 Drainage in Mid-sections Between Railway Stations

A drainage system between two railway stations consists of the following features:

Side drains   These should be provided along the track in cuttings and zero fill locations, where the cess level is not above the ground level. The typical cross section of a side drain is shown in Fig. 19.1. All drains must have an adequate gradient to enable the free flow of the collected water.

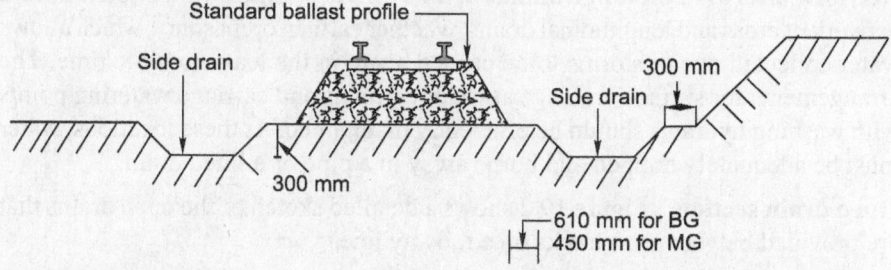

Fig. 19.1   Typical cross section of side drain

Lining of drains   Side drain lining is imperative on Group A routes and preferable on other routes, except those routes where the drains are in the rocky strata.

Adequate opening under level crossing   In order for all the water to flow out of the side drains in cuttings, adequate openings should be provided under level crossings, around trolley refuges, and around the overhead electric masts.

Catch water drain   Catch water drains should be provided wherever necessary, in cuttings. Their size should be worked out according to the volume of the water the drain is expected to carry.

No surplus ballast   Surplus ballast in the shoulders of the track retards drainage and encourages vegetation growth. All extra ballast should be taken out of the track and stacked in small heaps beside the track.

19.5.2 Drainage in Station Yards

When planning the drainage of station yards, the following guidelines should be kept in mind.

Open surface drains Surface drains should generally be left open to make cleaning and inspection convenient. When designing in-yard drains, a velocity range of 0.5 to 1.0 m/s may be allotted for earthen drains and 1.0 to 2.0 m/s for lined (or masonry).

Saucer-shaped drains As mentioned earlier, the longitudinal drains that lie between two tracks should be saucer-shaped with curved sides. However, drains with vertical sides may be provided wherever saucer-shaped drains are not practicable.

Drain top not to be above cess level Normally, the drain top should not be above the cess level for the effective drainage of the ballast bed. However, if it is essential that a drain with a higher top level be provided in order to retain the ballast, weep holes must also be provided at the assumed cess level and the drain so designed that the water it carrier does not flow the base level of the weep holes.

Outflow and slope Wherever there is a proper outfall available at either end of a yard, the longitudinal drains provided should have their slopes facing the direction opposite to the middle of the yard. This will ensure the minimum size and depth of the drains.

Position of ballast sections The ballast section in station yards should be the same as that on the main line.

Network of cross and longitudinal drains Every station yard should have a network of cross and longitudinal drains, whether earthen or masonry, which allows water collected due to storms to be carried away in the least possible time. The arrangements for surface drainage at water columns and carriage watering points with washing hydrants should be efficiently maintained. At these locations, water must be adequately trapped and borne away in a pipe or a lined drain.

Open drain section Figure 19.2 shows a detailed sketch of the open drains that are provided between two tracks on a railway line.

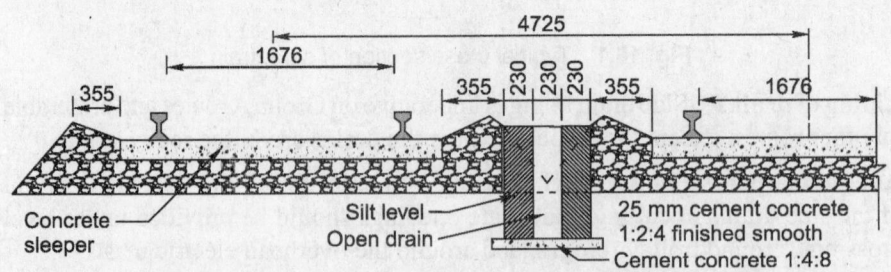

Fig. 19.2   Open drain between tracks (all dimensions in mm)

19.5.3 Drainage of Station Platforms

The following points should be taken into account when planning the drainage system of a platform:

Slopes away from track   Normally, all end platforms should be sloped away from the track. The details of the drainage system of a platform are illustrated in Fig. 19.3.

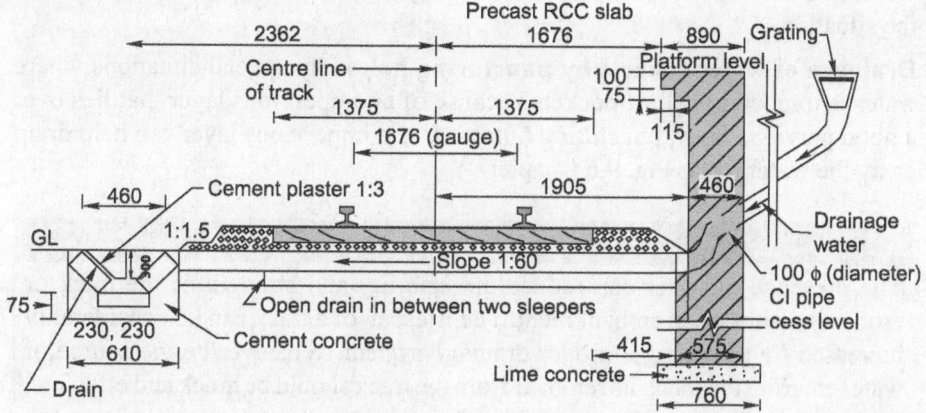

Fig. 19.3   Drainage of platform

Discharge on non-track side   All drains from platform shelters, tea stalls, toilets, water taps, etc., should be enclosed in pipes and should normally discharge or release their contents on the non-track side of the platform. Covered longitudinal drains should be provided on the platform, if necessary.

Discharge not towards run-through lines   In the case of island platforms, all drains should discharge their contents on the less important side of the track and not towards run-through lines.

19.6 SUB-SURFACE DRAINAGE

Water normally reaches the sub-surface due to capillary action, seepage from adjacent areas, or percolation of rainwater. When water reaches the sub-surface, it immediately affects the stability of the formation. The situation becomes worse particularly if the soil is bad. A variety of problems arise due to the proximity of the water to the subgrade. The nature of the problems faced and the remedial measures for the same have been discussed in detail in Chapter 9.

The various sub-drainage systems used under different conditions are briefly summarized here.

Provision of inverted filter   Some sub-drainage systems consist of a blanket of a non-cohesive graded material, which acts as a capillary break (Fig. 9.3 in Chapter 9).

Paving of catch water drains   The entrance of water into the subgrade can be checked by constructing effective catch water drains that are duly paved.

Provision of sand piling   Some drainage systems are provided with an adequate number of sand piles of diameter of about 20 cm. This improves sub-surface drainage (refer Fig. 9.4 Chapter 9).

Drainage of water pockets by perforated pipe Perforated pipes with a diameter of about 30 cm are provided at appropriate places to drain off the water pockets (refer Fig. 9.5, Chapter 9).

Cement grouting Cement can be used to grout the water pockets so as to ease the situation.

Drainage of water pockets by puncturing holes In special situations where water is trapped in ballast pockets because of an impervious layer that lies over a good pervious layer, puncturing holes into the impervious layer can help drain away the water (refer Fig. 9.6 Chapter 9).

SUMMARY

The presence of water can reduce the stability and strength of the track or erode the banks of an embankment. The lifespan of a track can be considerably increased by providing a proper drainage system. Whatever be the source of water entering the track, its removal from the track should be quick and effective. In the case of soft soils, sub-surface drainage may be necessary to keep the track safe.

REVIEW QUESTIONS

1. Why is it necessary to provide adequate drainage facility for a railway track? Suggest remedial measures to solve the following problems that occur due to poor drainage:
   (a) Clogging of the ballast because of wet earth
   (b) Sinking of ballast into the wet earth
2. What are the requirements of a good track drainage system? How can a good drainage systems be provided on railway tracks?
3. Describe the drainage systems required for the following areas in detail:
   (a) Drainage in station yards
   (b) Drainage of station platforms
   (c) Drainage in areas between stations
4. What are the various sources of percolated water in the track?
5. Write short notes on the following:
   (a) Inverted filter
   (b) Ballast pockets
   (c) Sub-surface drainage
   (d) Catch water drain

Choose the correct answer from the choices given.

6. The sources of percolated water in the track is by:
   (a) gravity   (b) capillary action
   (c) hydroscopic action from atmosphere
   (d) any of (a), (b) & (c)
7. Improper drainage of track causes the following problems:
   (a) settlement of embankment
   (b) reduction of bearing capacity of soild

    (c)   formation of ballast pockets

    (d)   all of the above (a), (b) & (c)

8. For lined open surface drains for drainage of yard, the section of drain should be designed based on velocity of water as:

    (a)   4 to 5 m/s   (b)   3 to 4 m/s

    (c)   2 to 3 m/s   (d)   1 to 2 m/s

Modern Methods of Track Maintenance

INTRODUCTION

Modern methods of track maintenance employ track machines and other modern track equipment for the maintenance of tracks as opposed to the traditional methods of manual packing. The methods used generally are the following:

(a) Mechanized maintenance of track with the help of track machines
(b) Measured shovel packing (MSP)
(c) Directed track maintenance (DTM), which is need based maintenance particularly if DTM is done by MSP

In these methods, the emphasis is not only on modern maintenance techniques but also on the identification and rectification of defects, the implementation of proper quality control in a systematic manner, and the use of proper tools and equipment.

For over a century, tracks on Indian Railways have been maintained by the method of 'beater packing' using manual labour and the performance has been excellent. It is now being felt that this method of manual maintenance, which has stood the test of time, may require a revision in view of all the recent technological and social advancements. The European Railways, which was also in a similar position till a few decades ago, has already switched over to modern methods of maintenance to suit the requirements of the modern track. Such a system of maintenance of track has given the men working on the permanent way a better social status, as it requires lesser physical strength and higher mechanical skill on their part. The system has also enabled them to maintain the modern track more economically and effectively in order to cater to the requirements of higher speeds and heavier axle loads. There is no doubt that the mechanized maintenance of track has become a technical necessity for the modern track structure with its long-welded rails and concrete sleepers. In spite of changes in our socio-technical background and the increasing emphasis on the modernization of the track to allow for higher speeds the following question arises: Is there really a need for the complete mechanization of the process of track maintenance, particularly when India has such a large labour force and is faced with the serious problem of unemployment?

20.1 MECHANIZED METHODS OF TRACK MAINTENANCE

The need to switch over from manual to mechanized methods of track maintenance is progressively being felt due to the following reasons:

(a) Beater packing is a very hard and strenuous job and thus the labourers have a tendency to shirk away from this type of work.

(b) It is difficult to ensure uniform quality of the compaction under the sleepers carried out by manual means due to the uncertainties associated with the varying physical strength of the labourers, commitment of the workers, varying weather conditions, and other allied factors.

(c) The intensity of the pressure and shock that the ballast is subjected to when the beater is being used is very high and in many cases exceeds the crushing strength of the stone. This results in the progressive clogging of the ballast section.

(d) Traffic densities, axle loads, and speeds have increased considerably on Indian Railways in the recent past. Beater packing does not enable track geometry to be maintained within the tolerances prescribed for a satisfactory length of time.

(e) The retention of the packing done through manual maintenance is not very good and the track geometry gets distorted in a short time due to high-speed traffic.

(f) Manual maintenance is not much suited to the modern track, which consists of LWR and heavy concrete sleepers.

(g) With the increase in traffic density, the time available between trains is becoming progressively short. It is, therefore, becoming increasingly difficult to maintain tracks by manual methods, which take a considerably long time.

(h) When a track is maintained manually, it takes considerable time for it to get fully consolidated and, therefore, speed restrictions exist for a long period after track renewal work has been completed.

(i) Manual methods do not emphasize on the identification of defects and monitoring of the work being done. These are, however, done in the case of modern methods of track maintenance, thereby giving more effective results.

Some of the important heavy track machines being used on Indian Railways (IR) are listed in Table 20.1.

Table 20.1 Track machines in use on Indian Railways

Type of track machine	Functions	Remarks and output
Plassermatic 06-16-USLC	Tamping, levelling, and aligning. Consists of 16 tamping tools and can tamp one sleeper at a time.	Output is about 700 sleepers per hour.
Plassermatic 08 Unomatic tamper	Tamping, levelling, and lining simultaneously. Consists of 16 tamping tools and can tamp one sleeper at a time.	Output is 1000 sleepers per hour.
Plassermatic 08 Dnomatic tamper	Tamping, levelling, and lining simultaneously. Consists of 32 tamping tools and can tamp two sleepers at a time.	Output is 2000 sleepers per hour.
Plasser 09 CSM tamper	Tamping, levelling, and lining simultaneously. The machine advances continuously, while tamping is done from sleeper to sleeper.	Latest track machine with an output of 2500 sleepers per hour.
Plasser 08-275 unimat P and C tamper	Tamping points and crossings. Consists of 16 independent tiltable tamping tools, which can tamp points and crossings as well as plain tracks.	Good machine for tamping points and crossings. Can tamp 1 turnout in a 4-hour block.
Plasser VDM-800 consolidator	Used for crib shoulder consolidation so as to improve the retention of packing.	Output is 1000 sleepers per hour.
DGS-62 N stabilizer	Dynamic track stabilizer that uses vibratory rollers to further stabilize the track.	Output is 500 sleepers per hour.
Matisa R-7 ballast regulator	Regulates ballast profile by transferring ballast from one place to another.	Can regulate ballast of 1 km of track per hour.
Plasser RM-80 BCM	Ballast cleaning machine used to excavate, clean, and put the screened ballast back in the track and to remove unwanted material.	Can handle 650 m^3 of ballast per hour.
Switch relaying machine WM-22	Removes and installs parts or complete assemblies of points and crossings.	Can relay about two sets of points and crossings everyday.
Track relaying trains (TRT)	Relays the entire track automatically with very little manual effort. Consists of four different units and other ancillary machines with a total cost of over ₹ 100 million.	Latest state-of-the art machine. Can accomplish the relaying of 1 km of track per day in a 4-hour block.

On railways, the mechanized maintenance of tracks normally involves the use of mechanical tampers, which are used to tamp or pack the ballast. Mechanical tampers are generally of two types, namely *off-track tampers* and *on-track tampers*.

20.2 OFF-TRACK TAMPERS

Off-track tampers are portable and can be quickly taken off the track by just two persons. These tampers work during the interval between the passage of trains and do not require any traffic blockage (Fig. 20.1). They consist of tools driven by compressed air, electricity, or petrol. There are generally two types of tampers, namely self-contained tampers and those that are worked from a common power unit. Tampers may be vibratory or of the percussion type or a combination of both.

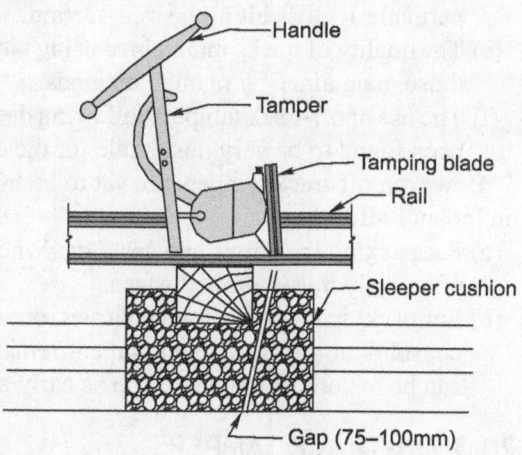

Fig. 20.1 Off-track tamper

In the vibratory type tamping is achieved by vibration as well as by the weight of the tamper itself, while in the percussion type. tamping is achieved by imparting blows. The important off-track tampers tried on Indian Railways are Cobra tampers, Jackson tampers, Shibaura tampers, and Kango tampers.

20.2.1 Use of Off-track Tampers

Off-track tampers are placed diagonally under the rail and worked in pairs from the opposite sides of the sleepers in order to ascertain the maximum consolidation of the ballast. Using beaters, the ballast is first loosened around the rail seat in the crib for a length of 450 mm (18") on either side of the foot of the raft. The tamper is then inserted vertically and the tamping tool blades are kept about 75 to 100 mm away from the sleeper so that enough ballast is available between the two as shown in Fig. 20.1. During its working, the head of the tamper should be moved slightly backward in the shape of an arc of a circle so that the surrounding ballast is well compacted. The operator should not exert force on the tamper while tamping is being done using either the vibratory system or the percussion system.

The average progress achieved by one set of off-track tampers is about 3 km per month, after taking the repairs, overhauling, etc., into consideration.

20.2.2 Limitations of Off-track Tampers

Off-track tampers have not been much of a success on Indian Railways because of the following reasons:
  (a) The maintenance of these tampers has been found to be extremely difficult because of the non-availability of spare parts, which are mostly imported.
  (b) Transporting off-track tampers along with their power units to the site of work in the mid-section is quite problematic.
  (c) Tamping with off-track tampers is very strenuous and a worker normally gets fatigued after 30–40 minutes. The quality of work done after this duration is likely to deteriorate.

(d) Intensive supervision is required to ensure the correct use of these tampers so that the work done is of the desired quality. This type of supervision becomes particularly difficult in the mid-section.

(e) The quality of tracks maintained using tampers is not very high compared to those maintained by manual methods.

(f) The use of off-track tampers following deep screening and relaying work has been found to be very unsuitable for the early restoration of normal speed.

However, off-track tampers are yet to be tried under the following conditions on Indian Railways.

(a) For packing the points and crossings where normal packing cannot be done effectively due to limited space.

(b) For packing newly realigned curves, on which the track requires immediate consolidation so that its alignment remains undisturbed and normal speed can be restored along its length as early as possible.

20.3 ON-TRACK TAMPERS

On-track tampers are self-propelled vehicles which facilitate automatic tamping of sleepers through the controls provided in the operator's cabin. Heavy on-track tampers weigh 20 to 30 tonnes and cannot be easily removed from the track. It is therefore, vital that the work be done after putting up the necessary traffic blocks. However, the tamper can be put off the track using special equipment, provided that there are adequate bank extensions available. These tampers can automatically and simultaneously perform the tasks of lifting, aligning, levelling, and tamping. On American railroads, a sleeper is referred to as a tie. Therefore, since on-track tampers are used for tamping sleepers (or ties), they are also called *tie tamping machines*.

20.3.1 Principles of Working of On-track Tampers

The principles of the working of an on-track tamper (tie tamping machine) are described below.

Tamping

Tamping is the most important application of tie tamping (TT) machines. Tamping consists of packing the ballast under the sleeper. This is achieved by vibrating the ballast, thereby making it fluid, and then compressing it by squeezing as shown in Fig. 20.2. Tamping is done with the help of either 16 or 32 tamping tools, depending

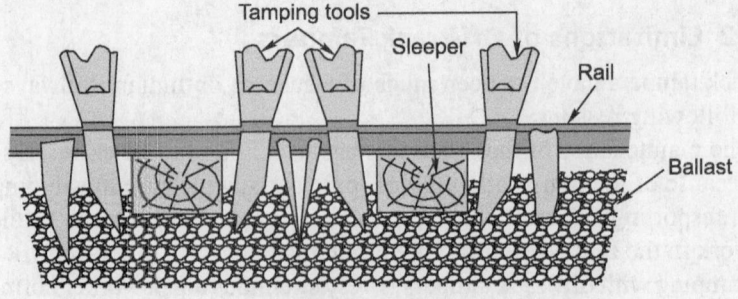

Fig. 20.2 Squeezing of ballast

upon whether single or double sleepers are to be packed at a time. Tamping is done by either the synchronous or the non-synchronous system of vibration. In the synchronous system, the movements of the two tamping tools on either side of a sleeper are similar and simultaneous, synchronizing with each other. In the case of non-synchronous systems, the two tools work independently.

Aligning the track

The alignment of the track is corrected by the two-chord system in the case of machines manufactured by Messrs Plasser and Theurer. In this method, two chords of lengths 24 m and 12 m are taken and placed in the area between two rails, where they are stretched parallel to the track and kept a certain distance apart from each other.

It can be geometrically proved that the versine BQ of the short chord AC measured at a quarter point B' is one-third of the ordinate BB' measured for a long chord at the same quarter point (Fig. 20.3). The ordinate at the quarter point of the 24 m chord H and the versine at the centre of the 12 m chord h are measured using a measuring bogie. In a circular curve, the ratio of $H : h$ is equal to 3:1. The two dimensions are measured at the same point by means of the measuring bogie and if the ratio is found to be equal to 3:1, it indicates that the curve is in order and that the alignment is correct. In the case of any defects, special rollers attached to the rail are used to slew the track at the location of the centre bogie till this ratio is achieved.

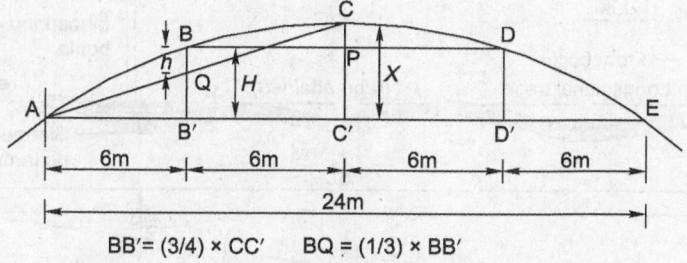

BB'= (3/4) × CC'   BQ = (1/3) × BB'

Fig. 20.3 Principle of alignment—two-chord system

This procedure is used for correcting alignment defects to a value that is one-sixth the value of the original fault because of the relative positions of the measuring bogie and the central bogie as shown in Fig. 20.4. The latest TT machines can be used to correct alignment defects using the three-chord and four-chord methods.

Levelling

Longitudinal levels are corrected on the basis of the principle of proportional levelling with the help of tampers along with an infrared transmitter, a shadow board, and photocells. The distance between these three units is fixed and is so arranged that any error in the longitudinal level is reduced to one-fifth of its value, as is clear from Fig. 20.5.

20.4 FUTURE OF TRACK MACHINES ON INDIAN RAILWAYS

Indian Railways has purchased over 430 tie tamping machines during the past few years, which have been giving satisfactory results. These machines have been found

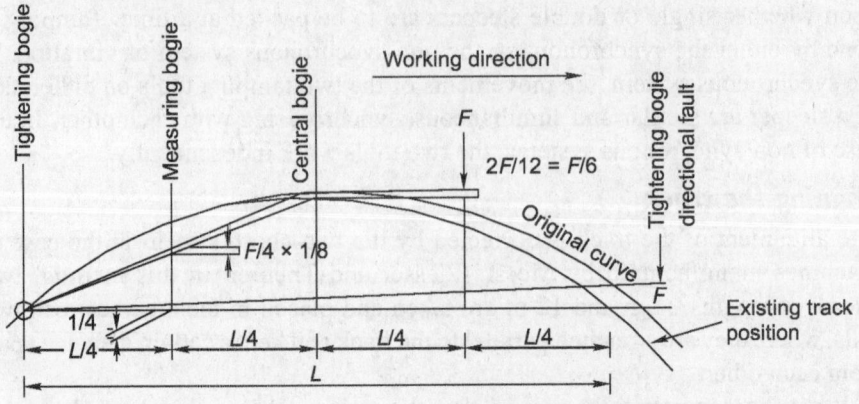

Fig. 20.4 Correction of alignment defect

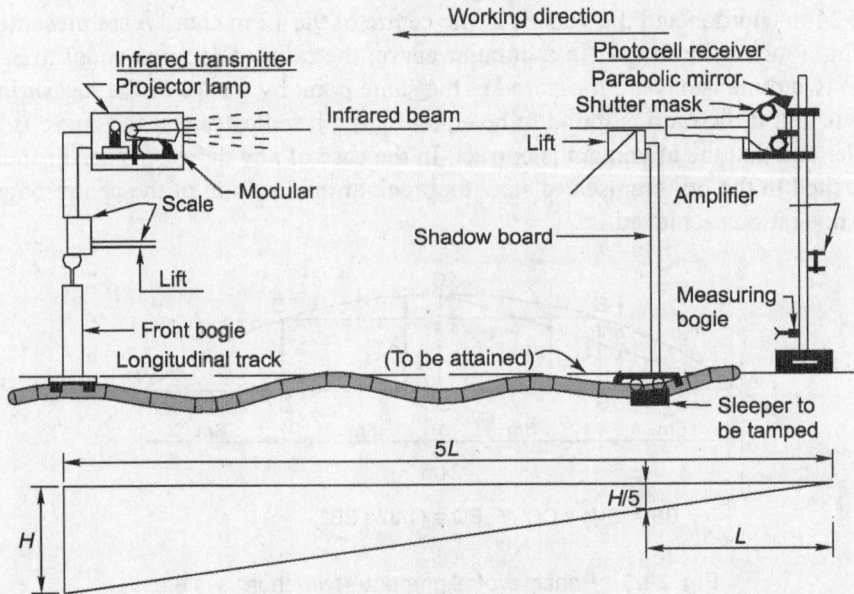

Fig. 20.5 Principle of levelling

most suitable for wooden and concrete sleepers and the condition of the sleepers maintained using these tie tampers is definitely superior to those maintained by the conventional method of beater packing. Indian Railways is systematically introducing LWR and concrete sleepers on high-speed routes and the tie tamping machine is ideally suited for the maintenance of such tracks. However, the following points require special consideration for the successful implementation of mechanical methods of maintenance on Indian Railways:

Blocks Adequate time blocks of about three to four hours per day should be made available so as to make mechanized maintenance effective and economical.

Ballast cushion A clean ballast cushion of a depth of about 20 cm (8") should be available below the sleeper bed of each track that is to be maintained mechanically.

Repair facilities   Adequate facilities for repair and overhauling should be made available at the site and in the workshop.

Spare parts   A good stock of spares should be kept ready at hand to ensure that the machines are always in good working order.

Training of staff   The staff employed for working the tie tamping machine should have thorough knowledge of the principles and method involved in its working. A full-fledged training school equipped with the necessary models and equipment should be made available to train the staff hired for the mechanical maintenance of tracks.

20.4.1 Three-tier System of Track Maintenance

With the introduction of concrete sleepers and on-track tamping machines on Indian Railways, the three-tier system of track maintenance is being adopted on all sections nominated for mechanized maintenance. This system consists of the following three tiers of track maintenance—on-track machine units (OMU), mobile maintenance units (MMU), and sectional gang.

Tier 1: On-track machine unit

The work of systematic mechanized maintenance of track is done with help of heavy on-track machines, which include tie tamping machines for plain tracks as well as points and crossings, shoulder ballast cleaning machines, ballast cleaning machines, ballast regulating machines, and dynamic track stabilizers. These machines are used when the following tasks are to be performed:
  (a) Systematic tamping of plain track as well as of points and crossings
  (b) Intermediate tamping of plain tracks as well as of points and crossings
  (c) Shoulder ballast cleaning, ballast profiling/redistribution, track stabilization, periodical deep screening

Tier 2: Mobile maintenance units

Picking up of slacks and other related works are done with the help of mobile maintenance units (MMU). The Railways uses two types of MMUs whose functions are as follows:

MMU-1 (Rail-cum-road vehicle based)   One such unit is provided to each PWI in charge of a 40–50 km-long double line or a 90–100 km-long single line with a view to carrying out the following work:
  (a) Need-based spot tamping
  (b) Casual renewal and repairs, except planned renewals; in-situ rail welding
  (c) Overhauling of level crossings
  (d) Replacement of glued joints as well as rail cutting or drilling and chamfering
  (e) Permanent repair of fractures
  (f) Creep or gap adjustments involving the use of machines; destressing of LWR/ CWR
  (g) Loading/unloading of materials and other miscellaneous functions

MMU-II (Road vehicle based)   One such unit should be kept aside for each subdivision. It is used for the reconditioning of turnouts and for the minor repairs of the various equipment that are a part of the MMUs. A list of all the equipment included in the two MMUs is given below.

Components of MMU-I These include spot tamping machines such as track tampers and lifting jacks, rail cutting and drilling equipment, rail welding equipment, destressing equipment for rail tension, etc., gas cutting equipment, material handling equipment, safety and protection equipment, inspection gadgets and communication equipment.

Components of MMU-II These include reconditioning equipment for points and crossings such as the welding generator, arc welding equipment, hand-held rail grinder, equipment such as spanner and gadgets for minor repairs.

Tier 3: Sectional gangs

Under the three-tier system of track maintenance, the section gangs are required to perform the following functions:
   (a) Patrolling of the track, namely the keyman's daily patrol, hot/cold weather patrolling, monsoon patrolling, patrolling of vulnerable locations
   (b) Dealing with emergencies by carrying out temporary repairs as in the case of fractures
   (c) Giving need-based attention to bridges, turnouts, SEJs, and approaches to level crossings
   (d) Greasing of ERCs, lubrication of joints, casual changing of rubber pads and other fittings
   (e) Minor cess repairs, cleaning of drains, and boxing of ballast
   (f) Paying attention to loops
   (g) Creep and gap adjustment not involving the use of machines
   (h) Cleaning of crib ballast for effective cross drainage
   (i) Pre and post-tamping attention
   (j) Assistance in the working of MMU and OMU as required

20.5 MEASURED SHOVEL PACKING

Measured shovel packing (MSP) is an improved form of manual packing which aims to provide a scientific method of track maintenance that does not use any sophisticated mechanical aid. This method, which was perfected on SNCF (French Railways) about 40 years ago, was the standard method of track maintenance in the UK prior to the introduction of mechanical maintenance. Even today, tracks on SNCF are mostly maintained by MSP. This method makes it possible to maintain fish-plated and LWR tracks for speeds of up to 160 km per hour in these countries. As such, this method has the potential of being used as a standard method of maintenance for high-speed routes, particularly for flat-bottomed sleepers. It is definitely an asset for controlling the overall economy by way of direct savings in labour and maintenance of the track, and long-term savings in terms of longer life of rails, sleepers, and fastenings due to improved track maintenance.

However due to technical and economic considerations, MSP has not been successful on Indian Railways and as such MSP is not being used on Indian Railways for the past two to three decades.

20.5.1 Essentials of MSP

MSP essentially consists of accurate measurements of track defects, particularly the unevenness and voids caused in the course of service and attending the same by placing of measured quantity of small sized stone chips under the sleeper to bring the track to predetermined levels; the compaction of chips is obtained by passage of traffic.

20.5.2 Equipment of MSP

Many different types of sophisticated equipment are used for maintaining the track by MSP. Details of some of these are given in the subsequent paras.

Canne-a-boule (Fig. 20.6)

This is used for assessing the extent of packing voids under the sleeper.

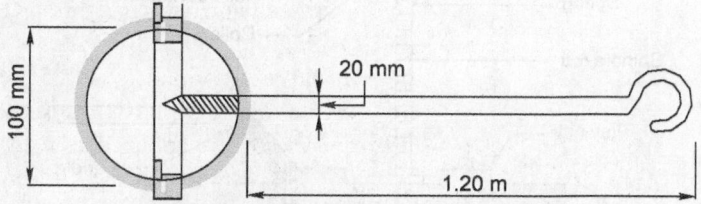

Fig. 20.6   Canne-a-boule

Dansometer

This is used for measuring of packing voids under the sleeper (Fig. 20.7).The tripod legs are fixed in the ballast bed while the dancing rod rests on the sleeper. The extent to which the friction sleeve can shift from its original position helps in determining the presence of voids under the sleeper in dynamic conditions.

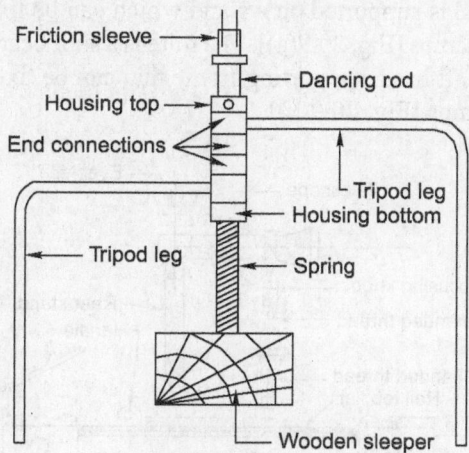

Fig. 20.7   Dansometer

Fleximeter

A fleximeter is used to measure the depression of the rail under the weight of plying traffic. It determines the degree to which voids occur in the packing together with the play in the fastenings, i.e., the gap between the rail foot and the sleeper. It is used in conjunction with the dansometer to check the tightness of fastenings. The difference between the fleximeter and dansometer readings indicates the extent to which the fittings between the rail and the sleeper have become loose (Fig. 20.8).

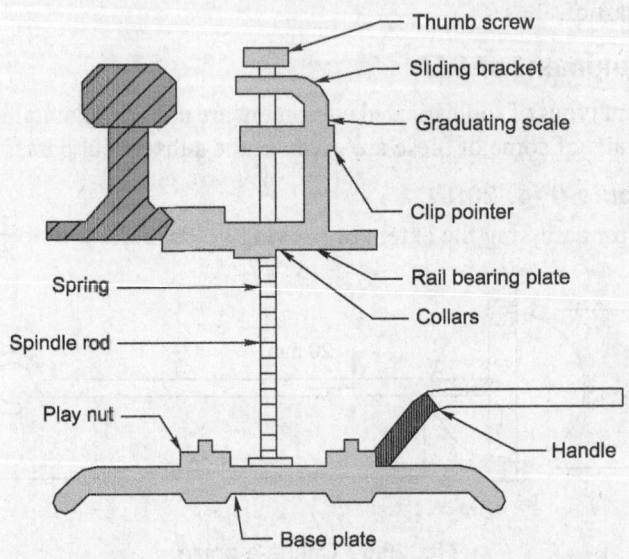

Fig. 20.8 Fleximeter

Viseur and mire

The viseur and mire are used to measure the unevenness of the rail top and for rectifying the alignment. The viseur is a type of telescope that has a magnifying power of about 12 and is supported on a stand which can be fixed to the rail seat with the help of two clamps [Fig. 20.9(a)]. The mire is a staff bearing five graduated scales, in millimetres. It has a supporting frame that can be fixed to the rail head by means of bent clamps [Fig. 20.9(b)].

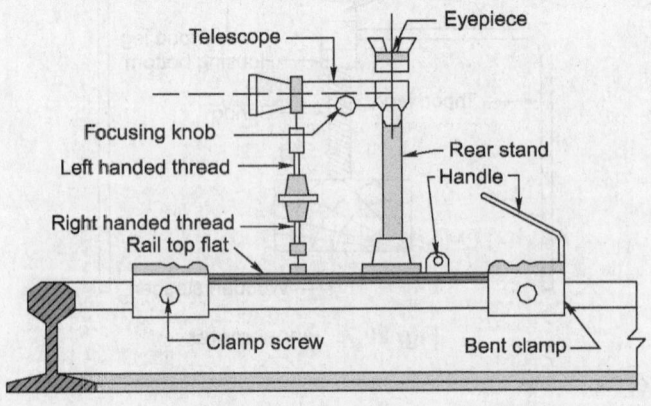

Fig. 20.9 (a) Viseur

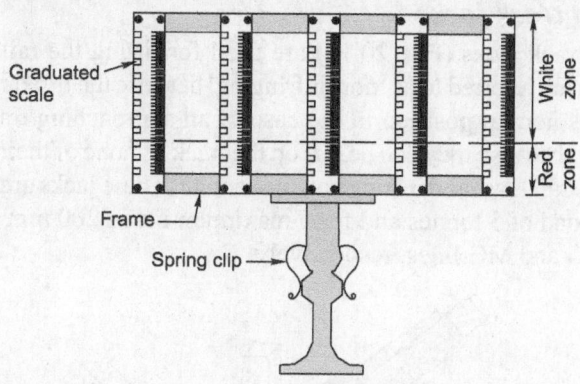

Fig. 20.9   (b) Mire

Gauge-cum-level

The gauge-cum-level is used for measuring the gauge of the track as well as the cross levels (Fig. 20.10). The cross level is measured with the help of an approximately 200 mm-long sensitive spirit level with a sensitivity of 2' 30". The cross level can be measured to an accuracy of 1 mm with the help of this instrument.

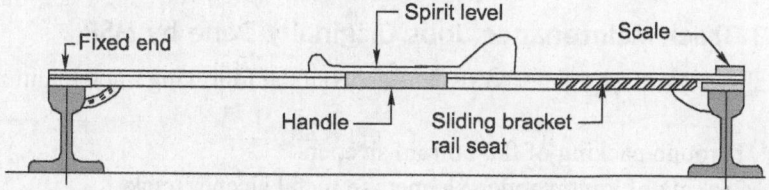

Fig. 20.10   Gauge-cum-level

Packing shovel

A packing shovel is used for placing stone chips over the full width of the sleeper under the rail seat. It is about 1 m long and has a pan for collecting the chips under the sleeper bed. The throw of the blade is 100 mm for BG lines and 85 mm for MG lines (Fig. 20.11).

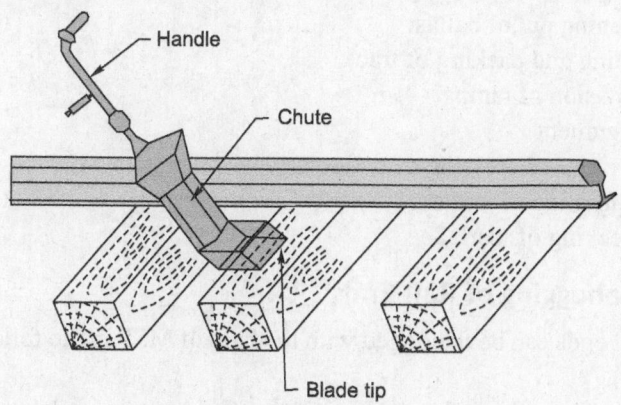

Fig. 20.11   Packing shovel

Non-infringing track jacks

Non-infringing track jacks (Fig. 20.12) are used for lifting the rail to a desired height. The jacks are referred to as 'non-infringing' because the lifted rail can easily be returned to its normal position in the case of an approaching train with little manipulation and because they can be left on the track as none of their components project above the rail level and infringe on movement. These jacks are designed for a safe working load of 5 tonnes and for a maximum lift of 200 mm and 160 mm in the case of BG and MG lines, respectively.

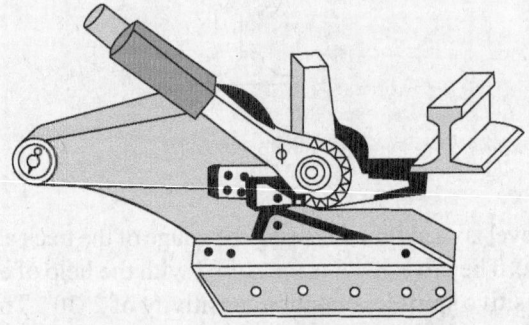

Fig. 20.12   Non-infringing track jack

20.5.3 Track Maintenance Jobs Originally Done by MSP

Originally it was thought that MSP can carry out the following track maintenance jobs.
- (i)   Through packing of flat bottom sleepers
- (ii)   Packing of joint wooden sleepers in metal sleeper track
- (iii)   Through packing of turnout with wooden or steel sleepers
- (iv)   Dehogging of rail ends

The following steps are involved while doing MSP of flat bottom sleeper track:
- (i)   Measurements of voids fixation of high points
- (ii)   Transferring high points to good points
- (iii)   Longitudinal levelling
- (iv)   Opening out of ballast
- (v)   Lifting and packing of track
- (vi)   Provision of ramps
- (vii)   Alignment
- (viii)   Boxing and dressing of ballast
- (ix)   Majoration of joints
- (x)   Checking of work

20.5.4 Dehogging of Rail Ends

Hogged rail ends can be dehogged with the help of MSP in the following manner (Fig. 20.13).
1. The dip at the joint sleeper (*a*) is measured by using a 1.5 m straight edge and a feeler gauge at a distance of 50 mm from the rail end.

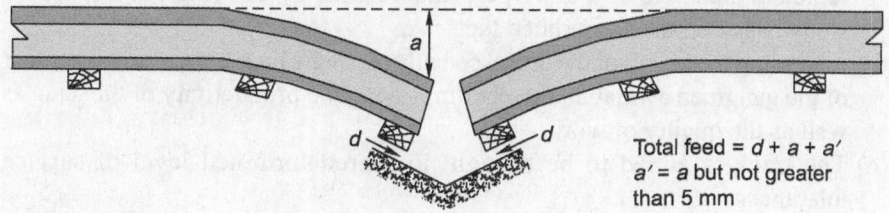

Fig. 20.13 Dehogging of rail ends

2. The dance at the joint sleepers (d), which is the gap between the sleeper and the ballast, is measured by a canne-a-boule or dansometer.

3. The joint sleepers are lifted and packed to a value equal to $d + a + a'$, where d is the value of dance, a is the amount of dip, and a' is equal to a or 5 mm, whichever is lower.

After allowing traffic to run on the track for a period of two days, the adjoining sleepers are beater or shovel packed depending upon whether it consists of metal or wooden sleepers. The dehogging of rail ends is achieved because of the train loads. It may be necessary to repeat this procedure in case the rails are not completely dehogged.

20.5.5 Future of MSP on Indian Railways

MSP was tried on Indian Railways for few years and subsequently it was observed that MSP is not technically and economically suitable for Indian conditions. MSP is, therefore not in use currently on Indian Railway except for dehogging of rail ends.

20.6 DIRECTED TRACK MAINTENANCE

As the name suggests, directed track maintenance (DTM) is a method of maintaining the track on the basis of directions that are given in this regard every day, and not as a prescribed routine. DTM essentially consists of need-based maintenance rather than routine maintenance. In the case of DTM, track maintenance is done by properly identifying any defects in track geometry and rectifying these defects by attending to the track at the affected locations under close supervision, thereby maintaining the track at predetermined standards.

20.6.1 Objectives

The two main objectives of DTM are as follows:

(a) To maintain high standards of track maintenance as per predetermined tolerances

(b) To reduce the cost of maintenance mainly by avoiding unnecessary work

In order to achieve the desired objectives, the following special features are incorporated in DTM, vis-à-vis the conventional system of maintenance.

(a) The level of supervision is improved by hiring a well-trained and qualified permanent way mistry.

(b) A thorough record of the track defects identified before and after the completion of work is maintained to assess the inputs and also to help devise remedial measures of a more permanent nature by carrying out a scientific study based on the assimilated facts.

(c) Increasing the length of the unit especially on single lines increases the number of the gangmen available, thereby improving the productivity of the gang as well as the quality of work.

(d) The track is aimed to be brought to a predetermined level of service tolerances.

20.6.2 Work Done Under DTM

The maintenance operations to be carried out in a section where DTM has been introduced can be placed into the following four categories.

Systematic overhauling

In DTM, while the emphasis is on need-based maintenance, the intention is not to completely dispense with routine maintenance works, such as systematic overhauling. Instead, the frequency of systematic overhauling is suitably increased, say by three to four years or as decided by the chief engineer, depending on local factors, such as the condition of the track and the formation, traffic density, permissible speed, and rainfall. A certain number of working days in the appropriate months of the year are earmarked for this work so as to cover one-third or one-fourth of the gang length by systematic overhauling, depending on the site conditions.

Periodic maintenance work

This includes works such as the lubrication of joints, cleaning of side drains, catch water drains, and repairs of the formation and cess. In the annual programme, an adequate number of working days should be set aside during the appropriate months for periodic maintenance work.

Occasional maintenance work

This includes other works such as scattered renewal of rails, sleepers, and other track components, adjusting creep, restoring correct spacing between sleepers, building damaged rail ends, realigning curves, overhauling level crossings as well as points and crossings, and properly removing any deficiencies in the ballast section. The PWI assesses the quantum of such works that are to be carried out periodically in the order of their priority and draws up a programme in consultation with engineers after taking into consideration the availability of track material, ballast, welding parties, etc.

Need-based maintenance

The remaining working days in the annual programme are devoted to need-based maintenance, which is a new concept and forms the main distinguishing feature of DTM as compared to the conventional system of maintenance. The operations involved in need-based maintenance are as follows:

(a) Location of defects by analysing the results of the track recording car/ oscillograph car/hallade recorder and by foot plate/rear vehicle/trolley/foot inspection

(b) Identification of defects by means of systematic inspection and by ground measurements taken by trained supervisors using precision instruments

(c) Recording of the observations

(d) Rectifying track defects by attending only to the defective portion followed by a post check of the same portion conducted by the supervisor to check its quality and output

20.6.3 Organization of Directed Track Maintenance

Beat of DTM Unit

The beat of the DTM unit should normally be six to eight route kilometres. The tool boxes should be so located, at manned level crossings, stations and gang huts, etc. that gang men with tools can reach the site of programmed work without undue loss of time and much effort.

Supervisors

The concept of DTM requires higher degree of skill in recording of measurements and assessing the maintenance requirements. It is, therefore, desirable that the units work under the supervision of an officer not lower in rank than a PWM. When the PWM is in charge of a unit, it is economical to combine two gangs into one and place them under the charge of the PWM. Such manning of gangs can be conveniently done in double line and multiple line sections where the total length of the unit will not be more than eight kilometres.

Training for DTM supervisors/PWM

PWMs in charge of DTM will be given detailed training in tack structure., conventional methods of maintenance, MSP, DTM, maintenance of LWR/CWR, track renewals, safety rules, etc. On satisfactory completion of the training, the PWMs will be awarded competency certificate by an authority nominated by the chief engineer.

20.6.4 Future Scope of Directed Track Maintenance

The concept of DTM is being used selectively at present on Indian Railways. The concept is generally good; the need-based method is economical and it also provides a higher standard of track maintenance.

SUMMARY

Manual methods of track maintenance are not suitable for modern tracks consisting of LWRs and heavy concrete sleepers. Mechanized maintenance has many advantages over manual maintenance. Measured shovel packing (MSP) has been tried on Indian Railways and has not been found successful. However, directed track maintenance (DTM) is another modern method of track maintenance that has been found suitable for high-speed tracks in India. These modern methods are cost-effective, time effective, and efficient. Indian Railways is slowly switching over to these methods of track maintenance.

REVIEW QUESTIONS

1. Explain the following methods of packing of tracks that are a part of the periodic correction of track geometry:
   (a) Beater packing
   (b) Measured shovel packing
   (c) Packing by track tamping machines

2. What is directed track maintenance (DTM)? In what way does it differ from through packing?

3. List the methods used for the maintenance of high-speed tracks and discuss the suitability of each method under Indian conditions.

4. What do you understand by modern methods of track maintenance? Why are modern methods of track maintenance required?

5. What is an on-track tamper? What kind of work is it used for? Describe the principles behind the working of an on-track tamper. What is the difference between an on-track and an off-track tamper?

6. What do you understand by measured shovel packing (MSP)? What are its advantages with respect to other methods of track maintenance? Briefly describe the main equipment used for MSP.

7. What are the various track components that are attended to under directed track maintenance? Name the equipment required to be kept with each unit of DTM.

8. Write short notes on:
   (a) canne-a-boule
   (b) viseur and mire
   (c) dehogging of rail ends
   (d) non-infringing track jack

Choose the correct answer from the choices given.

9. 'Off-track tampers' are so named because:
   (a) the tampers are assembled off the track
   (b) these tampers work off the track
   (c) these tampers can be taken off the track in a short time
   (d) none of these

10. The tamper that is not 'off-track tamper is:
    (a) Jackson tamper   (b) Shibaura tamper
    (c) Cobra tamper   (d) Matisa tamper

11. Small track machines are required to:
    (a) mechanize the work of sectional gangs
    (b) fill up the gap between the working of heavy track machines and manual working
    (c) tamp the isolated stretches
    (d) all of the above

12. The name 'On-track tamper' is given to the TT machine because:
    (a) TT machine is assembed 'on track'
    (b) the TT machine works while on track and cannot be taken off the track
    (c) both (a) and (b)

13. In plassermatic 06-16 SLC, the figure 16 represents:
    (a) the year of its design
    (b) the number of tamping tools
    (c) the weight of tamping tool
    (d) none of these
14. The track machine that is a ballast regulator is:
    (a) VDM-800
    (b) RM-76UHR
    (c) Matisa R-7
    (d) RM-80
15. The minimum clean cushion required for TTM working is:
    (a) 100 mm
    (b) 200 mm
    (c) 150 mm
    (d) 300 mm
16. The main functions of tamping machines include:
    (a) correction of Alignment
    (b) correction of levels
    (c) packing under sleepers
    (d) all of the above
17. MSP refers to:
    (a) method of short packing
    (b) measured shovel packing
    (c) maintenance system of packing
    (d) none of these
18. DTM refers to:
    (a) direct track maintenance
    (b) directed track maintenance
    (c) day-to-day track maintenance
19. With manual packing sectional speed can be restored on:
    (a) 30th day
    (b) 21st day
    (c) 16th day
    (d) 10th day
20. Dehogging of rail ends can be done by:
    (a) MSP
    (b) track machine
    (c) DTM
    (d) none of these
21. Viseur and Mire are used for measuring:
    (a) packing voids
    (b) unevenness at rail top
    (d) gauge cum-level
    (d) none of these

Rehabilitation and Renewal of Track

INTRODUCTION

The dynamic impact of running trains causes heavy wear and tear of the track. It becomes necessary to rehabilitate or renew the track periodically to ensure that it continues to be safe and efficient. Heavy track renewals are carried out on Indian Railways every year to keep the track safe and in a good running condition as well as to bring down the cost of maintenance as much as possible. The cost of track renewals carried out on Indian Railways runs into several millions. These track renewals consist of complete track renewal, through sleeper renewal, through rail renewal, and so on.

About 2000 to 3000 km of track renewal is done annually. As track renewals are costly proposals, they are formulated after a lot of deliberation and are well scrutinized at various administrative levels before they are finalized and included in the annual works programme.

Generally, there is heavy backlog of track renewal work on account of the shortage of permanent way material and the lack of adequate funds for track renewal programmes. Retaining the service of overaged tracks not only leads to an increased cost of maintenance, but also affects the safety and fluidity of the movement of traffic. Due to delays in track renewals, incidents of rail fractures have been progressively increasing. Long stretches of track have also been placed under speed restrictions owing to this backlog in track renewals.

21.1 CLASSIFICATION OF TRACK RENEWAL WORKS

Track renewals can be broadly classified as follows.

Complete track renewal   This implies the renewal of all the components of a track over a particular length. The necessary recoupment of the ballast and the provision of a full ballast cushion is also done along with complete track renewal (CTR).

Through rail renewal   Through rail renewal or TRR involves the renewal of all the rails between one point and the next.

Through sleeper renewals   This entails the renewal of the entire lot of sleepers from one point to another. The necessary recoupment of the ballast and the provision of a full ballast cushion is also done along with TSR (through sleeper renewal).

Casual renewal   This implies the replacement of some of the unserviceable rails or sleepers or both with serviceable, released rails and sleepers of a similar type and age.

Primary renewals   This refers to renewals that are done using new permanent way materials.

Secondary renewals   This refers to renewals that are done using released, serviceable permanent way materials.

21.2 CRITERIA FOR RAIL RENEWALS

The need for rail renewal arises because of any one of the following reasons:
   (a) Incidences of rail fractures or failures
   (b) Wear of rails
   (c) Maintaining tracks at the prescribed standards
   (d) Shortened service life of rails
   (e) Plan-based renewal
These criteria are discussed in detail in the subsequent sections.

21.2.1 Incidences of Rail Fractures or Failures

Since rail fractures or failures have a direct and vital bearing on the safety of the track, they should be given importance than all other factors when deciding on the rail renewals to be carried out. A spate of rail fractures or failures on a particular length of a track may necessitate rail renewal. In such cases, an ultrasonic test of the rails should be carried out, the results scrutinized, and the section is considered for rail renewal only if the rail fractures are high for a particular season (4 per cent in a year or 10 per cent overall).

21.2.2 Wear of Rails

Rail renewals may also become necessary because of excessive wear. This wear can be of various types as a result of the following.

Limiting loss of section

Rail renewal is called for when the loss in the weight of the section exceeds the prescribed limits given in Table 21.1. In such cases, the rails should be tested frequently using ultrasonic rail flaw detectors because such rails are more prone to rail failures.

Table 21.1   Limiting loss in weight of rail sections

Gauge	Rail section	Loss in weight
BG	52 kg/m	6.0 per cent
	90 R	5.0 per cent

(Contd.)

Table 21.1 Contd

Gauge	Rail section	Loss in weight
MG	75 R	4.20 per cent
	60 R	3.25 per cent

The loss of weight can be assessed either directly by measuring the actual weight or empirically by measuring the horizontal and vertical wear of the rail.

Wear due to corrosion

Rails get corroded due to the vagaries of weather, atmospheric conditions, and chemical reactions that take place when certain materials come in contact with rail metal, thereby resulting in wear. Such wear is more prominent in coastal areas. Corrosion beyond 1.5 mm in the web and the foot of the rail is specified as the criterion for rail renewal due to corrosion.

Vertical wear

When vertical wear (Fig. 21.1) causes a reduction in the depth of the rail head until a point beyond which there is a risk of the wheel flanges grazing the fish plates, the rails should be renewed. The limits of vertical wear for the different rail sections are given in Table 21.2.

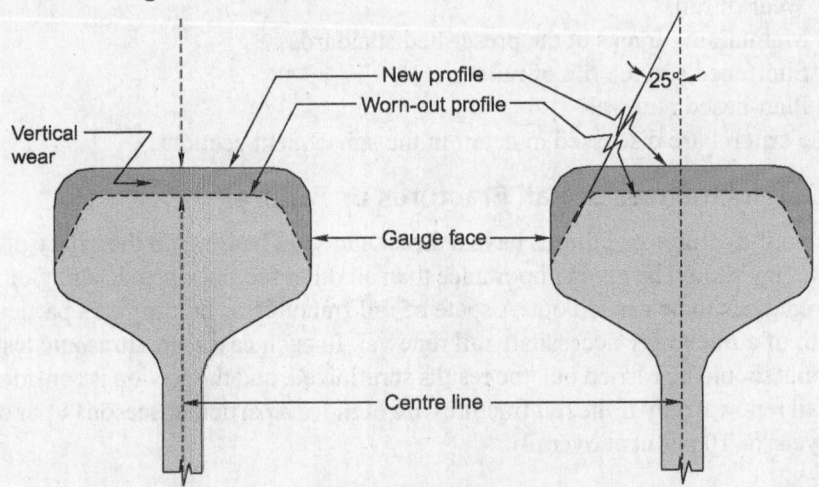

Fig. 21.1 Vertical wear

Table 21.2 Limits of vertical wear

Gauge	Rail section	Vertical wear (mm)
BG	60 kg/m	13.00
	52 kg/m	8.00
	90 R	6.00
MG	75 R	4.50
	60 R	3.00

Lateral wear

Lateral wear (Fig. 21.2) is measured at a distance of 13 to 15 mm below the rail top table. The profile of the worn-out rail is recorded and superimposed over a

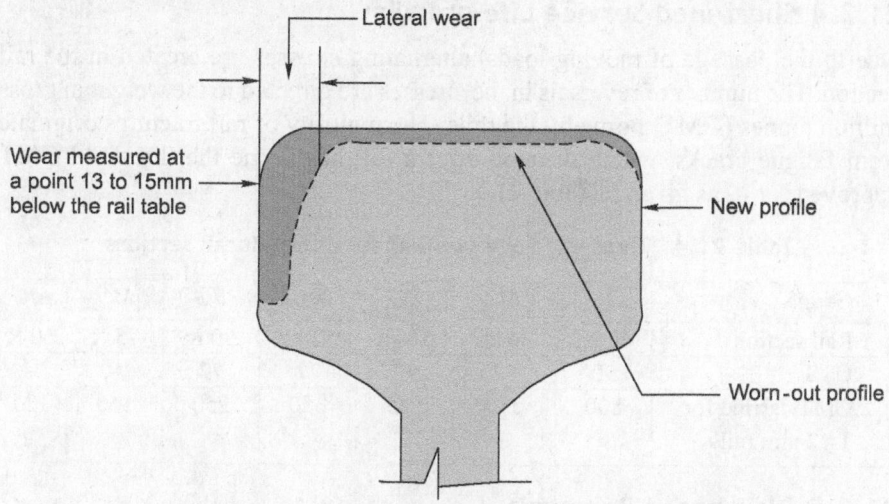

Fig. 21.2 Lateral wear

new profile to determine lateral wear. Excessive lateral wear causes lateral hunting of vehicles, resulting in an uncomfortable ride. This condition will manifest itself in the form of high values of the ride index during oscillograph car runs. When it becomes uneconomical to maintain such a track, through rail renewal becomes crucial. The permissible limits of lateral wear under various conditions on the Railways are presented in Table 21.3.

Table 21.3 Permissible limits of lateral wear

Section	Gauge	Category of track (route)	Lateral wear (mm)
Curves	BG	Group A and B	8
	BG	Group C and D	10
	MG	Q and R	9
Straight	BG	Group A and B	6
	BG	Group C and D	8
	MG	Q	6
	MG	R	8

21.2.3 Maintaining Tracks at the Prescribed Standards

There may be situations in which renewals become necessary even though the service life of a rail has not yet expired, due to local factors, such as curves, steep gradients, high speeds, heavy axle loads, burrs, scabbing, and wear of rails that have the following adverse effects on the track.

(a) Poor running quality of track in spite of the extra maintenance labour engaged on the section

(b) Disproportionate cost of maintaining the track under safe conditions

(c) Poor condition of the rail due to hogging, battering, scabbing, wheel burn, etc., and other causes, such as the excessive corrugation of the rail and which make track maintenance difficult and uneconomical and affect riding quality

21.2.4 Shortened Service Life of Rails

Due to the passage of moving loads, alternating stresses are created in the rail section. The number of reversals in the stresses are directed to the weight in gross million tonnes (GMT) borne by the rails. The majority of rail fractures originate from fatigue cracks, which develop after a rail has borne the threshold GMT approved for it, as given in Table 21.4.

Table 21.4 Threshold values of GMT for different rail sections

Gauge	BG	BG	BG	BG	BG	MG	MG
Rail section	60 kg	60 kg	52 kg	52 kg	90 R	75 R	60 R
UTS	90	72	90	72	72	–	–
GMT carried for T-12 mm rails	800	550	525	350	250	150	125

21.2.5 Plan-based Renewals

Plan-based renewals are planned with the objective of modernizing the track structure on selected routes in the quickest possible time. It may sometimes involve a premature renewal of the track also.

21.3 THROUGH SLEEPER RENEWALS

Through sleeper renewals should be done when the percentage of unserviceable sleepers is equal to 30 per cent on trunk routes/main lines and 35 per cent on branch lines. Various conditions warrant the renewal of each type of sleeper, as explained in the following.

Wooden sleepers These become unserviceable only when they are split, broken, burnt, or rounded on the underside with deeply cut rail seats. They are also considered unserviceable when the spikes that fix the rails on them lose their gripping power or when there is general deterioration in the structure of the sleepers as a whole due to an insect attack, etc.

Steel sleepers These become unserviceable when there is excessive corrosion and also when the rail seats or lugs crack or give way. The life of a steel sleeper is reckoned to be about 40 to 50 years.

CI sleepers These sleepers become unserviceable when there is excessive corrosion of the tie bar and the rail seat, when the sleeper plate breaks or develops a crack, or when the wear of the lug and rail seat is so excessive that the sleeper no longer grips the rail firmly and the keys become loose resulting in creep. The average life of a CI sleeper is 40 to 50 years. Generally, these sleepers can be considered unserviceable when they are unable to either hold the gauge or retain the packing. The renewal of worn-out fish plates should be considered when the gap between the fishing plates of the rails and the fish plates exceeds 1.5 mm at the centre.

21.4 EXECUTION OF TRACK RENEWAL OR TRACK RE-LAYING WORK

Track renewal or track re-laying work can be done either manually or by mechanical means. These works involve the following steps:

(a) Survey of the section

(b) Preliminary works

(c) Actual execution of track renewal works

21.4.1 Survey of the Section

The section to be re-laid should be surveyed in advance. The survey includes the preparation of the longitudinal sections of the track, including the existing rail levels, at every 100 m and at obligatory points, such as level crossings, road over bridges, and overhead crossings. The cross sections are taken at every 50 m in station yards and on platform lines, at the site of foot overbridges, overhead watering arrangements, etc.

21.4.2 Preliminary Works

Preliminary works are the minor works undertaken prior to the actual work involved in track renewal. These works include the following tasks:

(a) Unloading track materials

(b) Ballasting the section

(c) Fixing level and centre-line pegs

(d) Arranging traffic blocks

Each one of these preliminary works is briefly discussed here.

Unloading track materials   It should be ensured that the relevant materials are unloaded at the precise location where they are to be laid, as explained in the following.

Rails   These should be unloaded in pairs on the cess on either side of the track in the case of single-line sections and between two tracks in the case of double-line sections.

Sleepers   The sleepers are taken directly to the site in wagons and unloaded on the section. They are piled in small stacks and subsequently laid at their final location. Alternatively, they are laid 'end on' with respect to their future location, wherever possible.

Fittings   Small fittings are likely to be lost or stolen. These should, therefore, be unloaded in depots, gang huts, gate lodges, etc., and taken to the site as and when required. New fish plates are graphited and fish bolts and nuts are oiled and lubricated before use.

Ballasting the section   An adequate quantity of ballast is provided in advance and is unloaded at the site, preferably with the help of ballast trains. Some of the ballast can also be collected, or even procured, at the cess to avoid ballast deficiency. The ballast should preferably be deep screened in advance.

Fixing centre-line and level pegs   Centre-line and level pegs are provided at every 50 m.

Arranging traffic blocks   The arrangement of traffic blocks should be finalized in consultation with the operating department. A joint memorandum of understanding should be prepared for drawing up the exact programme of traffic blocks.

21.4.3 Actual Execution of Track Renewal Work

Once the preliminary works are completed, the actual work of track renewal can start. This involves the following tasks:

Setting up traffic block A two to four-hour traffic block is set up in order to execute the work of track re-laying.

Dismantling old track The old track is dismantled by removing the fish-bolts and fish plates as well as the sleeper fastenings. The old rails and sleepers are then removed and stacked on the side of the track opposite the new material.

Deep screening The ballast should be deep screened if this was not done earlier along with the other preliminary works. The ballast should be dressed and properly levelled in accordance with the correct profile.

Linking new track After the ballast has been laid in keeping with the correct level, the sleepers are laid almost in their correct places, and the rails are linked over them using proper expansion liners.

Packing new track The track is then lifted to the correct level, aligned, and packed either manually or with the help of machines.

21.4.4 Progress and Output

On an average, the renewal of about 0.4 km of track can be achieved by 15 to 30 gangs in a three to four-hour traffic block. The deep screening of the ballast should, however, have been completed ahead of the re-laying work.

21.4.5 Rail Dolly

The rail dolly (Fig. 21.3) is an equipment used for the quick and easy transportation of rails. It consists of a tabular frame fitted with a chain that ends in a rail clamp. There are two double-flanged wheels with ball bearings, which run on rails. A leverage handle is available for operating and moving the rails. The equipment, weighing about 55 kg, is extremely useful for carrying rails and sleepers to the site of work. In the event of an approaching train, the rail as well as the equipment can be cleared off the track within 10 seconds.

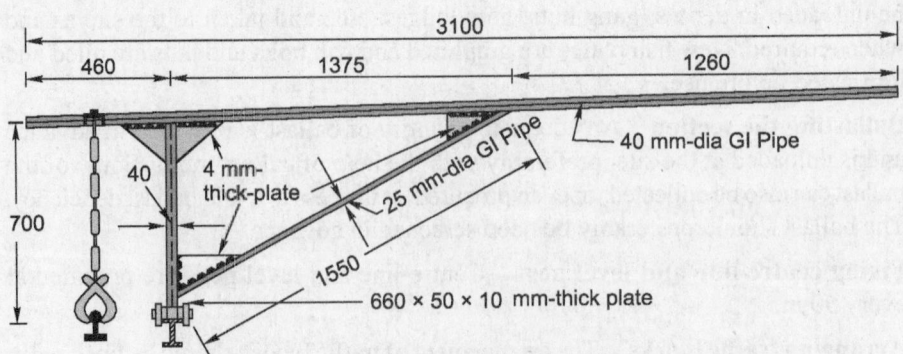

Fig. 21.3 Rail dolly (all dimensions in mm)

21.5 MECHANIZED RE-LAYING

A concrete sleeper weighs more than 200 kg and, therefore, unlike wooden or metal sleepers, cannot be handled easily by manual labour. Apart from being heavy, a concrete sleeper is brittle in nature and is likely to get damaged if dropped from some height or handled roughly. Furthermore, the conventional method of re-laying, which employs manual labour, is quite slow and the quality of the work done is also not very good.

The need for mechanized re-laying has been increasingly felt on account of these limitations, particularly in view of the large-scale use of concrete sleepers on Indian Railways.

Mechanical re-laying can be done by either one of the following methods.

(a) Semi-mechanized re-laying using prefabricated panels, in which the main work of re-laying is done mechanically with the help of PQRS (Plesser quick re-laying system) equipment but all other minor works are done manually.

(b) Fully mechanized re-laying with the help of track re-laying trains.

21.5.1 Mechanical Re-laying using Prefabricated Panels

On Indian Railways, mechanical re-laying is mostly done with the help of PQRS equipment. This equipment is manufactured by an Austrian firm, Messrs Plasser and Theurer in a factory near Faridabad, India. The equipment basically consists of two self-propelled portal cranes, each capable of lifting 5 to 9 tonnes. The portal cranes run on an auxiliary track with a 3.4 m gauge. Track panels consisting of concrete sleepers, rails, and fittings are prefabricated in the base workshop and used for re-laying the track. These prefabricated panels are loaded in special BFR-type wagons and taken to the worksite along with portal cranes, etc. [BFR is a classification scheme for freight wagons on Indian Railways. B indicates bogie wagon, F indicates flat car, and R indicates rapid (forced) discharge]. The old track is dismantled and new panels are laid with the help of portal cranes. A fully-equipped train called the PQRS rake that consists of the following parts is then taken to the site of work:

(a) Engine

(b) Coal wagon (for recoupment of coal in the engine)

(c) Crew rest van (for the staff to stay and rest in)

(d) Guard compartment (for controlling the movement of the train)

(e) Mobile workshop-cum-rest van (for urgent repairs to be done at the site)

(f) Special wagon called BFR loaded with two portal cranes

(g) Two empty BFRs for loading released panels

Semi-mechanized track renewal work is then carried out in the following sequence.

1. The auxiliary track with a 3.4 m gauge is first laid at the location where re-laying is to be done. The rails of the existing track are also replaced with 13 m service rails.

2. A fully-equipped train comprising the various components and loaded with prefabricated panels is taken to the work site after setting up a traffic block, and also a power block in the case of an electrified section.

3. The portal cranes are then unloaded mechanically on the auxiliary track of a 3.4 m gauge.

4. The portal cranes are utilized to dismantle the two panels of the existing track and load them onto empty BFRs.
5. The ballast bed is prepared manually and levelled up to a predetermined point.
6. On the return trip, the portal cranes are utilized to lift the pre-assembled panels of concrete sleepers from the BFRs and lay them on the levelled ballast bed.
7. The procedure is repeated so that in the forward trip, the portal cranes lift the dismantled panels and place them in empty BFRs, and on the return trip, the portal cranes lift the two preassembled panels from the loaded BFRs and lay them on the prepared bed (see Fig. 21.4).

Fig. 21.4 Portal cranes lifting concrete sleeper panels

8. The new panels are then connected using fish plates, fittings, and fastenings.

This method of re-laying is quick and independent and is best suited to Indian conditions. It is possible to achieve an average output of about 200 m per effective hour of working. About 0.8 km of re-laying can be accomplished in a four hour block.

One 5 tonne crane can be used to lift a panel of 13 m rail assembled on CST-9 sleepers. Two cranes are, however, able to lift a similar panel of concrete sleepers, which is comparatively heavier. One crane can normally lift concrete sleeper panel of 9.1 m length only because of the crane capacity being limited to 5 tonnes only. Indian Railways recently acquired portal cranes of 9 tonnes capacity and one crane of this type can lift 13 m rail panel having concrete sleepers.

21.6 TRACK RENEWAL TRAINS

Indian Railways has purchased a few track renewal trains (TRTs) (Fig. 21.5) of modern design, which can carryout track re-laying work automatically with minimum manual effort. These trains are of the P-811 model and are manufactured by the Tamper Corporation of the US. The cost of a TRT along with all ancillary units is about ₹ 100 million.

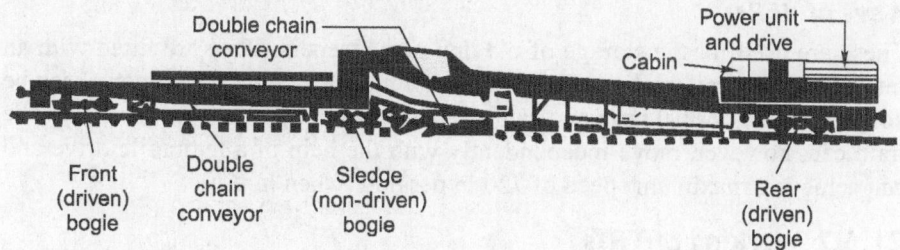

Fig. 21.5 Track re-laying train

21.6.1 Structure of Track Re-laying Train

A track re-laying train is designed to simultaneously and continuously perform all the operations involved in the replacement of rails and sleepers. The train carries out multifarious functions such, as removal of old rails, removal and stacking of old sleepers, levelling and compacting of the ballast bed, placing new sleepers in their proper positions, laying of new rails, and removal of released rails.

The train consists of three main units, which are described in the subsequent paragraphs.

Main vehicle

The main vehicle weighs about 110 tonnes and has an overall length of about 45 m. It further comprises the following parts.

  (a) A sleeper handling device, which removes the old sleepers and replaces them with new ones with the help of several conveyors
  (b) A power vehicle fitted with a *rail lifting and guidance system* for dissecting the old rails at the side of the track
  (c) A *triangular smoothing plough* and a compaction plate to prepare the ballast for new sleepers

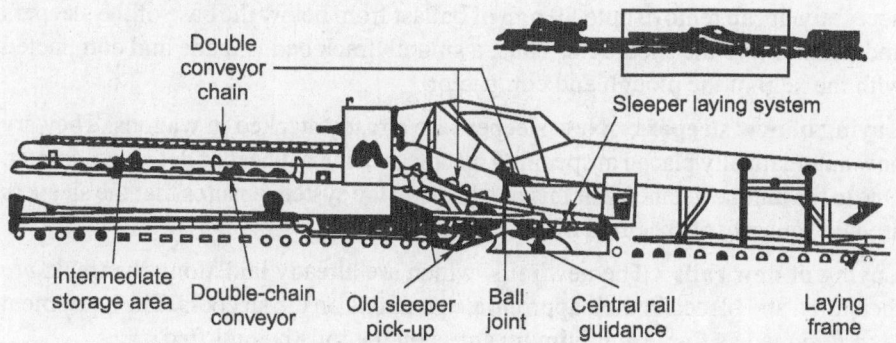

Fig. 21.6 Details of main vehicle of a TRT

Handling gantry

The train has a handling gantry for transporting old and new sleepers. The gantry weighs about 6.6 tonnes and has a lifting force of about 50 kN. It can lift about 20 concrete sleepers at a time and travel at a maximum speed of 15 km per hour.

A set of BFRs

These are used for the storage of old and new sleepers. They are fitted with an interconnecting rail track on which the handling gantry runs. The renewal can be coupled with a normal train and can reach a speed of upto 100 km per hour. The train can, however, move independently with the help of a hydraulic drive and can achieve a maximum speed of 720 m per hour when in use.

21.6.2 Working of TRTs

The following steps are involved in the working of track relaying trains.

Removal of fastenings   The fastenings joining the old rails to the sleepers are removed ahead of the commencement track renewal work.

Placement of new rails   The duly welded rails are placed in advance either beside the track duly welded or fish plated, at a distance of 1.5 m from the track.

Positioning on the track   The main vehicle is now positioned on the track to be renewed. The design of the machine is such that with the help of a *guidance sled* it is possible to acquire an accurate reproduction of the old track layout without the aid of an external guidance system.

Lifting of old rails   The old rails, which are already free from their fastenings, are lifted with the help of the rail lifting and guidance frame. The machine continuously lifts the old rails and deposits them on either side of the track. The ballast can be simultaneously screened if necessary, while the rails rest on the ends of the sleepers.

Picking up of old sleepers   The old sleepers are picked up with the help of a sleeper pick-up system and placed upright on the conveyor system.

Levelling and compaction of ballast bed   The TRT also consists of a vibratory plough and compactor. The plough levels the ballast and make it ready to accept the new sleepers. The plough can be adjusted vertically to remove the ballast as necessary. It can remove upto 80 mm of ballast from below the base of the sleepers and transfer it to the shoulders. Thus, a smooth track bed is made and compacted with the help of the plough and compactor.

Laying of new sleepers   New sleepers are already stacked in wagons. They are now automatically placed at specified distances using a wheel of definite diameter. Due to the unique geometry of this mechanism, the system ensures that the sleepers are laid squarely at specified distances.

Laying of new rails   The new rails, which are already laid along the track, are then lifted and placed in their appropriate positions on the sleepers. The equipment is so designed as to cause minimum stress on the rails being lifted.

Fixing insulators and elastic rail clips   Elastic rail clips and insulators are then fixed to the rail with the help of a small mechanical appliance. These items are normally carried in a separate wagon.

Picking up of old rails   The final operation is the removal of old rails. It is important that prior to rail pick up, the rails are cut into lengths of 39 m to ensure that they are handled properly. The wagons, which are towed by a utility vehicle

(UTV), contain a mobile crane that picks up the old rails and places them into the wagons for transportation to the depot.

The manufacturers claim that a TRT can lay 16 sleepers/minute and give an average output of about 300 to 400 m per hour. In an average traffic block of four hours per day, it is expected that this train will give an average output of 200 to 250 km per year.

21.7 REQUIREMENT OF TRACK MATERIAL

It is important that the exact quantities of track material are conveyed to the site. Any excess material gets wasted, whereas shortage of material results in difficulties and delays in carrying out renewal work. It is normally desirable to have a small spare stock of extra material ready in case of unforeseen contingencies. For requirement of track materials students are advised to refer to details as given in Chapter 28.

SUMMARY

Track renewals are required every year to keep the track safe and in good running condition. The tracks on Indian Railways require heavy renewal since they are very old and have been in use for a very long time. Track renewal is done to rectify defects in rails, fish plates, and sleepers. The track renewal train is used on Indian Railways for mechanized renewal of the track. It can lay 16 sleepers per minute and is able to give an average output of 200–250 km per year

REVIEW QUESTIONS

1. Categorize the different types of rail wear and enumerate the steps to reduce the same.
2. Discuss the standard method of re-laying a track. Also discuss the various aspects to be kept in mind when replacing a track.
3. What is meant by spot and through renewal of track components? Discuss the reasons for the same along with the methods employed.
4. Why is re-laying of a track required? Describe the standard method of re-laying a track on single-line sections.
5. What is the criterion laid down on Indian Railways for track renewal works?
6. What do you understand by mechanized re-laying? How is it done on the Railways?
7. In brief give the various steps involved in carrying out the manual re-laying of track.

Choose the correct answer from the options given.

8. The permissible vertical wear in 52 kg rails is:
   - (a) 6 mm
   - (b) 8 mm
   - (c) 10 mm
   - (d) 12 mm
9. The limiting loss of section in 52 kg rail is:
   - (a) 4 %
   - (b) 6 %
   - (c) 8 %
   - (d) 10 %

10. The service life of 60 kg 90 UTS rail in GMT is:
    (a) 800 (b) 1000
    (c) 1500 (d) None of these.

11. A rail panel of a BG 13m rail with concrete sleepers can be lifted by a portal crane of capacity of:
    (a) 5.0 tonnes (b) 6.0 tonnes
    (c) 7.5 tonnes (d) 9.0 tonnes

12. The gauge of the auxiliary track for working of PQRS equipment is:
    (a) 3400 mm (b) 3700 mm
    (c) 3000 mm (d) 1676 mm

13. The track is considered for renewal after it has carried a threshold GMT of ...... for BG having a rail section of 60 kg with 90 UTS.
    (a) 250 (b) 350
    (c) 525 (d) 800

14. The criteria for rail renewals is:
    (a) incidence of rail fractures or rail failures
    (b) wear of rails
    (c) maintainability of track to the prescribed standards
    (d) service life of rails
    (e) all of the above

15. Rail renewal becomes due when the corrosion in web and foot exceeds:
    (a) 1.5 mm (b) 2.5 mm
    (c) 4.0 mm (d) 5.0 mm

16. It becomes necessary to carry out through rail renewal, when lateral wear exceeds .... mm for BG Group A and B routes for a straight track:
    (a) 6 mm (b) 8 mm
    (c) 9 mm (d) 10 mm

17. The limiting loss of section for track renewal of 60 kg rails is:
    (a) 4 % (b) 6 %
    (c) 8 % (d) 10 %

Railway Accidents and Disaster Management

INTRODUCTION

Any occurrence that does or may affect the safety of the railways, its engines, rolling stock, permanent way, works, passengers, or personnel or cause delays to trains or losses to the railways is termed as an accident. It is the duty of every railway personnel to take the following actions whenever any accident comes to his/her notice:

(a) Take immediate steps to stop the train if it is still in motion, since any further movement is likely to worsen the situation.

(b) Provide all possible assistance in order to relieve the injured and stranded passengers.

(c) Take immediate steps to remove any obstructions or remedy the situation, that is, carryout the necessary actions competently.

(d) Report the incident to the nearest station master by the quickest possible means.

22.1 TYPES OF TRAIN ACCIDENTS

In pursuance of the recommendations made by the Railway Accident Inquiry Committee (Sikri Committee), 1978, the Railway Board has decided that all incidents that result in mishaps should be termed train accidents. Train accidents can broadly be classified under two distinct categories.

Consequential train accidents

These include collisions, derailments, accidents at level crossings, train fires, and similar accidents that have serious repercussions in terms of casualties and damage to property.

Train accidents under the following classification will be termed as consequential train accidents:

Collision	All cases under categories A-1 to A-4
Fire	All cases under categories B-1 to B-4
Level crossing	All cases under categories C-1 to C-4
Derailment	All cases under categories D-1 to D-4
Miscellaneous	All cases under category E-1

Other train acidents

All other accidents which are not covered under the definition of consequential train accidents are to be treated as 'other train accidents'. These include accidents under categories B-5, B-6, C-5 to C-8, D-5, and E-2.

Yard accidents All accidents that take place in a yard and do not involve a train are termed as yard accidents. These include accidents falling under categories A-5, B-7, C-9, and D-6.

Indicative accidents In the real sense of the term they are not accidents but are serious potential hazards and include all cases of trains passing a signal at danger, averted collision, breach of block rule coming under classification F, G, and H.

Equipment failures These include all failure of railway equipment, i.e., failure of locomotive, rolling stock, permanent way, overhead wire, signalling and telecommunication equipment and include cases falling under classification J, K, L, and M.

22.1.1 Sounding of Hooters

It is the practice on the Railways that different number of hooters are sounded according to the type of accident, as indicated in Table 22.1.

Table 22.1 Types of accidents and corresponding hooters

Description of accident	No. of hoots to be sounded*
Accident in the locomotive shed or traffic yard adjoining the locomotive shed	Two long
Accident at the out-station where the main line is clear	Three long
Accident at the out-station where the main line is clear and the relief train is to be accompanied by the medical car	Three long and one short
Accident at the out-station where the main line is blocked and the relief train is not accompanied by the medical car	Four long
Accident at the out-station where the main line is blocked and the relief train is to be accompanied by the medical car	Four long and one short

* The duration of the long hoot is 30 sec and that of the short hoot is 5 sec.

Whenever there is an accident resulting in damage to any part of the permanent way and affecting the free passage of trains, the PWI and AEN of the section should proceed to the accident site as soon as possible. If possible, the men and material required to deal with the emergency should be collected on the way to the site.

In case the traffic is interrupted, the divisional engineer should also proceed to the site if the circumstances of the case at hand require his personal supervision and guidance. In the event of an accident taking place in a division at a point where assistance could be more expeditiously rendered by officials of an adjacent division, the concerned officials should be alerted. These officials should proceed to the scene of the accident immediately and render all possible assistance.

22.2 CLASSIFICATION OF RAILWAY ACCIDENTS

Railway accidents are categorized into different classes according to the type of accident (Table 22.2).

Table 22.2 Detailed classification of accidents

Class	Type of accident	Detailed classification
I & II	**TRAIN ACCIDENTS**	
A	Collisions	A-1 to A-5
B	Fire or explosion in trains	B-1 to B-7
C	Train mingling into road traffic, and/or traffic running into trains, at level crossings	C-1 to C-4
D	Derailment	D-1 to D-6
E	Other train accidents (Train running over or against any obstruction including fixed structure)	E-1 to E-2
III	**INDICATIVE ACCIDENTS**	
F	Averted collisions	F-1 to F-4
G	Breach of block rules	G-1 to G-4
H	Train passing signal at danger	H-1 to H-2
J	Failure of engine and rolling stock	J-1 to J-10
K	Failure of permanent way, viz. buckling of track, weld failure, rail fracture; failure of railway tunnel, bridge, viaduct/formation/cutting and culvert, etc.	K-1 to K-7
L	Failure of electric equipment	L-1 to L-4
M	Failure of signalling and telecommunication	M-1 to M-7
N	Unusual incidents and like train wrecking	N-1 to N-3
O	Causalities	P-1 to O-3
P	Other incidents like accidental or natural death or grievous hurt. Murder or suicide in a train, robbery, attempted robbery, theft, fire or explosion, blockade to train services due to agitation	Q-1 to P-6
Q	Miscellaneous like vehicle running away, train running over cattle, floods, breaches and landslides.	R-1 to-5

22.2.1 Serious Accident

An accident of a train carrying passengers which is attended with loss of life or with grievous hurt to a passenger or passengers in the train, or with serious damage to railway property of the value exceeding ₹ 2.5 million and any other accident which in the opinion of the Chief Commissioner of Railway Safety or Commissioner of Railway Safety requires the holding of an inquiry by the Commissioner of Railway Safety shall also be deemed to be a serious accident. However, the following shall be excluded:

(a) cases of trespassers run over and injured or killed through their own carelessness or of passengers injured or killed through their own carelessness;

(b) cases involving a railway servant or holding valid passes/tickets or otherwise who are killed or grievously injured while travelling outside the rolling stock of a passenger train, such as on foot board or roof or buffer but excluding the inside of vestibules between coaches, or run over at a level crossing or elsewhere on the railway track by a train; and

(c) level crossing accidents where no passenger or railway servant is killed or grievously hurt unless the Chief Commissioner of Railway Safety or Commissioner of Railway Safety is of the opinion that the accident requires the holding of an inquiry by the Commissioner of Railway Safety.

22.3 TYPES OF ACCIDENTS AND REMEDIAL MEASURES

The various types of accidents and remedial measures are explained in this section.

22.3.1 Collisions

Collision refers to one train colliding with another train or any other rolling stock. There are three major types of collisions, these are head-on-collisions, rear-end collisions and side collisions. The main causes of these collisions are failure of station staff, failure of drivers by an act of omission, failure of technical staff and miscellaneous causes.

Remedial measures

The remedial measures for avoiding collisions are as follows:

(i) Development of human resources
(ii) Modern signalling technologies and electronic gadgets

22.3.2 Derailments

Derailments are those occurrences where a train gets derailed due to some obstruction on track or due to some other abnormality. The main causes of derailments may be track defects, vehicle defects, signalling and interlocking defects, human failure including defective operating features and miscellaneous causes.

Remedial measures

The remedial measures for avoiding derailments are as follows:

(i) Improved track structure and track maintenance
(ii) Modern rolling stock with planned maintenance schedules
(iii) Modern signalling and electronic equipments with innovative techniques
(iv) Development of human resources

22.3.3 Accidents at Level Crossing

These are accidents that occur at manned and unmanned level crossings. These accidents are caused mostly due to human error.

Remedial measures

The remedial measures for avoiding accidents at level crossing are as follows:
(i) Better alertness of railway drivers and other staff
(ii) Road drivers to be more vigilant and careful
(iii) Better publicity campaigns

22.3.4 Fire Accidents in Trains

Fire accidents in trains, particularly passenger trains, take a heavy toll of life and property. Main causes are short circuiting in the wiring of coaches, carriage of inflammable materials such as petrol, kerosene oil, gas cylinders and irresponsible behaviour of passengers in lighting cigarettes and *bidies*.

Remedial measures

The remedial measures for avoiding fire accidents in trains are as follows:
(i) Public awareness
(ii) Better publicity to avoid fire accidents

22.3.5 Actions to be Taken by Engineering Officials in Case of Derailments

In case of a derailment, the first engineering official arriving at the site of accident should take the following actions:

Protect the train Any engineering staff available at the site of accident shall assist the guard and driver to protect the train. The inspector should ensure that protection has been afforded to the train in front and in rear. In the case of double line, if the other line is also affected by the accident, steps should be taken to protect both the lines. If no infringement exists, trains must be controlled and passed cautiously on the unaffected track.

First aid and rescue The inspector should arrange for first aid to injured passengers and railway staff and rescue of trapped persons. If there is any medical practitioner on the train, his assistance should be obtained.

Examine the entire site The entire site inclusive of the track should be examined over which the train has passed immediately before derailing, noting down any unusual features observed, especially any part of vehicles or other materials lying on or near the track.

Advice to the nearest station master After a rapid survey of the position, particulars should be sent to the nearest station master, about the accident. In case of controlled sections, field telephone should be got commissioned at once.

Detailed examination A detailed examination should be carried out of the wheel marks on the rail head, fastenings, sleepers and ballast, the wheel trail marks and the corresponding marks on the wheel sets of derailed and other vehicles to identify the wheel set which derailed first and to establish the initial point of derailment.

Dimensioned sketch A dimensioned sketch should be prepared giving the full particulars of the site of accident including the track leading upto the point of derailment, path followed by derailed vehicles, place of mounting and drop, which pair of wheels and which vehicles were first derailed and the position at which the derailed vehicles came to a stand.

Preserve clues Clues which bear evidence of wheel marks especially at the point of mount drop or any other such clues which may help in analysing the cause of the accident should be preserved. In particular, it should be ascertained whether the derailment has occurred due to the flange climbing the rail or due to the wheel set suddenly jumping the rail, leaving no mounting marks.

Recording of track geometry Details of track and measurement of track geometry in the prescribed manner should be recorded in the presence of representatives of other departments.

Gang charts The gang chart should be examined to ascertain the date when the track was last attended.

Vehicle measurements The measurement of the vehicle and its deficiency, if any, should be recorded in the prescribed manner in the presence of representatives of two other departments. Also, the history of repair of the above vehicle should be established.

Operating features Various operating features, such as speed, train formation, loading condition of the vehicle, etc., and the factors which might have contributed to the derailment should be investigated and recorded.

Preliminary report A preliminary report should be prepared to be passed on to the immediate superior. The report should contain the following information:

1. The nature of the accident
2. Cause, if known
3. Particulars of loss of life, injuries to passengers and staff
4. Extent of damage to way and works
5. Steps taken for resumption of traffic
6. Probable date and time when normal working is likely to be resumed
7. Whether transhipment is necessary and if so, when it is likey to be opened
8. Details of any assistance required, such as additional labour, ballast, train bridging material, etc.
9. Follow-up action Take action to collect men and material to repair the track and restore the traffic. The actual repair work should be started only when the measurements of the track have been completed.

22.3.6 Actions to be Taken in Case of Sabotage

In all cases of accidents, the cause of which might possibly be due to sabotage, it is essential that the clearance and restoration operations are not commenced till the police officials arrive at the site and intimate their agreement to the commencement of clearance and restoration work after making thorough investigations.

A factual note of the conditions obtained at the site prior to restoration work should be prepared and signed jointly by the senior most police and railway officials. Difference of opinion, if any, may be recorded on the joint factual note.

This should not, however, be allowed to interfere with rendering of first aid to the injured, which is the first essential in all accidents.

22.4 DERAILMENT MECHANISM

The motion of a vehicle on a track is a compelx phenomenon. A very large number of factors are at play having bearing on the safety and stability of the track vehicle and the dynamic track vehicle system. It is for this reason that at times it becomes difficult to establish with certainty the cause of derailment.

Whenever a derailment takes place, it generally happens that one or more factors have crossed the safety limits. At the time of derailment, the actions of the factors and their values, cannot be precisely known. Since, it is almost impossible to reproduce all conditions that prevailed at the time of derailment exactly, it may happen that the same vehicle when moved again over the same track, at the same speed under the same operating conditions may not derail.

After the track and rolling stock particulars have been taken and operating features have been identified, a sound theoretical understanding of the whole phenomenon of vehicle-track interaction helps one to analyse all the evidences logically and systematically and to arrive at the probable cause of derailment.

22.4.1 Derailment Categories

There are two categories of derailments:
 (a) Sudden derailment by wheel sets jumping off the track
 (b) Derailment by flange climbing, that is, by wheel mounting the rail in a relatively gradual manner

22.4.2 Nadal's Formula for Derailment Analysis

At the time of derailment, when the wheel flange is in the process of climbing onto the rail, certain forces act at the part of contact between the rail and the wheel, which lie at positive angularity to each other as shown in Fig. 22.1.

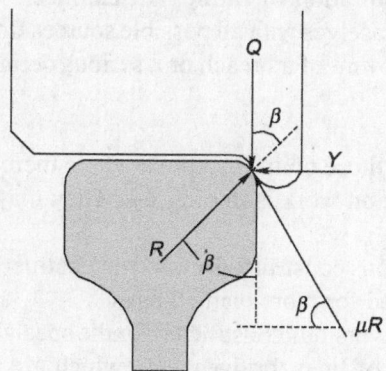

Fig. 22.1 Forces at rail–wheel contact with positive angularity

Nadal's formula for safety against derailment is as follows:

$$\frac{Y}{Q} > \frac{\tan \beta - \mu}{1 + \mu \tan \beta}$$

(22.1)

where Y is the flange force, Q is the instantaneous wheel load, R is the normal reaction of the rail, μR is the frictional force acting upwards, μ is the coefficient of friction between the wheel flange and the rail, and β is the flange angle. R and μR in figure 22.1 are normal reactions of the rail and frictional force, respectively Y/Q is known as the *derailment coefficient*. Nadal's formula provides an important criterion for the assessment of the stability of rolling stock. It also has practical use in the investigation of derailments.

Based on Nadal's formula, the factors that contribute to derailments, whether due to track defects, vehicular defects, or unfavourable operational techniques, are as follows:

(a) Increase of flange force Y

(b) Decrease of instantaneous wheel load Q

(c) Increase of coefficient of friction between the wheel flange and the rail (μ). This is normally the consequence of a rusted rail, a newly turned wheel standing on the rail, sharp flanges, etc.

(d) Increase in positive angle of attack condition (β)

(e) Increase in positive eccentricity. This increases primarily due to the slope of the wheel flange becoming steeper

(f) Persistent angular running of the axle

22.5 RESTORATION OF TRAFFIC

After the necessary information has been collected, immediate and prompt action should be taken to restore traffic on the line. In the case of causeways, etc., it may be necessary to restore traffic by making temporary arrangements, such as the insertion of a rail cluster or the provision of a temporary diversion. Arrangements should be made for adequate labour and permanent way material and the damaged track should be attended to as soon as possible.

It may be emphasized here that adequate labour is the most predominant requirement for the restoration of traffic. All engineers and inspectors should thoroughly acquaint themselves with all possible sources from where labour can be readily obtained in the event of a breach or a serious derailment in their section.

22.5.1 Diversions

Diversions are set up at those points in a track where there is some obstruction or where some reconstruction work is in progress. They may be classified into two main categories:

(a) Temporary diversions constructed around an obstruction to traffic, which are not likely to be used for more than 10 days.

(b) Semi-permanent diversions constructed for the special purpose of facilitating the reconstruction of lines, bridges, etc., which are likely to be used for a period of more than 10 days.

All trains must stop before entering a temporary diversion and proceed at very low speeds. On a semi-permanent diversion, trains may proceed at a non-stop but reduced speed but only after an adequate period of diversion.

Normally, semi-permanent diversions are provided on the Indian Railways with an initial speed requirement of zero, though this may be relaxed after the consolidation of the bank. A semi-permanent diversion should be laid according to the specifications presented in Table 22.3.

In a difficult terrain, it may be necessary to lay curves of radius not less than 225 m on BG and 125 m on MG and adopt the steepest gradient on the section.

In addition to the standards specified in Table 22.3, it is desirable that a semi-permanent diversion has the following features:

(a) The existing bank should not be cut to lay the diversion.
(b) The gradient should be suitably compensated for curvature in order to ensure that in spite of the curvature, the gradient does not exceed the limiting value specified in Table 22.3.

Table 22.3 Geometric elements for semi-permanent diversions

Item	BG	MG	NG
Radius of curve	Not less than 450 m	Not less than 300 m	Not less than 45 m
Gradient	Not steeper than 1 in 100	Not steeper than 1 in 80	Not steeper than 1 in 40
Superelevation	Normally there is no superelevation, but if it is provided, it should be limited to suit the speed permitted on the diversion and should run out at a rate not exceeding 12 mm (1/2") in 10 m (30 ft)		

(c) If the diversion includes a bridge with cribs, a straight and level portion of 30 m (100 ft) should be provided on either approach to the bridge.
(d) To guard against the possible settling of the diversion, a blanket of coal ash should be provided on the new embankment and rolled, if possible.

Plan and longitudinal section of diversion

The plan laid to ensure that the diversion settles in its place is illustrated in Fig. 22.2. The formulae used in connection with the same are the following:

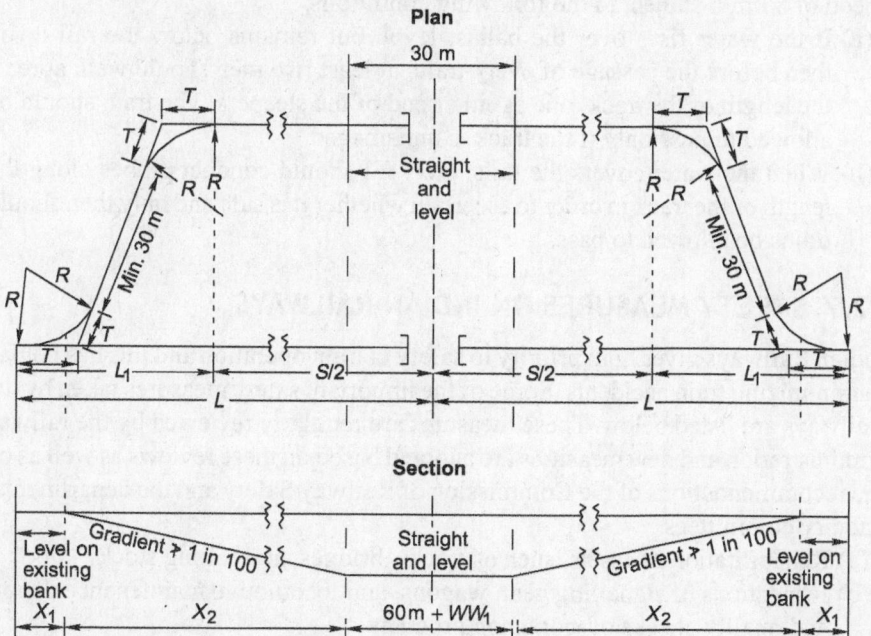

Fig. 22.2 Plan and longitudinal section of diversion

$$L = \sqrt{C^2 + 4RD - D^2 + \frac{S}{2}} \tag{22.2}$$

$$T = \frac{RD}{L - \dfrac{S}{2} + C} \tag{22.3}$$

Here L is the length of half the diversion measured along the original alignment, D is the maximum distance of diversion from the original alignment, C is the straight line between two reverse curves connecting the original alignment to the diversion, which is not less than 30 m (100 ft), S is the straight portion of the diversion that is parallel to the original alignment, R is the radius of the curve, and T is the length of tangent.

22.6 FLOOD CAUSEWAY

When a causeway gets flooded and the velocity of water is not too high, trains may be allowed to pass, subject to the following conditions:
   (a) The PWI has examined the track on foot and is satisfied as to the safety and integrity of the permanent way.
   (b) The depth of water does not exceed the following values:

	Passenger trains	Goods trains
BG	300 mm	450 mm
MG	225 mm	225 mm
NG	225 mm	225 mm

When a track gets flooded, trains should be made to halt and then proceed at a speed of 8 km/h subject to the following conditions:
   (a) If the water rises over the ballast level, but remains below the rail level, then before the passage of every train, at least two men should walk abreast the length of the track, one at either end of the sleepers. The train should be allowed to pass only if the track is undamaged.
   (b) When the water covers the rails, the PWI should conduct probes along the length of the track in order to ascertain whether it is safe and only then should trains be allowed to pass.

22.7 SAFETY MEASURES ON INDIAN RAILWAYS

Indian Railways gives high priority to safety in train operation and this has helped in minimizing train accidents. Some of the important safety measures taken by the Railways are listed below. These measures are regularly reviewed by the railway administration and new measures are adopted based on these reviews as well as on the recommendations of the Commission of Railway Safety and the departmental enquiry committees.
   (a) Rehabilitation of assets, such as tracks, bridges, and rolling stock
   (b) Inspections of signalling gear, wagons, and locomotive maintenance depots and quality checks of workshop products
   (c) Monitoring the training and performance of the operational staff
   (d) Counselling the drivers about their driving techniques

(e) Extension of technical aids, such as, auxiliary warning system, axle counters, route relay interlocking, and track circuiting

(f) Surprise checks to verify that no inflammable or explosive substances are being carried on passenger trains

(g) Providing whistle boards/speed breakers and road signs at unmanned level crossings and making provisions to improve visibility for drivers

(h) Audio–visual publicity to educate road users on how to safely use level crossings

(i) Holding disaster management courses for staff and officers in training institutes

(j) Making the safety of the Railways a multidisciplinary affair for effectiveness

22.8 DISASTER MANAGEMENT

A disaster is a sudden and great calamity which causes deep distress to passengers, staff, and their families. In the context of railways, disaster management envisages expeditious, orderly, effective, and adequate relief measures in the case of a disaster. The term 'disaster management' first came to be used on Indian Railways in 1986, and has been in vogue ever since. In some respects, this term redefines an accident as a disaster.

22.8.1 Classification of Disasters

Figure 22.3 gives a broad classification of disasters on the Railways.

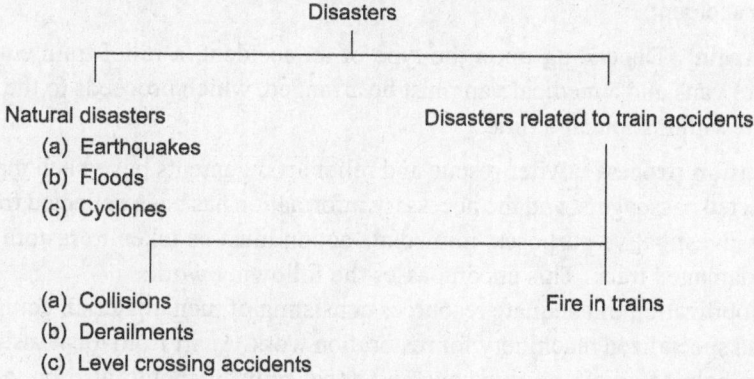

Fig. 22.3 Classification of disasters

22.8.2 Details of Disaster Management

Disaster management on railways may be broadly divided into the following stages:

Disaster mitigation and prevention

This stage consists of strengthening the basic infrastructure, such as the track, bridges, rolling stock, and signalling equipment as well as introducing systems and procedures to ensure that accidents are minimized, if not prevented.

Disaster management planning and preparedness

The successful management of a disaster depends on the ability to foresee and control it in time. Planning is thus vital for minimizing disaster effects, quick recovery, and resumption of work. The key to developing an effective disaster plan is to anticipate every possible vulnerability and taking the appropriate action to tackle the same effectively. In this context it is necessary to identify the available resources and utilize them most efficiently. The duties of various officials should be defined in detail so that work can proceed as per the planned strategy. The management should be fully prepared to face a disaster and should possess a proper action plan for dealing with the same.

Post-disaster management including rescue and relief arrangement

This comprises all the work that is undertaken by the railways following a train accident.

First responsibility of the railways   The first responsibility of the railways in the case of accidents is to reach and extricate accident victims and organize effective trauma care. The basic principle of trauma management is speed and expediency, as most trauma patients can be saved within the first hour. This hour is called 'the golden hour'.

Accident relief measures   The Railways should have full-fledged arrangements for rescue and relief operations in the case of railway accidents. Steps must be taken to provide prompt and effective relief to the affected passengers in the event of train accident.

Relief train   Depending upon the type of an accident, a relief train equipped with tool vans and a medical van must be arranged, which proceeds to the site of accident within stipulated time.

Restoration process   After rescue and relief arrangements have been made for the affected passengers, and the necessary information has been collected from the site for investigative purposes, immediate action must be taken to restore traffic on the damaged track. This encompasses the following work:

(a) Mobilization of adequate resources consisting of men, material, equipment, and specialized machinery for restoration work. Apart from local assistance, the help of adjoining divisions and zonal railway should also be taken. If necessary, contact can also be established with villagers, local administration, civil authorities, and defence establishments for the necessary aid.

(b) Restoration work should be started immediately by way of removal of debris and establishment of new track connections after the necessary earthwork has been done. In case a bridge is affected, either a diversion or a temporary bridge can also be planned. Once the damaged track has been repaired, trains are allowed to traverse it at a restricted speed, which is slowly restored to the normal speed in subsequent stages.

(c) All efforts must be made to ensure that restoration work is done expeditiously and that trains start using the affected section as soon as possible.

22.8.3 Actions to be Taken by Engineering Officials

The first engineering officials to arrive at the site of the accident should be very alert and take proper action (For details refer to Section 22.3.5)

22.8.4 Availability of Resources

The resources available in the case of a major accident may be grouped into four different units, depending on the time frame within which they can be made available after an accident. These groups are the following.

Resource unit-I Railway and non-railway resources available on the train and in the nearby surroundings

Resources unit-II Railway resources available on the accident relief train with medical van/accident relief train (ARMV/ART) depots and elsewhere within the division

Resources unit-III Railway resources available at ARMV/ART depots and elsewhere on adjoining zones and divisions

Resources unit-IV Non-railway resources available within or outside the division

22.8.5 Unified Command Centre/Combined Assistance and Relief Enclosure

There are two aspects of disaster management work carried out at an accident site, as described below.
 (a) Rescue, relief, and restoration operations, which are carried out by one set of functionaries. This work is done under the control of the Unified Command Centre (UCC) on Indian Railways.
 (b) The second aspect pertains to the rehabilitation of the accident affected passengers, taking care of dead bodies, dealing with relatives of the dead victims, etc., for which a different set of functionaries is required. This work is done by the Combined Assistance and Relief Enclosure (CARE) on Indian Railways.

22.8.6 Review of Disaster Management

A high-level committee was constituted in September 2002 to review disaster management on Indian Railways and to modernize it. The committee presented its recommendations on a report in April 2003, which was subsequently approved and accepted. The implementation of these recommendations would require an estimated budget of ₹ 4000 million.

The various aspects that would help in the modernization of disaster management are given below.
 (a) Faster response
 (b) Better facilities and equipment
 (c) Expanding resources to meet requirements in major accidents
 (d) Better customer focus
 (e) Training and preparedness

SUMMARY

Accidents and derailments result in heavy loss of life and property, and disruption of normal traffic. Accidents are classified into different groups depending upon their nature and severity; the more severe accidents are collision, derailment, and fire accidents. Although it is practically impossible to eliminate accidents completely, attempts should be made to reduce their probability as much as possible. Disaster management helps in dealing with the aftermath of a train accident.

REVIEW QUESTIONS

1. Define derailment coefficient, and explain how it indicates the proneness of a section to derailment. Describe the various defects in the permanent way that may contribute to derailments.
2. A portion of a track is washed away during floods. What are the emergency measures adopted under such circumstances?
3. A temporary diversion has to be laid for releasing a single-span 12 m girder over a canal bridge. The rail level is 3 m above the water level of the canal, which is not navigable. 4° curves and suitable gradients are to be provided for the diversion. Draw a dimensioned sketch of the diversion and show the positions of the temporary signals.
4. What is derailment? Discuss the reasons for train derailments. What is Nadal's formula and how does it help in analysing the cause of derailments?
5. Discuss and analyse the various types of derailments and their probable causes.
6. What is a diversion? What is the difference between a temporary and a semi-permanent diversion? Give the specifications of a semi-permanent diversion.
7. When is a diversion required? What are the various features of a diversion?

Choose the correct answer from the choices given.

8. For taking track measurement at an accident site stations shall be marked at.... apart.
   (a) 1 m
   (b) 2 m
   (c) 3 m
   (d) 5 m
9. At an accident site while taking track measurement the versine shall be recorded on a ...... chord for BG track upto 600 m radius.
   (a) 10 m
   (b) 12 m
   (c) 15 m
   (d) 20 m
10. Derailment coefficient is indicated by:
    (a) Y
    (b) Q
    (c) Y/Q
    (d) none of these
11. Formula for Nadal's equation for safety against derailment is:
    (a) $\dfrac{Y}{Q} > \dfrac{(\tan\beta - \mu)}{1 + \mu\tan\beta}$
    (b) $\dfrac{Y}{Q} < \dfrac{(\tan\beta - \mu)}{1 + \mu\tan\beta}$
    (c) $\dfrac{Y}{Q} \ngtr \dfrac{(\tan\beta - \mu)}{1 + \mu\tan\beta}$
    (d) none of these

12. An accident will be termed as a serious accident when the loss of a railway property is more than:
    (a) 3 lakhs
    (b) 1.0 million
    (c) 2.5 million
    (d) 1.5 million
13. Collision as per new the classification is classified as:
    (a) A class
    (b) B class
    (c) C class
    (d) none of these
14. CARE is abbreviated for:
    (a) Centre for Accident Relief Equipment
    (b) Combined Assistance and Relief Enclosure
    (c) Combined Accident and Relief Enclosure
    (d) Care and relief Engineer
15. As per the new classification of accidents, a level crossing accident comes under the category of:
    (a) consequential train accidents
    (b) other train accidents
    (c) indicative accidents
    (d) unusual incidents

CHAPTER
23

Level Crossings

INTRODUCTION

Level crossings are provided on railway lines to allow road traffic to pass across the track. As the level of the passing road traffic is the same as that of the railway track, the crossing is referred to as a level crossing. Other types of crossings are road overbridge or road underbridge, where road traffic passes over or under the railway track. In both these cases, the necessary clearance between the road bed and the railway track is kept as prescribed in the schedule of dimensions.

23.1 CLASSIFICATION OF LEVEL CROSSINGS

Level crossings are classified in different categories in consultation with road authorities depending upon the class of the road, visibility conditions, volume of road traffic, and number of trains passing over the level crossing. The classification of level crossing is made as (i) special class (ii) A class (iii) B class (iv) C class (v) D class. Table 23.1 gives the criteria for classification of these level crossings. Brief details about each of these level crossings are given below.

Indian Railways as on 31 March 2011 maintains 32,735 level crossings, out of which 17,839 have gatekeepers and the remaining 14,896 crossings are unmanned.

Special class

These are the busiest level crossings in terms of road traffic. Most of the busy level crossings on the national highway are special class level crossings. Normally the gates are open to road traffic but whenever a train passes by, the gates are closed to road traffic. The gates of the level crossings are interlocked with signals. They are manned round the clock by three gatemen working in eight-hour shifts.

Table 23.1 Criteria for different classes of level crossings

Class of level crossing	Criteria
Special Class	TVUs greater than 50,000
A Class	TVUs between 30,000 to 50,000. Line capacity utilization > 80% (on single line) and road vehicles > 1000
B Class*	TVUs > 20,000 and road vehicles > 750
C Class	All other level crossings for roads not covered above.
D Class	For cattle crossing

Notes:

1. *Position of gates: The normal position of gates will be open to road traffic, when TVU> 25,000 and gates will be interlocked. In case TVU<25,000 the gates will normally be closed to road traffic.*
2. *The following facilities will be provided in the following order of priority under revised classification of gates as the case may be: (a) Provision of telephones, (b) Provision of lifting barriers, (c) Inter-locking of manned level crossing gates.*
* *B Class is further subdivided as B1 for TVU between 30,000 to 25,000 and B2 for TVU between 25,000 to 20,000.*

'A' class

These level crossings are also busy in terms of road traffic. All level crossings on important roads are mostly A class level crossings. In this case also, the gates are normally open to road traffic. All other provisions are the same as for special class level crossings except that these level crossings are provided with only two gatemen who work in 12-hour shifts, as these crossings are not as busy as special class level crossings.

'B' class

These level crossings are relatively less busy. Normal B class level crossings can be found on metalled roads. The gates are normally closed to road traffic, but can be kept open to road traffic provided that the gates are interlocked with signals. They are provided with two gatemen working 12-hour shifts.

'C' class

These level crossings are mostly provided on unmetalled roads. Some of these level crossings are unmanned because of low volume of road traffic.

'D' class

These level crossings are provided for cattle; they are normally used by cattle or pedestrians.

23.1.1 Square or Skew Level Crossing

All level crossings should preferably cross the railway line at right angles. In special cases, when modification is required in unavoidable situations to suit the road approaches, the angle of crossing should not be less than 45°.

At all such skew level crossings, the gate posts shall be fixed square to the road. A typical layout of a square crossing is given in Figure 23.1.

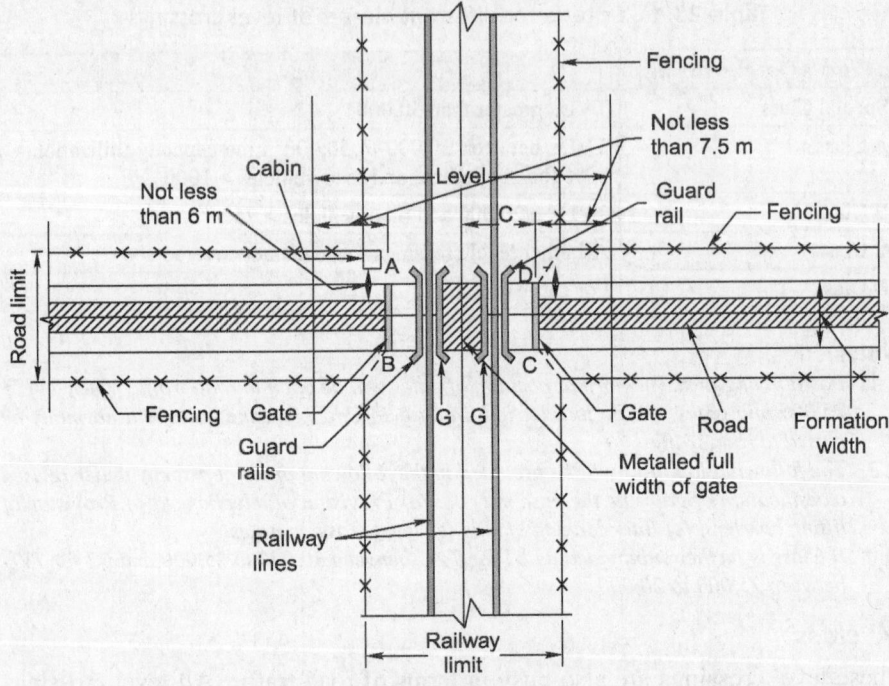

Fig. 23.1 A typical layout of a level crossing

23.2 DIMENSIONS OF LEVEL CROSSINGS

Some of the important dimensions of square A class and B class level crossings are given in Table 23.2. It may be mentioned here that the angle of crossing between the centre line of the railway and the road should not be less than 45°.

Table 23.2 Dimensions of A class and B class level crossings

Description of item	A class level crossing on a class-I road	B class level crossing on a class-II road
Maximum width of gates	9 m	7.5 m
Minimum length of guard rails	11 m	9.5 m
Position of gates when open to road traffic	Always from or towards, but not across, the line	Same as for A class
Maximum distance of gate post from the centre line of the track	3 m	3 m
Maximum distance of gate lodge from the centre line of the nearest track	6 m	6 m
Minimum width of metalled road between gates	Same as the width of the gate	Same as the width of the gate
Level and gradients		
(a) between gates	Level	Level

Contd.

Table 23.2 *(Contd.)*

Description of item	A class level crossing on a class-I road	B class level crossing on a class-II road
(b) beyond gates	Level up to a distance of 8 m; a grade of 1 in 30 beyond that	Level up to a distance of 8 m; a grade of 1 in 20 beyond that
Minimum length of road outside the gates	22.5 m	15 m
Minimum fencing on lines parallel to the track	15 m	15 m
Provision of light indicators at the gates for the benefit of road users	Red when the gate is closed and white when it is open to road traffic	Same as for A class

23.3 ACCIDENTS AT LEVEL CROSSINGS AND REMEDIAL MEASURES

Level crossings are vulnerable to accidents due to a number of reasons, but the main reason behind these accidents is the negligence of road users. Whatever may be the cause of an accident on a level crossing, the railways come under severe criticism whenever such an accident occurs, and there is concern at all levels to determine the corrective action that should be taken to minimize these accidents. Some of the safety measures that can be taken to reduce accidents on level crossings are as follows:

Manning of level crossings Busy unmanned level crossings, particularly at locations where buses ply, should be manned in a phased manner.

Providing lifting barriers Lifting barriers should be provided in preference to gates to quickly clear road traffic.

Level crossing indicators At the approaches to all unmanned C class level crossings or manned level crossings where the view is obstructed, bilingual whistle boards should be erected along the track at a distance of 600 m from the level crossing to enjoin the drivers of approaching trains to give an audible warning of the approach of the train to road users. The drivers of approaching trains should blow the train whistle continuously from the time they pass the whistle boards till the time they have negotiated the level crossing.

Stop signs for level crossings Stop signs should be provided on either side of the road approaches to all unmanned C class level crossings at suitable points within the railway boundary.

Speed breaker at level crossings A series of three speed breakers (bumps) spaced 100 m apart should be provided within the railway boundary on either side of all unmanned level crossings with metalled approaches, where motor vehicles ply. The speed breakers should have a smooth parabolic profile so that the vehicles can slow down without experiencing extra jerks.

Rumble strips at level crossings Rumble strips should be provided by road authorities on the road approaches on either side of all manned level crossings where road travellers face poor visibility conditions with a view to making them alert and vigilant. Rumble strips consist of intermittent raised bituminous overlays of 15–20 mm height and 200–300 mm width across the roadway. About 15 to 20 such strips spaced at about 1 m centre-to-centre distance are provided in the approaches to level crossings to caution the motorist.

Following extant rules Normally level crossings should not be kept open to road traffic in contravention of extant orders. On each railway, all cases where the gates of the level crossing are open normally to road traffic should be reviewed immediately and it should be ensured that such instances are permitted only in strict conformity to extant orders.

Job analysis Wherever the gates of a level crossing are kept open to road traffic, the work of the gatemen must be analysed and, where required, an extra gateman may be provided as per the rules for intensive work, thereby increasing the number of gatemen from two to three.

Visibility Approaching trains should be adequately visible from the road. Whenever an obstruction is discovered on the track, immediate action must be taken to remove it.

Surprise checks Surprise checks should be made to ensure that the gatemen are alert and that road vehicles follow the regulations properly.

23.4 RECORDS TO BE MAINTAINED AT LEVEL CROSSINGS

Following are the records to be maintained at level crossings:
   (i) Private number book and train passing log register on level crossings provided with a telephone
  (ii) Level crossing inspection book
 (iii) Complaint book for users
  (iv) Duty roster
   (v) Latest working instructions in Engligh and local language .
  (vi) Competency certificates of gatemen
 (vii) Due date of next periodical medical examination and vision test of gatemen
(viii) Due date of next refresher course
   (x) Tools list (with columns drawn for checking of tools)

23.5 TRACK STRUCTURE AT LEVEL CROSSINGS

   (i) At level crossings, concrete sleepers of special design should preferably be used. Wherever, it is not feasible to provide concrete sleepers, 'U' category wooden sleepers should preferably be used.
  (ii) 'U' category wooden sleepers used in level crossings shall be provided with suitable bearing plate; at each rail seat, four spikes should be provided.
 (iii) There should be no combination joint on level crossing proper and its approaches upto 50 m.
  (iv) Rail joints should preferably be avoided in check rails and on the running rails within level crossing and 3 m on either side. In case of SWR the fish plated joint should be avoided on the level crossing and within 6 m from the end of level crossing.

(v) The level crossing should not fall within the breathing length of LWR, as per the provisions contained in the *Manual of Instructions of Long Welded Rails*.

23.6 MAINTENANCE AND INSPECTION OF LEVEL CROSSINGS

Clearing of trees and bushes   All trees, bushes or undergrowth that interfere or tend to interfere with the view from the railway or roadway when approaching level crossings, should be cut down.

Overhauling of level crossing   The level crossings having sleepers other than concrete sleepers must be overhauled at least once a year or more frequently, as necessary.

However level crossings laid on concrete sleepers should be overhauled with each cycle of machine packing or more frequently as warranted by conditions and in no case, should be delayed by more than two years.

During overhauling, the condition of sleepers and fittings, rails and fastenings should be examined. The wooden sleepers be given a coat of coal tar.

Maintenance of rail track   In all cases, rails and fastenings in contact with the road shall be thoroughly cleaned with wire brush and a coat of coaltar applied. Flange way clearances, cross level gauge and alignment should be checked and corrected as necessary, and the track packed thoroughly before reopening the level crossing for road traffic.

Maintenance of whistle boards and stop boards   There should be proper upkeep and maintenance of whistle boards and 'STOP' boards provided on the approaches to level crossings.

Maintenance of check rails   Check rails of level crossings are required to be removed for tamping, overhauling, destressing, track renewals and should be refixed as quickly as possible preferably before leaving site. In case, check rail cannot be fixed, a speed restriction of 30 km per hour will follow, and a stationary watchman shall be posted to ensure safety besides arranging for diverting passage of road traffic.

Table 23.3   Specified clearances of check rails at level crossings

Item	BG mm	MG mm	NG mm
Minimum clearance of check rails at level crossings	51	51	51
Maximum clearance of check rails at level crossings	57	57	57
Minimum depth of space for wheel flange from the rail level	38	35	31

Indication post for detonators   This should be provided so as indicate the location and number of detonators to be placed in case of an obstruction to level crossings.

For BG   Indication post with one dot at 600 m and another post with three dots at 1200 m from the rear of the level crossing.

For MG and NG   Indication post with one dot at 400 m and another post with three dots at 800 m from the rear of the level crossing.

Height gauges These should be erected on either side of level crossing equipment or other equipment at every level crossing so as to ensure that all vehicles can pass with adequate clearance. The height gauges shall be located at a minimum distance of 8 m from the gate posts.

Duty huts These should be so located that a clear and unobstructed view is obtained of all approaching trains and road vehicles. The minimum distance of duty hut from centre lane of nearest track edge of road metalling shall be 6 m.

23.6.1 Checking of Equipment and Knowledge of Gateman

The sectional PWIs should inspect each level crossing once in a month. During their inspection, they shall:
 (a) Inspect the equipment with the gateman
 (b) Ensure that the gateman has a correct knowledge of rules. Besides this, they should endeavour to examine them periodically during their trolley inspection particularly on appointment, promotion or transfer by conducting practical demonstration of protection of level crossing in case of emergency.
 (c) Ensure presence and alertness of the gateman by carrying out surprise night inspection of level crossing once in a month.

23.6.2 Visibility Requirements for Unmanned Level Crossings

For new unmanned level crossings, the visibility requirements for road users along with the track is 600 m, with single or double-line track. Where this is not feasible, the distance may be reduced suitably with the approval of the chief engineer, provided the maximum permissible speed is less than 100 km per hour and there is only a single track to be crossed on the level crossings. The visibility of trains for road users at unmanned level crossings may be assessed from a distance of 5 m from the centre of the track.

23.6.3 Census of Traffic at Level Crossings

A periodical census of traffic at all level crossings whether unmanned or manned should be taken at least once in five years to review the classification in the case of manned level crossing and need for manning in the case of unmanned level crossing. The total train vehicle unit (TVU) per day (train units × vehicle units) is worked out taking the census for a week. Trains, motor vehicles, bullock carts, and tongas are considered as one units; cycles, rickshaws, auto-rickshaws are considered as half units.

When the train vehicle unit exceeds 6,000 TVU or visibility is poor as in the case of an unmanned level crossing, the same is considered as accident prone and action should be taken to man such level crossings in a phased manner.

In the case of an accident at an unmanned level crossing, a census should be conducted immediately to determine whether manning is required.

In the case of manned level crossings, the quinquennial census should be substituted by a census cum job analysis so as to avail of the opportunity of checking up the adequacy of men on consideration of hours of employment regulations.

23.6.4 Equipment of Level Crossings

The following are the various types of equipment of level crossings:

(i)   Two hand signal lamps, (tricolour), provided with bright reflectors

(ii) One green and two red hand signal flags

(iii) One staff suitable for exhibition of the red lamp and the red flag

(iv) Two long spare chains with 'stop' marked at the centre to cover the full width of the gate in case the gate is damaged

(v)   Two spare small chains and pad locks for locking gates in case the locking arrangement of the gates become defective

(vi) Twelve detonators, which are still active

(vii) Two gate lamps

(viii) One tin case for detonators, one tin case for muster sheet, and one can for oil

(ix) Tools and equipment, viz., one tommy bar, one mortar pan, one phowrah, one hammer, and one bucket

(x)   Relevant document, viz. one tools list, one book of safety rules, duty roster, complaint book for users, one inspection register, and level crossing instructions book

23.7 INSPECTION OF LEVEL CROSSINGS BY PWI AND AEN

The PWI should inspect level crossings once a month, check the equipment to be used by the gatemen, examine their knowledge of safety rules, and carry out other routine inspections. She or he should also carry out surprise inspections of level crossings to ensure that the gateman is in attendance and is alert. In addition to the aforementioned duties, the PWI should inspect the rails, sleepers, and fastenings at least once a year after getting the level crossings fully opened out.

The AEN should inspect all the equipment at every manned level crossing that comes under her or his subdivision once every six months and quiz the gateman regarding rules during the inspection. She or he should scrutinize the manuscript registers of level crossings that are maintained by the PWI and inspect as many level crossings as possible when they are opened completely during the year.

SUMMARY

The level crossing is a point where a railway line and a road cross each other at the same level. These are classified into different categories depending upon the importance of the road and the track. Level crossings are often prone to accidents and sufficient care must be taken to avoid the same. This includes proper maintenance and regular inspection by the appointed railway personnel.

REVIEW QUESTIONS

1. What is a level crossing? Draw a neat sketch of an A class level crossing giving its typical dimensions and details.

2. How are level crossings classified in India? Discuss their important features.

3. 'Level crossings are potential accident sites'. Comment.

4. What are the special precautions to be taken regarding the safety of level crossings? Describe the inspection schedule of PWIs for level crossings.

Choose the correct answer from the choices given.

5. Traffic passing through 'A' class level crossing should be:
   (a) 10000 to 20000 TVU
   (b) 20000 to 30000 TVU
   (c) 30000 to 50000 TVU
   (d) above 50000 TVU

6. Traffic passing through 'B' class level crossing should be:
   (a) 10000 to 20000 TVU
   (b) 20000 to 30000 TVU
   (c) 30000 to 50000 TVU
   (d) above 50000 TVU

7. TVU is the abbreviation of:
   (a) Track vulnerable unit
   (b) Train vehicle unit
   (c) Total vehicle unit
   (d) none of these

8. The minimum depth of space for wheel flange from rail table on a level crossing for BG should be:
   (a) 45 mm
   (b) 38 mm
   (c) 25 mm
   (d) 15 mm

9. The total number of level crossings on Indian Railways is about:
   (a) 25000
   (b) 30000
   (c) 34000
   (d) 38000

10. For an unmanned level crossing the visibility requirement for roads users along the track shall be .... with single-or double-line track.
    (a) 200 m
    (b) 300 m
    (c) 500 m
    (d) 600 m

11. The minimum distance of gate posts from the centre line of the nearest track for a BG section should be at least:
    (a) 1.5 m
    (b) 2.5 m
    (c) 3.0 m
    (d) 4.0 m

12. The minimum distance of gate posts from the centre line of the nearest track for a MG section should be at least:
    (a) 1.5 m
    (b) 2.5 m
    (c) 3.0 m
    (d) 4.0 m

13. Rumble strips should be provided on approaches of all manned level crossings with a height width of:
    (a) 15 to 25 mm and 200 to 300 mm respectively
    (b) 10 to 20 mm and 150 to 250 mm respectively
    (c) 20 to 30 mm and 250 to 350 mm respectively
    (d) 15 to 25 mm and 250 to 350 mm respectively

Locomotives and Other Rolling Stock

INTRODUCTION

The locomotive is a powerhouse mounted on a frame that produces the motive power needed for traction on railways. There are three distinct locomotives used on the railways, each drawing its power from a different energy source. In a steam locomotive, the motive power is the steam generated in a pressure vessel called the boiler. Thus the thermal energy of fuel is converted into the mechanical energy of motion. In a diesel locomotive, the motive power is an internal combustion engine, which uses high-speed diesel oil as its source of energy. An electric locomotive derives its power from an electric conductor running along the track.

24.1 TYPES OF TRACTION

There are three types of traction on Indian Railways.
  (a) Steam traction by steam locomotives
  (b) Diesel traction by diesel locomotives
  (c) Electric traction by electric locomotives

  Diesel and electric locomotives are comparatively more efficient than steam locomotives. They have greater hauling capacity, permit better acceleration and deceleration, and are capable of carrying heavy loads at higher speeds. Table 24.1

Table 24.1 Details of rolling stock on Indian Railways

Type of rolling stock	Number	Details
Locomotives	9213	Steam 43, diesel 5137 and electric 4033
Coaching vehicles	53555	EMU 7334, passengers coaches 45123, DMU 763 and other coaching vehicles 6493
Goods, wagons	229381	Covered 26.5%, others 73.5%

presents details of rolling stock on Indian Railways. Diesel and electric locomotives are fast replacing steam locomotives, as can be seen from Table 24.2.

Table 24.2 Different types of locomotives on Indian Railways

Year	Number of locomotives				Average tractive effort per locomotive (kg)	
	Steam	Diesel	Electric	Total	BG	MG
1950–51	8,120	17	72	8,209	12,801	7,497
1960–61	10,312	181	131	10,624	14,733	8,201
1970–71	9,837	1,169	602	11,608	17,303	9,607
1980–81	7,469	2,403	1,036	10,908	19,848	10,429
1990–91	2,915	3,759	1,743	8,417	21,088	12,438
2000–01	54	4,702	2,810	7,566	29,203	18,537
2008–09	43	4,964	3,586	8,593	33,499	18,452
2010–11	43	5,137	4,033	9,213	34,380	18,304

It may be noted here that though the total holdings of locomotives on Indian Railways have been decreasing since the past 25 years, the average tractive effort has increased progressively due to the provision of more efficient diesel as well as electric traction in place of steam traction.

24.2 NOMENCLATURE OF STEAM LOCOMOTIVES

A steam locomotive is normally identified by the arrangement of its wheels. This locomotive generally has two types of wheels.

Driving wheels These wheels are coupled with each other and are directly connected with the pistons. The higher the diameter of the driving wheels, the greater the tractive power, but the lower the speed.

Idle wheels These wheels distribute the load of the locomotive on the track.

A locomotive is normally designated by a three-digit code. The first digit indicates the number of idle wheels in the front, the middle digit indicates the number of driving wheels, and the last digit denotes the number of idle wheels that come after the driving wheels. A locomotive with a nomenclature of 4-6-2 indicates four idle wheels in the front, six driving wheels, and two idle wheels after the driving wheels (Fig. 24.1).

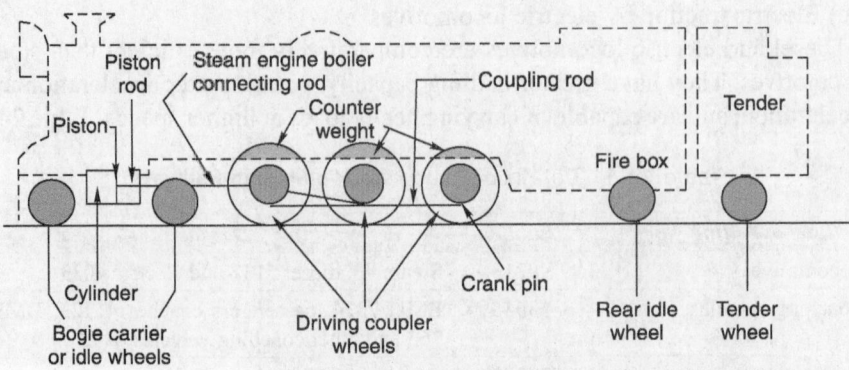

Fig. 24.1 Steam locomotive

It may be noted here that the middle digit indicating the driving or coupled wheels is the most important of all. The hauling capacity of the locomotive is decided by the number of driving wheels.

24.3 CLASSIFICATION OF LOCOMOTIVES

The first locomotive, which hauled the first train in India in 1853 from Bombay (now Mumbai) to Thane, was imported from England and was named the Lord Falkland locomotive. The locomotive was of the 2-4-0 type. Since then, more than 5000 types of steam locomotives have been imported or manufactured for different gauges.

In 1929, Indian Railways standardized the classification of various locomotives. Locomotives are classified by a two-digit code as given in Table 24.3.

Table 24.3 Classification of locomotives

First digit, representing gauge		Second digit, representing service	
Digit	*Meaning*	*Digit*	*Meaning*
X or W	Broad gauge	A or L	Light passenger
Y	Metre gauge	B or P	Standard passenger
Z	Narrow gauge	C	Heavy passenger
		D or G	Standard goods
		E	Heavy goods
		M	Mixed goods and passenger
		U/M/T/W	Shuttle or shunting

In the case of diesel or electric locomotives, one additional alphabet—D, A, or C—is used to indicate diesel, ac electric, or dc electric engines, respectively. Thus, YP indicates a steam locomotive for MG for a standard passenger train. WAG indicates a broad gauge ac electric locomotive used to haul a goods train. Table 24.4 enlists the salient features of some of the typical locomotives in use on Indian Railways.

Table 24.4 Salient features of some typical locomotives

Class	Service	Type	Axle load (tonnes)	Overall length (m)	IPH (hp) (tonnes)	Max. speed (kmph)
				BG locomotives		
XA	Light passenger	4-6-2	13.2	19.2	933	75
XB	Standard passenger	4-6-2	17.3	23.1	1257	75
XC	Heavy passenger	4-6-2	20.1	23.2	1379	95
XD	Standard goods	2-8-2	17.3	23.3	1115	75
XE	Heavy goods	2-8-2	22.9	24.0	1668	65
				MG locomotives		
YB	Standard passenger	4-6-2	16.2	18.3	700	75
YC	Heavy passenger	4-6-2	12.2	18.3	700	60
YD	Standard goods	2-8-2	10.2	18.2	717	55

(Contd.)

Table 24.4 *(Contd.)*

Class	Service	Type	Axle load (tonnes)	Overall length (m)	IPH (hp) (tonnes)	Max. speed (kmph)
				NG locomotives		
ZB	Standard passenger	2-6-2	6.1	12.8	689	40
ZE	Heavy goods	2-8-2	8.1	14.6	574	50

24.3.1 Steam Locomotives

Figure 24.1 shows the sketch of a typical steam locomotive and the various parts of a steam locomotive are described in detail in the following subsections.

Boiler

A boiler consists of a fire box, a barrel, and a smoke box. Coal is fired in the fire box, where water is converted into steam. The smoke and gases from the fire box pass through the smoke pipe and finally get discharged through the chimney. The chemical energy of the fuel is thus converted into the heat energy of the steam.

Steam engine

The steam engine consists of a cylinder, a piston, and other moving parts. It converts the heat energy of the steam into rotary energy.

Wheel and frame

The locomotive has idle as well as loaded wheels. The frame is supported on the wheels; it carries the boiler and also holds the steam engine and other parts of the locomotive.

Tender

A tender is provided behind the locomotive to store fuel. The tender is a type of mini-wagon that is supported on wheels and is coupled with the main locomotive.

General

Indian Railways has phased out all steam locomotives and no steam locomotive is currently being manufactured by or used on the railways.

24.3.2 Diesel Locomotives

The diesel locomotive works on the principle of a diesel engine. It uses diesel oil as fuel and combustion takes places inside a cylinder. The air inside the cylinder is compressed, raising the temperature of the air. The fuel is injected inside the cylinder, causing spontaneous combustion. The diesel engine mostly comprises four-stroke cycles consisting of suction, compression, ignition, and exhaust. The energy thus generated is utilized for driving the locomotive.

The horsepower generated in a diesel locomotive is transmitted to its wheels in the following manner:

(a) Mechanical transmission in the case of conventional diesel locomotives
(b) Hydraulic transmission in the case of diesel-hydraulic locomotives
(c) Electric transmission in the case of diesel-electric locomotives

Diesel-electric locomotives have become quite popular because of economy in operation and high tractive efforts even at low speeds. All the standard main line BG and MG locomotives on Indian Railways function through electric transmission. The different types of diesel locomotives in use on Indian Railways are as follows:

Broad gauge   WDM1, WDM2, WDM4, WDG4, WDP4, WDM3, WDG2, WDP2, WDS4 and WDM1.

Metre gauge   YDM1, YDM2 and YDM4

Indian Railways has, however, recently standardized WDM2 diesel locomotives for BG lines and YDM4 diesel locomotives for MG lines. Both these locomotives are of the ALCO design and are currently being manufactured in DLW Varanasi. The WDM2 locomotives have a horsepower of 2600 and are capable of hauling 3600 tonnes of train load at a speed of 72 m per hour (115 km per hour) on a level track. The maximum speed that can be achieved using this locomotive is 120 km per hour.

The latest series of locomotives are the WDG2 locomotives for goods traffic and WDP1 and WDP2 locomotives for passenger traffic. These locomotives have greater horsepower and superior performance capabilities.

WDP4 Diesel Locomotive Baaz, which is now at New Jalpaiguri (Courtesy: Metasur. This file is licensed under the Creative Commons Attribution-Share Alike 2.5 Generic license; http://creativecommons.org/licenses/by-sa/2.5/deed.en)

WAP-7 numbered 30255 hauls train number 18237 Chattisgarh Express (Courtesy: Shan H. Fernandes. This file is licensed under the Creative Commons Attribution-Share Alike 3.0 Unported license; http://creativecommons.org/licenses/by-sa/3.0/deed.en)

24.3.3 Electric Locomotives

In electric locomotives, movement is brought about by means of electric motors. These motors draw power from an overhead distribution system through pantographs (joined frameworks conveying current to an electric train from overhead wines) mounted on the locomotives. There are different systems for feeding power to these locomotives, namely 1500-V dc, 750-V dc, 25-kV ac single-phase, and ac three-phase. The most economical way of transmitting and distributing power is to directly transmit ac power at industrial frequency, whereas the motors most suited for electric traction are dc series motors. Indian Railways has locomotives that have the best of both systems, i.e., locomotives with dc series motors that collect power at 25 kV ac single-phase at an industrial frequency of 50 cycles. The voltage is stepped down to a lower voltage by means of a transformer present inside the locomotive, converted from 1500 V or 750 V ac to dc by means of silicon rectifiers, and this power is then fed to the dc series motors inside the electric locomotives. This system has been adopted in India since 1957 and is based on the system adopted by French National Railways.

Earlier, Indian Railways had an electric traction system that worked using 1500-V dc and 3000-V ac. The electrification of traction in India was first undertaken between 1925 and 1932 at 1500 V dc. This consisted of 304-route kilometre on Central Railway from Bombay VT to Igatpuri and Pune, 62 km on Western Railway between Churchgate and Virar, and 32 km between Madras beach and Tambaram. These sections still use a 1500 V dc system to power their locomotives. In addition, 142 route km around Calcutta were electrified using a 3000-V dc system in 1954–58, which has subsequently been converted into a 25-kV ac system.

The details of the four different types of electric locomotives in use on Indian Railways (IR) are given in Table 24.4.

Table 24.4 Details of electric engines on Indian Railways

Gauge	ac or dc	Type of electric locomotives
BG	ac	WAM1, WAM2, WAM3, WAM4, WAP1 WAG1, WAG2, WAG3, WAG4, and WAG5
BG	ac	WCM1, WCM2, WCM3, WCM4, WCG1
BG	ac/dc	WCG2, WCM1
MG	ac/dc	YAM1

The WAM1, WAG1, WAG3, and WAG4 locomotives are based on European design whereas the WAM4, WAP1, WCM1, and WAG5 locomotives have been designed in India with a six-wheel arrangement for their bogies, which is similar to diesel WDM locomotives. Some of these locomotives are capable of hauling the latest type of BOXN wagons with air brake arrangement.

It may be mentioned here that the WAP1 is a passenger locomotive that has a vacuum as well as an air brake system and is designed to run at speeds of upto 130 km per hour. The Rajdhani Express on the Delhi–Howrah route is hauled by such locomotives. The horsepower, weight, and maximum service speed of different types of locomotives are given in Table 24.5.

Table 24.5 Horsepower, weight, and service speed of various locomotives

Type of locomotive	Horsepower	Weight of locomotive (tonnes)	Maximum service speed (km/h)
WAP1	3800	108.80	130
WAM1	2840	75.20	112
WAM2	2812	76.03	112
WAM4	3640	112.60	120
WAG1	2840	85.20	80
WAG2	3200	85.20	80
WAG4	3150	87.60	80
WAG5A	3850	118.80	80
YAM1	1600	52.00	80

24.4 PREVENTIVE MAINTENANCE OF LOCOMOTIVES

The efficient operation of railways depends on the reliability and availability of locomotives. The proper and efficient maintenance of locomotive is the basis for economical train operation. Maintenance practices have evolved on the basic principle that necessary attention should be paid to all assets before deterioration through wear and tear makes them prone to failure. This is known as *preventive maintenance*. The preventive maintenance of locomotives can be broadly classified into periodical overhaul and schedule repairs.

24.4.1 Steam and Diesel Locomotives

The prescribed life of steam and diesel locomotives is 40 and 36 years, respectively. In view of the long lifespan, it is imperative that both these assets are properly and periodically maintained to ensure that there is no deterioration in their condition. The preventive maintenance of steam and diesel locomotives is done in the following manner:

Periodical overhaul

Periodical overhaul (POH) of these locomotives is undertaken in railway repair workshops. The frequency of the periodical overhauling of steam and diesel locomotives is as follows:

Diesel locomotives   6 years
(or 0.8 million km for BG and 0.6 million km for MG lines)
Steam locomotives
(a) Passenger   0.3–0.35 million km
(b) Goods   0.2–0.25 million km
(c) Inferior services   5 years (or about 0.2 million km)

During POH, the loco is completely stripped and all its parts and components are repaired and/or replaced, as their condition warrants. After POH, the locomotives are in an 'almost new' condition.

Schedule shed maintenance

Various examination schedules have been drawn up as a part of preventive maintenance so that specific components and parts of the locomotives may be given need-based attention at intervals. The examination schedules list all the various aspects that should be taken care of at the time of maintenance, which are dependent on the wear and tear of the components. The total number of kilometres that should have been covered before these examinations have also been specified.

Diesel locomotives

A trip (or weekly) schedule, a fortnightly schedule, a monthly schedule, a three-month schedule, a six-month schedule, an yearly schedule, a three-year schedule, and a six-year schedule (POH) have been drafted for diesel locomotives.

These preventive maintenance schedules have been worked out on the basis of experience and informed investigations into the causes leading to the breakdown of locomotives. Compared to a steam locomotive, the diesel traction unit is far more complex, usually incorporating an indirect control system with much more elaborate ancillary equipment. As a result of this greater complexity, it is more liable to be rendered unserviceable by the failure of a particular component. Therefore, in order to ensure a high standard of reliability, various schedules have been instituted for its inspection and maintenance. The 'servicing' of minor schedules, i.e., trip, fortnightly, monthly, etc., involving routine inspections, minor attention, and the like, takes about 4 to 12 hours. The major schedules take a longer time, lasting from 6 to 14 days, since they require repair and replacement of major components.

24.4.2 Electric Locomotives

The schedules of inspection and checks given in Table 24.6 are generally followed on Indian Railways to ensure the proper maintenance of electric locomotives.

Table 24.6 Schedule of maintenance of electric locomotives

Frequency	Code	Time period alloted for completion	Nature of attention
Weekly/trip (within 1500 km)	Trip	1 hr	Weekly or trip inspection
15 days	Fortnightly	1 to 2 hrs	Fortnight inspection
Monthly	IA	2 hrs	Incidental inspection
Two months	IB	8 hrs	Complete/partial inspection
Four months	IC	8 hrs	Complete inspection
Every third IC	AOH	16 hrs	Annual inspection
At 300,000 km or after 3 years	IDH	15 days	General inspection or intermediate overhauling
At 600,000 km or after 6 years	POH	30 days	Thorough overhauling or periodical overhauling

The frequency and duration of various schedules have been defined after taking the several failures of locomotives, the existence of indigenous components, and the atmospheric levels of heat and dust in the country into consideration. Compared to steam and diesel locomotives, electric locomotives include more contactors, relays, and auxiliaries/equipment. Electrical quantities and changes in the condition of the locomotive are not physically visible. Therefore, it is essential to check the equipment, contactors, and relays periodically to ascertain the condition of the various equipment without too much of dismantling. Most of the equipment gets inspected during four monthly inspections and, therefore, these are excluded from the annual inspections. During such inspections, the equipment is disconnected from mechanical fixtures, assemblies are stripped and cleaned, and worn-out or damaged parts are replaced and reconnected.

Trip or weekly and fortnightly inspections are conducted at outstation running sheds, and all other schedule inspections, except POH, are carried out at sheds. POH is, however, carried out in the workshops.

24.5 ROLLING STOCK

Rolling stock includes locomotives, passenger coaches, goods wagons, and all other types of coaches and wagons, such as electric multiple units (EMUs), diesel rail cars, and special wagons, such as BOX wagons. This section gives some of the details of passenger coaches and goods wagons.

24.5.1 Coaching Stock

The different types of passenger coaches include the electric multiple units that are part of suburban trains and conventional coaches, such as II class, I class, II sleeper, ac three tier, ac two tier, and ac I class coaches.

In 2011, Indian Railways had a stock of about 7334 EMU coaches and 45,123 conventional coaches. These coaches have three basic structural designs.

(a) Integral coaches built by the Integral Coach Factory (ICF), Perambur, Chennai
(b) Integral coaches built by Bharat Earth Movers Ltd (BEML), Bengaluru
(c) Non-integral wooden body coaches made in accordance with the Indian Railways standard design (IRS)

Some of the salient features of these three types of coaches are given in Table 24.7.

Table 24.7 Salient features of IRS, ICF, and BEML coaches

	IRS	*ICF*	*BEML*
Type of body	Wooden	Light metal	Light metal
Type of structure	Wooden body resting on rigid steel underframe	All metal, all welded integral coach of a tabular construction	Similar to an ICF coach except that here load is taken by the floor
Coach dimensions			
Overall width	3250 mm	3250 mm	3250 mm
Overall length	21,031 mm	21,337 mm	21,336 mm
Overall height	3886 mm	3886 mm	3926 mm
Damage during collisions and derailments	Extensive damage	Less damage because of the presence of an anti-telescopic structure in the end walls to absorb shocks	Same as ICF

24.5.2 Goods Wagons

Goods wagons are primarily meant for the carriage of goods traffic. Indian Railways currently has a stock of about 0.229 million goods wagons with a haulage capacity of about 10 million tonnes. These goods wagons mostly consist of covered and open wagons as well as special wagons, such as BOX wagons for carrying coal and other bulk traffic.

Until the middle of the last century, practically all the goods traffic was transported in general-purpose wagons or in covered, open high-sided, and open low-sided wagons. The standard wagon on the broad gauge was a four wheeler with a 22.19 tonnes haulage capacity, while the standard wagon on the metre gauge weighed 5.69 tonnes and had the capacity of carrying 18.69 tonnes of goods. Recently, a number of new bogie wagons has been designed and put into service, which lay emphasis on a higher payload and on the provision of facilities for the loading and unloading of special type of traffic. These include the BOX, BCX, BOBX, BOY, BOXN, CRT wagons. It has been decided that only bogie wagons will be put into service on the Railways, as the four-wheeler wagon is a non-viable unit in the present context of the bulk movement of commodities. In the above-mentioned classification of wagons, B stands for bogie wagon, C for centre discharge, O for open wagon, X for high-sided (also for both centre and side discharge), and Y for low-sided walls. N is used for air braked, C for covered wagon, R for rail-carrying wagon, and T for transition coupler. The B indication is sometimes omitted as all new wagons are bogie stock.

24.6 BRAKE SYSTEMS

Brakes in the locomotive are applied to stop a moving train. There are two basic types of brakes.

Compressed air brakes

This brake system is made up of a brake cylinder containing a piston and lever arrangement, which is provided under each vehicle. A brake pipe running is also provided under each vehicle, which extends from the main reservoir to the locomotive. The reservoir is provided in that part of the locomotive where compressed air is fed in through the air pump (Fig. 24.2). When compressed air is admitted into the system, the movement of the piston results in the application of the brakes.

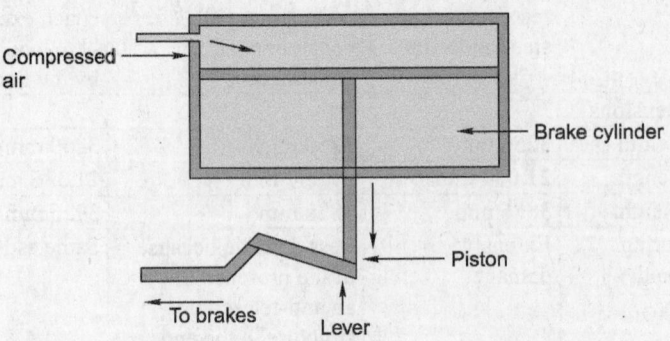

Fig. 24.2 Compressed air brake system

Vacuum brakes

The equipment consists of a vacuum brake cylinder with a piston and lever arrangement provided under each vehicle. The cylinder is connected to the train pipe running from one end of the vehicle to the other. A direct admission valve provided with each coach is also used for applying brakes in the case of an emergency. A vacuum of about 20 inches of mercury is maintained in the vacuum cylinder on one side of the piston in order to operate the brake system. This vaccum exerts an effective working pressure of about 10 psi on the piston when the brakes are applied. The vacuum cylinders are designed to supply the required amount of brake power at the wheels by making use of this vacuum (Fig. 24.3).

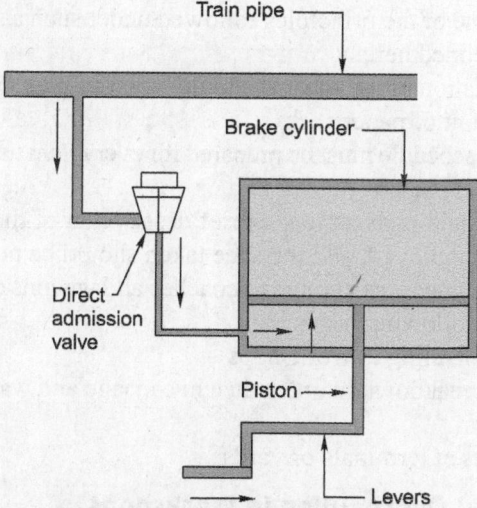

Fig. 24.3 Vacuum brake system

24.6.1 Vacuum Brakes Versus Air Brakes

It has become necessary to run longer, heavier, and faster trains on account of the constant increase in traffic on the Railways. Therefore, whereas the prevalent automatic vacuum brake system is continuously being improved, the feasibility of using air brakes is also being explored. The fundamental considerations which govern the performance of a railway braking system are summarized here.

(a) The fast propagation of air along the entire train length to ensure the uniform application of brakes
(b) A very rapid initial filling up of the cylinder so that the brake blocks are quickly brought in contact with the wheel to start the process of deceleration
(c) A slower subsequent filling up of the cylinders to allow a gradual retardation of the wagons
(d) A fast release of brakes throughout the train length so that trains can restart quickly after having come to a stop

The air brake system claims superiority over the existing vacuum brake system in all these spheres. However, air brake system has a few comparative disadvantages, which are as follows.

(a) A long train gives a sluggish response to the control of the driver, as the time required for the compressed air to reach the last vehicle is considerable.

(b) If for some reason the train gets divided into two parts, the front portion of the train receives the power to operate the air brakes whereas the rear portion does not receive it. This can result in an accident.

24.7 MAINTENANCE OF COACHES AND WAGONS

The prescribed economic lifespan of coaches and wagons is 30 years and 40 years, respectively. Coaches and wagons go through cycles of periodic maintenance in order to ensure that they are in good working order throughout their lifespan. As stated earlier, the basic principle followed under the preventive maintenance system is that the equipment should be paid the necessary attention so that there is no wear and tear and consequent deterioration and the wagons do not become prone to failure. Some of the principles followed under such a scheme of planned maintenance are outlined here.

(a) The implementation of the scheme should be as simple as possible involving minimum amount of paperwork.
(b) A maintenance schedule must be prepared for every item that requires planned preventive maintenance.
(c) When the schedule is executed, a brief description of the action taken, the type of labour employed, and the time taken should be noted down.

Preventive maintenance, as applied to coaches and wagons on the Railways, is classified under the following heads.

(a) Periodical overhauling in workshops
(b) Schedule and breakdown maintenance in carriage and wagon depots or sick lines
(c) Running repairs at terminals or yards

24.7.1 Periodical Overhauling in Workshops

Coaching stock is periodically overhauled in workshops at prescribed intervals based on the type of service for which it is used. The frequency of POH varies from 12 to 24 months. Such an overhaul consists of lifting, which involves a thorough examination of all components, and the execution of such repairs as may be necessary to enable the vehicle to remain in service until the next periodical overhaul. The prescribed intervals for the periodical overhauling of goods stock are presented in Table 24.8.

Table 24.8 Periodicity of POH of goods stock

Type of wagon	Interval of first POH (yr)	Interval of subsequent POH (yr)
BOX wagons	4.5	4.0
Brake vans	4.0	1.0
Containers	1.5	1.0
Departmental stock, etc.	4.0	4.0
All other wagons	4.0	3.5

24.7.2 Schedule Maintenance

The maintenance of different types of rolling stock as per approved schedules is described below.

Coaching stock

Between consecutive overhauls, coaches are serviced in coaching depots on the basis of a schedule known as *intermediate lifting*. This is normally done after six months of the date of the last POH or lifting. During the process of lifting, trolleys, underframe members, the body and floor of the coaches, etc., are thoroughly examined and the necessary repair and replacement of the components of the running gear are carried out. The side bearer/centre pivot oil is also replenished. Similarly, coaching stock fitted with plain-bearing axle boxes (as distinguished from roller-bearing axle boxes) undergo the process of 'repacking' at regular intervals as follows:

Mail and express train coaches	Two months
Passenger train coaches	Three to four months
Other coaches	Four to six months

Goods stock

All axle boxes with bearings of the entire goods stock are periodically repacked at maintenance depots according to prescribed schedules. In addition, the axle boxes are packed whenever a wagon is lifted off its wheels. Roller-bearing axle boxes are attended to in workshops during overhaul maintenance.

Coaching and goods stock

There are a few schedules that are common to both coaching and goods stock. Some of the important maintenance work done in compliance with these schedules are the following:

(a) *Vacuum brake cylinders* The vacuum brake fitted on the stock must be overhauled at intervals of 18 months in the case of coaching stock and 18 to 24 months in the case of goods stock. Normally there is a separate section for vacuum cylinder overhauling in the maintenance depot.

(b) *Axle boxes* The axle boxes (plain-bearing) of coaching and goods stock are oiled periodically. The coaching stock on passenger trains is oiled every 15 days. A goods stock with permanently secured face plates must be lubricated with at least 225 g of oil per axle box at intervals of one month, either in maintenance depots or in the originating/terminating yards.

(c) There are many other scheduled maintenance works, such as alarm chain apparatus testing, tank wagon valve testing, water tank painting, and which are implemented in the maintenance depots at the prescribed frequencies.

24.7.3 Breakdown Maintenance

The maintenance depots attend to all the breakdowns or failures of the rolling stock. Breakdowns are unusual incidents resulting either from human failure or due to the failure of the equipment. Some of these breakdowns even result in accidents.

The most common failure of a coach or a wagon is the one that occurs as a result of the axle box heating up in the course of the run, popularly known as 'hot box'. Every journal (the part of the axle at its ends, which bears the weight of the wheels) that becomes hot during the run of the train, thereby necessitating the detachment of the coach or wagon from the train before it can reach its booked destination, is considered a hot box.

There are many other types of breakdowns that must be attended to in maintenance depots to ensure the safety of the wagon or box. These are as follows:

(a) Cracked wheel tyre
(b) Expanded axle guard
(c) Bent sole bar
(d) Dead buffer
(e) Leaking water tank
(f) Wheel defects, etc.

24.7.4 Running Maintenance at Terminals and Yards

The rolling stock is on the move often. Since the major scheduled maintenances, such as overhauling are done only when due, and call for the temporary immobilization of the stock, the normal maintenance of the stock is done in the course of its service at terminals and yards.

This maintenance mainly involves an examination or inspection of the coaches or wagons of the train and is carried out in two stages.

Stage I: At the approach to the terminal or yard when the train is still in motion

Stage II: At the terminal or yard where the train normally stops

The first stage of the inspection helps in discovering those defects of the under-gear that are indicated by unusual sounds or disjointed/broken hanging components. For example, a flat tyre is indicated by the heavy thudding of the train as it passes over the rails. In the second stage, each coach or wagon is thoroughly examined and all defects, both exterior and interior, are recorded for rectification. Coaches or wagons found with defects that can be attended to only in maintenance depots are marked for handling there. The remaining coaches and wagons are attended to at the terminal (washing line) or in the yard itself.

24.8 DESIGN FEATURES OF MODERN COACHING AND GOODS STOCK

The introduction of high-speed trains on Indian Railways has necessitated some changes in the design of the rolling stock. The important design features of modern coaches and goods wagons are described in the following sections.

24.8.1 Coaching Stock

The need for increasing the number of passengers in every coach along with providing them with greater comfort, introducing more coaches in every train, and running trains at progressively increasing speeds has necessitated the modernization of the rolling stock.

The coaches manufactured by the Integral Coach Factory, Chennai are of the latest design. The body of the coach is made of a lightweight metal and has a tubular structure. The trough underflooring, the sides, and the roof are welded together to form an integral unit. The floor consists of cross-beams, specially designed head stocks, sole bars, etc., all of which are welded together. The steel trough flooring is run under the cross-beams and is welded to the head stocks and the sole bars. The advantage in this is that a trough section can withstand greater longitudinal loads and still help in keeping a check on the weight of the coach. The curved body of the coach reduces wind resistance and helps stability. The ends are also specially designed to absorb heavy impact shocks. Further, at the ends, the under-flooring is designed to be collapsible so that in the case of an accident, it absorbs the maximum force of the impact. The tubular structure ensures that the coach does not collapse inwards and thus ensures maximum passenger safety against heavy impacts. ICF coaches are, therefore, called 'anti-telescopic' coaches.

Like the coach body, the rest of the ICF bogie is also of a welded, lightweight design with a 9' 6" rigid wheel base running on roller bearings and mounted on pendulum type axle boxes. The axle box guides are fitted with dash pots for absorbing the effects of the lateral bumps. Lateral shock absorbers prevent or absorb lateral shocks. All these fittings and the coiled spring design of the bogies ensure that they provide a very smooth side.

24.8.2 Goods Stock

The BOX-type open wagon is the most significant of all the new wagons. It has a tare weight of 25 tonnes and a haulage capacity of about 55 tonnes. It has a robust structure and is fitted with centre buffer couplers at the ends that provide automatic coupling. It runs on two bogies fitted with roller-bearing axle boxes. These wagons are most suited for long block rakes (consisting of 35 to 45 box wagons per train) carrying bulk commodities at high speeds.

The covered version of this type of wagon is the BCX wagon measuring 14.5 m in length, about 3 m in width, and about 2.5 m in height from the inside and with a haulage capacity of 52 to 78 tonnes. It has two wing-cum-flap doors on each side and is provided with centre-buffer couplers.

The effort behind the design of these wagons has been directed towards increasing their size. The average haulage capacity of a wagon has progressively increased in a bid to absorb the growth in traffic as much possible without increasing the line capacity, which is costly in terms of both money and time.

24.9 MANUFACTURE OF ROLLING STOCK

Indian Railways has developed adequate capacity for manufacture of most of the rolling stock (locomotives, coaches, and wagons) indigenously.

Manufacture of locomotives

Chittaranjan Locomotive Works (CLW)   Chittaranjan manufactured 230 BG electric locomotives including 70 state-of-the-art 3-phase 6000 HP electric locos.

Diesel Locomotive Works (DLW)   Varanasi manufactured 267 BG diesel locomotives including 150 indigenous high power 4000 HP GM locomotives. Out of these, 35 diesel locomotives were supplied to non-railway customers (NRCs). It also exported spares worth ₹ 21.8 million.

Manufacture of coaches

Integral Coach Factory (ICF) Chennai   Manufactured 1,503 coaches including 535 Electric Multiple Units (EMUs) and 52 coaches for NRCs and exported spares worth ₹ 281.6 million.

Rail Coach Factory (RCF)   Kapurthala manufactured 1,576 coaches including 316 lightweight LHB coaches and 140 hybrid stainless stell coaches with high passenger comfort and amenities. It supplied 17 coaches to NRCs.

Manufacture of wagons

The entire requirement of wagons of Indian Railways is met by manufacturing units both in public, private sectors, railway workshops and PSUs under the administrative control of the Ministry of Railways. In 2010–11, the production of wagons totalled 16,638 in terms of Vehicle Units (VUs). Out of these, 1,570 VUs (including BLC wagons) were manufactured at railway workshops and the remaining 15,068 VUs including 1,347 VUs against WIS, BLC, and other private wagons were manufactured by wagon industry.

24.10 NEW TECHNOLOGY FOR DESIGN OF LOCOMOTIVES, COACHES, AND WAGONS

Some of the major research studies done for new technology by Research Design and Standards Organisation (RDSO) for desgin of rolling stock are as follows:

For improvement of safety
- Development of a new crashworthy design of 4500 HP WDG4 locomotive incorporating new technology to improve dynamic braking and attain significant fuel savings.

For passenger amenities improvement
- Development of 4500 HP Locomotive to provide clean and noise-free power supply to coaches from locomotive to eliminate the existing generator car of Garib Rath express trains
- Field trials conducted for electric locomotive hauling Rajdhani/Shatabdi express trains with head on generation (HOG) system to provide clean and noise-free power supply to end-on coaches

For operational efficiency
- Design and development of 5000 HP WDG5 diesel locomotive for faster, longer, and heavier trains
- Development of improved rubber side bearer for locomotive bogies to improve fatigue life and reliability
- Framing of technical specification of Guidance for Optimised Loco Driving (GOLD)—an in-cab service system to help loco pilots to save fuel
- Design and development of high-performance fuel-efficient 2300 HP freight/industrial/shunting locomotive (WDS7A)

SUMMARY

There are three types of traction on Indian Railways, namely steam traction, diesel traction, and electric traction. Steam traction has gradually been replaced by diesel and electric traction. Electric traction has many advantages over diesel and steam traction and is most suitable for high-speed and super-high-speed tracks. Locomotives, coaches, and wagons are the three components of rolling stock. Rolling stock for the modern tracks on Indian Railways is designed and manufactured by Integral Coach Factory, Chennai. Special attention is paid to the maintenance of these coaches, which have been modified to suit the growing needs of the Railways.

REVIEW QUESTIONS

1. With the help of a neat sketch, describe a vacuum brake and its working principle.
2. What are the requirements of a locomotive? Briefly describe the merits of the different types of tractions commonly in use in India.
3. What is a locomotive? What do you understand by a locomotive with a nomenclature 4-6-2? Describe the various components of a steam locomotive.
4. What do you understand by preventive maintenance of locomotives? What are the various schedules prescribed for the maintenance of steam and diesel locomotives?
5. Where are the coaches for Indian Railways manufactured? Compare the structural design of coaches manufactured at different sources. What do you understand by anti-telescopic coaches?
6. Describe the salient features of the vacuum brakes and compressed air brakes used on Indian Railways. Discuss their relative advantages/disadvantages.
7. What are the special design features of the modern coaching and goods stock on Indian Railways?

Choose the correct answer from the choices given.

8. The first steam locomotive that hauled the first train in India in 1853 from Bombay to Thane was named as:
   (a) Golden Locomotive (b) Lord Falkland
   (c) Queen Mary (d) Lord Wellingtrans
9. A steam locomotive does not have:
   (a) driving wheels (b) ideal wheels
   (c) pony wheels (d) boggie carriers
10. The continuous automatic brakes are of:
   (a) mechanical type (b) hydraulic type
   (c) vacuum type (d) electric type
11. POH is the abbreviation of:
   (a) preventive overhead equipment (b) periodic overhauling
   (c) preventive overhauling (d) none of these
12. High-sided bogie open wagons for movement of bulk commodities like coal and iron ore are:
   (a) BCX wagons (b) BOX wagons
   (c) BOXN wagons (d) BTPN wagons

CHAPTER
25

Train Resistance and Tractive Power

INTRODUCTION

Various forces offer resistance to the movement of a train on the track. These resistances may be the result of movement of the various parts of the locomotives as well as the friction between them, the irregularities in the track profile, or the atmospheric resistance to a train moving at great speed. The tractive power of a locomotive should be adequate enough to overcome these resistances and haul the train at a specified speed.

25.1 RESISTANCE DUE TO FRICTION

Resistance due to friction is the resistance offered by the friction between the internal parts of locomotives and wagons as well as between the metal surface of the rail and the wheel to a train moving at a constant speed. This resistance is independent of speed and can be further broken down into the following parts.

Journal friction This is dependent on the type of bearing, the lubricant used, the temperature and condition of the bearing, etc. In the case of roll bearings, it varies from 0.5 to 1.0 kg per tonne.

Internal resistance This resistance is consequential to the movement of the various parts of the locomotive and wagons.

Rolling resistance This occurs due to rail-wheel interaction on account of the movement of steel wheels on a steel rail. The total frictional resistance is given by the empirical formula

$$R_1 = 0.0016W \tag{25.1}$$

where R_1 is the frictional resistance independent of speed and W is the weight of the train in tonnes.

25.2 RESISTANCE DUE TO WAVE ACTION

When a train moves with speed, a certain resistance develops due to the wave action in the rail. Similarly, track irregularities such as longitudinal unevenness and differences in cross levels also offer resistance to a moving train. Such resistances are different for different speeds. There is no method for the precise calculation of these resistances but the following formula has been evolved based on experience:

$$R_2 = 0.00008\,WV \qquad\qquad (25.2)$$

where R_2 is the resistance (in tonnes) due to wave action and track irregularities on account of the speed of the train, W is the weight of the train in tonnes, and V is the speed of the train in kmph.

25.3 RESISTANCE DUE TO WIND

When a vehicle moves with speed, a certain resistance develops, as the vehicle has to move forward against the wind. Wind resistance consists of side resistance, head resistance, and tail resistance, but its exact magnitude depends upon the size and shape of the vehicle, its speed, and the wind direction as well as its velocity. Wind resistance depends upon the exposed area of the vehicle and the velocity and direction of the wind. In Fig. 25.1, V is the velocity of wind at an angle θ. The horizontal component of wind, $V\cos\theta$, opposes the movement of the train. Wind normally exerts maximum pressure when it acts at an angle of 60° to the direction of movement of the train.

Wind resistance can be obtained by the following formula:

$$R_3 = 0.000017\,AV^2 \qquad\qquad (25.3)$$

where A is the exposed area of vehicle (m^2) and V is the velocity of wind (kmph).

Studies also support the fact that the important factors that affect wind resistance are the exposed area of the vehicle and the relative velocity of the wind, vis-à-vis that of the vehicle. In fact, wind resistance depends upon the square of the velocity of the wind. The following formula has been empirically established on the basis of studies.

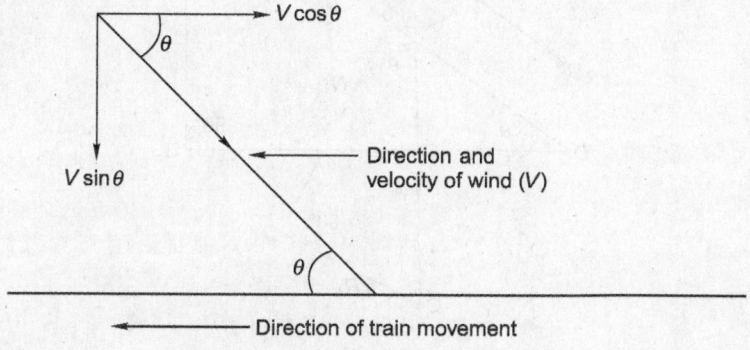

Fig. 25.1 Resistance due to wind

$$R_3 = 0.0000006 \, WV^2 \tag{25.4}$$

where R_3 is the wind resistance in tonnes, V is the velocity of the train in km per hour, and W is the weight of the train in tonnes.

25.4 RESISTANCE DUE TO GRADIENT

When a train moves on a rising gradient, it requires extra effort in order to move against gravity as shown in Fig. 25.2.

Assuming that a wheel of weight W is moving on a rising gradient OA, the following forces act on the wheel.

(a) Weight of the wheel (W), which acts downward

(b) Normal pressure N on the rail, which acts perpendicular to OA

(c) Resistance due to rising gradient (R_4), which acts parallel to OA

These three forces meet at a common point Q and the triangle QCD can be taken as a triangle of forces. It can also be geometrically proved that the two triangles QCD and AOB are similar. From $\triangle$QCD,

$$R_4 = W \sin\theta$$

From $\triangle$OAB,

$$R_4 = W \times \frac{AB}{OA}$$

In actual practice, the gradients are very small and, therefore, OA is approximately equal to OB. Therefore,

$$R_4 = W = \frac{AB}{OB} = \frac{W \times \% \text{slope}}{100}$$

$$R_4 = \frac{W \times \% \text{slope}}{100} \tag{25.5}$$

This means that if the weight of the train is 50 tonnes and the slope is 1 in 50 (2%), the resistance due to gradient is

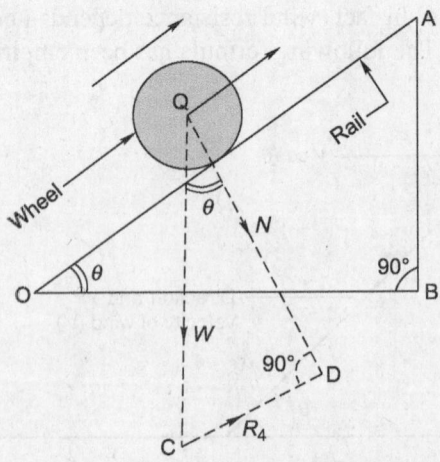

Fig. 25.2 Resistance due to gradient

$$R_4 = \frac{W \times \% \text{slope}}{100} = \frac{50 \times 2}{100} = 1 \text{tonne}$$

It may be noted here that when a train ascends a slope, extra effort is required to overcome the resistance offered by the gradient. The position is, however, reversed when the train descends a slope and the resistance offered by the gradient helps in the movement of the train.

25.5 RESISTANCE DUE TO CURVATURE

When a train negotiates a horizontal curve, extra effort is required to overcome the resistance offered by the curvature of the track. Curve resistance is caused basically because of the following reasons (Fig. 25.3):

(a) The vehicle cannot adapt itself to a curved track because of its rigid wheel base. This is why the frame takes up a tangential position as the vehicle tries to move in a longitudinal direction along the curve as shown in Fig. 25.3. On account of this, the flange of the outer wheel of the leading axle rubs against the inner face of the outer rail, giving rise to resistance to the movement of the train.

(b) Curve resistance can sometimes be the result of longitudinal slip, which causes the forward motion of the wheels on a curved track. The outer wheel flange of the trailing axle remains clear and tends to derail. The position worsens further if the wheel base is long and the curve is sharp.

(c) Curve resistance is caused when a transverse slip occurs, which increases the friction between the wheel flanges and the rails.

(d) Poor track maintenance, particularly bad alignment, worn-out rails, and improper levels, also increase resistance.

(e) Inadequate superelevation increases the pressure on the outer rail and, similarly, excess superelevation puts greater pressure on the inner rails, and this also contributes to an increase in resistance.

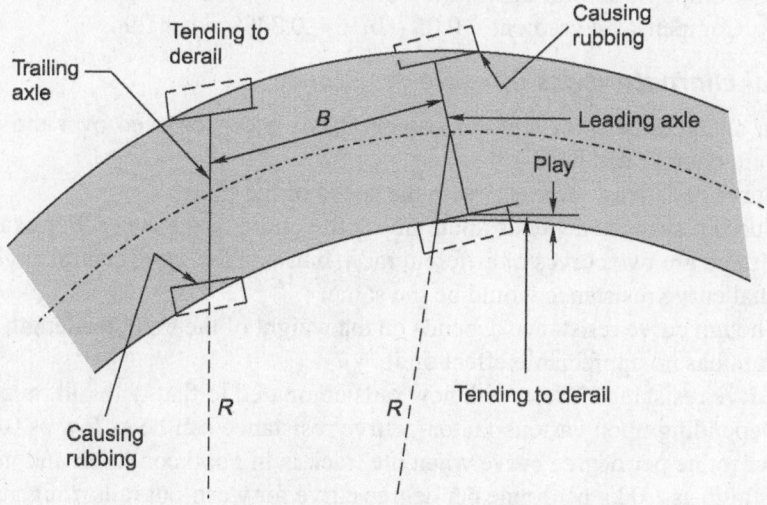

Fig. 25.3 Resistance due to curvature

The value of curve resistance can be determined by the following equation:

$$Curve\ resistance = C\frac{FG}{R} \tag{25.6}$$

where F is the force of sliding friction, G is the gauge of the track, R is the mean radius of the curve, and C is the constant, which is dependent on various factors. This equation indicates that
 (a) curve resistance increases with increase in gauge width and
 (b) resistance is inversely proportional to the radius, i.e., it increases with an increase in the degree of the curve.

Empirical formulae have been worked out for curve resistance, which are as follows:

Curve resistance for BG (R_5) = 0.0004 WD (25.7)
Curve resistance for MG (R_5) = 0.0003 WD (25.8)
Curve resistance for NG (R_5) = 0.0002 WD (25.9)

where W is the weight of the train in tonnes and D is the degree of the curve. It means that for a 4° curve on a BG line, the curve resistance for a train weighing 250 tonnes would be $0.0004 \times 250 \times 4 = 0.4$ tonne.

Compensated gradient for curvature

Curve resistance is quite often compensated or offset by a reduction in the gradient. In this way, the effect of curve resistance is translated in terms of resistance due to gradient. The compensation is 0.04 per cent on BG, 0.03 per cent on MG, and 0.02 per cent on NG lines for every 1° of the curve. This will be clear through the solved example given below.

Example 25.1 Calculate the compensated gradient on a BG track with a 4° curvature on a ruling gradient of 1 in 200.

Solution
 Ruling gradient = 1 in 200 = 0.5%
 Compensation for curvature = 0.04 × 4 = 0.16%
 Compensated gradient = 0.05 – 0.16 = 0.34% = 1 in 294

Special characteristics of curve resistance

Certain characteristics of curve resistance have been observed over the years. These are enumerated here.
 (a) Curve resistance increases with the speed of the train.
 (b) Curve resistance depends upon the central angle of the curve. For example, if there are two curves of different radii, but with the same central angle, the total curve resistance would be the same.
 (c) Though curve resistance depends on the weight of the train, the length of the train has no appreciable effect on it.
 (d) Curve resistance is less with new rails compared to that with old rails.
 (e) Depending upon various factors, curve resistance can be as low as 0.20 kg per tonne per degree curve when the track is in good condition and may go as high as 1.0 kg per tonne per degree curve for worn-out rails, rough tracks, and other unfavourable conditions of the track.

25.6 RESISTANCE DUE TO STARTING AND ACCELERATING

Trains face these resistances at stations when they start, accelerate, and decelerate. The values of these resistances are as follows:

Resistance on starting, $R_6 = 0.15\,W_1 + 0.005\,W_2$ (25.10)

Resistance due to acceleration, $R_7 = 0.028\,aW$ (25.11)

where W_1 is the weight of the locomotive in tonnes, W_2 is the weight of the trailing vehicles in tonnes, W is the total weight of the locomotive and vehicle in tonnes, i.e. $W_1 + W_2$, and a is the acceleration, which can be calculated by finding the increase in velocity per unit time, i.e., $(V_2 - V_1)/t$, where V_2 is the final velocity, V_1 is the initial velocity, and t is the time taken.

Table 25.1 summarizes the various resistances faced by a train.

Table 25.1 Details of various resistances

Nature of resistance	Value of resistance	Remarks
Resistance due to friction (R_1)	$0.0016\,W$	Resistance independent of speed
Resistance due to wave action, track irregularities, and speed (R_2)	$0.00008\,WV$	Resistance dependent on speed
Resistance due to wind, or atmospheric resistance (R_3)	$0.0000006\,WV^2$	Resistance dependent on square of speed
Resistance due to gradient (R_4)	$\dfrac{W \times \%\,\text{of slope}}{100}$	Resistance dependent on slope
Curve resistance for BG (R_5)	$0.0004\,WD$	Resistance dependent on degree of curve
Resistance due to starting (R_6)	$0.15W_1 + 0.005\,W_2$	Resistance dependent on weight
Resistance due to acceleration (R_7)	$0.028\,aW$	Resistance dependent on weight and acceleration

25.7 TRACTIVE EFFORT OF A LOCOMOTIVE

The tractive effort of a locomotive is the force that the locomotive can generate for hauling the load. The tractive effort of a locomotive should be enough for it to haul a train at the maximum permissible speed. There are various tractive effort curves available for different locomotives for different speeds, which enable the computation of the value of tractive effort. Tractive effort is generally equal to or a little greater than the hauling capacity of the locomotive. If the tractive effort is much greater than what is required to haul the train, the wheels of the locomotive may slip.

A rough assessment of the tractive effort of different types of locomotives is provided in the following sections.

25.7.1 Steam Locomotive

The tractive effort of a steam locomotive can be calculated by equating the total power generated by the steam engine to the work done by the driving wheels.

Assume P to be the difference in steam pressure between the two sides of the cylinder, A the area of the piston of the engine, d the diameter of the piston of the engine, L the length of the stroke of the engine, D the diameter of the wheel of the locomotive, and Te the mean tractive effort of the locomotive. Work done by a two-cylinder steam engine

$$= 2 \times \textit{difference in steam pressure} \times \textit{area of the piston} \times$$
$$\quad 2 \times \textit{length of the stroke}$$
$$= 2P \times A \times 2L$$
$$= 2P \times \frac{(\pi d^2)}{4} \times 2L = \pi P d^2 L \qquad (25.12)$$

Work done in one revolution of the driving wheel of the locomotive:

$$= \textit{tractive effort} \times \textit{circumference of the wheel}$$
$$= T_e \times \pi D \qquad (25.13)$$

On equating Eqs (25.12) and (25.13),

$$\pi P d^2 L = \text{Te} \times \pi D$$

we get $\quad Te \quad = \dfrac{P d^2 L}{D} \qquad (25.14)$

It is clear from Eqn (25.14) that tractive effort increases with an increase in steam pressure difference and the diameter and length of the piston, but decreases with an increase in the diameter of the driving wheel of the locomotive.

25.7.2 Diesel Locomotive

Tractive effort of a diesel-elective locomotive can be assessed by the following empirical formula.

$$Te = \frac{308 \times \text{RHP}}{V} (\text{kg}) \qquad (25.15)$$

where Te is the tractive effort of a diesel-electric locomotive, RHP is the rated horsepower of the engine, and V is the velocity in km per hour.

25.7.3 Electric Locomotive

The tractive effort of an electric locomotive varies inversely with the power of speed. The empirical formulae for calculating the approximate value of tractive effort are as follows:

For an dc electric locomotive: Te $= a/V^3$ $\qquad (25.16)$
For an ac electric locomotive: Te $= a/V^5$ $\qquad (25.17)$

where a is a constant depending upon the various characteristics of the locomotive.

The important characteristics of three types of tractions are compared in Table 25.2.

25.8 HAULING POWER OF A LOCOMOTIVE

The hauling power of a locomotive depends upon the weight exerted on the driving wheels and the friction between the driving wheel and the rail. The coefficient of

Table 25.2 Comparison of steam, diesel, and electric traction

Characteristics	Steam locomotive	Diesel locomotive	Electric locomotive
Design characteristics			
Source of energy and basic design characteristics	Coal or oil is burnt to generate steam; a steam engine converts the heat energy of the steam into the rotary energy of the moving wheels.	Diesel oil is used for the generation of power with the help of a diesel engine, the generated power is transmitted by means of a mechanical, hydraulic, or electrical transmission system for propelling the wheels.	Electric energy is supplied from a stationary prime mover, which is converted into mechanical energy for propelling the wheels.
Simplicity of design	The design of the engine is simple. The engine itself is heavy and bulky.	The design is not as simple and compact, the engine weighs less.	The design is complicated but the weight is comparatively low.
Tractive effort	Tractive effort is low because torque is not uniform.	Tractive effort is higher because torque is uniform.	An even higher tractive effort is obtainable.
Adhesion (ratio of tractive effort to weight on wheels, beyond which slipping occurs)	0.20–0.25 because torque is not uniform.	0.25 for electric transmission, 0.33 for hydraulic transmission.	0.25 dc 0.33 ac
Ratio of horse-power and weight	75 kg per horsepower.	45 kg per horsepower	25 kg per horsepower
Overload capacity	10%–25% overload capacity.	Only 6%–10% overload capacity, which is the overload capacity of the diesel engine.	Over 50% overload capacity, as energy is drawn from an outside source.
Thermal efficiency	About 7%.	About 25%	About 90%
Technical experience on Indian Railways	Simple machinery, over hundred years of experience.	More complicated machinery, very limited experience in our country.	Some experience on 1500 V dc systems, experience being acquired on 25 kV ac systems.
Cost of locomotive	About ₹ 5 million	About ₹ 15 million	About ₹ 18 million

(Contd)

Table 25.2 (Contd)

Characteristics	Steam locomotive	Diesel locomotive	Electric locomotive
Reversing arrangement	A steam locomotive requires a turntable for reversing its direction.	Reversing of engine is not required. Only the driver and guard have to change their positions.	Reversing of locomotive is not required.
Life of locomotive	About 40 years	About 30 years	Over 30 years
Utilization of power	Fuel is consumed from the moment it is lighted, whether the locomotive is in use or not.	There is no wastage of power when idle if the engine has been switched off.	There is no wastage of power when idle.
Operational characteristics			
Requirement of staff for operation	One driver and two firemen.	One driver	One driver
Smoke and fire	Both fire and smoke present.	No fire and little smoke.	No smoke and no fire.
Promptness of service	Takes time for igniting coal and raising enough steam for the engine to start.	Service is readily available.	Ready service without any wastage of time.
Importance of driving skill	Driving skill is important because the driver is required to control each regulating factor separately.	There is not much variance in the regulating factor and so the driver's skill is not important.	Driving is simpler and normal driving skills are sufficient.
Normal working hours	About 12 hours a day	About 21 hours a day	About 21 hours a day.
Monthly kilometrage	3500 km/month	9000 km/month	10,000 km/month
Speeds	Only low speeds are possible; on gradients, speed gets reduced further.	Higher speeds can be achieved; even on gradients, better speeds possible.	Very high speeds; can negotiate steep grades because of high are overload capacity.
Rate of acceleration	Very low	Better	Best
Flexibility in haulage	Can haul only a limited number of coaches to ensure the economic utilization of power.	Can haul a greater number of coaches.	Can haul a large number of coaches.

(Contd)

Table 25.2 *(Contd)*

Characteristics	Steam locomotive	Diesel locomotive	Electric locomotive
Condition of track	The track gets damaged due to hammer blow action.	No such damage is caused.	Movement is smooth and there is no damage.
Shed arrangement and time required for service	Shed should have arrangement for turning. General service and boiler maintenance require about 70 hours per month.	Shed is simpler than for a steam locomotive. Maintenance requires about 20 hours a month.	Shed is simple like the diesel ones. Maintenance required is about 4 hours a month.
Transport of fuel	Carries its own fuel supply of water and coal, this consumes a lot of power.	The quantity of oil required is about one-eighth that of coal; cost of transport of fuel is much lower.	Does not require fuel. Overhead electric lines are required for power transmission.
Fuelling points	Intermediate fuelling points required for replenishing the supply of coal and water.	Intermediate fuelling points required for filling diesel oil.	No intermediate fuelling points are required.
Repairs and renewals	Frequent	Comparatively less	Minimum
Suitability	▲ Wherever traffic density is not heavy	▲ For heavy traffic	▲ For heavy loads and on gradients. For underground railways.
	▲ Wherever coal is available at cheap rates	▲ As an intermediate stage between steam operation and electric traction.	▲ For high speed
	▲ Wherever water is available	▲ As intermediate traction	▲ For suburban traffic and for quick acceleration/deceleration
Future prospects	▲ Steam locomotives have been phased out on Indian Railways.	▲ Bright prospects; more diesel locomotives are being procured.	▲ Best prospects. An increasing number of electric locomotives are being procured.

friction depends upon the speed of the locomotive and the condition of the rail surface. The higher the speed of the locomotive, the lower will be the coefficient of friction, which is about 0.1 for high speeds and 0.2 for low speeds. The condition of the rail surface, whether wet or dry, smooth or rough, etc., also plays an important role in deciding the value of the coefficient of function. If the surface is very smooth, the coefficient of friction will be very low.

Hauling power = number of pairs of driving wheels × weight exerted on each driving axle × coefficient of friction

Thus, for a locomotive with three pairs of driving wheels, an axle load of 20 tonnes, and a coefficient of friction equal to 0.2, the hauling power will be equal to 3 × 20 × 0.2 tonne, i.e., 12 tonnes.

Example 25.2 Calculate the maximum permissible load that a BG locomotive with three pairs of driving wheels bearing an axle load of 22 tonnes each can pull on a straight level track at a speed of 80 km/h. Also calculate the reduction in speed if the train has to run on a rising gradient of 1 in 200. What would be the further reduction in speed if the train has to negotiate a 4° curve on the rising gradient? Assume the coefficient of friction to be 0.2.

Solution

(a) Hauling power of the locomotive = number of pairs of driving wheels × wt exerted on each pair × coefficient of friction = 3 × 22 × 0.2 = 13.2 tonnes

(b) The total resistance negotiated by the train on a straight level track at a speed of 80 kmph:

R = Resistance due to friction + resistance due to wave action and track irregularities + resistance due to wind

= $0.0016\,W + 0.00008\,WV + 0.0000006\,WV^2$

Substituting the value of $V = 80$ kmph

$R = 0.01184\,W$

Assuming total resistance = hauling power,

$W \times 0.01184 = 13.2$ tonnes

or

$$W = \frac{13.2}{0.01184} = 1114.86 \text{ tonnes, Approx. 1115 tonnes}$$

(c) On a gradient of 1 in 200, there will be an additional resistance due to gradient equal to $W \times$ % of slope. Since hauling power = total resistance,

$$13.2 = 0.0016W + 0.00008WV + 0.0000006WV^2 + W\frac{0.5}{100}$$

$$= W(0.0016 + 0.00008V + 0.0000006V^2 + 0.005)$$

Since $W = 1114.8$ tonnes,

$$13.2 = 1114.8\,(0.0016 + 0.000008V + 0.0000006V^2)$$

On solving the equation further,

$V = 48.13$ kmph

Reduction in speed = 80 – 48.13 = 31.87 kmph = 32 kmph

(d) On a curve of 4° on a rising gradient of 1 in 200, curve resistance will be equal to

$R = 0.0004 \times$ degree of curve × wt

$= 0.0004 \times 4 \times W = 0.0016\,W$

Hauling power of locomotive = total resistance. Therefore,

$$13.2 = 0.0016W + 0.00008WV + 0.0000006WV^2 + 0.005W + 0.0016W$$

By substituting the value of $W = 1114.8$ tonnes in the equation and solving further,

$$V = 43.68 \text{ kmph}$$

Further reduction in speed = $48.13 - 43.68 = 4.45$ kmph. Therefore,

Maximum permissible train load = 1115 tonnes

Reduction in speed due to rising gradient = 31.87 kmph

Further reduction in speed due to curvature = 4.45 kmph

Example 25.3 Compute the steepest gradient that a train of 20 wagons and a locomotive can negotiate given the following data: weight of each wagon = 20 tonnes, weight of locomotive = 150 tonnes, tractive effort of locomotive = 15 tonnes, rolling resistance of locomotive = 3 kg/tonnes, rolling resistance of wagon = 2.5 kg/tonnes, speed of the train = 60 kmph.

Solution

(a) Rolling resistance due to wagons = rolling resistance of wagon × weight of wagon × number of wagons

$$= 2.5 \times 20 \times 20 = 1000 \text{ kg} = 1 \text{ tonne}$$

(b) Rolling resistance due to locomotive

= rolling resistance of locomotive × wt of locomotive

$$= 3 \times 150 = 450 \text{ kg} = 0.45 \text{ tonne}$$

(c) Total rolling resistance = rolling resistance due to wagons + rolling resistance due to locomotive = $1.00 + 0.45$ tonnes = 1.45 tonnes

(d) Total weight of train = weight of all wagons + wt of locomotive

$$= 20 \times 20 + 150 = 550 \text{ tonnes}$$

(e) Total train resistance = rolling resistance + resistance dependent on speed + resistance due to wind + resistance due to gradient

$$= 1.45 + 0.00008WV + 0.0000006WV^2 + W/g$$
$$= 1.45 + 0.00008 \times 550 \times 60 + 0.0000006 \times 550 \times 60^2 + (550/g)$$
$$= 1.45 + 2.64 + 1.19 + (550/g) = 5.28 + (550/g)$$

where g is the gradient.

(f) Tractive effort of locomotive = Total train resistance

$$15 = 5.28 + (550/g)$$

or

$$g = 56.5$$
$$= 1/56 = 1 \text{ in } 56$$

Therefore, the steepest gradient that the train will be able to negotiate is 1 in 56.

Example 25.4 Calculate the maximum permissible train load that can be pulled by a locomotive with four pairs of driving wheels with an axle load of 28.42 tonnes each on a BG track with a ruling gradient of 1 in 200 and a maximum curvature of 3°, travelling at a speed of 48.3 kmph. Take the coefficient of friction to be 0.2.

Solution

(a) Hauling capacity of locomotive

= Number of pairs of driving wheels × axle load × coefficient of friction

$$= 4 \times 28.42 \times 0.2 = 22.736 \text{ tonnes}$$

(b) Total resistance of train = resistance due to friction + resistance due to speed + resistance due to wind + resistance due to gradient + resistance due to curve

$$= 0.0016W + 0.00008WV + 0.0000006WV^2 + W(1/g) + 0.0004WD$$
$$= 0.0016W + 0.00008W \times 48.3 + 0.0000006W \times (48.3)^2 + W \times$$
$$(1/200) \times 0.0004 \times W \times 3$$

(c) Hauling capacity = total resistance

$$22.73 = 0.01306W$$

or

$$W = 1740 \text{ tonnes}$$

Therefore, the maximum weight of the train is 1740 tonnes.

SUMMARY

There are various types of resistances or forces that oppose the movement of a train on a track. These resistances may be due to the atmosphere, track condition, gradient, curvature, or any other factor. The tractive effort of a locomotive should be sufficient to overcome these resistances so that the desired speed can be maintained. Steam locomotives have gradually been replaced with diesel and electric locomotives on Indian Railways. There are numerous advantages of electric traction over steam and diesel traction.

REVIEW QUESTIONS

1. Differentiate between the hauling capacity and the tractive effort of a locomotive.
2. (a) A BG locomotive has three pairs of driving wheels with an axle load of 20 tonnes. If this locomotive runs at a speed of 120 kmph, what is the train weight in tonnes that the locomotive can pull on a straight level track?
   (b) What is the train weight that the same locomotive will be able to haul on a 2° curve and a 1-in-100 gradient?
3. (a) List and explain the various resistances that a locomotive in motion has to overcome.
   (b) Determine the maximum permissible train load that a locomotive with four pairs of driving wheels of a 22.86 tonnes axle load each can pull on a level BG track at a speed of 90 kmph. Also determine the reduced speed of the train if it has to ascend a gradient of 1 in 200 with the same train load. (Assume the hauling capacity of the locomotive to be one-sixth of the load on the driving wheels).
4. What are the requirements of a locomotive? Briefly describe the merits of the different types of traction commonly used in India.
5. A train with 20 wagons, each weighing 18 tonnes, is supposed to run at a speed of 50 kmph. The tractive effort of a 2-8-2 locomotive with a 22.5 tonnes load on each driving axle is 15 tonnes. The weight of the locomotive is 120 tonnes. The rolling resistance of the wagons and locomotive are 2.5 kg/tonnes and 3.5 kg/tonnes, respectively. The resistance, which depends upon the speed, is computed to be 2.65 tonnes. Find out the steepest gradient for these conditions.

6. Discuss how the hauling capacity of a locomotive is worked out. Compute the steepest gradient that a train of 20 wagons and a locomotive can traverse. Use the following data: weight of each wagon = 20 tonnes, weight of locomotive (with tender) = 150 tonnes, tractive effort of locomotive = 15 tonnes, rolling resistance of wagons = 2.5 kg/tonne, speed of the train = 60 kmph.

7. What will be the gradient for a BG track when the gradient resistance together with curve resistance due to a 3° curve is equal to the resistance due to ruling gradient of 1 in 200? What would be the resistance when an 8° curve is provided on an MG line and a train with a total weight of 914.85 tonnes passes over it.

8. What resistances does a locomotive have to overcome for hauling a train in hilly terrains? A goods train with 80 wagons weighing 30 tonnes each is to run at a speed of 50 kmph, while ascending a 0.25% gradient with 2° curves. The train is hauled by a 2-8-4 locomotive with 18.5 tonnes load on each driving axle. Find out whether the locomotive will be able to haul the load at the desired speed. Assume the coefficient of rail–wheel friction to be 0.2.

9. Name the different train resistances that a locomotive has to overcome in hauling a train under adverse circumstances. Explain the factors that would affect speed-dependent resistances. What do you understand by gradient compensation on curved alignment?

10. Compare the various characteristics of steam, diesel, and electric traction.

Choose the correct answer from the choices given.

11. The resistance that is not dependent on speed is:
    (a) resistance due to wind
    (b) resistance due to fraction
    (c) resistance due to wave action

12. The starting resistance depends upon:
    (a) speed of the train   (b) degree of curve
    (c) weight of the train   (d) slope of track

13. A train is hauled by a 2-8-2 locomotive with 22.5 tonnes load on each driving axle. Assuming the coefficient of rail–wheel friction to be 0.25, the hauling capacity of the locomotive will be:
    (a) 15.0 tonnes   (b) 22.5 tonnes
    (c) 45.0 tonnes   (d) 90.0 tonnes

14. The load on each axle of a locomotive is 22 tonnes. If the coefficient of friction is 0.2, then the hauling capacity due to 3 pairs of driving wheels will be:
    (a) 26.4 tonnes   (b) 19.8 tonnes
    (c) 13.2 tonnes   (d) 6.6 tonnes

Railway Stations and Yards

INTRODUCTION

A railway station is that place on a railway line where traffic is booked and dealt with and where trains are given the authority to proceed forward. Sometimes only one of these functions is carried out at a station and accordingly it is classified as a flag station or a block station. In the case of a flag station, there are arrangements for dealing with traffic but none for controlling the movement of the trains. In the case of a block station, a train cannot proceed further without obtaining permission from the next station and traffic may or may not be dealt with. However, most railway stations perform both the functions indicated above.

26.1 PURPOSE OF A RAILWAY STATION

A railway station is provided for one or more of the following purposes:
   (a) To entrain or detrain passengers
   (b) To load or unload goods or parcels
   (c) To control the movement of trains
   (d) To enable trains to cross each other in the case of a single-line section
   (e) To enable faster trains to overtake slower ones
   (f) To enable locomotives to refuel, whether it be diesel, water, or coal
   (g) To attach or detach coaches or wagons to trains
   (h) To collect food and water for passengers
   (i) To provide facilities for change of engines and crew/staff
   (j) To enable sorting out of wagons and bogies to form new trains
   (k) To provide facilities and give shelter to passengers in the case of emergencies, such as floods and accidents, which disrupt traffic

26.2 SELECTION OF SITE FOR A RAILWAY STATION

The following factors are considered when selecting a site for a railway station.

Adequate land   There should be adequate land available for the station building, not only for the proposed line but also for any future expansion. The proposed area should also be without any religious buildings.

Level area with good drainage   The proposed site should preferably be on a fairly level ground with good drainage arrangements. It should be possible to provide the maximum permissible gradient in the yard. In India, the maximum permissible gradient adopted is 1 in 400, but a gradient of 1 in 1000 is desirable.

Alignment   The station site should preferably have a straight alignment so that the various signals are clearly visible. The proximity of the station site to a curve presents a number of operational problems.

Easy accessibility   The station site should be easily accessible. The site should be near villages and towns. Nearby villages should be connected to the station by means of approach roads for the convenience of passengers.

Water supply arrangement   When selecting the site, it should be verified that adequate water supply is available for passengers and operational needs.

26.3 FACILITIES REQUIRED AT RAILWAY STATIONS

The passenger station is the gateway through which people find their way into a town or community. A first impression is a lasting one and, hence, a well-designed station building with well-maintained surroundings is important. Whilst service is the main consideration, the type and finish of a station building should be, as far as practicable, in keeping with the best standards of civic amenities available in that area. A large passenger station should provide for facilities corresponding to the anticipated demands of at least the first 20 years of its life, with provisions for future expansion. The facilities required at stations are broadly classified into the following main groups.

Passenger requirements

This includes waiting rooms and retiring rooms, refreshment rooms and tea stalls, enquiry and reservation offices, bathrooms and toilets, drinking water supply, platform and platform sheds, and approach roads.

Traffic requirements

This includes goods sheds and platforms, station buildings, station master's office and other offices, signal and signal cabins, reception and departure lines and sidings, arrangements for dealing with broken-down trains, and station equipment.

Locomotive, carriage, and wagon requirements

This includes the locomotive shed, watering or fuelling facilities, turntable, inspection pits, ashpits, and ashtrays.

Staff requirements

This includes rest houses for officers and staff, running rooms for guards and drivers, and staff canteens.

26.4 REQUIREMENTS OF A PASSENGER STATION YARD

The main requirements of a passenger yard are the following.

(a) It should be possible to lower the signals for the reception of trains from different directions at the same time. This facility is particularly necessary at junction stations so that all the trains that are to be connected with each other may be received at the same time.

(b) Unless all trains are booked to stop at the station, it should be possible to run a train through the station at a prescribed speed.

(c) In the case of an engine changing station, an engine coming from or going to a shed should cause minimum interference in the arrival and departure of trains.

(d) An adequate number of platforms should be provided so that all trains can be dealt with at the same time.

(e) There should be convenient sidings where extra carriages can be stabled after having been detached from trains or before their attachment to trains.

(f) There should be provision of facilities for dealing with special traffic such as pilgrim and tourist traffic, parcels in wagon loads, livestock, and motor cars.

(g) Stabling lines, washing lines, sick lines, etc., should be provided as per requirement.

26.5 CLASSIFICATION OF RAILWAY STATIONS

Railway stations can broadly be classified into various classes on the basis of three main considerations, viz. operational, functional, and financial.

Operational considerations

As per the general and subsidiary rules of Indian Railways, stations are classified into block stations and non-block stations. Block stations are further classified into A class, B class, and C class stations. Non-block stations are classified into D class or flag stations.

Functional considerations

Stations are classified based on the functions they are required to perform. Under this category, stations are classified into halt stations, flag stations, crossing stations or wayside stations, junction stations, and terminal stations.

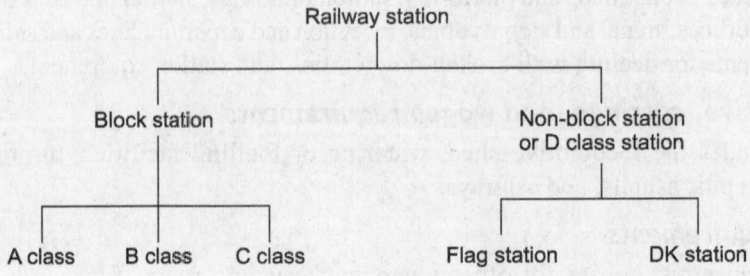

Earning considerations

Depending upon the passenger earning of the railway stations and some other important considerations, the stations have been classified into six categories, viz. A,B,C,D,E and F which is also an indicator of the passenger traffic.

The following factors are taken into consideration when classifying a railway station.

(a) Least expenditure with regard to the provision of the least number of signals
(b) Flexibility in shunting operations
(c) Increasing the line capacity
(d) Faster movement of trains
(e) Passenger earnings of the station

26.5.1 Block Stations

A block station is a station at which the driver has to obtain an 'authority to proceed' in order to enter the next block section. In a railway system that is inclusive of block stations, the entire railway line is divided into convenient block sections of 5 to 10 km and a block station is provided at the end of each block. This system ensures that a suitable 'space interval' is provided between running trains so that there are no collisions and accidents. There are three types of block stations.

A class station

A class stations are normally provided on double-line sections. At such stations a 'line clear' signal cannot be granted at the rear of a station unless the line on which a train is to be received is clear and the facing points are set and locked. No shunting can be done after a line clear has been granted.

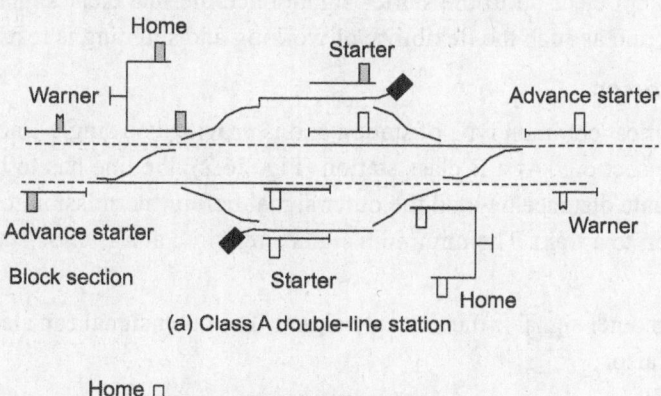

(a) Class A double-line station

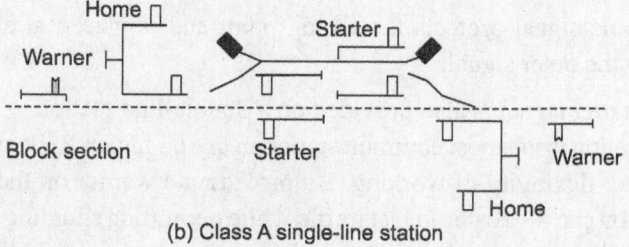

(b) Class A single-line station

Fig. 26.1   A class station

A class stations are suitable for sections where traffic passes rapidly. It is essential for the driver of the train to have an advance knowledge of the layout of the block station. The typical layout of an A class station with two-aspect signalling is shown in Fig. 26.1.

The signals required at an A class station are as follows.

Warner A warner signal is placed at a warning distance from the home signal, the main function of which is to indicate whether the section beyond is clear or otherwise.

Home A home signal is the first stop signal.

Starter A starter signal is placed at an adequate distance from the home signal and marks the point up to which the line should be clear so that the train can be given permission to approach.

Advance starter This signal is optional and is provided to allow the drivers to further increase the speed of the trains.

Advantages

(a) More economical vis-à-vis B class stations because of the use of fewer signals.
(b) Ensures the safety of the train because of the provision of a warner signal ahead of a home signal.
(c) Trains normally stop within the station limits.

Disadvantages

(a) No shunting is possible once line clear has been granted.
(b) Another clear disadvantage of A class stations is that a line at the station has to be kept clear up to the starter signal once the line clear signal has been given, and as such the flexibility of working and shunting is restricted.

B class station

This is the most common type of station and is provided on single-line as well as double-line sections. At a B class station (Fig. 26.2), the line has to be clear up to an adequate distance beyond the outer signal before 'permission to approach' can be given to a train. The minimum signals required at a B class station are as follows.

Outer An outer signal is the first stop signal. The outer signal can also be below the warner also.

Home A home signal protects the facing point and is placed at an adequate distance from the outer signal.

Starter A starter signal is also provided on a double-line section.

The B class station is the most common station in use on Indian Railways because it offers greater flexibility of working. By providing a warner on the outer arm post, this station can also cater to fast traffic while permitting shunting of vehicles even when a clear signal has been given.

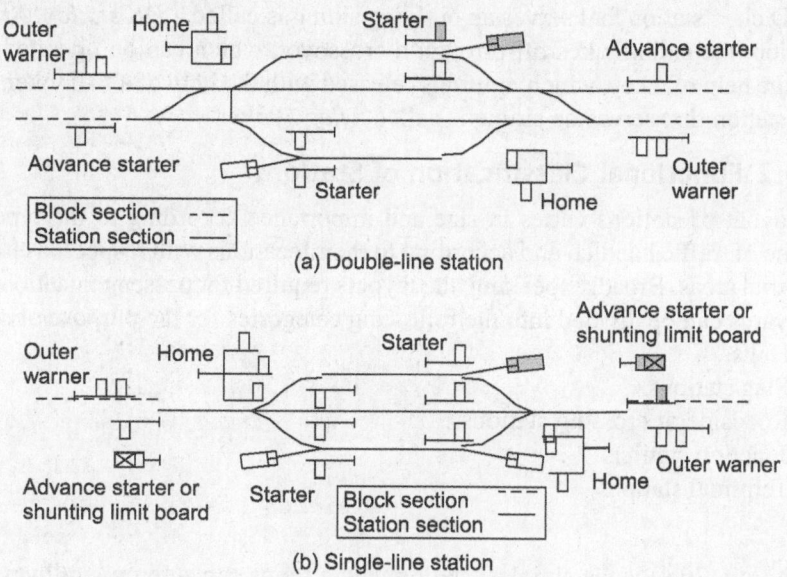

Block section ---
Station section ——

(a) Double-line station

Block section ---
Station section ——

(b) Single-line station

Fig. 26.2   B class station

C class station

The C class station (Fig. 26.3) is only a block hut where no booking of passengers is done. It is basically provided to split a long block section so that the interval between successive trains is reduced. No train normally stops at these stations. The minimum signals required are as follows.

Warner   A warner signal placed at an adequate warning distance from the home signal to indicate whether the section ahead is clear or not.

Home   A home signal is the first stop signal.

The advantage of a C class station is that it ensures the faster movement of trains and increases line capacity. The disadvantage, however, is that no shunting is possible and trains cannot stop at these stations.

Non-block Stations or D Class Stations

D class or non-block stations are located between two block stations and do not form the boundary of any block section. No signals are provided at D class stations.

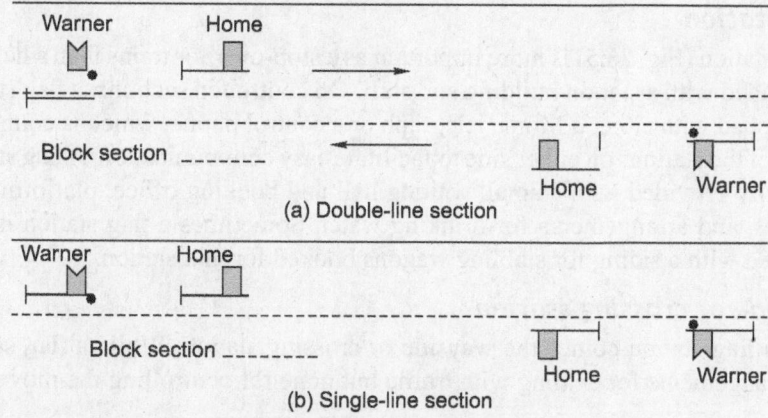

Block section ___

(a) Double-line section

Block section ___

(b) Single-line section

Fig. 26.3   C class station

A D class station that serves an outlying siding is called a *DK station*. At such a station, the siding takes off through a crossover, which can be operated only with the help of a key, which in turn is released with the help of a ball token. A D class station that serves no siding is called a *flag station*.

26.5.2 Functional Classification of Stations

The layout of stations varies in size and importance according to the type and volume of traffic handled and according to their locations with respect to cities or industrial areas. Broadly speaking, the layouts required for passenger stations and their yards can be divided into the following categories for the purpose of study:

(a) Halts
(b) Flag stations
(c) Roadside or crossing stations
(d) Junction stations
(e) Terminal stations

Halt

A halt (Fig. 26.4) is the simplest station where trains can stop on a railway line. A halt usually has only a rail level platform with a name board at either end. Sometimes a small waiting shed is also provided, which also serves as a booking office. There is no yard or station building or staff provided for such types of stations. Some selected trains are allotted a stoppage line of a minute or two at such stations to enable passengers to entrain or detrain. The booking of passengers is done by travelling ticket examiners or booking clerks. A notable example of the halt is a Gurhmukteshwar bridge halt, which is situated on the bank of river Ganga.

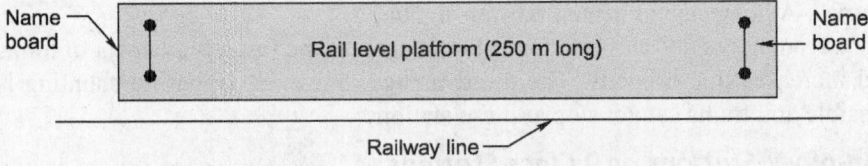

Fig. 26.4   Layout of a halt station

Flag station

A flag station (Fig. 26.5) is more important as a stop-over for trains than a halt and is provided with a station building and staff. On controlled sections, a flag station is equipped with either a Morse telegraph or a control phone, which is connected to one of the stations on either side to facilitate easy communication. A flag station is usually provided with a small waiting hall and booking office, platforms and benches, and arrangements for drinking water. Sometimes a flag station is also provided with a siding for stabling wagons booked for that station.

Wayside or crossing station

After a flag station comes the wayside or crossing station. While a flag station has arrangements for dealing with traffic but none for controlling the movement

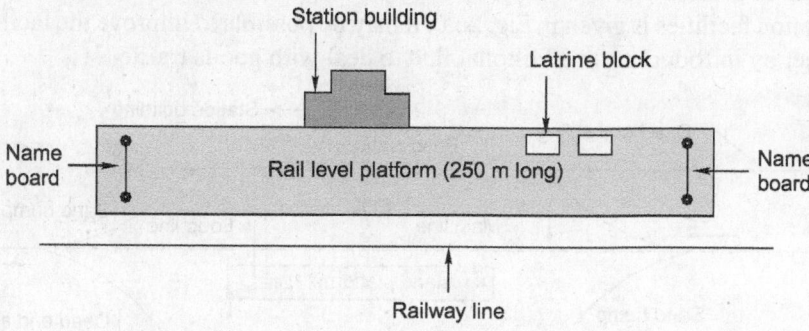

Fig. 26.5   Layout of a flag station

of the trains, a crossing station has arrangements for controlling the movement of trains on block sections. The idea of a crossing station was initially conceived for single-line sections, to facilitate the crossing of trains going in opposite directions so that there may be a more rapid movement of trains.

Crossing stations may be further classified into (a) roadside small and medium sized stations and (b) major stations. Some of the important tasks dealt with these stations are the following.

Operating work   The main operations performed at these stations include attending to the passing and crossing of trains, giving precedence to important trains, and other miscellaneous works done for stopping passenger trains. Slow passenger trains mostly stop at small stations whereas mail and express trains stop at major stations.

Goods traffic   These stations mostly deal with parcel traffic only. Piecemeal wagon load goods traffic is now being accepted on roadside stations as per the new policy of the Railway Ministry with effect from December 1994.

Operation of points and signals   The operation of points and signals is controlled either by a central cabin or two cabins at either end of the station.

Reception and dispatch of trains   The reception and dispatch as well as shunting of trains is handled as per the instructions laid down in the 'station working order'. Block instruments are provided either in the station master's office or in the cabin, but the entire responsibility of carrying out these operations lies with the station master.

Station master's duty for run-through trains   When a train runs through the station, the station master should stand opposite his office in proper uniform and exchange 'all right' signals with the driver and guard of the train. He should watch the running train carefully and if there are any unusual occurrences such as the incidence of a hot box, he should instruct the station officials in advance to stop and examine the train.

Wayside or crossing station on a single-line section   Increasing traffic on a single-line section necessitates the construction of a three-line station, which provides an additional line as well as more facilities for passing traffic. A typical layout of a three-line station providing one additional line and simultaneous reception facilities is given in Fig. 26.6. It may be possible to improve the facilities further by introducing an additional line to deal with goods traffic.

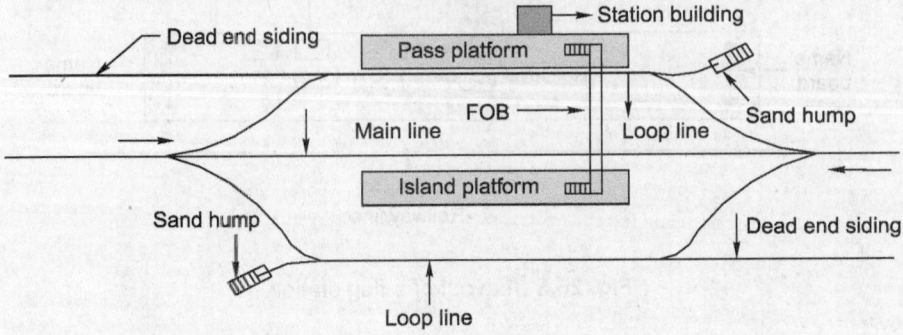

Fig. 26.6 A wayside or crossing station on a single-line section

The following are some of the important features of this track layout.

(a) It is a three-line station and provides facilities for the simultaneous reception of trains from both sides because of the proximity of sand humps in each direction.

(b) There are two platforms, namely an island platform and a platform near the station building. The island platform can deal with two stopping trains simultaneously. Also, if a goods train has to be stopped at an island station, it can be accommodated on the loop line of the platform, thus keeping the main line free for run-through traffic. Important trains can be made to halt on the platform near the station building.

(c) There is a dead end siding at either end of the station to accommodate wagons that are marked sick.

(d) The foot overbridge (FOB) helps the passengers to reach the island platform from the station building and vice versa.

Double-line crossing station with an extra loop In the case of a double-line section, which consists of separate up and down lines to deal with traffic moving in either direction, the layout of a station yard is somewhat different.

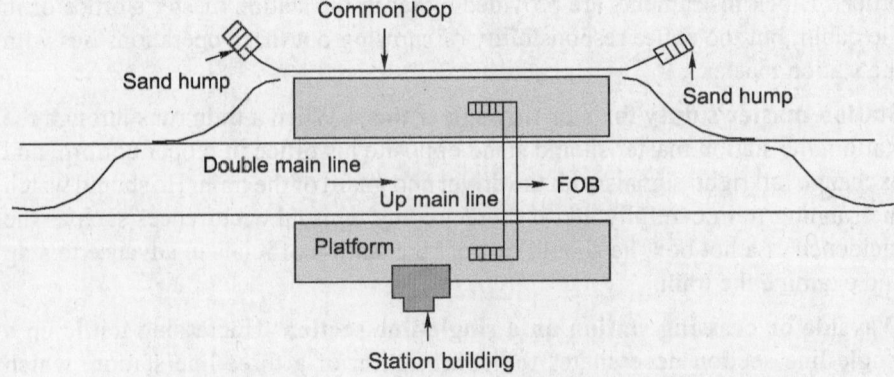

Fig. 26.7 Double-line crossing station with three lines

Figure 26.7 shows a double-line station with three lines receiving, with one common loop for trains coming from both sides. Some of the important features of this layout are as follows:

(a) This is a wayside station for a double-line section with almost minimum facilities.

(b) In addition to two main lines an up line and a down line, there is a common loop that can receive trains from either direction. There is a total of three lines only.

(c) It consists of two platforms, one an island platform and the other a platform beside the station building.

(d) There is a foot overbridge to connect the station building to the island platform and back.

(e) There are emergency crossovers provided on either side of the station so that it can be converted into a single-line station in case of an emergency.

Double-line crossing station with four lines The more common layout of a station yard on a double-line section has four-line station as shown in Fig. 26.8. The important features of this layout are as follows:

(a) This is a four-line station, where, apart from two up and down main lines, there are two extra loops. These loops are directional loops, i.e., one is known as a down loop as it is meant for down trains while the other is an up loop and is meant for up trains.

(b) There are two platforms provided with connection loops. One of these platforms can also be an island platform.

(c) There is provision of a foot overbridge to connect the two platforms.

(d) Two emergency crossovers are provided on either side of the station so that it can be converted into a single-line station in case of an emergency.

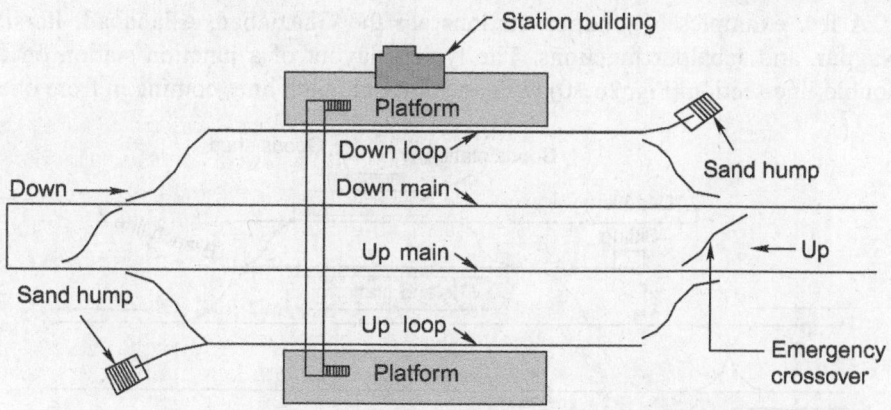

Fig. 26.8 Double-line crossing station with four lines

Junction stations

A junction station is the meeting point of three or more lines emerging from different directions. Normally at junctions, trains arrive on branch lines and return to the same station from where they started or proceed to other stations from where they again return to their originating stations.

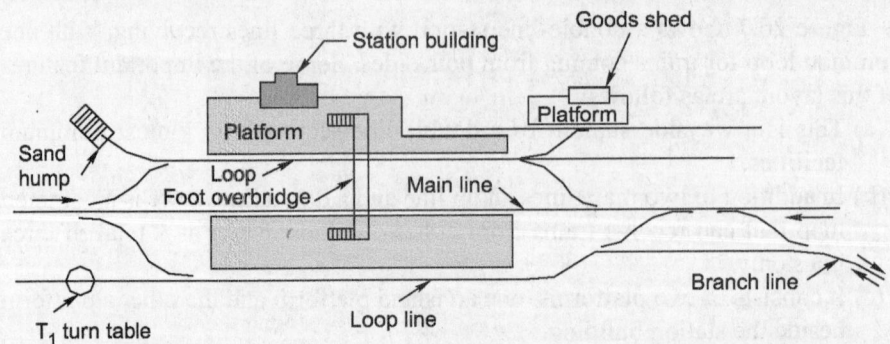

Fig. 26.9   Junction station with single main line and single branch line

The typical layout of a junction station with a single main line and a single branch line is shown in Fig. 26.9. The important features of junction stations are as follows:

(a) There are two platforms—the main line platform and the island platform. In case the timings of two trains match, both the trains can be received and made to wait on either side of the island platform. This helps in the easy trans-shipment of passengers and luggage. Also, main line as well as branch line trains can be received on the main platform.

(b) A foot overbridge is provided for passengers to move between the station platform and the island platform.

(c) It is provided with a small goods siding and a goods platform to deal with goods traffic.

(d) A turntable is provided for reversing the direction of an engine, if required.

(e) The emergency crossover provided on either side of the station helps in switching to a single-line set-up in case of an emergency.

A few examples of junction stations are the Ghaziabad, Allahabad, Itarsi, Nagpur, and Jabalpur junctions. The typical layout of a junction station on a double-line section (Fig. 26.10) with one or two branch lines coming in from one

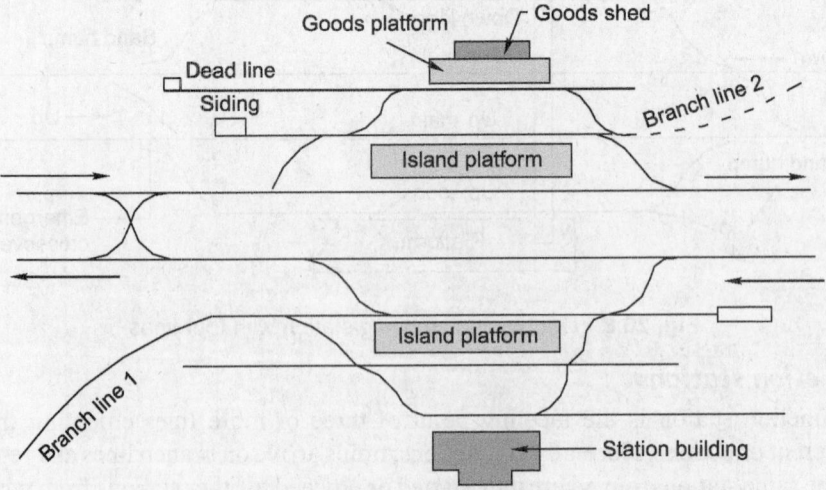

Fig. 26.10   Junction station with double main line and two branch lines

or two different directions is shown in Fig. 26.10. The most important feature of this layout is that such a station receives traffic from four different directions, i.e., up main line, down main line, branch line 1, and branch line 2. Most of the facilities provided at this station are almost the same as described for the layout shown in Fig. 26.9.

Terminal station

The station at which a railway line or one of its branches terminates is known as a terminal station or a terminal junction (Fig. 26.11). The reception line terminates in a dead end and there is provision for the engine of an incoming train to turn around and move from the front to the rear of the train at such a station. In addition, a terminal station may need to be equipped with facilities for watering, cleaning, coaling, fuelling, and stabling the engines; storing, inspecting, washing

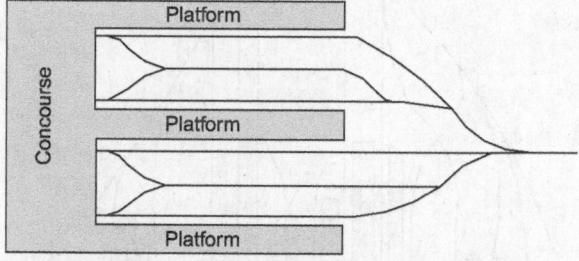

Fig. 26.11 Terminal station with run round line

and charging the carriages; and such other works.

On unimportant branch lines, the terminal station will have only one platform, but there are big terminal stations, such as the Howrah and Mumbai stations, which are provided with elaborate facilities. The general layout of a big terminal station is shown in Fig. 26.12.

It may be noted that access from one platform to another is via a concourse and that there are no foot overbridges provided for this purpose.

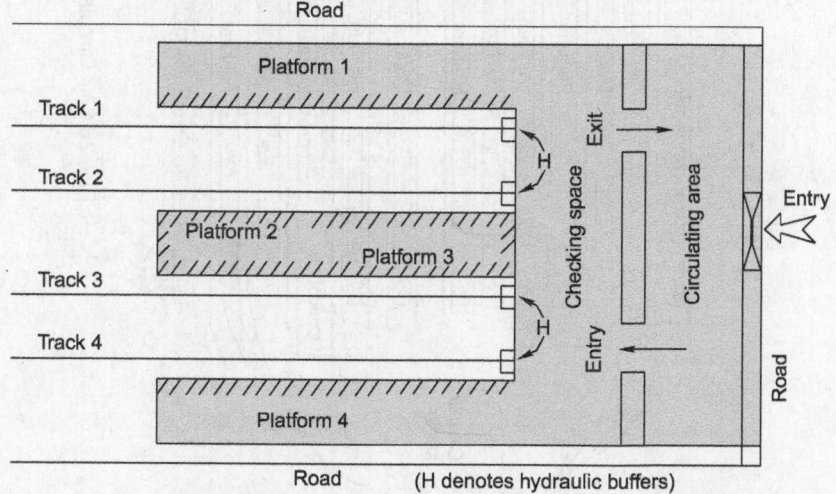

Fig. 26.12 Layout of a big terminal station

Grand Central Station at New York   Figure 26.13 depicts the circular loop provided at the Grand Central Station, New York. The provision of circular loops enables the trains to pass through a terminal station without any delay. A further advantage attached to the loop system is that it enables the provision of special stations for dealing with suburban traffic at underground locations away from the congested area of the main terminal and in close proximity to business districts, thus affording direct connections with other stations.

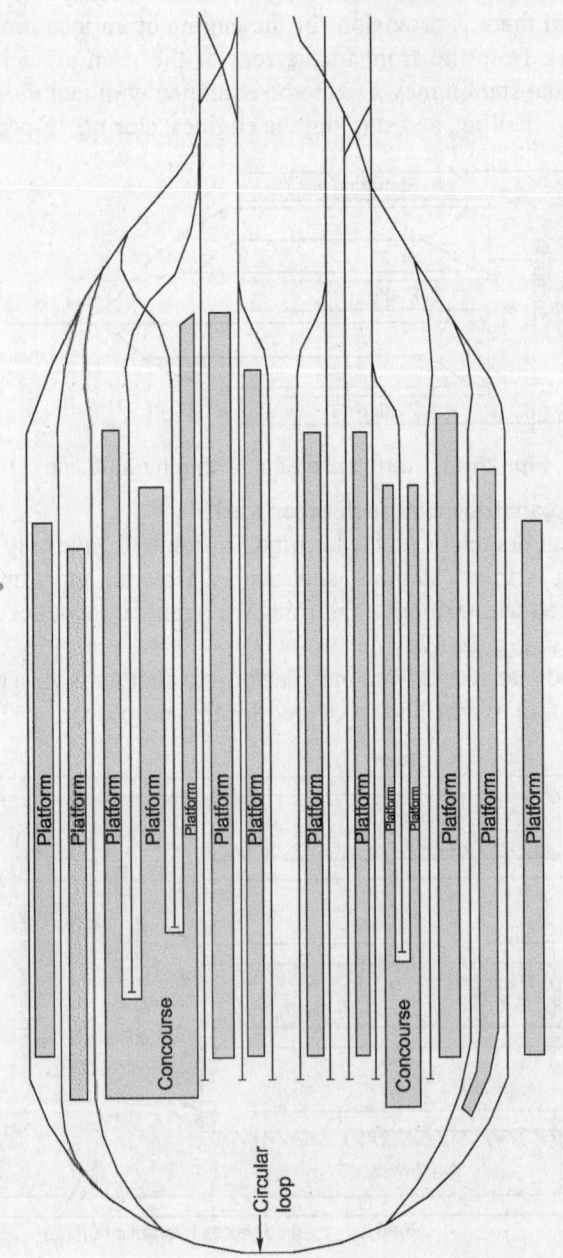

Fig. 26.13   Circular loop at Grand Central Station, New York

26.5.3 Classification Based on Earning Considerations

(I) Category 'A'
- (i) Non-suburban stations with annual passenger earnings of ₹ 60 million and above
- (ii) Divisional and zonal railway headquarters

(II) Category 'B'
- (i) Non-suburban stations with annual passenger earnings between ₹ 30 and ₹ 60 million
- (ii) Stations serving places of tourists interest and having a very significant tourist traffic and junction stations

(III) Category 'C'
   All suburban stations

(IV) Category 'D'
- (i) Non-suburban stations with passenger earnings between ₹ 10 and ₹ 30 million
- (ii) District headquarters and stations of local importance to be classified by the Railway

(V) Category 'E'
   Non-suburban stations with earnings less than ₹ 10 million

(VI) Category 'F'
   Halt stations

Minimum Essential Amenities

- (i) When a station is constructed, certain minimum amenities should be provided at each category of station. These were earlier termed as basic amenities/infrastructural facilities, but are now called 'Minimum Essential Amenities'.

Railway Board has prescribed the norms or scale of essential facilities for passengers depending upon the type of railway station and availability of resources. These are listed in Table 26.1.

Table 26.1 Norms for quantum of minimum essential amenities at various categories of stations

Amenities	Station categories					
	A	*B*	*C*	*D*	*E*	*F*
Booking facility (No. of counters)	15	10	7	4	2	1
Drinking water** (No. of taps)	12 taps on each PF	12 taps on each PF	6 taps on each PF	6 taps on each PF	1 tap/HP on each PF	1 hand pump at station
Waiting hall/shed@	100 m²	50 m²	0'	30 m²	15 m²	10 m² booking office cum waiting hall
Seating arrangement (No. of seats/PF)	100	75	10	50	10	-
Platform shelter shade trees (on each PF)	400 -	200 m² -	200 m² -	50 m² -	- shade trees	- shade trees
Urinals#	10	6	4	4	1	-
Laterines#	10	6	2	4	1	-

Contd.

Table 26.1 (*Contd.*)

Amenities	Station categories					
	A	B	C	D	E	F
Platforms Level[+]	High	Medium	High	Medium	-	-
Lighting and Fans[++]	Yes	Yes	Yes	Yes	Yes	-
Foot overbridge[+]	Yes	Yes	Yes	-	-	-
Time table display	Yes	Yes	Yes	Yes	Yes	Yes
Clock[+]	Yes	Yes	Yes	Yes	Yes	Yes

++ Stations falling in water scarcity zones or at station, where water source dries up in summer, drinking water facility should be ensured at every platform by means of sintex tanks/CANS/*Matkas/Piaus,* etc., to be decided by GM of the Railways. At less important stations, may be provided as per Minimum Essential Amenities. Drinking water facility would include all necessary units whether donated by private parties or provided by the Railway themselves.

@ If the variation is marginally on the lower side (upto-5 sqm), then it can be taken to be adequately provided.

1. Number of latrines/urinals includes provision in waiting room/halls. 1/3rd of the toilet may be reserved for ladies. In case of 2 toilets existing, one each should be earmarked for ladies and gents.

2. Number of latrines/urinals can be reduced in water scarcity areas by the Railway with the approval of GM.

$ At A, B, C & D category of stations, the booking counters to operate round the clock except at stations where there is no night working.

Notes: (i) *At stations where only one ASM is posted, only one booking window will be provided. In respect of 'E' category stations, where the earnings is less than ₹ 200 million per annum, the quantum of amenitieis to be provided at such stations could be decided by general managers based on actual requirements.*

(ii) *Scale of all the amenties prescribed above are the bare minimum to be provided at the appropriate class of stations. Amenities over and above the prescribed minimum scales will continue to be provided as per norms for provision of amenities at 'Recommended Level'.*

+ *Final decision about Level of Platforms, Foot-overbridge, Time Table display and clock to be taken by general manager.*

++ *As per Railway Board order No. 95/Elec/G/138 dated 19.3.1996.*

(Authority: Rly. Board Letter No. 94/LMB/2/175 dated 24-6-2003)

26.6 MODEL STATIONS

(i) For the purpose of upgradation of amenities, some stations have been selected as model stations. Such stations would be provided with the level of 'Desirable Amenities' specified for the category as given in Table 26.2.

Table 26.2 Desirable amenities

Amenities	Station category					
	A	B	C	D	E	F
Retiring room	Yes	Yes	-	Yes	-	-
Waiting room withbathing facilities				Yes	-	-

Contd.

Table 26.2 (*Contd.*)

Amenities	Station category					
	A	B	C	D	E	F
• Common	Yes	Yes	Yes	-	-	-
• Separate for upper second class	Yes	-	-	-	-	-
• Separate for ladies and gents	Yes	-	-	-	-	-
Cloak room	Yes	Yes	-	-	-	-
Enquiry and computer-based announcement	Yes	Yes	Yes	Yes	-	-
Multimedia information Kiosk (MMIK)	Yes	-	-	-	-	-
Interactive voice response system (IVRS)	Yes	Yes	-	-	-	-
Public address system	Yes	Yes	Yes	-	-	-
Book stalls/other-stalls of essential goods	Yes	Yes	Yes	-	-	-
Refreshment room	Yes	Yes	-	-	-	-
Parking/circulatory area	Yes	Yes	Yes	Yes	-	-
Washable apron with jet cleaning	Yes	Yes	-	-	-	-
Train indicator board	Yes	Yes	Yes	-	-	-
Public phones and Internet	Yes	Yes	Yes	Yes[@]	Yes[@]	-
Touch screen MMIK	Yes	-	-	-	-	-
Watering vending machines	Yes	Yes	-	-	-	-
Water coolers	Yes	Yes	Yes	-	-	-
Signages (Standardized)	Yes	Yes	Yes	-	-	-
Modular catering stalls*	Yes	Yes	Yes	Yes	-	-
Automatic vending machines	Yes	Yes	Yes	-	-	-
Pay and use toilets	Yes	Yes	Yes	Yes	Yes	-
Self-printing ticket machine (SPTM) and unreserved ticket system (UTS)	Yes	Yes	Yes	-	-	-
Computerization of complaints	Yes	-	-	-	-	-
Circulating area lights	Yes	Yes	Yes	-	-	-

\#   Washable apron may be provided in a planned manner to cover only stations from where trains terminate/originate or stop for longer duration in the morning hours.

@   Only public phones would be adequate.

*   In end platforms, all stalls be embedded in walls

  (i) The number of model stations identified for development of Indian Railways is 319 out of which 110 stations have already been developed.

 (ii) The amenities requiring less expenditure should be provided first and completed at all model stations as early as possible.

26.7 STATION PLATFORMS

Station platforms are provided for the entraining and detraining of passengers. Platforms can be rail-level, low-level, or high-level platforms depending upon the expected passenger traffic at each station. The general policy of Indian Railways is to provide high-level platforms at all important main line stations, low-level platforms at less important main line stations, and rail-level platforms at unimportant wayside stations.

The height of rail-level platforms coincides with the rail level, low-level platforms lie at a height of 455 mm (1'–6″), and high-level platforms lie at a height of 760 mm to 840 mm (2' – 6″ to 2' – 9″) in the case of BG lines and 305 mm to 405 mm (1' – 0″ to 1' – 4″) in the case of MG lines. Other details of these platforms are given in Table 26.3.

Table 26.3 Important features of passenger and goods platforms

Item	*Details*	
Passenger platforms		
Height of platform	High-level platforms (for all important main line stations)	0.76–0.84 m on BG, 0.305–0.405 m on MG
	Low-level platforms (less important than these for main line stations)	0.455 m on BG
	Rail-level platforms (for unimportant wayside stations)	At rail level
Length of platform	Enough length to accommodate the longest passenger train on the station. Minimum length of platform to be 180 m. Normally, a platform of a length of 450 m is provided on a main line to accommodate 20 bogies.	
Width of platform	Platform to be wide enough to accommodate the entire train load of passengers. The suggested yardstick for the width of a platform is 1.5 m^2 per passenger for main line and 1.0 m^2 per passenger for suburban trains. Minimum width of platform to be 3.66 m.	
End of platform	A ramp is provided with a slope of 1 in 6.	
Platform cover	Platform to be covered as per passenger requirement. Minimum length of platform cover to be 60 m.	
Water supply	The number of taps approved is two taps per 100 passengers.	
Toilets, urinals, and bathrooms	The prescribed scale is four toilet seats per 100 passengers, one urinal per 100 passengers, and one bathroom per 200 passengers.	
Station name boards	Two station name boards to be placed, one on each side of the platform, perpendicular to the track. Name of the station to be written in Hindi, English, and the regional language. Height of underside of boards to be 1.8 m.	
Goods platforms		
Height of platform	1.07 m for BG, 0.69 m for MG, and 0.61 m for NG	
Length of platform	Adequate enough to deal with goods received or dispatched; normally not less than 60 m.	
Width of platform	Depends upon volume of traffic, minimum width specified is 3.1 m.	
Other facilities	Weighing facilities, direct access road, paved platform, etc.	

26.8 MAIN BUILDING AREAS FOR DIFFERENT TYPES OF STATIONS

The main facilities provided in the case of a small station are a waiting hall, booking hall, assistant station master's (ASM) office, and storeroom. Different designs have been standardized for each type of station by the various railways, which provide all the facilities required by small and medium-size stations. When considering big stations, however, the design of an individual station building has to be drafted based on the requirement of passenger traffic with due regard to its architectural features.

26.8.1 Scale for Passenger Amenities

Booking office   A booking window should be provided at the rate of 800 tickets per window per seat. Electrically operated machine and computerized ticketing machine have different yardsticks. Additional windows are to be provided for reservations and cancellations and for inquiries.

Waiting hall and benches are to be provided for 45 per cent of the maximum passenger dealt at a time (excluding mela/fairs) for small stations and 30 per cent for large stations. For waiting hall, per passenger minimum area of 1.4 m^2 is to be provided. Number of benches to be provided at the rate of 40 seats per 100 passenger, for total number of passengers determined in the manner mentioned above.

Station platforms   These are provided for the passengers to entrain or detrian the trains. The platforms can be rail-level, low-level or high-level platforms depending upon the expected passenger traffic at each station. The policy generally adopted on the Indian Railways is to provide high-level platform at all important main line stations, low-level platforms, at less important main-line stations, and rail-level platforms at unimportant wayside stations as given in Table 26.4.

Table 26.4   Details of Platforms

Class of station	BG line	MG line
Important	High level (760 mm above rail level)	High level (405 mm above rail level)
Large	Low level (455 mm above rail level)	Low level (305 mm above rail level)
Small	Rail level	Rail level

Note: *In case of suburban stations and stations in cutting, the high-level platform should be 840 mm above rail level.*

The platform surface should be of such material, which is free from dust and mud during the dry and rainy seasons. A slope of 1 in 60 away from the coping is to be provided in case of a single face platform. In case of a two face (island platform) a slope of 1 in 60 away from the centre of platform is to be given. End of the platform is to be given a ramp of 1 in 6.

A platform is to be provided on all single face platforms the height of which should be 1800 mm.

Shade trees on platforms   Where there is no platform shelter adequate number of shade trees are to be planted. Even if platfrom shelter is to be provided some shade trees at suitable locations can be planted. It should be ensured that shade trees do not infringe the moving dimensions and should be away from any type of live conductor.

Drinking water   This is also a basic requirement. In case there is no facility for running water supply such as taps at least two hand pumps on each platform should be provided. On big stations in addition to adequate number of taps, water coolers should also be provided.

Latrines and urinals   These should be provided for 45 per cent of the maximum number of passengers dealt at a time for small stations and at the rate of 30 per cent for large stations. Latrines should be provided at a rate of minimum 4 seat per cent 100 passenger subject to minimum of 4 seats per platform. At large station urinal should be provided at the rate of one urinal per 100 passengers. The total number of urinals required will be calculated based on 30 per cent of the maximum number of passegers dealt at a time.

Platform cover   Depending upon the climatic condition, number of passengers and nature of traffic, platform covers should be provided at the rate of 6. sqm. per passenger, preferably to accommodate 50 per cent at the maximum passenger dealt at a time.

Foot overbridge or subway   A foot overbridge or subway as per the requirement should be provided keeping in view the following factors:
- inter connection between high-level or low-level platform
- maximum number of passengers dealt with at a time
- frequency of train services
- blocking of line platforms by freight trains

26.9 TYPES OF YARDS

A yard is a system of tracks laid out to deal with the passenger as well as goods traffic being handled by the Railways. This includes receipt and dispatch of trains apart from stabling, sorting, marshalling, and other such functions. Yards are normally classified into the following categories.

Coaching yard

The main function of a coaching yard is to deal with the reception and dispatch of passenger trains. Depending upon the volume of traffic, a yard provides facilities, such as watering and fuelling of engines, washing of rakes, examination of coaches, charging of batteries, and trans-shipment of passengers.

Goods yard

A goods yard provides facilities for the reception, stabling, loading, unloading, and dispatch of goods wagons. Most goods yards deal with a full train load of wagons. No sorting, marshalling, and reforming is done at goods yards except in the case of 'sick' wagons or a few wagons booked for that particular station. Separate goods sidings are provided with the platforms for the loading and unloading of the goods being handled at that station.

Marshalling yard

A goods yard which deals with the sorting of goods wagons to form new goods trains is called a marshalling yard. This is discussed in detail in Section 26.9.1.

Locomotive yard

The locomotive yard houses the locomotives. Facilities for watering, fuelling, examining locomotives, repairing, etc., are provided in this yard. The yard layout is designed depending upon the number of locomotives required to be housed in the locomotive shed. The facilities are so arranged that a requisite number of locomotives are serviced simultaneously and are readily available for hauling the trains. Such yards should have adequate space for storing fuel. The water supply should be adequate for washing the locomotives and servicing them.

Sick line yard

Whenever a wagon or coach becomes defective, it is marked 'sick' and taken to sick lines. The sick line yard deals with such sick wagons. Adequate facilities are provided for the repair of coaches and wagons, which include examination pits, crane arrangements, and train examiner's office and workshop. A good stock of spare parts should also be available with the TXR (train examiner) for repairing defective rolling stock.

26.9.1 Marshalling Yard

The marshalling yard (Fig. 26.14) is the yard where goods trains are received and sorted out, and new trains are formed and finally dispatched to various destinations. It receives loaded as well as empty goods wagons from different stations for further booking to different destinations. These wagons are separated, sorted out, properly marshalled, and finally dispatched bearing full trainloads to various destinations. The marshalling of trains is so done that the wagons can be conveniently detached without much shunting en route at wayside stations.

Functions

A marshalling yard serves the following functions at the specified locations within the yard itself.

Reception of trains   Trains are received in the reception yards with the help of various lines.

Sorting of trains   Trains are normally sorted with the help of a hump with a shunting neck and sorting sidings.

Departure of trains   Trains depart from departure yards where various lines are provided for this very purpose. Separate yards may be provided to deal with up and down traffic as well as through trains, which need not be sorted out.

Principles of design

A marshalling yard should be so designed that there is minimum detention of wagons in the yard and as such sorting can be done as quickly as possible. These yards should be provided with the necessary facilities, such as a long shunting neck, properly designed hump, braking arrangement in the shape of mechanical retarders, etc., depending upon the volume of traffic. The following points should be kept in mind when designing a marshalling yard.

(a) Through traffic should be received and dispatched as expeditiously as possible. Any idle time should be avoided.
(b) There should be a unidirectional movement of the wagons as far as possible.
(c) There should be no conflicting movement of wagons and engines in the various parts of the yard.
(d) The leads that permit the movement of wagons and train engines should be kept as short as possible.

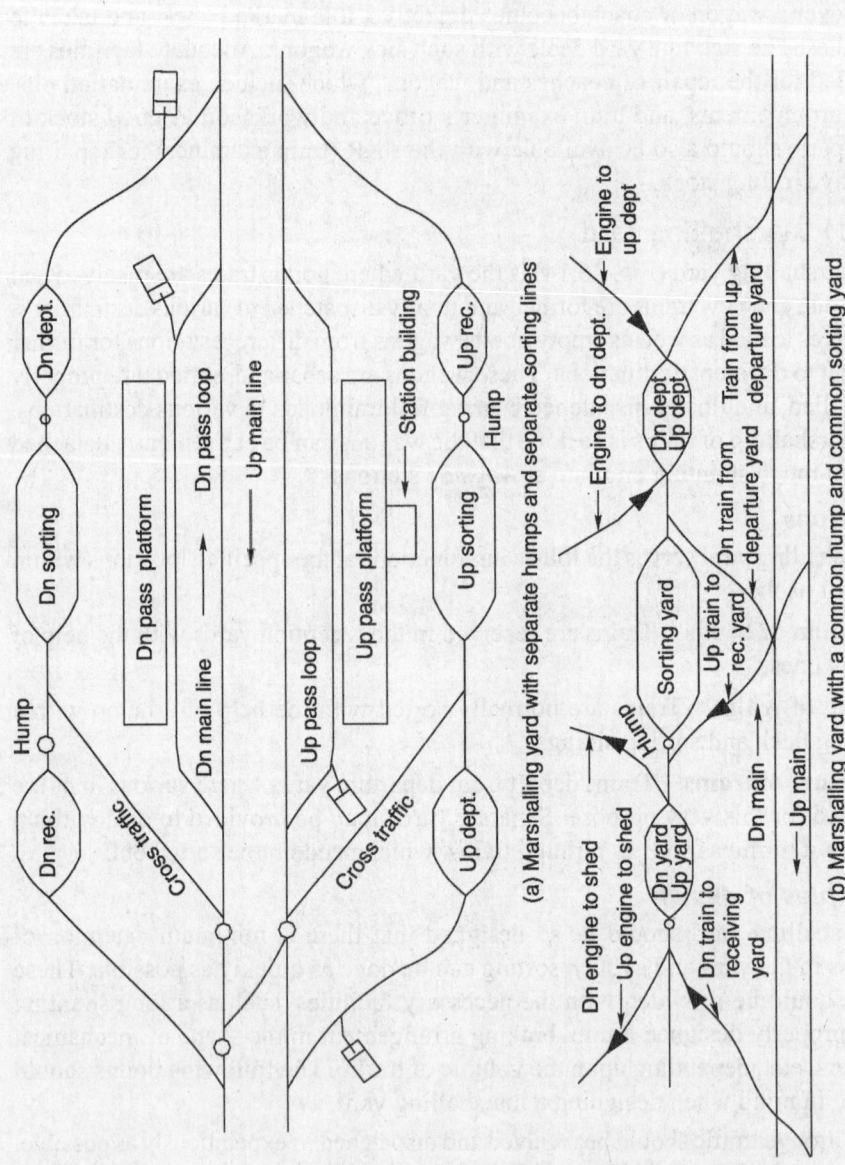

(a) Marshalling yard with separate humps and separate sorting lines

(b) Marshalling yard with a common hump and common sorting yard

Fig. 26.14 Layout of marshalling yards

(e) The marshalling yard should be well lighted.

(f) There should be adequate scope for further expansion of the marshalling yard.

Types

Marshalling yards can be classified into three main categories, namely flat yards, gravitation yards, and hump yards. This classification is based on the method of shunting used in the marshalling yard.

Flat yard   In this type of yard, all the tracks are laid almost level and the wagons are relocated for sorting, etc., with the help of an engine. This method is costly as it involves frequent shunting, which requires the constant use of locomotive power. The time required is also more as the engine has to traverse the same distance twice, first to carry the wagons to the place where they are to be sorted and then to return idle to the yard. This arrangement, therefore, is adopted when

(a) there is limitation of space,

(b) there is a severe limitation of funds, or

(c) the number of wagons dealt with by the marshalling yard is very low.

Gravitation yard   In this yard, the level of the natural ground is such that it is possible to lay some tracks at a gradient. The tracks are so laid that the wagons move to the siding assigned for the purpose of sorting by the action of gravity. Sometimes, shunting is done with the help of gravity assisted by engine power. However, it is very seldom that natural ground levels are so well suited for gravitation yards.

Hump yard   In this yard, an artificial hump is created by means of proper earthwork. The wagons are pushed up to the summit of the hump with the help of an engine from where they slide down and reach the sidings under the effect of gravity. A hump yard, therefore, can be said to be a gravitation yard as shunting is done under the effect of gravity. The gradients normally adopted in this regard are listed in Table 26.5. These are, however, only recommended gradients and the final gradient for a particular yard is decided after a test run of the trains over the humps, taking into consideration the rolling quality of different types of wagons and the spacing between successive groups of wagons. The topography of the location of the yard also plays an important role in deciding the gradient.

Table 26.5   Gradients in marshalling yards

Item	Gradients to be adopted for	
	Mechanical yards	*Non-mechanical yards*
Rising gradient of approach	1 in 50 to 1 in 125	1 in 50 to 1 in 100
Top of hump	Level	Level
First falling grade after apex of hump	1 in 17 to 1 in 20	1 in 25 to 1 in 35
Intermediate grade up to the point where the trains start	1 in 50 to 1 in 60	1 in 80 to 1 in 200

Contd.

Table 26.5 *(Contd.)*

Item	Gradients to be adopted for	
	Mechanical yards	*Non-mechanical yards*
Final falling gradient up to clearance of points	1 in 200 to level	1 in 80 to 1 in 200
Gradient of the sidings	Down-gradient eased off and then an up-gradient given to stop wagons at the end	Falling gradient 1 in 400 to 1 in 600

Regulation of speeds in hump yards The speed of the wagons is regulated to ensure that they are kept in a stable condition in the siding where they are to be sorted, so that there is least damage to them. The regulation of speed is done as follows.

Mechanical method In this method, wagons are slowed down automatically with the help of 'retarders' (Fig. 26.15). Retarders normally in the shape of bars fixed on either side of the track, operate electrically or electromechanically and offer resistance to the movement of wagons by pressing against the sides of the moving wheels. This finally stops the wagons at the appropriate place. Such mechanical retarders are used extensively in Germany and on other developed railways.

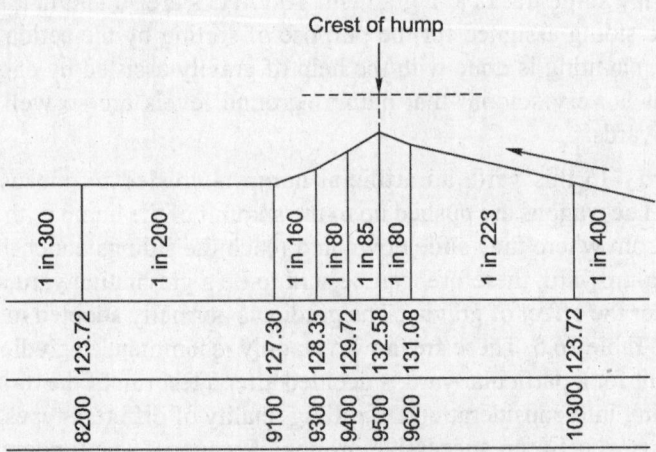

(a) New Katni—grades in hump yard

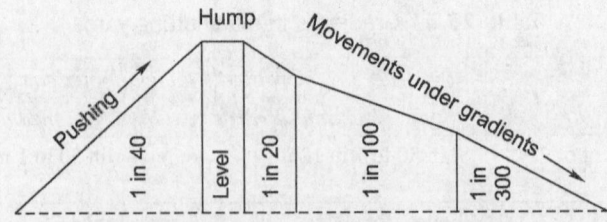

(b) Typical gradients in mechanized hump yard

Fig. 26.15 Gradients in mechanical hump yard

Non-mechanical yard In a non-mechanical yard, the speed of the wagon is regulated manually with the help of hand brakes or skids. A shunting porter runs alongside the wagons and applies a hand brake to the wagon at an appropriate place, making the wagon slip and stop. Skids are also used to slow down the wagons. Skids are placed on the track; they get dragged by the rolling wagon and the friction thus developed reduces the speed of the wagon and stops it at the desired location.

Design of various constituents

The design details of the various components of a marshalling yard are discussed below.

Spacing of marshalling yards   This depends upon the average distance that a long-distance train can go. If the lead is 500 km and the section train can go up to 100 km, the approved spacing of a marshalling yard is 400 km.

Siting of marshalling yard   A marshalling yard is normally sited at a junction point, a depot yard to a group of collieries, a feeder yard for a big terminal point, or a steel plant, etc.

Reception yard   The number of lines to be included in a reception yard depends upon the number of trains to be received and on the frequency of their arrival. Normally one reception line is provided for every three to four trains. The approved length of a siding is normally 700 m for BG and 650 m for MG.

Shunting neck   The length of the shunting neck should be longer than the longest train.

Hump   The hump should be designed to meet the following objectives.
  (a) It should be such that even the wagon whose movements are affected the worst by the most adverse weather conditions can clear the fouling mark, when sent to the outermost siding.
  (b) It should be such that a successive group of wagons are separated from each other to the extent that it enables the point between them to be operated upon so that the wagons can be sent to various sidings.
  (c) The hump should be such that the speed of the wagons is so regulated that there is no damage to the wagons when they bump against each other in the sorting lines. The figures given in Table 26.6 can be taken as a rough guide for choosing the design of the humps.

Table 26.6   Design of humps

Design element	Suggested value
Average gradient from the hump to the end of switching zone	2% for empty and 1.5% for loaded wagons
Average height of ordinary hump	2.5–3 m (8–10 ft)
Average height of mechanized hump	3.5–6 m (12–20 ft)

Sorting yard   The number of lines to be included in the sorting yard depends upon the number of destinations for which the trains are to be assembled. The length of each sorting line is about 15 to 20 per cent more than that of a normal train so that there is provision of some space behind the wagons. The layout of the sorting yard may be of the ladder or the balloon type. The speed of the wagons is controlled by hand brakes while the skids and the mechanical retarders are controlled by manual and mechanical means, respectively.

Departure yard The number of lines to be included in a departure yard depends upon the number of trains proposed to be dispatched from the yard and on the frequency of their departure. Some engineers feel that there is no need for a separate dispatching yard because it unnecessarily increases the length of the marshalling yard. According to them, trains should be dispatched straight from the sorting lines. This arrangement, however, runs into problems if the departure of the trains is delayed on account of operational reasons.

The pattern of transportation of goods traffic has changed drastically in the recent past. Now, most goods traffic is carried as trainloads from point to point. The loading of piecemeal wagons has also been drastically reduced. Consequently, the need and importance of goods marshalling yards has reduced considerably.

26.10 CATCH SIDINGS AND SLIP SIDINGS

Catch sidings are provided in the case of hilly terrains, where the gradients near railway stations are very steep. The purpose of catch sidings is to arrest the movement of the vehicles if they start to roll down the grade, which may eventually foul up the running lines. A separate siding is provided outside the station yard so that the vehicles can be collected there.

In Fig. 26.16, DEF is a running line and AB is a dead end siding. BC is the catch siding connected to the dead end siding preferably by the means of a spring-operated point. The catch siding lies on a rising gradient and its length is so designed that the vehicle loses its kinetic energy when it reaches the dead end. Thus the vehicle is protected from damage and the safety of the trains on the running line is ensured. There is a sand hump provided at the end of the catch siding to prevent any minor damage to the vehicle.

In the case of hilly terrains, normally one siding is provided at each end of the station as explained here.

Catch sidings These are provided at the higher level or upper end of a station when it starts to slope downwards along the track in an unauthorized manner from the previous station.

Slip sidings These are provided at the lower level on the lower end of the station. If by chance the vehicle is not caught in a catch siding and enters the station premises, the same will be caught and shipped into the slip siding.

Clapham Junction (London) of the Southern Region on British Railways is probably the largest junction station in the world. It has 17 platforms but 12 of these lines

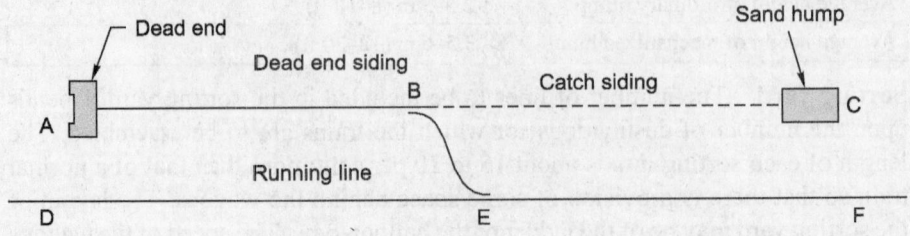

Fig. 26.16 Catch siding

are used by trains that either make only an ordinary station stop or do not stop at all. The other five platforms are used for miscellaneous purposes and are chiefly provided for trains transporting milk and other similar articles, which require a significant stop before the lines can be cleared. Thus, as many as 2500 trains can run daily with ease from this station, as very few trains occupy the platform for more than a minute or so.

SUMMARY

Stations and yards are provided to control the movement of trains, passengers, and goods. Stations are classified based on their operational and functional characteristics. The facilities to be provided at a station depend upon the type of station. Similarly, yards are also classified into coach yards, goods yards, marshalling yards, or locomotive yards depending upon their purpose. The efficiency of a station largely depends on the efficiency of its yards.

REVIEW QUESTIONS

1. (a) What are marshalling yards and where are they usually located?
   (b) Enumerate the principal types of marshalling yards and the basic facilities that should be provided with each of them.
2. Differentiate the following.
   (a) Flag station and block station
   (b) Island platform and dock platform
   (c) Junction and terminal
3. What is the purpose of providing marshalling yards? What are the points to be considered in the design of marshalling yards? What are the main siding features of marshalling yards?
4. What are the functions of a railway station? Explain briefly the various requirements of a railway station at an important city.
5. Draw a diagrammatic and dimensioned layout of a BG three-lines crossing station with the minimum provisions for goods handling. Also, mark the signals at either end at the appropriate distances. Assume the station to be a B class station with standard III interlocking and 70-wagons loop capacity.

Choose the correct answer from the choices given.

6. A railway station is provided for the following purpose:
   (a) to entrain or detrain the passengers
   (b) to load or off load the goods or parcels
   (c) to enable the faster trains to overtake the slower trains.
   (d) all of the above
7. For Category "A" stations, the annual passenger earning should be:
   (a) ₹ 100 million of more
   (b) ₹ 60 million or more
   (c) ₹ 30 million or more
   (d) less than ₹ 30 million

8. The number of taps on each platform of 'A' category stations as per norms for minimum essential facility criteria is:
   (a) 6   (b) 8
   (c) 12   (d) 15

9. The number of latrines on 'A' category stations as per norms for essential amenities is:
   (a) 2   (b) 4
   (c) 6   (d) 10

10. For A category stations it is necessary to provide:
    (a) high-level platform
    (b) medium-level platform
    (c) rail-level platform
    (d) all of the above

11. The height of high-level platform for passengers for BG is:
    (a) 0.455 m   (b) 0.560 m
    (c) 0.650 m   (d) 0.760 m

12. The height of low-level platform for passengers for BG is:
    (a) 0.455 m   (b) 0.560 m
    (c) 0.650 m   (d) 0.760 m

13. The number of model stations that have already been developed on Indian Railways is about:
    (a) 55   (b) 110
    (c) 265   (d) 320

Equipment at Railway Stations

INTRODUCTION

Many different types of equipment are required at railway stations and yards for the efficient working of the railway system. These serve the following purposes.

(a) Providing facilities for the convenience of passengers—platforms, foot overbridges, and subways.

(b) Receipt and dispatch of goods traffic—cranes, weigh bridges, loading gauges, and end loading ramps.

(c) Equipment for locomotives and coaches—locomotive sheds, examination pits, ashpits, water columns, turntables, and triangles.

(d) Isolation of running lines—derailing switch, scotch block, sand humps, buffer stops, and fouling marks.

The details of some of the important railway equipment are given in this chapter.

27.1 PLATFORMS

The following types of platforms are provided on Indian Railways.

Passenger platforms

These are provided to facilitate the movement of passengers and to help them entrain and detrain. Passenger platforms can be of three types, namely rail-level platforms, low-level platforms, and high-level platforms. These have been discussed in detail in Chapter 26.

Goods platforms

These platforms are provided for the loading and unloading of goods and parcels onto and from wagons. The essential features of these goods platforms are the following.

(a) A goods platform is normally surfaced with bituminous carpet or concrete. In the case of light traffic, moorum or water-bound macadam can also be used.

(b) The height of a goods platform is 1.07 m for BG, 0.69 m for MG, and 0.61 m for NG lines. The height of the platform is measured from the rail level and is such that the platform surface is flush with the floor of the wagon for the easy loading and unloading of goods.

(c) Adequate storage accommodation is provided on goods platforms for the storage of goods and parcels.

(d) Mobile cranes or fixed overhead gantry cranes are provided for the handling of bulky and heavy parcels.

27.2 FOOT OVERBRIDGES AND SUBWAYS

Foot overbridges are provided for the movement of passengers and light baggage from one platform to another. Bulky or heavy goods are taken from one platform to another by means of a handcarts, which are carried across the tracks near the end of the platform in order to reach the requisite platform. Some stations are also provided with subways for the movement of the passengers and goods between platforms.

27.3 CRANES

Cranes are normally provided in goods sheds to load and unload bulky or heavy material, such as heavy machines and logs from wagons. These are normally of three types.

Fixed jib crane   This crane is fixed at a convenient location on the goods platform for the purpose of loading and unloading bulky and heavy goods from wagons.

Mobile crane   This crane is mounted on a wagon or a truck and can be moved anywhere on the platform as per requirement to load or unload bulky parcels.

Overhead gantry cranes   It consists of two horizontal girders or beams supported on a number of vertical posts. A travelling platform is fixed in between the two girders, which is fitted with equipment for hoisting goods and is capable of moving to and fro on the girders. Wagons or road vehicles are brought under the gantry for loading and unloading materials.

27.4 WEIGH BRIDGE

A weigh bridge is used to weigh a loaded wagon so as to get an idea of the weight of its contents. It is basically a small length of track on a platform that is supported on beams. The beams are located in a pit under the track and rest on knife edges attached to levers. When a wagon is placed on the weigh bridge, its weight is indicated by a pointer on a graduated disc located in an adjoining structure. A weigh bridge is normally provided on a siding and not on a through track.

27.5 LOADING GAUGE

The loading gauge is the gauge or profile up to which a vehicle can be loaded in order to maintain a minimum clearance between the loaded top of the wagon and the underside of a structure, such as a bridge, tunnel, or signal post. The dimensions of the load of the wagon are kept within the fixed limits by erecting a loading gauge across the track in the shape of a steel frame, which causes an obstruction if the wagon is loaded to an extent that exceeds the loading gauge (see Fig. 2.2 in Chapter 2).

Loaded wagons whose dimensions exceed the normal permissible loads are known as *oversized dimensioned consignments* (ODCs). ODC parcels can also be sanctioned up to a certain level after taking the following precautions.

(a) Trains carrying ODCs should move at a restricted speed, which is specified for the purpose.
(b) Such trains should not move at night.
(c) Train carrying ODCs should be accompanied by a supervisor whose duty is to ensure the safety of the train along the route.

Construction Gauge

The construction gauge is decided by adding necessary clearance to the loading gauge so that vehicles can move safely at prescribed speed without any infringement. Various fixed structures on the rails such as bridges, tunnels, and platform sheds are built as per construction gauge so that the side and top remain clear of the loading gague.

27.6 END LOADING RAMPS

End loading ramps (Fig. 27.1) are provided to allow the unloading of the wagons at their rear end. Such ramps are also used for unloading cars and other mobile vehicles. An end loading ramp has the following essential features.

(a) It has a dead end siding with a buffer stop and a platform with a ramp.
(b) The platform is at a height of 1.3 m for BG and 0.86 m for MG lines.

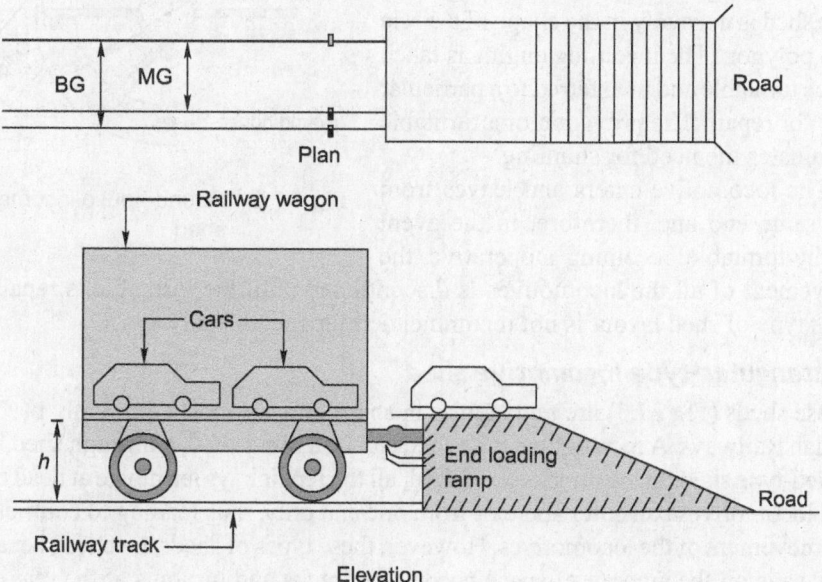

Fig. 27.1 End loading ramp

(c) A small gap is maintained between the buffer stop and the ramp platform to minimize the damage to the platform. This gap is covered by the hinged plates of the wagon while it is being unloaded.

27.7 LOCOMOTIVE SHEDS

Locomotive or running sheds are meant for the maintenance and servicing of locomotives. The location and design of a locomotive shed depends on the volume and pattern of traffic, the layout of the terminal station and the marshalling yard, and other allied factors. Locomotive sheds are normally spaced at about 250 to 300 km apart in order to avoid the idle movement of locomotives and crew. Locomotive sheds are basically of two types.

Homing sheds These are provided to house locomotives and attend to their maintenance and servicing. Equal stress is laid on the servicing and maintenance of locomotives in these sheds. These sheds are normally designed to house about 80 to 100 locomotives.

Turn round sheds These are provided for servicing locomotives and bringing them back to the homing sheds. They may also be provided for attending to certain minor repairs. These sheds are normally designed to hold about 30 to 50 locomotives.

27.7.1 Layout of Locomotive Sheds

The layout of locomotive sheds is normally of three different types namely, round house type, rectangular type, and mixed type.

Round house locomotive shed

The round house shed (Fig. 27.2) consists of a number of locomotive repair and stabling lines radiating from a turntable in the centre. The shed is normally in the shape of a circle or a polygon. The incoming engine is taken to the turntable and transferred to a particular line for repair. The provision of a turntable eliminates the need for shunting.

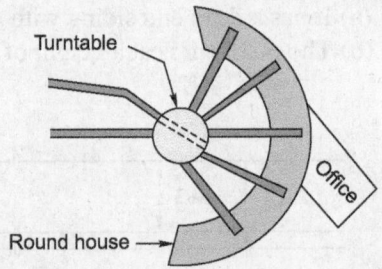

Fig. 27.2 Round house locomotive shed

The locomotive enters and leaves from the same end and, therefore, in the event of the turntable becoming inoperative, the movement of all the locomotives is discontinued until the turntable is repaired. This type of shed layout is not recommended for Indian Railways.

Rectangular-type locomotive shed

These sheds (Fig 27.3) are rectangular in shape and are most commonly used on Indian Railways. A rectangular-type shed can be a blind shed, a through shed, or a mixed-type shed. In the blind type of shed, all the repair bays terminate at dead ends and locomotives can enter and exit from one end only, thus leading to conflicts in the movement of the locomotives. However, these types of sheds are cheap because they save on the amount of space needed for tracks and turnouts at the rear end. Such rectangular sheds can be planned for small locomotive holdings.

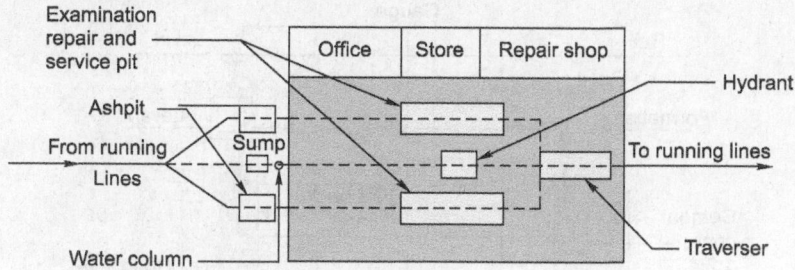

Fig. 27.3 Rectangular-type locomotive shed

In a through rectangular shed, the locomotives move from one end to the other in the correct sequence of operations that are to be performed for servicing. Such a shed offers considerable scope for future expansion and tenders the maximum flexibility in shed operations.

Mixed-type shed

A mixed-type shed has the characteristics of both a blind and a through type. These sheds are suitable for locations that require expansion in the near future but lack the requisite.

27.7.2 Essentials of a Well-laid Locomotive Shed

A locomotives shed should satisfy the following conditions.

(a) As far as possible, the design of the shed should allow the engines to move only in one direction. The multidirectional movement of the engine, apart from creating unsafe conditions, would invariably slow the engine down and cause them to be unduly detained.

(b) The design should be such that the time taken by the engine in passing through the shed is reduced to the minimum.

(c) The layout should be such that it is possible to skip one or more stages of servicing, as all engines do not require the entire cycle of servicing.

(d) There should be adequate stabling accommodation, both covered and open, for all engines.

(e) There should be adequate facilities for the servicing and repairs of engines.

(f) There should be a separate shed for carrying out heavy repairs involving the lifting of engines or dropping of wheels.

(g) It should be provided with an adequate water supply for the servicing of locomotives as well as for domestic use.

(h) There should be provision for stores and a tool room for general maintenance and repair work. Adequate office accommodation should also be made available for the supervisors and the staff.

(i) The sequence of operations normally followed in locomotive sheds is as follows: inspection → cooling → turning → fire cleaning → placement → repairs → outgoing inspection → departure to traffic yard.

27.8 ASHPITS

Ashpits (also called de-ashing pits) (Fig. 27.4) are provided to collect the ashes falling from the locomotives.

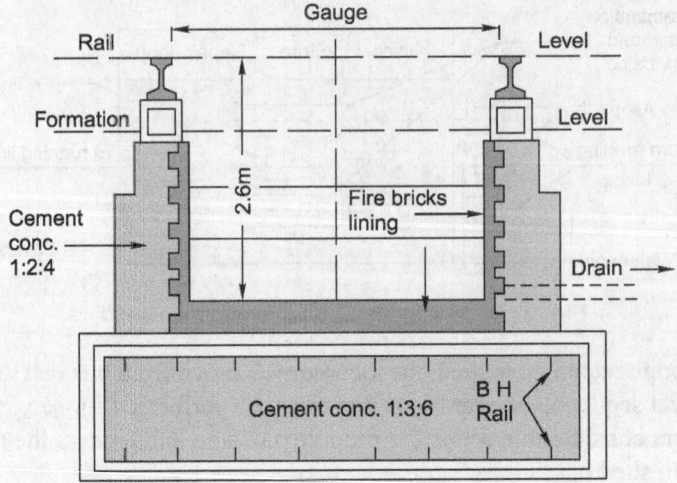

Fig. 27.4 Ashpit

Ashpits are normally provided at those points in the locomotive sheds where the locomotives turn for cleaning or dropping of fire. These are also provided in big stations at places where the locomotives collect water for de-ashing.

27.8.1 Ash Pans

Ash pans are also used for the de-ashing of locomotives. These consist of U-shaped precast reinforced concrete units placed side by side for the retention of ashes. Ash pans are normally provided in station yards. Though these pans have a very low capacity, they still have the following advantages.

(a) Easy to construct

(b) Very economical

(c) No speed restriction necessary on the main line when ash pans are provided

27.8.2 Examination Pits

Examination pits (also called outgoing pits) are used both for fire de-ashing before the locomotives leave the sheds and for outgoing engine examination and repairs. These pits should have a minimum length of 25 m with stairs at the ends to enable the staff to go underneath the locomotives for inspection and repair. The pit should be about 1 m deep and lined with fire bricks for about 6 m in the centre where fire cleaning is to be carried out. A water column should be provided by the side of each pit.

27.8.3 Drop Pits

Drop pits are provided in order to enable the wheels of the locomotives to be removed for examination, repairs, and renewals. These pits are normally provided at right angles to the track. Mobile jacks are installed to enable the wheels and axles to be removed.

27.9 WATER COLUMNS

Water columns (Fig. 27.5) are provided to supply water to the locomotives.

Water columns are provided in locomotive yards as well as at various stations, where engines are required to be watered and fuelled.

Note: As steam locomotives have already been phased out on Indian Railways, these types of equipment (ash pits, water column, etc.), meant for servicing steam locos are of no relevance today.

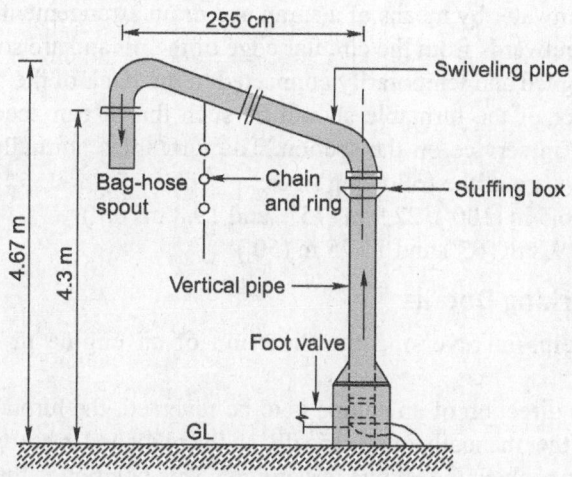

Fig. 27.5 Water column

27.10 TURNTABLE

A turntable (Fig. 27.6) is a device used for changing the direction of a locomotive. It is normally provided at terminal stations, locomotive yards, and marshalling yards.

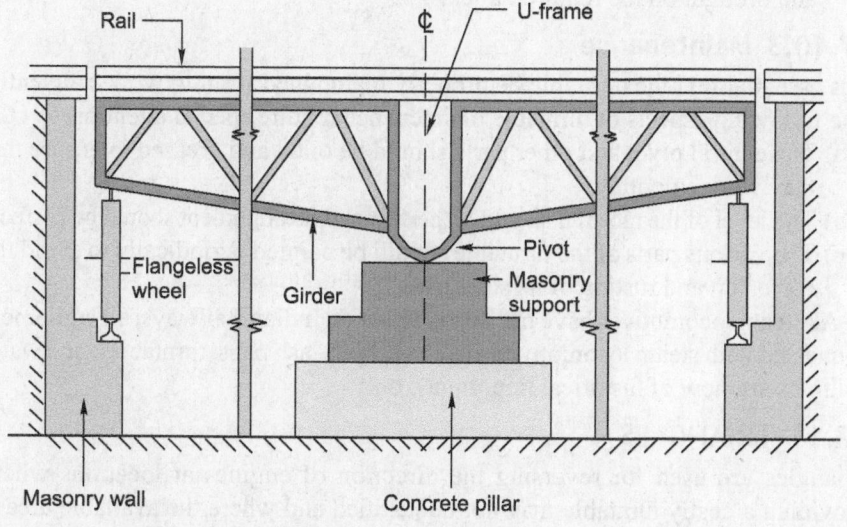

Fig. 27.6 Turntable

27.10.1 Main Features

A turntable basically consists of a track supported on two parallel fish-bellied girders, which are braced together and supported on or suspended from a central pivot. Sets of flangeless wheels are attached to the ends of the girders, which roll over the rails along the circumference of the pit, called race rails. The turntable is installed in a circular masonry pit. The girders supporting the track move around the rail and can be fixed in the desired position with the help of locking bolts. The bottom of the masonry pit slopes towards the centre and a sump is provided for draining rainwater by means of a sump and drain arrangement. Two or more tracks radiate outwards from the circular edge of the pit and are so designed that they can be aligned and temporarily connected to the track of the turntable.

The diameter of the turntable should be such that it can accommodate the longest engine in service on the section. The turntables normally provided on Indian Railways are of the following sizes:

BG: 30.5 m (100′), 22.9 m (75′), and 19.8 m (65′)

MG: 19.8 m (65′) and 15.75 m (50′)

27.10.2 Working Details

A turntable helps in reversing the direction of an engine in the following manner.

1. When the direction of an engine is to be reversed, the turntable is made to revolve either manually or electrically on the pivot and is brought in line with the track on which the engine is standing. This position is then locked with the help of locking bolts.
2. The engine is then brought on the track of the turntable and the locking bolts are removed.
3. The turntable is rotated again till the turntable track aligns with the track where the engine is required to go.
4. This position is locked with the help of locking bolts and the engine is shunted and brought on the required track.

27.10.3 Maintenance

It is essential that the turntable be properly maintained for it to work efficiently. The following aspects of turntable maintenance require special attention.

(a) The central pivot and other parts should be oiled and greased to ensure that they work smoothly.
(b) The level of the race rail should be perfect and its alignment should be correct.
(c) The various parts of the turntable should be painted periodically to avoid the corrosion and rusting of its steel parts.

As steam locomotives have been phased out on Indian Railways, all equipment connected with steam locomotives, such as ashpits, ash pans, turntables, and water columns are now of historical importance only.

27.11 TRIANGLES

Triangles are used for reversing the direction of engines at locations where providing a costly turntable may not be justified and where the available area is adequate for the provision of a triangle. Triangles are normally provided at the terminals of short lines. The details of a triangle are given in Chapter 15.

27.12 TRAVERSER

Traverser is a device for transferring vehicles from one track to a parallel one without the use of a turnout or a crossover. It is quite a costly arrangement and is preferred only in workshops where space is limited and a coach or locomotive is required to be shifted from one shop to another on a nominated line.

A traverser (Fig. 27.7) consists of a platform with a track that is mounted on small wheels or rollers, which can traverse to and fro and can fall in line with the track on either side of the traverser. The following steps are involved in transferring a vehicle standing on track 3 on the left-hand side of the traverser to track 2 on the right-hand side.

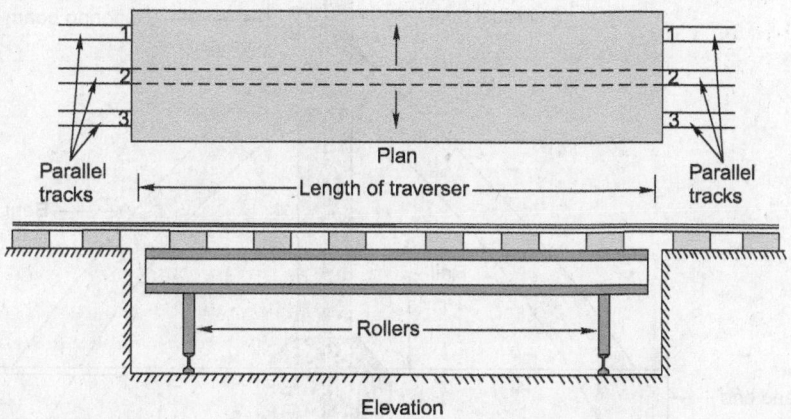

Fig. 27.7 Traverser

1. The traverser is aligned with track 3 on the left-hand side.
2. The vehicle is then transferred to the traverser track.
3. The traverser is then shifted so as the align it with the track on the right-hand side.
4. The vehicle is then transferred to track 2 on the right-hand side.

27.13 CARRIAGE WASHING PLATFORMS

Terminal stations are provided with the facility for washing carriages so that passenger bogies can be cleaned and washed properly. This consists of two or three long sidings that can accommodate the entire length of the rake and washing platforms that are provided between the sidings. The salient features of these washing platforms are the following.

(a) Washing platforms are long platforms of a height equal to the height of the carriage floor that are generally made of cement concrete. In the case of BG lines, the width is about 0.61 m.

(b) The washing platform is equipped with a number of hydrants for washing the carriages. An adequate water supply is made available to ensure that the pressure of the water inside the hydrant is sufficient.

(c) A washing platform is normally provided between two tracks so that two carriages can be washed simultaneously.

(d) The tracks are supported on masonry structures and have adequate arrangement for the drainage of water.

27.14 BUFFER STOP

Buffer stops (Fig. 27.8) or 'snag dead ends' are provided at the end of a siding to ensure that the vehicles stop while still on the track and do not go off it. The buffer stop is a type of barrier placed across the track, which stops the vehicles from going beyond the selected point. Its essential features are the following.

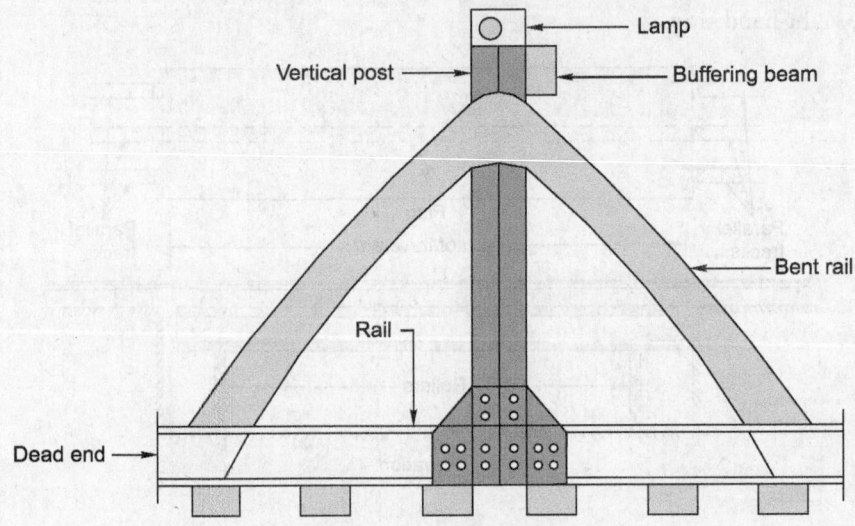

Fig. 27.8 Buffer stop

(a) The buffer stop should be structurally strong to take the impact of a rolling vehicle.
(b) It should have a buffer disc with a cross-sleeper, which is normally painted red. A red lamp should be provided at its centre for night indication.
(c) Normally the track should be straight for some distance near the buffer stop.
(d) It should be visible from a long distance.

27.15 SCOTCH BLOCK, DERAILING SWITCH, AND SAND HUMP

It is a normal practice to isolate a through running line from a siding so that a vehicle standing on the siding does not accidentally roll onto the running line and foul the same. A scotch block or derailing switch is provided on a siding or shunting neck to ensure that the vehicle does not go beyond a particular point and if that happens, the vehicle gets derailed.

Scotch block It is a wooden block (Fig. 27.9) placed on the rail and properly held in its place with the help of a device to form an obstruction. Once it is clamped in position, the scotch block does not allow a vehicle to move beyond it.

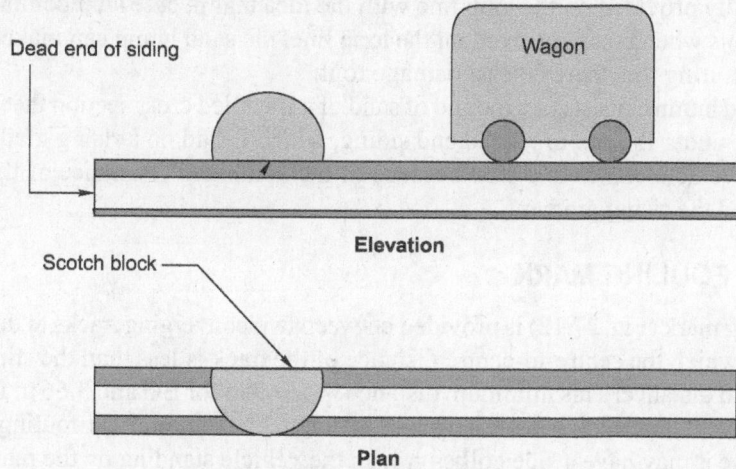

Fig. 27.9   Scotch block

Derailing switch   It consists of a half-switch (Fig. 27.10), i.e., only a tongue rail, which in its open position faces away from the stock rail, leaving a gap in between, and this causes a discontinuity in the track. A vehicle cannot go beyond this point and gets automatically derailed if it does manage to do so. The switch can be closed with the help of a lever and a vehicle can then traverse it normally. This is also called a *trap switch*.

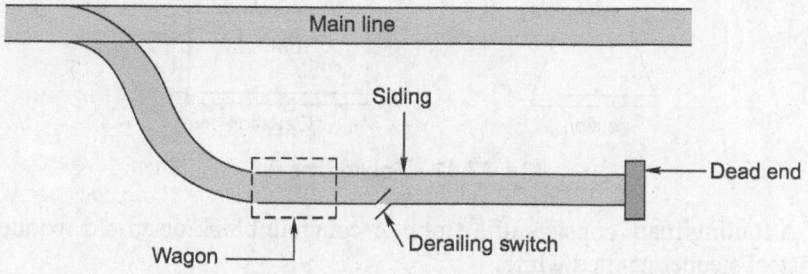

Fig. 27.10   Derailing switch

Sand hump   This is possibly the most improved method of isolating and stopping a moving vehicle without causing much damage to it. The sand hump (Fig. 27.11)

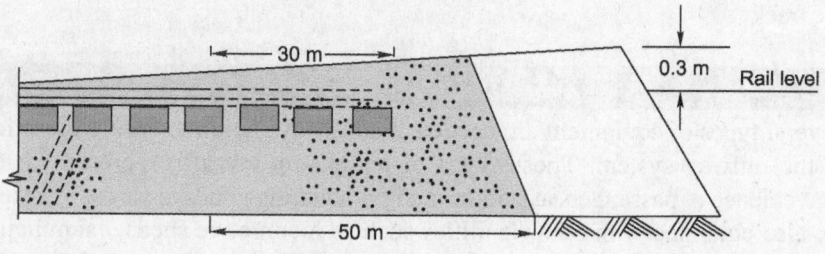

Fig. 27.11   Sand hump

is normally provided on the loop line with the idea that in case an incoming train overshoots when being received on the loop line, the sand hump can make it stop while ensuring that there is least damage to it.

A sand hump consists of a mound of sand of a specified cross section that covers the track under the end of a dead end siding, which is laid on a rising gradient. A moving vehicle comes to a stop because of the combined resistance of the sand hump and the rising gradient.

27.16 FOULING MARK

A fouling mark (Fig. 27.12) is provided between two converging tracks at the point beyond which the centre-to-centre distance of the track is less than the stipulated minimum distance. This minimum distance is 4.265 m for BG and 3.66 m for MG lines. A vehicle standing on a loop line is not stabled beyond the fouling mark, otherwise it may have a side collision with the vehicle standing on the main line. The salient features of a fouling mark are as follows.

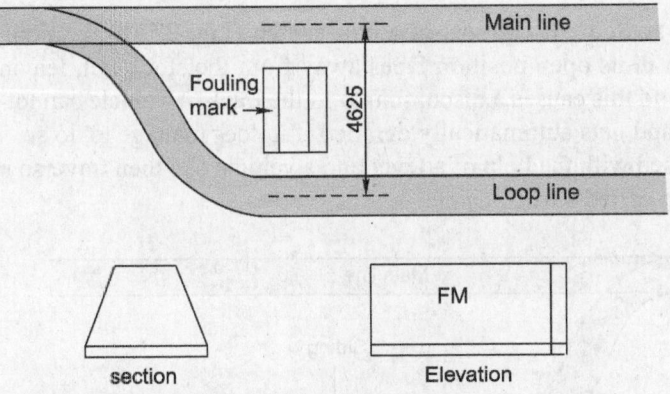

Fig. 27.12 Fouling mark

(a) A fouling mark consists of a stone or concrete block or an old wooden or steel sleeper painted white.
(b) The fouling mark should be visible from a distance. Therefore, it is painted white and has the letters FM marked on it in bold using black paint.
(c) The top of the fouling mark should be in line with the top of the ballast section.
(d) The fouling mark should be fixed firmly on the ground at right angles to the track.

SUMMARY

Several types of equipment are needed at a railway station to ensure the safety of the railway system. These types of equipment are also required for the convenience of passengers at stations and for handling goods in yards. The yards are also equipped with other facilities such as locomotive sheds, examination pits, and turntables. The safety and efficiency of a station are greatly influenced by the quality of the equipment.

REVIEW QUESTIONS

1. What are the various devices and types of equipment used in station yards?
2. What is a turntable? In which railway yard is this located? Explain its function.
3. Give a list of the equipment required by an ordinary railway station.
4. What are the essentials of a well-laid locomotive shed? Describe with the help of sketches the layout of locomotive sheds on Indian Railways.
5. Write short notes on: (a) turntable, (b) ashpit, (c) loading gauge, (d) fouling mark, (e) derailing switch

Choose the correct answer from the choices given.

6. ODC is the abbreviation of:
   (a) Overtime Difficult Construction
   (b) Oversized Dimensional Consingnment
   (c) Overhead Direct Consignment
   (d) none of these
7. In an end-loading ramp, the platform for BG is at a height of:
   (a) 1.00 m   (b) 1.15 m
   (c) 1.30 m   (d) 1.45 m
8. Ashpit is used in the working of a:
   (a) steam locomotive   (b) diesel locomotive
   (c) electric locomotive   (d) none of these
9. The maximum diameter of turntables for BG is:
   (a) 15 m   (b) 22 m
   (c) 30.5 m   (d) 35.5 m

New Lines, Doublings, and Gauge Conversion Projects

INTRODUCTION

The construction of a new railway line in an area is normally required for one or more of the following reasons.

(a) Strategic and political considerations
(b) Development of backward areas
(c) Connecting new trade centres
(d) Shortening of existing rail lines

Doubling of an existing single railway line to a double line is also done in a few cases to cope up with the additional requirement of traffic.

A large number of gauge conversion projects have also been taken up recently for converting the existing MG lines into BG lines in order to have a uniform gauge for smooth flow of traffic.

28.1 CONSTRUCTION OF NEW LINES

The main tasks involved in the construction of a new line are as follows.

(a) Land acquisition
(b) Earthwork and bridges
(c) Construction of station building, staff quarters, and other allied facilities, including platforms and sheds
(d) Laying of plates including ballasting of track
(e) Opening of section to traffic

28.1.1 Land Acquisition

The work of land acquisition should start well in advance so that all the legal and financial formalities are completed in time and the possession of the land is acquired for the start of the construction work. Land acquisition is done with the help of the State Government as per the procedure laid down in the Land Acquisition Act.

Normally, Sections 4 and 6 of the Land Acquisition Act are applied for the acquisition of land. Whenever land is to be acquired, it is generally done after giving a certain notice and paying the requisite compensation. In the case of an emergency, land can also be acquired urgently by the application of Special Sections 9 and 17 of the Land Acquisition Act.

The land being acquired should be sufficient for the formation, berm, and borrow pits. It should also have adequate provision for any future expansion. Even when a single line is to be constructed, it should be ascertained that the land made available is suitable for future conversion into a double line. Normally a strip of 15 to 25 m of land is acquired for the construction of a railway line. An extra width of land is acquired for station yards. In the case of small stations, the width normally adopted is 150 × 1000 m.

The minimum measurements of the selected land should be such that it can cater to the following requirement.

Width of formation The land should be adequate so as to accommodate the width of the formation.

Side slope This depends on the nature of the soil and is normally taken as 2:1 (horizontal:vertical).

Width of berm The width of the berm is usually kept as 3 m.

Borrow pits If the land is not very costly, adequate land should also be made available for borrow pits. Borrow pits may be provided on one side of the track for low banks and on both sides of the track for medium and high banks.

When the land is expensive, borrow pits need not be provided and instead earth can be borrowed from adjoining areas. Extra land is, however, required for station yards, level crossings, and bridge approaches.

28.1.2 Earthwork for Formation

Depending upon the rail level and general contour of the area, the formation may be laid in an embankment or in a cutting. A formation laid in an embankment is normally preferred because it affords good drainage. The height of the embankment also depends on the high flood level (HFL) of the area and a reasonable free board should be given above the HFL.

The standard widths of the formation for BG and MG lines are given in Table 28.1.

Table 28.1 Standard width of formation for BG and MG lines

Gauge	Type of formation	Width of formation in m (ft)	
		Single line	*Double line***
BG	Embankment	6.85 m (22′6″)	12.155 m (41′2″)
	Cutting	6.25 m (20′6″)	11.155 m (37′0″)
MG	Embankment	5.85 m (19′1″)	9.81 m (32′2″)
	Cutting	5.25 m (17′1″)	9.21 m (30′2″)

** *Notes:* (i) Doube line in BG are at spacing of 4.725 m (15′6″) track centres and in MG are at spacing of 3.96 m (13′0″) track centres

(ii) The difference between width provided in BG and MG is mainly on account of the difference in length of concrete sleepers and their end thickness.

Some of the points to be kept in mind with regard to earthwork are given below.

(a) Earthwork is normally done in 30 cm layers so that the soil is well compacted.

(b) Mechanical compaction is normally done after each layer of earthwork with the help of a sheep foot roller to obtain 90 per cent maximum dry density at an optimum moisture content.

(c) A shrinkage allowance of 5 per cent is made for the consolidation of the final cross section in the case of mechanical compaction. The shrinkage allowance is increased to 10 per cent if no mechanical compaction is involved.

(d) A blanket of a thickness of about 30 cm is provided at the top of the embankment where the soil is not of good quality.

(e) In areas where there are both cuttings and embankments, the soil from the cuttings should be used for the embankments up to an economical lead. The economical limit of moving the earth in the longitudinal direction is determined by the *mass-haul curve*.

(f) For the early execution of earthwork, the section is normally divided into convenient zones, with each zone requiring earthwork costing ₹ 1.5 to 3 million approximately. Tenders are separately invited for each zone so that work can progress simultaneously in all the zones.

28.1.3 Bridges

Bridges should be designed to bear the load of the heaviest locomotive likely to pass the section. Depending upon the topography of the location and the type of stream to be crossed, hume pipe culverts, RCC slab bridges, plate girders, PRC girder bridges, or steel bridges are designed. Bridges, being important structures, are normally constructed to accommodate a double line even in those sections where only a single line is being set up so that future expansions can be planned.

Separate tenders are invited for the construction of important bridges so that the bridge design and construction can also be included in the tender documents. Minor bridges and culverts are normally included in the earthwork zones mentioned earlier.

28.1.4 Service Buildings and Staff Quarters

Service buildings include buildings, such as the station master's office or telegraph office, which are basically required for providing assistance in the running of trains. Apart from this, staff quarters and other passenger amenities such as platforms, foot overbridges, waiting halls, and retiring rooms are also provided at the stations. Many other ancillary facilities such as water, drainage, telephone lines, and electricity are also made available at every station.

All these constructions are taken up simultaneously by civil engineers, electrical engineers, and signal engineers so that all work can progress together.

28.1.5 Plate Laying or Track Linking

Once the formation is ready, plate laying or track linking should be done, which consists of laying rails, sleepers, and fastenings. The following methods can be adopted for plate laying.

Tram line method

In this method, a temporary line known as a *tram line is* laid by the side of the proposed track for transporting track material to the site. This method can be useful in flat terrains, where laying of a tram line on the natural ground may be comparatively easier. This method is, however, seldom used in actual practice.

A modification of the above method, called *side method*, is also in practice, where track and bridge material such as steel girders and RCC slabs is carried to the site in trucks on a service road that runs parallel to the track. These materials are then unloaded near the work site. This method is used only in cases where the terrain is comparatively flat.

American method

In this method, rails and sleepers are first assembled in the base depot, and the pre-assembled track panels are then conveyed to the site along with the necessary cranes, etc. The track panels are then unloaded at the site of work either manually or with the help of cranes and laid in their final position.

This procedure is used in many developed countries, particularly where concrete sleepers are laid, which are quite heavy and not very easy to handle manually.

Telescopic method

This method is widely used on Indian Railways. In this method, the rails, sleepers, and other fittings are taken to the base depot and unloaded. The track material is then taken to the rail head, where the tracks are linked and packed. The rail head is then advanced up to the point where the track has been laid. The track materials are then taken up to the extended rail head with the help of a dip lorry and the track is linked and packed again. Thus, the rail head goes on advancing till the entire track has been linked.

28.2 FIELD EXECUTION OF TRACK LINKING

Field execution of track linking can be done either by conventional methods using manual labour or by mechanical means using modern track machines.

28.2.1 Conventional Methods Using Manual Labour

Unloading of materials

The track materials are taken to the base depot and unloaded with the help of material gangs. The first base depot lies at the junction of the existing line and the new line to be constructed. All the track material is taken from the base depot to the rail head with the help of a dip lorry (a special type of trolley). The rail head goes on advancing till the track is sufficiently linked. After that, a subsidiary depot is established at a distance of about 5 km and track material carried to this depot with the help of a material train. Alternatively, track material is transferred from base depot with the help of a dip lorry up to a distance of about 2 km and by means of a material train beyond this distance. The base depot has arrangements for advanced processes, such as adzing and boring of sleepers as well as for matching materials, etc., to ensure the speedy linking of the track at the site.

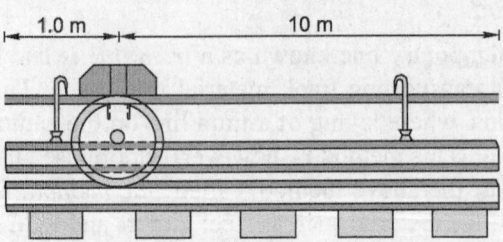

Fig. 28.1 Anderson rail carrier

Linking of track

Once the track material is unloaded, the track is linked with the help of linking gangs. The following procedure is normally adopted for this purpose.

1. A string is first stretched along the central line of the alignment and the sleepers are laid with their centres on the string. The sleepers are laid roughly at the desired spacing, keeping the total number of sleepers per rail intact.
2. The rails are carried using rail tongs and laid on the cess of the bank almost near the final position. Carrying rails is a strenuous job, as about 12 to 15 gangmen are required to carry each rail (each rail weighs about 0.6 tonnes or so). A special type of rail carrier known as the Anderson rail carrier, shown in Fig. 28.1, can be used for carrying rails with lesser strain.
3. Next the sleepers are distributed over the length of the formation. The rails have markings to indicate the final position of the sleepers as shown in Fig. 28.1.
4. Small fittings such as fish plates and bolts are kept near the joints. The fittings required for each sleeper are kept near the ends of the sleepers.
5. The rails are then placed on the sleepers and fixed with the help of fittings, which are chosen depending upon the type of sleeper. For example, rail screws are used for fixing rails to wooden sleepers. In the case of steel sleepers, rails are fixed with the help of keys. Bearing plates are also provided wherever required, as per the prescribed track standards.
6. The rails are joined with each other after ensuring that there is sufficient gap between them. Normally, the initial laying of the tracks is done using three rail panels. Adequate expansion gaps should be provided in the case of single-rail as well as three-rail panels. The recommended expansion gaps are provided with the help of steel liners or shims of appropriate thickness (1 mm to 4 mm), which are fixed between the two rail ends.

Packing of track

The track is then thoroughly packed with the help of beaters by the packing-in-gangs. The following aspects should be examined during this process.

(a) The track should have a proper gradient.
(b) If the track is on a curve, it should have proper curvature.
(c) The cross levels should be even. If a track is to be provided with the recommended superelevation, this can be achieved by raising the outer rail.
(d) The track should be thoroughly packed and should be free of hollow spaces.

Ballasting of track

The railway line is normally covered with the ballast after the embankment has settled and has endured at least two monsoons. Ballasting is generally done with the help of a ballast train, which has special hoppers that are used for automatically unloading the ballast onto the track. Alternatively, the ballast is taken to the cess and then placed on the track manually. Either method ensures that the ballast is thoroughly packed and inserted properly under the track.

28.2.2 Mechanical Means Using Modern Track Machines

The following mechanized track linking system can be used.
1. Linking of track with the help of portal cranes using pre-assembled panels.
2. Deployment of automatic track relaying train, where all the track linking items are done automatically with single set of machines expeditiously and at great precision.

Details of modern track machines which are used for track linking, viz. portal cranes and track relaying train (TRT) are given in Chapter 20.

28.2.3 Requirement of Track Materials

The requirement of track material for 1 km length of BG track with (M + 7) sleeper density is calculated as follows:

Rails The standard rails are of 13 m length for BG and 12 m length for MG.
- Number of sleepers per rail $= 13 + 7 = 20$
- Number of rails per km for BG $= \dfrac{1000}{13} \times 2 = 154$
- Wt of 52 kg rails per km $= 52 \times 154$
  $= 8008$ kg

Number of sleepers The number of sleepers depend upon the sleeper density.
For a sleeper density of M + 7,
Number of sleepers per rail $= 13 + 7 = 20$
Number of sleeper per km $= 77 \times 20 = 1540$

Fittings and fastenings with wooden sleeper track
(i) Number of fish plates per km $= 2 \times$ number of rails per km
$= 2 \times 154 = 308$
(ii) Number of fish bolts $= 4 \times$ number of rail per km
$= 4 \times 154 = 616$
(iii) Number of bearing plates $=$ Number of sleepers $\times 2$
$= 1540 \times 2 = 3080$
(iv) Number of screw spikes $=$ Number of sleepers $\times 4$
$= 1540 \times 4 = 6160$

Fittings and fastenings with concrete sleeper track
(i) Number of concrete sleepers per km of track $= 20 \times 77 = 1540$
(ii) Number of grooved rubber pads (2 per sleeper) $= 1540 \times 2 = 3080$
(iii) Number of nylon liners (4 per sleeper) $= 1540 \times 4 = 6160$
(iv) Elastic rail clips (4 per sleeper) $= 1540 \times 4 = 6160$

The requirement of track material for MG track can also be calculated using the same method.

28.3 DOUBLING OF RAILWAY LINES

The doubling of a railway line refers to the construction of an additional line. This is normally done when a single line is no longer capable of carrying all the trains that ply on the section. The main steps involved in the doubling of a railway line are described here.

Engineering-cum-traffic survey This survey is conducted to examine the traffic prospects of the section and to roughly identify the quantum of work and the cost involved. The Railway Board finally sanctions the work and authorizes the railway to execute the same.

Specification of work During the execution of the work, it is necessary to clearly understand and follow the standards of construction and the specifications that have been laid down. The track standard should also be decided during the survey so that the same can be followed as the work progresses. The ruling gradient and maximum curvature are normally the same as those of the existing line.

Land acquisition If necessary, extra land can be acquired for the double-line station yard, bridges, etc., in case it has not already been done.

Earthwork The earthwork for a double line is done by widening the existing formation to suit the double line. The land where earthwork is to be done is divided into different zones and a separate contract is awarded for each zone. The earthwork is properly consolidated and is normally made to endure two rains. If the soil is of a poor quality, a blanket of an adequate thickness is provided just below the final level of the formation. The soil can also be consolidated by using a sheep foot roller or vibratory roller to achieve the desired compaction. When this is done, traffic can be allowed to ply on the track immediately after the work has been completed.

Bridges The existing bridges are extended to suit the double line. This extension is done after drawing up a plan that causes the minimum dislocation or disruption of traffic. Separate zonal contracts are awarded for major bridges, whereas minor bridges are included in the zones defined for earthwork. The work is normally so planned that it is completed in the working season lasting from October to June.

Plate laying Rails, sleepers, and fastenings are collected in track depots specifically established for the project. Normally, such depots are situated at every railway station. The work of track linking is usually done by the telescopic method.

Opening for goods traffic and ballasting As soon as construction work is completed, the line is opened to goods traffic plying at restricted speeds. Ballast trains also run on the section and unload the ballast on the track, which is then packed manually with the help of beaters.

Opening the line for passenger traffic Once the line is properly packed, the same is opened to passenger traffic after obtaining the sanction of the commissioner of railway safety. The speed restrictions on the section are slowly relaxed till the trains can ply at the maximum permissible speed once again.

28.4 GAUGE CONVERSION

A number of projects have recently been taken up by Indian Railways for converting railway lines from MG to BG. *Gauge conversion* projects (Fig. 28.2) basically aim at providing a uniform gauge for the smooth and fast flow of traffic, which may be necessitated either due to strategic reasons or on account of operating considerations.

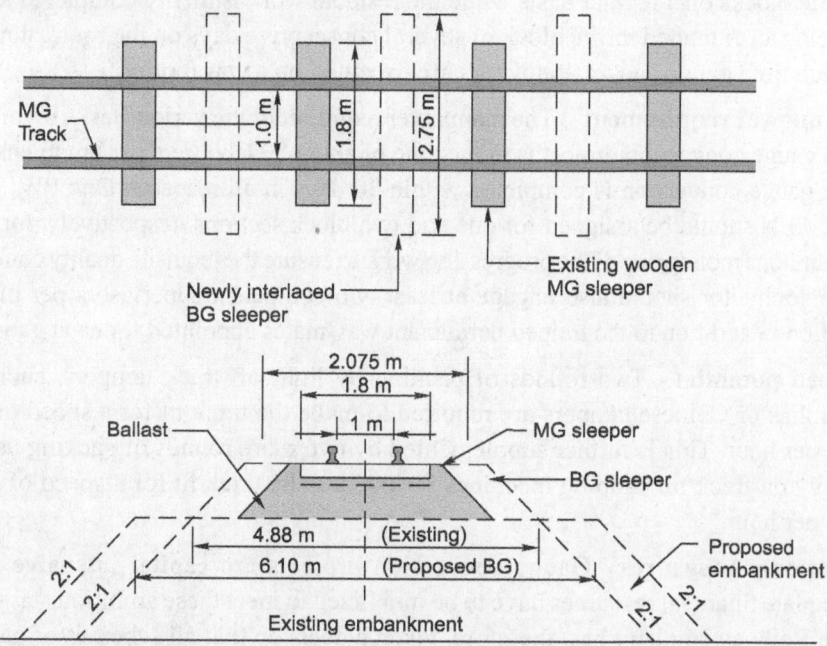

Fig. 28.2 Gauge conversion from MG to BG

The main advantages of gauge conversion as well as its details are discussed in Chapter 2.

28.4.1 Execution of Gauge Conversion Projects

The execution of gauge conversion projects involves the following:

Details of works The fieldwork normally involved in gauge conversion projects is presented in Table 28.2.

Table 28.2 Details of fieldwork for gauge conversion projects

Nature of work	Brief details of works
Civil engineering works	Engineering-cum-traffic survey; land acquisition; earthwork for widening the formation; extension of minor bridges; extension of major bridges of PSC slabs or steel girders; permanent way work consisting of supplying and spreading of ballast and laying of rails, sleepers, and fastenings, including points and crossings, and all connected works
Electrical works	Conversion to high tension (HT) crossings; augmentation of power supply; wiring of new structures; modifying LT (low tension) and HT installations; shifting of floodlight masts
Signalling works	Erection of signals at new locations; interlocking of points; track circuiting; relay track installation; wiring and testing works

Planning of works Gauge conversion projects require detailed planning as they involve expansion, which needs to be carried out in a tight time schedule and multifarious activities, such as assigning of contracts; posting and management of manpower, including supervisors and artisans; supply of track materials; and coordination between the engineering, service and telegraph, electrical, operating, and various other departments. Much of the engineering work is done by placing traffic blocks on a regular basis while the residual work is finally completed after placing an extended traffic block of several consecutive days on the route, during which time gauge conversion works are executed on a war footing.

Manpower requirement The manpower required for the various tasks involved in a gauge conversion project is to the tune of about 80 labourers per km to ensure that gauge conversion is completed within 30 days in all respects. One PWI and one AEN should be assigned for one and two block sections, respectively, for the continuous monitoring of the progressing work to ensure the requisite quality control. The contractor should also engage at least two competent supervisors per block section in addition to the trained permanent way mates appointed for each gang.

Speed potential Two rounds of packing by light off-track tampers, such as phooltas or Chinese tampers are required to make the track fit for a speed of 50 km per hour. This is further supplemented by two more rounds of packing using heavy on-track tie tamping machines for making the track fit for a speed of 100 km per hour.

Economy measures Gauge conversion projects are capital intensive and adequate financial resources have to be mobilized to meet these ambitious targets. The Railway Ministry has, therefore, taken a decision that all schemes of gauge conversion are to be carried out in the most economical way. By cutting down the cost of gauge conversion in various ways, the Railways plan to carry out gauge conversion works at an economical cost of ₹ 6 to 7 million per km as against the normal cost of ₹ 9 to 10 million per km, thereby saving about 30 to 40 per cent on the initial cost. The guidelines to be followed in this regard are the following.

(a) The maximum permissible speed on the converted BG section need not be more than what was previously specified for the MG section.

(b) The facilities required at the stations should not be more than what has already been provided except that the loop length should be 686 m.

(c) The length of the platform at those stations that currently serve as stops for mail/express trains should be just enough for accommodating 16 coaches. If the existing platform needs to be extended, the extended portion should be constructed at rail level. On suburban sections, the length of the platform should be enough for 12 coaches. At important stations, the platform length should be sufficient for no more than 22 coaches with added provisions for future expansion to include 26 coaches.

28.4.2 Civil Engineering Works of Gauge Conversion Projects

Engineering-cum-traffic surveys The survey is done to examine the technical feasibility of the proposal and the economical aspects of the same.

Land acquisition The land requirement for converting the track from MG to BG is assessed and, if necessary, extra land is acquired as per the existing procedure laid down in the Land Acquisition Act.

Earthwork It is done on either side of the existing formation of a single-line track to increase its width from 4.88 m (16 ft) to 6.10 m (20 ft) in the case of embankments and from 4.27 m (14 ft) to 5.49 m in the case of cuttings.

Bridges and culverts The existing bridges and culverts are extended to suit the BG formation and to conform to the standards adopted for BG sections. The extension of these bridges is properly planned to ensure the minimum dislocation of traffic.

Track linking Track linking involves the following operations.

Unloading of track material Track material such as rails, sleepers, and fastenings are taken from the base depot to the work site and unloaded at the final location. Preliminary work such as boring and adzing of sleepers, proper matching of rails, and drilling of holes in the rails are planned in advance.

Interlacing of sleepers Wooden sleepers are best suited for gauge conversion projects, if available. They are interlaced with the existing sleepers of the track and the two are spiked together.

Arrangement of men and materials The deployment of manpower, including the supervising staff, is planned in detail so that the conversion of the track from MG to BG is achieved with minimum interruption of traffic. Normally 50 to 100 men per km are required on the track during a traffic block. In addition, the work of skilled artisans such as blacksmiths is also required for track conversion. All track material, including fittings, are properly organized and arranged at the site. Tools and equipment such as augers and drills, Jim Crows, crowbars, rail tongs, sleeper tongs, and spanners are also arranged in adequate quantities.

Traffic blocks Traffic is suspended on the railway line for a period of 15 to 30 days depending upon the length of the track to be dealt with, so that the work of gauge conversion can be carried out at one stretch. During a traffic block, traffic is diverted on alternate routes or trans-shipped onto road vehicles.

Linking the new track During the traffic block, the existing rails are removed and new rails placed in what is to be their final position on the BG track. The new sleepers are also placed in their proper position and the rails are spiked to the sleepers as per BG requirements. The track is properly levelled and aligned and traffic is allowed on the new line at a restricted speed after obtaining the sanction of the commissioner of railway safety.

Ballasting and packing An adequate quantity of ballast is then put in the track and the track is properly packed. The speed restrictions are gradually relaxed as the stability of the track improves.

28.5 MODERN CONSTRUCTION OF RAILWAY LINES

Modern construction of railway lines includes the following:

 (i) Use of sophisticated modern techniques for carrying out survey work of railway lines
 (ii) Use of modern geo-technology for preparation of subgrade (formation)
 (iii) Deployment of sophisticated modern track machines for laying of track

Details of these techniques are discussed briefly in the following sections.

28.5.1 Modern Surveying Techniques

Modern survey technology uses satellite imageries, aerial photographs, and digital terrain modelling in deciding the initial alignment as well as final location of the new railway line. Satellite imageries as well as topo maps are available for almost all places around the world. These are very useful in inital marking out of any new alignment of railway line. Satellite imageries can also help in generating topographical data of the route. By this technique the intial rough alignment of railway line can be checked out in the topographical topo sheet in the drawing office itself.

Subsequently, aerial photographs can be taken of the area and with the help of topographical sheets, satellite imageries, and aerial photographs, digital terrian models can be generated. These models will be able to help in further refining the alignment to arrive at the most optimal solution, which can meet all the requirements of the new line.

Once the tentative alignment is checked out, field units can be made to get actual field data including hydrological and geological data for planning and execution of the project. Modern survey techniques can not only expedite the survey work, but also considerably reduce the cost of the project.

28.5.2 Modern Geotechnology for Preparation of Subgrade (Formation)

Geotechnology has recently made great advancements. This technology can be very successfully used in the design of subgrade for railway lines to provide a stable and trouble-free formation. Sometimes if the natural soil is not good, the technology can be used for providing the right treatment to subgrade so as to get sound and trouble-free formation.

In case of doubling and gauge conversion projects, if the existing formation is giving trouble, the same can be improved while executing these projects.

Some of these technologies or remedial measures are as follows:

(i) Provision of pre-fabricated vertical drains and pre-loading
(ii) Removal of soft soils of limited depth to be replaced by good material fill
(iii) Provision of an inverted filter
(iv) Cement grouting by a slurry of cement and sand to be pumped by pressure
(v) Provision of sand piles of about 30 cm diameter to a depth of 2 m to 3 m
(vi) Soil stabilization by the geo-textile method

Details of some of these soil treatment methods can be seen in Chapter 9 of the book.

28.5.3 Mechanized Track Linking

As per conventional method, track linking used to be done conventionally by spreading ballast over the formation, placing of sleepers at specified distance and then linking the track. This method however, has recently been improved. Apart from the cost factor and difficulty in the availability of labour, modern track requires use of heavy concrete sleepers and work to be done expeditiously with good precision.It is briefly described in Section 28.2.2. For more details please refer to Chapter 20.

SUMMARY

Both passenger and goods traffic is increasing at a very fast rate on Indian Railways. Therefore, the construction of new lines or the doubling of the existing single-line tracks becomes important. The latter is brought about through extensive gauge conversion projects, which involve land acquisition, earthwork, construction of bridges and station buildings, and many other activities. The project should be planned and executed following the standard norms to ensure that the trains run safely.

REVIEW QUESTIONS

1. Using a sleeper density of $n + 6$, determine the number of sleepers required for constructing a BG railway track that is 760 m long, if the length of the rail is 13 m.
2. Describe the principal operations involved in the plate laying method used extensively in India.
3. Define plate laying and give an elaborate description of the same.
4. You are placed in charge of a plate laying operation for the construction of a BG railway track. Give a detailed and step-by-step account of how you would carry out the work to achieve the desired result.
5. A 5 km-long BG main line straight track with CST-9 sleepers and duplex joint sleepers is to be constructed. Estimate the total quantity of track material required with respect to 90 R rails and a $N + 5$ sleeper density.
6. Discuss briefly the factors affecting track alignment and describe in brief the methods one would adopt to align a railway track in a region where a heavy gradient is unavoidable. Using a sleeper density of $N + 5$, determine the number of sleepers required for constructing a railway track of length 640 m on a BG section.
7. Work out the quantities of the various track materials required for laying a single-line BG track for a length of 1 km. Make suitable assumptions for all the other data that may be required.
8. Briefly discuss the various steps involved in the construction of a new BG railway line.
9. Write notes on (a) doubling of a railway line (b) gauge conversion project.

Choose the correct answer from the choices given.

10. The formation width for banks for a BG single-line section is:
    (a) 6.10 m (b) 6.50 m
    (c) 6.85 m (d) 7.0 m
11. The formation width for cuttings for a BG single-line section is:
    (a) 5.40 m (b) 6.20 m
    (c) 6.25 m (d) 6.50 m
12. The formation width for bank for a BG double-line section is:
    (a) 10.82 m (b) 11.0 m
    (c) 11.58 m (d) 12.155 m
13. The formation width for bank for a MG double-line section is:
    (a) 8.50 m (b) 8.84 m
    (c) 9.65 m (d) 9.81 m

14. On Indian Railways the linking of track is done mostly by the:
    (a) tram line method   (b) American method
    (c) telescopic method   (d) none of these
15. The number of standard rails (13 m) for 1 km BG single-line section is:
    (a) 120   (b) 154
    (c) 168   (d) 196
16. The number of sleepers required with a sleeper density of M + 7 for 1 km of BG single line is:
    (a) 1360   (b) 1540
    (c) 1660   (d) none of the these
17. Gauge conversion means:
    (a) converting MG to BG   (b) converting NG to MG
    (c) converting NG to BG   (d) converting MG/NG to BG
18. The number of grooved rubber pads required per km for laying a new BG track of 13 m rail length with concrete sleepers laid at M + 7 sleepers is:
    (a) 1540   (b) 3080
    (c) 4620   (d) 6160
19. For the bank of a newly constructed railway line, the cross slope generally given should be:
    (a) 1 in 10   (b) 1 in 20
    (c) 1 in 30   (d) 1 in 50

Suburban Railways in Metro Cities

INTRODUCTION

The term 'metropolitan city' is commonly used for major or important cities. Most metropolitan cities in India have grown in an unplanned and haphazard manner. Even in places where city masterplans were available, the actual lands barely resemble what was envisaged in the plans. Delhi is one such example. The rapid growth in the population and economy of metropolitan cities has resulted in several social and economic problems. The imbalance in the distribution of population and economic activity in these cities has led to large-scale intra-city movement resulting in a serious transportation problem.

There are generally many limitations in the movement of people in these metropolitan cities due to the following reasons.

(a) The traffic capacity of the roads in the major metropolitan cities has not kept pace with the growing demands of traffic and this has resulted in severe congestion on the roads, particularly during peak hours and in central business districts.

(b) The average vehicular speed in these cities is about 20 to 30 km per hour and in some of the congested parts of the core areas, speeds have been reduced to as low as 5 to 10 km per hour. A heavy, often unidirectional, peak load is required to be carried through certain routes during specific hours everyday. The design and capacity of these roads are unable to meet the requirements of traffic and because of the several architectural structures that have come up on their either side; the further widening of these roads is not possible.

(c) The number of road vehicles has increased considerably in the past few years, of the order of 7 to 10 per cent per year in the four metropolitan cities. In Delhi, the number of registered vehicles has increased manifold in the last 10-year period. These cities keep expanding in all directions at an alarming rate, placing additional demands on the existing transport system.

(d) Environment pollution is widespread in metropolitan cities on account of the increase in vehicular traffic and all-round congestion. This has led to increased levels of noise and dust, increased vehicular emissions, and a loss of sunlight and daylight.

(e) The congestion on the roads in metropolitan cities has resulted in a large number of accidents. India, with 142,000 road deaths and more than 0.5 million serious injuries, suffers an annual loss of more than ₹ 10,00,000 million.

(f) There is a considerable wastage of time of a large number of people staying in metropolitan cities on account of the slow movement of vehicles and the formation of long queues on the roads.

A possible solution to these problems is to establish a proper mass transport system. The existing transport facilities must be suitably augmented and expanded to meet the growing traffic demands. All types of road transport such as two wheelers, autorickshaws, cars, and buses have a maximum load capacity. Depending on the traffic density, one possible solution would be to strengthen and develop electric rail services, which, besides providing high-capacity transit facilities, also help substantially in energy conservation and environmental preservation. Further, from the point of view of relieving the roads of excess traffic and also of conserving energy, there is a need for urgent and deliberate measures that will discourage commuters from using personal modes of transport and promote the use of public conveyance instead.

29.1 URBAN TRANSPORT

The basic objective of urban transport is to provide residences with access to facilities such as workplaces, schools, and shopping centres. The different forms of urban transport in use in most cities of the world are the following.

(a) Buses
(b) Trolley buses
(c) Tramways
(d) Surface railways
(e) Underground railways
(f) Elevated railways
(g) Monorails
(h) Tube railways

29.1.1 Buses

Buses are the most convenient form of transport and are used extensively in metropolitan cities. These run mostly on diesel oil and their exhaust emissions have an adverse effect on the environment. Moreover, buses, though very convenient for transporting passengers, have very limited seating capacity.

29.1.2 Trolley Buses

Trolley buses derive their energy through overhead electric transmission. These are superior to buses as they do not pollute the environment. On the other hand, huge expenses are incurred in providing overhead traction for supplying power to trolley buses.

29.1.3 Tramways

Tramways require a track on which the trams can run and as such require the infrastructure of a proper railway track. Their initial cost is quite high. They cause minimal air pollution; however, they contribute significantly to noise pollution. Tramways are almost obsolete now and are used only in some parts of the country such as in Kolkata.

29.1.4 Surface Railways

Surface railways are the cheapest and most extensively used form of railway service in the world. In such a system, the track is laid on a ground that has a suitable embankment or cutting, depending upon the topography of the area. Metropolitan railways use electric traction because of the following advantages.

(a) Electric traction does not pollute the environment.
(b) The acceleration and deceleration of trains is faster.
(c) Electric traction ensures the availability of power for improved and modern signalling.
(d) An electric locomotive can haul a train with the same efficiency in both the directions and there is no need for reversing the direction of the locomotive.
(e) This system uses special type of coaches called electric multiple units (EMUs), which can carry more traffic than conventional coaches.

Cost considerations

The construction of surface railways with an overhead electric transmission system is quite costly and comes to about ₹ 10 to 15 million per km (excluding the cost of the land) in normal terrain. The cost, however, increases considerably in dense and thickly populated areas, where the cost of the land to be acquired is quite steep. In addition, the railway line may have to be constructed across roads, some of which may be quite busy.

Type of crossings

The points where the roadways and the railways cross each other are provided with a level crossing, a road underbridge, or a road overbridge, depending upon the volume of traffic being carried by rail or road, availability of financial resources, road level, vis-à-vis rail level, proximity of land, and other allied factors. The special characteristics of each of these crossings are given in the subsequent paragraphs.

Level crossing At this crossing, the railway track and the road cross each other at the same level. In urban areas, most level crossings are manned by two to three watchmen who are responsible for operating the same. Normally the level crossing remains closed to road traffic and is opened only when road vehicles need to cross to the other side. Though quite cheap and convenient, level crossings have the following limitations.

(a) Gatemen have to be provided for manning level crossings, which proves to be quite costly.
(b) Level crossings cause delay in the movement of road traffic, particularly when trains are crossing them.

(c) Level crossings pose a significant safety hazard and the number of accidents at level crossings is quite substantial.

Road overbridge The road overbridge is a type of bridge where the road passes over the railway line. It is an improvement over the level crossing and overcomes most of the limitations of a level crossing. However, the road overbridge has certain limitations, which are as follows.

(a) The road overbridge is a costly arrangement, as apart from the cost of construction of the main bridge, heavy expenditure is involved in constructing the road approaches and acquiring land.

(b) The length of the detour provided for crossing the track is considerable, particularly for cyclists and pedestrians. In some sections, therefore, a foot overbridge is provided so that at least pedestrians and cyclists can use it to cross the track.

Road underbridge In this type, the road passes under the railway line. A road underbridge is preferred in places where the general ground level is low and the railway line is at a comparatively higher level. A road underbridge is also preferred in areas where enough land is not available. Normally, a road underbridge costs less, but its construction is quite complicated. The cost may also increase if excavation is required in rocky areas. If the water table is high, there may be an additional cost involved for lowering the water table. A road underbridge normally presents drainage problems, particularly during monsoons.

29.1.5 Underground Railways

In underground railways, the railway line is constructed below the ground level. The requisite construction work is done mostly by the 'cut and cover method'. The area is excavated in the shape of trenches and once the formation is ready, the track is laid, the necessary overhead structures are provided, and finally the trenches are covered and the ground is restored to its original state.

An underground railway system normally uses 'electric traction', as steam and diesel tractions produce smoke and lead to the pollution of the environment, which in this case becomes particularly hazardous since these railways are underground. Proper arrangements are also made for the drainage of underground railways as the low-lying areas in which they are constructed are likely to get flooded during the rains. Such underground railways have been constructed in Kolkata and Delhi and in other countries around the world.

The main advantages and limitations of underground railways are as follows.

Advantages

(a) Trains can run fast and unobstructed in an underground railway system as there are no road crossings or other similar problems.

(b) As the trains move at incredible speeds, underground railways can deal with a very high concentration of human traffic.

(c) There is no wastage of land and a large area of the city, which would have otherwise been used for surface railways, remains available for other utilities.

(d) Provides safety from aerial attacks, particularly during war.

Limitations

(a) The underground railway system is a very costly arrangement and a heavy financial backing is required. The cost may vary anywhere from ₹ 30 million to ₹ 100 million per km, depending upon the geographical features and other conditions.

(b) Special attention needs to be given to the drainage as well as proper ventilation of underground railways.

(c) During construction, the residents of the city are greatly inconvenienced as excavation work is normally carried out throughout the city. The water supply, electricity supply, and sewerage system of the city are also affected, as the diversion of many of these services is required during the constructional phase.

Cross section of an underground railway

An underground railway may have either one of the following cross sections depending upon the method used in its construction.

Cut and cover In this case, excavation is done by the cut and cover method. This method affects all public services, such as water supply, sewerage mains, and electric and telephone lines, which have to be diverted or suspended temporarily. The typical cross section of an underground railway constructed using the cut and cover method is given in Fig. 29.1.

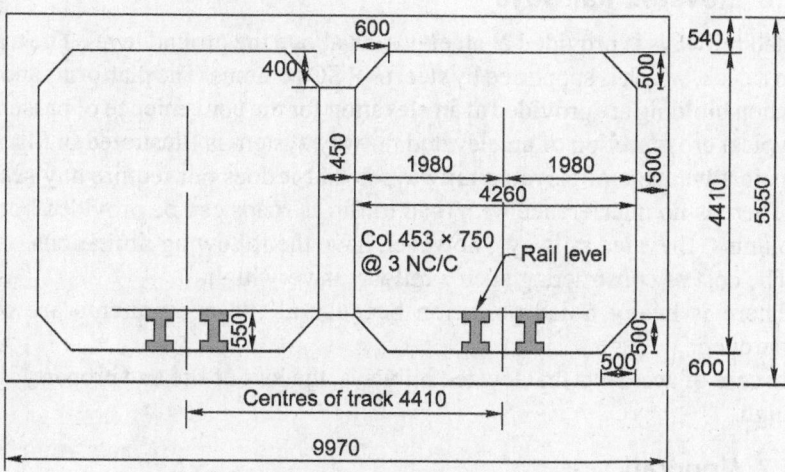

Fig. 29.1 Typical cross section of an underground railway (cut and cover method); all dimensions in mm

Tunnel section The tunnel for the underground railway is dug very deep into the ground, much below the level at which the ground is dug for water and sewerage mains and for telephone and electric lines. The circular cross section of the tunnel is sometimes made by the application of modern techniques that involve pushing a big pipe of an adequate size through the ground. The necessary infrastructure is then provided within this circular tunnel. The typical cross section of such an arrangement is shown in Fig. 29.2.

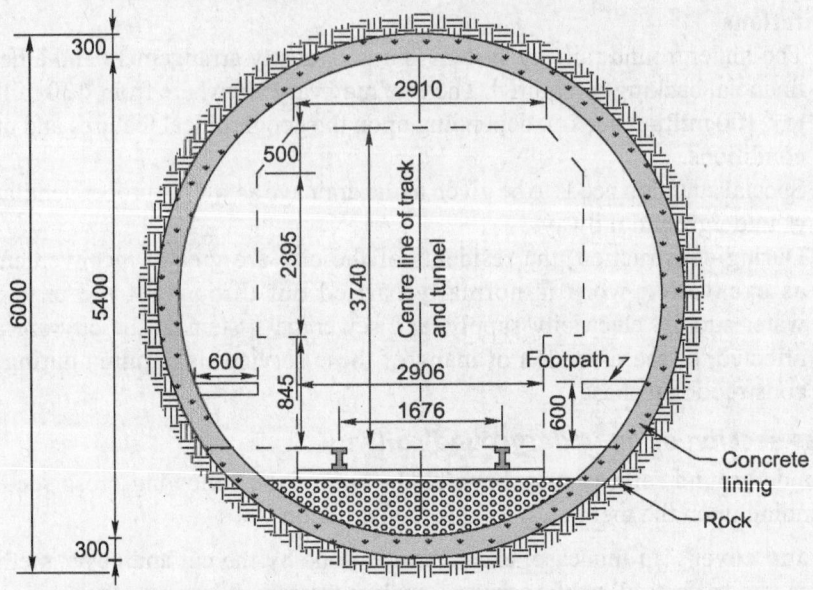

Fig. 29.2 Typical cross section of an underground railway (tunnel section); all dimensions in mm

29.1.6 Elevated Railways

Elevated railways is provided at an elevation above the ground level. The track is laid on a deck, which is supported by steel or RCC columns. The platforms and even the station building are provided at an elevation for the convenience of passengers. The typical cross section of an elevated railway system is illustrated in Fig. 29.3. The main advantage of elevated railways is that it does not require any separate land. There is no interference with road traffic as roads can be provided between the columns. Elevated railways, however, have the following limitations.

(a) The cost of constructing such a railway is very high.
(b) There is heavy noise pollution because all its components are out in the open.
(c) In case of accidents on elevated railways, the loss of life and property is very high.

29.1.7 Monorail

Monorail is a form of elevated railway that is provided with only one rail on which trains run. The trains can be suspended on the monorail as in Montreal, Canada, or can be mounted on pylons as in Tokyo, Japan. The monorail system is recommended only in exceptional cases where operating the conventional systems is difficult.

29.1.8 Tube Railways

In tube railways, the underground railways are generally provided at a depth of more than 25 m. The railway line is constructed in a tunnel that is circular or tubular. The main reason for taking the railway so deep into the ground is to avoid

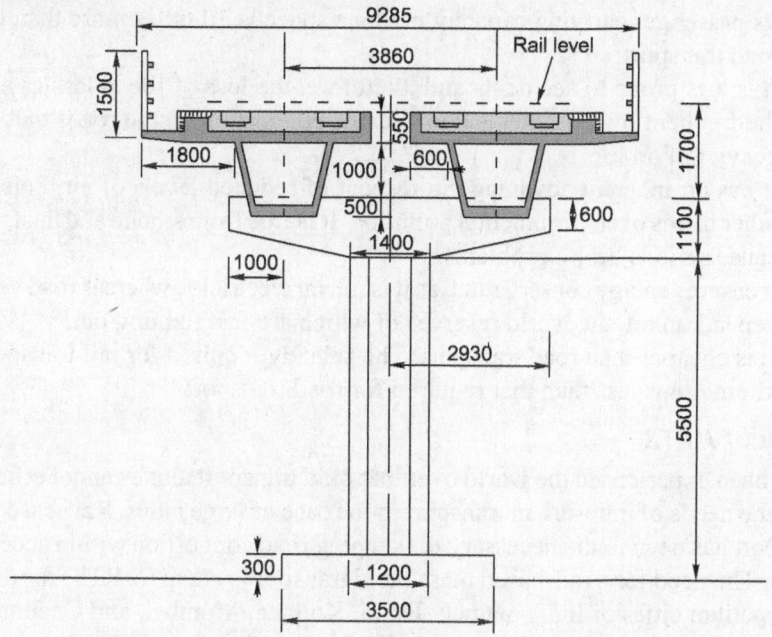

Fig. 29.3 Typical cross section of an elevated railway (all dimensions in mm)

it interfering with the water supply mains, sewerage system, telephone lines, gas lines, etc., which are normally located within 10 m of the natural ground. Some of the salient features of the tube railway are given below.

(a) The railway stations of the tube railway are generally cylindrical in shape.
(b) Only electric traction is used for hauling trains in order to avoid smoke and the resulting environmental pollution.
(c) For the convenience of passengers, escalators are provided for accessing and exiting the stations.
(d) An automatic signal is provided for the faster and effective movement of trains. The trains stop automatically when the signal indicates impending danger. Human error is thus avoided to the maximum possible extent.
(e) In order to ensure the safety of passengers, it is essential that the doors of the compartments are closed before the train can start. This is done by means of a centrally operated switch installed in the driver's cabin.
(f) Automatic ticket issuing machines are installed in the concourse in order to reduce the need for manpower and to ascertain that tickets are issued speedily for the convenience of passengers.
(g) Adequate indication and warning boards are displayed at significant locations so that passengers have ready information regarding the timing of the trains and the platform number for each train without making too many enquiries.

Superiority of rail transport over road transport

Rail transport is becoming increasingly popular all over the world as far as its relation to urban transport is concerned. This is due to the following advantages that rail transport has over a road transport system.

(a) Its passenger carrying capacity per lane space is 10 times more than that of road transport.
(b) It is less prone to accidents and, therefore, the loss of life is less as against the frequent road accidents due to acute congestion on the roads that take a heavy toll on life.
(c) It has an inherent advantage in respect of reduced levels of air, noise, and other forms of environmental pollution. It is free from smoke and dust, which cause serious health problems.
(d) It ensures energy conservation, as it is run on electricity, whereas road vehicles depend on oil, the world reserves of which are fast running out.
(e) It is cheaper than road transport. The subsidy required for rail transport per kilometre is less than that required for road transport.

Study of MRTS

It has been experienced the world over that road transport alone cannot efficiently meet the needs of intra-urban transport in the case of large cities. Rail-based mass transport has been found necessary for keeping road congestion within acceptable limits. The need for a rail-based mass rapid transport system (MRTS) for the four metropolitan cities of India, namely Delhi, Kolkata, Mumbai, and Chennai, was recognized in the early 1970s. Accordingly, the Planning Commission set up MRTS study teams for each of the four cities under the guidance of the Ministry of Railways. The study reports were finalized in the mid-1970s. On account of the massive investments that were required and the simultaneous paucity of funds, the project remained on hold except for a corridor in Kolkata, which was taken up subsequently. For various reasons, progress of the Kolkata project was slow and was commercially operated for the first time in October 1984.

29.2 THE DELHI RING RAILWAYS

Most of the traffic requirements of Delhi commuters are met by the road transport system, which is managed by the Delhi Transport Corporation (DTC). In addition to this, electric multiple unit services were introduced on the Ring Railway in August 1982 in the suburban area of Delhi. These services originate at Nizamuddin and cover Safdarjang, Patel Nagar, Dayabasti, and New Delhi before terminating at Nizamuddin. The total distance covered is 36 km and the EMU services take about 55 minute for one complete trip. Recently, EMU services have also been initiated between the Delhi and New Delhi stations and the Ghaziabad station.

29.3 MASS RAPID TRANSIT SYSTEM IN DELHI

At the time of Independence, the population of Delhi was a mere 0.6 million. Starting from this humble figure, the population of the metropolis grew to 5.7 million in 1981, 12 million in 1998, and 16.7 million in 2011. In the absence of an efficient mass transport system, the number of motor vehicles in Delhi has increased from 3 million in 1998 to about 6.5 million in 2011. The large number of vehicles lead to congestion on roads, resulting in a decline of vehicular speed, thereby resulting in fuel wastage, increased air pollution, and an increase in road accidents.

It is a well-known fact that out of all the cities in the country, the process of urbanization has been the fastest in Delhi. The city witnesses about 11.7 million transit trips per day, of which no less than 62 per cent are by public transport. Among the various public transport options available, 99 per cent are road based and only 1 per cent are rail based, despite the fact that Delhi has 144 route kilometres of rail tracks converging into the city from five different directions.

In 1989, a study for an MRTS network for Delhi was undertaken at the instance of the Delhi Government. The report brought out the urgent need for a rail-based transit system comprising a network of 181 route kilometres that consisted of 29.5 km of underground alignments, 40.5 km of elevated route, and 111 km of surface/ elevated route length. Nearly 99 km of the route length was analogous with the existing rail network and was proposed as a surface (or at-grade) corridor. This scheme, which was the basis of the Delhi Master Plan 2001, however, remained unimplemented.

Corridors for MRTS

There are two distinct types of corridors planned for the MRTS in Delhi.

Rail corridor The system is basically located on the surface (at-grade) or on elevated ground and works on 25 kV ac traction and consists of a rolling stock with sealed doors and windows that remain closed while in motion.

Metro corridor The system is underground except for depot connections and lines, which may be partly at-grade.

29.3.1 Phase I of Delhi Metro System

Once the government decided to take up the Delhi metro project, the Delhi Metro Rail Corporation (DMRC) was set up in 1995 for the implementation and subsequent operation of the Delhi MRTS. The DMRC, however, became effectively functional in 1998, after the appointment of general consultants and a team to execute the project.

Before starting with the implementation of the project, the corridors included in phase I were reviewed and it was finally decided to include three corridors covering a route length of 62.1 km, which included 12.1 km of underground, 38.2 km of elevated, and 11.8 km of at-grade alignment (Fig. 29.4). The first phase of the Delhi Metro system consists of three lines, as presented in Table 29.1.

Table 29.1 Details of phase I of the Delhi Metro

Lines	Total length (km)	Underground (km)	Elevated (km)	At-grade (km)	No. of stations
Line no. 1 Shahdara–Barwala	28.0	Nil	17.5	10.5	21
Line no. 2 Vishwavidyalaya– Central Secretariat	11.0	11.0	Nil	Nil	10
Line no. 3 Connaught Place– Dwarka	23.1	1.1	20.7	1.3	22
Total	62.1	12.1	38.2	11.8	53

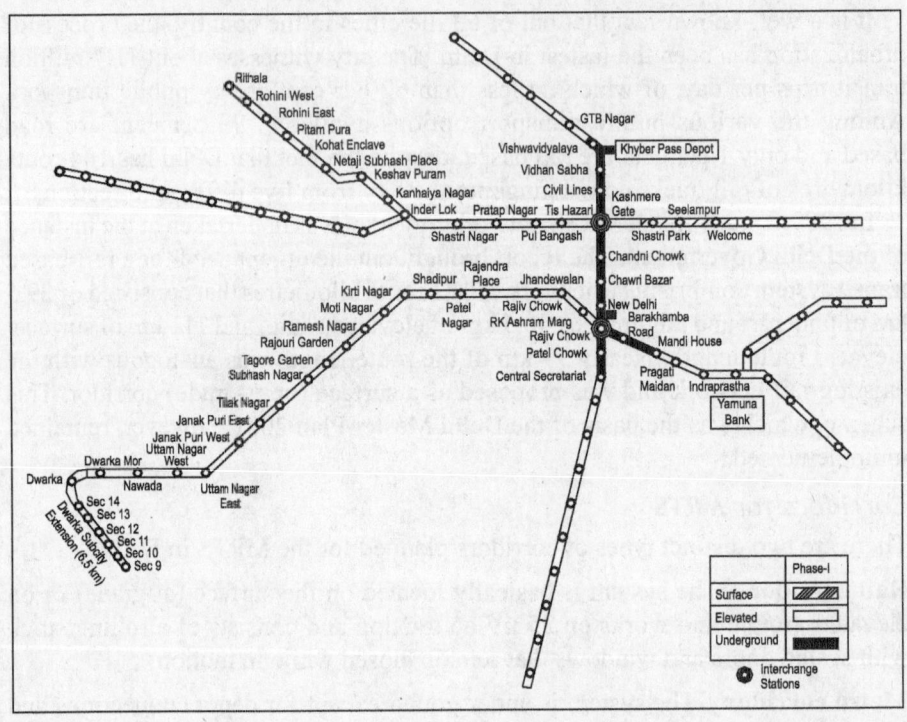

Fig. 29.4   Map of phase I of Delhi Metro

The government decided to take up the first phase of the Metro system in August 1996. The project was started in right earnest in 1998 and was completed in December 2005.

29.3.2 Phase II of Delhi Metro Rail Project

Phase II of Delhi Metro consits of several lines as given in Table 29.2 (see Fig. 29.5).

Table 29.2   Phase II of Delhi Metro

Line	Length (km)	No. of stations
Shahdara–Dilshad Garden	3.09	3
Indraprastha–Noida Sector 32 City Centre	15.07	11
Yamuna Bank–Anand Vihar ISBT	6.17	5
Vishwavidyalaya–Jahangir Puri	6.36	5
Inderlok–Kirti Nagar–Mundka	18.46	15
Central Secretariat–HUDA City Centre	27.45	19
Dwarka Sector 9–Dwarka Sector 21	2.76	2
Airport Express Line	22.70	6
Anand Vihar–KB Vaishali	2.57	2
Central Secretariat–Badarpur	20.04	15
Total	124.67	83

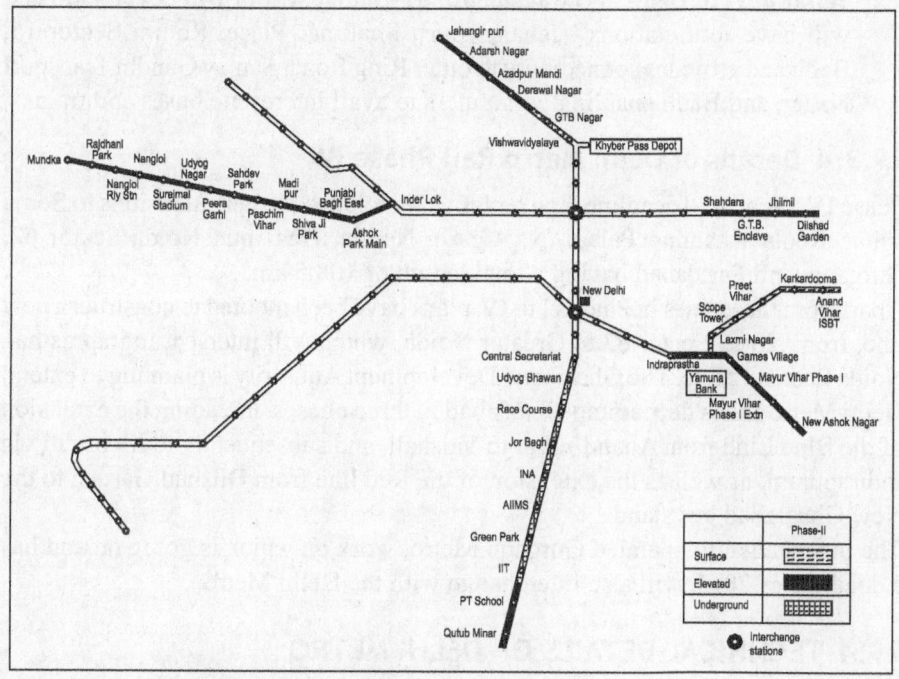

Fig. 29.5 Map of phase II of Delhi Metro

29.3.3 Delhi Metro Rail (DMRC) Phase-III

DMRC Phase-III will connect 103 km to the existing network of about 190 km of phase I and phase II.

The Phase-III which will be completed by 2016, will have a total of 67 stations with 15 interchange points that will facilitate free movement of people. There are four corridors of DMRC Phase III.

Mukundpur to Yamuna Vihar The Mukundpur corridor will be the longest with 56 km length and 35 stations and will provide connectivity to the northern, western, and eastern parts of Delhi. The line will also have the third bridge on Yamuna Vihar.

Janakpuri West to Kalindi Kunj The Janakpuri West–Munirka–Kalindi Kunj will cover 34 km and the corridor will have 22 stations, providing connectivity to the southeastern parts of Delhi with the western part of Delhi. The route provides integration with the existing stations at Janakpuri west (Yamuna Bank to Dwarka Sector 21 Line), Hauz Khas (Jahangirpuri to HUDA City Centre Line), and Nehru Place (Central Secretariat–Badarpur Corridor) enabling commuters to go to Dwarka, Gurgaon, Central Secretariat, Badarpur, and Qutab Minar. A major portion of this corridor is along the outer ring road.

Central Secretariat to Kashmiri Gate The Central Secretariat–Kashmiri Gate corridor will be an extension of the Badarpur–Central Secretariat corridor. When the Faridabad corridor comes up, it will be integrated with this corridor.

Janakpuri to Badli   The Jahangirpuri–Badli corridor will cover 4 km and will have four stations—Jahangirpuri, Shalimar Place, Rohini Sector 15, Badli and provides connectivity to outer Ring Road, Sanjay Gandhi Transport Nagar, and Badli enabling commuters to avail interestate buses and trains.

29.3.4 Details of Delhi Metro Rail Phase IV

Phase IV has a 2020 deadline, and tentatively includes further extensions to Sonia Vihar, Reola Khanpur, Palam, Najafgarh, Narela, Ghazipur, Noida Sector 62, Gurgaon, and Faridabad having a total length of 108.5 km.

Apart from these lines in Phases I to IV, plans have been mooted to construct a new line, from Noida Sector 62 to Greater Noida, which will intersect Indraprastha–Noida Sector 32 line. The Ghaziabad Development Authority is planning to extend Delhi Metro Lines deeper into Ghaziabad in three phases, including the extension of the Blue Line from Anand Vihar to Vaishali, and subsequently to Mehrauli via Indirapuram, as well as the extension of the Red line from Dilshad Garden to the New Ghaziabad bus stand.

The independently operated Gurgaon Metro, work on which is going on and has a deadline of 2013, will also interchange with the Delhi Metro.

29.4 TECHNICAL DETAILS OF DELHI METRO

The Delhi metro is planned on the lines of a world class metro rail system and is equipped with modern communication and train control systems. In the Delhi metro system, trains are available at a three-minute frequency during peak periods. The entrances and exits of metro stations are monitored by flap doors operated by smart cards.

Table 29.3   Details of underground lines and all lines of Delhi Metro

Phase	Underground lines		All lines of Delhi Metro		Date of completion
	Length (km)	No. of stations	Length (km)	No. of Stations	
Phase-I	12.10	13	65.00	53	October 2006
Phase-II	34.89	18	124.63	89	August 2011
Phase-III	41.044	28	103	67	2016
Phase-IV	NA	NA	108.5	NA	2020
Total	88.034	59	401.13	209	2011 – 2020

For the convenience of commuters, an adequate number of escalators are installed at the metro stations. A special feature of the Delhi Metro will be its integration ultimately with other modes of public transport, enabling the commuters to interchange from one mode to another. The technical details and design parameters of the Delhi Metro system are briefly described below.

29.4.1 Rolling Stock

The rolling stock is the mainstay of a metro rail system. Metro coaches are lightweight, air conditioned with stainless steel bodies that are equipped with features, such as three-phase ac motors. All systems in the coach are monitored by a microprocessor-based train integrated management system (TIMS). All coaches are of the same type and each coach is able to accommodate nearly 380 passengers. The coaches have a width of 3.2 m and an overall length of 22 m (including sitting and standing passengers).

Some of the important features of the metro rolling stock are listed below.

(a) Automatic electric door closing mechanism.
(b) All doors fully open within 2.5 seconds and fully close within 2.5 to 3.5 seconds.
(c) The train cannot move unless all the doors are properly closed and automatically locked.
(d) The train can be halted with the application of the emergency brake, should a door open during running.
(e) Emergency evacuation facility in the form of an emergency front door.
(f) Emergency illumination and ventilation in case of a power failure.

The initial plan involved running four-coach trains, each consisting of two motor coaches that are fitted with propulsion equipment and two driving trailer coaches, with a total carrying capacity of 1500 passengers. Subsequently, six-coach trains were introduced to meet traffic demands at rush time. Plans are on the anvil to introduced eight-coach trains in some busy lines. These trains are planned to run at an average speed of 32 to 35 km per hour over an average interstation distance of 1.1 to 1.3 km.

29.4.2 Signalling and Train Control

Delhi Metro trains are provided with the latest signalling technology and control system as described of in the subsequent paragraphs.

Signalling It has been decided that the trains plying underground and in surface corridors will run at a 3 minute interval in peak hours. The signalling system is designed for an ultimate headway of two minutes, with a continuous automatic signalling system comprising the following special features.

Automatic train protection This consists of cab signalling, whereby the drivers will get information as regards the condition of the line beforehand and so that they can control the speed of the trains in advance as per the track status or obstruction. Normally, the driver will apply the brakes in case of any perceived obstruction on the track. In case a driver fails to do so, emergency brakes are applied automatically and all these events are recorded in the appropriate system device for the purpose of establishing accountability.

Automatic train supervision (ATS) This is a sophisticated computer-based supervisory system, which will takeover the critical functions of the controller in the control room and will drastically reduce the workload of the station master. The assistance of ATS is necessary in cases where trains run at a close headway of 2 minutes. The train describer system of ATS displays the positions of the trains while they are in motion in visual form in the control room as well as provide precise information regarding passengers.

Automatic train operation This feature is required to be provided for underground corridors only. It enables the train to be driven without any human intervention, thereby providing a high level of operational efficiency.

Interlocking of yards An advanced interlocking system with high-speed (50 kmph) turnouts has been recommended for the yards. The entire section is to be circuited with audio frequency track circuits, which should eliminate any failures on account of the insulated joints giving way, thus requiring replacement.

Telecommunication The telecommunication network between various stations and services is like a backbone of optic fibre, which will have an enormous capacity for channelling data and voice communication. It will carry all train control information through telemetry links, which will bear the details regarding the running of the train and also the data collected by the passenger information system.

Passengers information system In the system that has been proposed, the central train describer will provide precise information to the passengers at every platform on a real-time basis by furnishing online information through the central computer. All the stations will be provided with centrally synchronized clocks on the various platforms to maintain time standards. The public address system will enable centralized as well as local announcements on all stations platforms.

CCTV system Closed circuit TVs are provided at all the critical locations in the station premises to monitor every safety aspect.

29.4.3 Power Supply

To ensure the continuous availability of quality power for the running of metro trains, utmost efforts have been made to plan and design a power supply system with the degree of reliability found in other world class metros. The Delhi Metro system derives its operating power in the form of 25kV ac traction.

29.4.4 Tracks

The design standards for the track of the metro system are briefly described in the following paragraphs.

Spacing of tracks The metro routes consist of two broad gauge tracks spaced 4.10 m apart, which is less than the 4.725 m standard spacing on Indian Railways. The reduced spacing has been feasible due to the fact that the doors of the coaches are closed and none of the doors open outward as is the case with the goods wagons of the Indian Railways. The points and crossings are 1 in 12 and 1 in 8.5, as in the case of the Indian Railways.

Gradients The steepest gradient permissible is 3.0 per cent. Gradients steeper than 2.5 per cent are adopted only in exceptional cases.

Curves Curves of radius less than 450 m are adopted sparingly on running lines. No curve on a running line may have a radius less than 300 m.

Tracks on stations On stations, tracks are not to be laid on gradient steeper than 0.1 per cent (1 in 1000) and they should also not be laid on curves with radius less than 1000 m. Vertical curves normally have a radius of 2500 m at points where there is a change in the gradient. In the case of transition curves, the rules followed are the same as those on Indian Railways. The other features of the metro track are as follows.

(a) The metro track consists of 60kg UIC rails on level tracks, which rest on concrete sleepers and are fixed using elastic fastenings (Pandrol clip mark 11). The sleeper density is 1660 sleepers per km.
(b) The ballast cushion is prepared using hard stone ballast and measures 300 mm in thickness.
(c) 60 kg head-hardened rails have been used to enhance the service life of the rails, particularly on sharp curves and steep gradients.
(d) To minimize the need for track maintenance and reduce the dimensions of structures that are consequential to the quality of the run, a 'ballastless track' is being laid on elevated sections, viaducts, and tunnels. These stretches are provided with 60 kg rails that are fixed with the help of Vossloh fastenings.
(e) To improve the standard of maintenance and ensure a comfortable ride, rails are mostly welded as LWRs and efforts have been made to ascertain that the entire track is almost 'joint-less'. For this purpose, even the turnouts have been integrated into the LWRs. Specially-designed turnouts are used with thick web switches and CMS crossings by means of welded leg extensions.

29.4.5 Metro Stations

There are about 53 metro stations, including 12 underground stations on phase 1 of the Delhi Metro. Metro stations are normally two-line stations with side or island platforms. The platforms are 185 m long, so as to accommodate eight-coach trains. The platform surface lies 1.08 m above the rail level, so that the floor of the coaches is almost level with the platform. The width of the platform is normally 6 m in the case of side platforms and 10 m in the case of island platforms.

Tracks at stations with side platforms are placed 4.1 m apart, while those with island platforms, are placed around 13.3 m apart. Stations located in well-populated areas are normally spaced 1 km apart. The interstation distance can, however, vary marginally to suit site conditions.

Stations situated on an elevated corridor along the central verge of the road can be provided with tracks at a height of about 12 m above the road level. Most of the underground metro stations are generally two-level stations between 270 m to 300 m in length. A typical station is 20 m wide and 15 m deep. The first level boasts of various passenger facilities together with ventilation and electronics and communication equipment rooms and the lower level comprises the platform with electric equipment rooms at each end of the box. Island platforms are provided with stations that are adjacent to board tunnel sections while side platforms are provided with cut and cover tunnels. All stations have been designed with the option to accommodate platform screen doors.

29.5 CIVIL WORKS

Some of the new technologies and innovations in construction techniques that are being deployed for the Delhi MRTS are briefly mentioned here.

29.5.1 Construction of Bridges

Several innovative construction techniques were used in the construction of the bridges. For example, a new Yamuna bridge (12 spans of 46.2 m each) was constructed using modern techniques during the construction of the Delhi Metro. The substructure of this bridge consists of capsule-shaped piers that rest on well (caisson) foundations of 10 m diameter with a steining thickness of 1.0 m, which have been sunk up to a depth of about 39 m. The 'jack-down' method, supplemented with air/water jetting, has been used for sinking the wells. This method involved pushing the well assembly down into the ground by applying pressure to counter the resistance arising due to skin friction around the periphery of the wells and below the cutting edge. Soil dredging has been carried out inside the well simultaneously. The new techniques resulted in faster sinking of the wells. Also, the wells have been sunk plumb into the water with the help of minimum tilts or shifts.

The superstructure of the bridge comprises a single box girder of a constant depth of 3.5 m that has been launched using an incremental launching technique. This technique involved casting the girder on the bridge approach behind the abutment in segments of length 23.1 m, which is half the length of one span. Each segment has been cast behind the previous unit. After a sufficient concrete strength was attained, the new unit was post-tensioned to the previous one. The assembly of units was pushed forward to permit the casting of the succeeding segment. This technique resulted in a single continuous girder of length 554 m with no joints.

The innovative features of the Yamuna bridge are summarized below.

(a) Second incrementally launched bridge in India, the first being the Panvel Nadi bridges of the Konkan Railway.
(b) First box girder in India carrying dual unballasted tracks.
(c) An innovative technique for sinking the wells; the jack-down method has been adopted for sinking 15 wells. Ground anchors have been installed near the well staining, which have been used for taking the reaction of the specially designed hydraulic jacks at the top.

29.5.2 Elevated Viaduct

The elevated viaduct comprises supported spans with lengths varying from 21.1 m to 29.1 m. It has been constructed by the precast segmental technique with epoxy bonded joints and traditional internal pre-stressing. The span lengths have been determined mainly on the basis of the site constraints at the ground level for the location of the foundations and piers. All precasting has been done at a centralized casting yard.

An aesthetic form of the substructure has been evolved so as to harmonize with the flow of the forces. In this structure the pier gradually tapers outward at the top to support the bearings under the box webs. All the piers have been cast in place using rigid steel, which was poured in a single attempt in order to avoid any construction joints. The use of bolts in the concrete has not been permitted. Bored piles of a diameter of 1.2 m that have been cast in situ have been included by employing modern hydraulic rigs. To ensure a standard span, groups of nine piles with a length of 23 m to 28 m have been stacked in the soil. In rocky terrains, piles have been stacked in the rock in groups of six up to a depth equal to 1.5 to 3.0 times the diameter of the pile, depending on the type of rock encountered. A temporary casing has been employed for the stabilization of the bore holes.

29.5.3 Cut and Cover Tunnels

It is possible to construct tunnels by excavating the ground surface, hence the cut and cover method has been employed for tunnel construction in these stretches.

Cut and cover tunnels are generally designed as a single structural unit with a dividing wall, and walkways for each track giving an average box width of 10 m. The over-run tunnels are used as sidings and contain three tracks while at the depot spur, the tunnels contain four tracks. The construction methodology for cut and cover tunnels is similar to that of cut and cover stations.

Cut and cover construction, though economical, results in a lot of public inconvenience if planned improperly. As such, it requires exhaustive planning with regard to utilities, traffic management, and environmental concerns, such as cutting of trees, and air and noise pollution.

29.5.4 Bored Tunnel Construction

The bored tunnel construction technique has been used at many places in the metro corridor tunnels. The internal diameter of such tunnels is about 5.6 m and they are generally lined with precast reinforced concrete segments. Tunnel boring machines (TBMs) have been used in place of conventional tunnelling methods that employ various hand mining techniques, since conventional tunnelling methods suffer from major handicaps with regard to slower progress and risks to existing structures in the vicinity of tunnel excavation.

Since the bored tunnel construction for the Delhi Metro involved tunnelling through both quartzite and soft ground, two types of TBMs have been used, namely rock TBM for rocky ground and earth pressure balance machine (EPBM) for soft ground. In the case of an EPBM, the excavated face is supported by pressurizing the soil that has been dug up, together with additives if necessary, to give it a plastic texture. A screw auger is used to remove the excavated material. The earth pressure within the cutter head is maintained by balancing the rate at which it advances, with the rate at which the spoil material is removed and adjusting the parameters as required. The gap between the excavation surface and the precast concrete lining is filled by means of an automatic injection from the tail of the machine, which minimizes ground settlement. Watertightness at the tail is ensured by injecting grease into the three rows of brush seals.

29.6 UNDERGROUND VERSUS ELEVATED ALIGNMENT

Underground construction is a much costlier option than elevated alignment. The current average cost per route kilometre for construction on an elevated alignment, including stations, rolling stock, signalling, and electrical equipment as well as land and incidental works, is ₹ 1000 million as compared to ₹ 3000 million for underground construction. On account of the vast cost difference, the choice of underground construction is restricted to areas where elevated alignment is not suitable. In the case of Delhi Metro, underground construction is restricted to the core areas of Delhi on the North–South and East–West corridors, which intersect at the Connaught Place area. Much of the route length outside the core area is likely to be constructed as an elevated route.

In the case of elevated segments of the MRTS, the alignment on city roads usually follows the central verge of the road. Piers of about 2 m diameter are spaced 20–30 m apart along the central verge. The arrangement of girders above the pier cap is shown in the general arrangement given in Fig. 29.3. Tracks are generally laid 8.5–9.5 m above the road level on elevated routes outside the station. The desirable minimum width of the roads for locating an elevated MRTS on the same is 30 m, although a road 40–45 wide is preferable. The geometry of the road should be favourable for the adoption of curves adhering to the minimum MRTS standards. In the case of underground construction on stretches outside the station, the rail level lies at a minimum depth of 9 m below the road surface. This is done with the object of 2–3 m of space below the road for electrical cables, telephone cables, and drains.

A corridor need not be wholly underground or elevated. An underground corridor in the core city area could get converted into an elevated corridor in the outer parts, where the conditions are favourable for such constructions. It is crucial that an extra stretch of aligned road surface be available between the point where the underground section ends and the elevated section starts. The alignment in this stretch should rise from about 9 m below the ground to 9 m above the road level. In order to ensure that the length of the at-grade stretch is not unduly long, the tracks are provided with a steep gradient for negotiating this difference in the level. This stretch can measure 400–500 m in length depending on the site characteristics.

In the at-grade stretch, the MRTS tracks are walled on both sides, their outer faces being nearly 10 m apart. This implies that the road width is effectively reduced by about 11 m on this stretch. This factor is of great importance in choosing the location for an at-grade ramp.

Schedule of dimensions

The Delhi Metro Rail Corporation has drawn up a schedule of the dimensions for its system (Table 29.4). These dimensions should normally be adhered to when undertaking new works and finding alternatives to existing works on BG lines (1676 mm), except in exceptional cases where the sanction of the commissioner of railway safety has to be specifically obtained.

Table 29.4 Delhi Mass Rapid Transit System (MRTS)—Salient Features

Parameter	Details	Remarks
Gauge	1676 mm	BG
Stations at a distance	1 km each	–
Train frequency during peak hours	3 minutes	5 minutes frequency in normal times
Speeds	(i) Maximum 80 kmph (ii) Scheduled 30 kmph	
Traction system	25 kv ac traction	
Train composition	8 coaches	Coaches are air cooled
Carrying capacity	380 passengers	Each rake will carry about $(380 \times 8) = 3040$ passenger
Type of signalling and interlocking	Cab signalling and solid state interlocking	Continuous automatic train control
Telecommunication	Fibre optic transmission system	Train mobile radio and closed circuit TV system
Platform: Length height	185 m 1.08 m	Operation and supervision

(Contd.)

Table 29.4 *(Contd.)*

Parameter	Details	Remarks
Rulling gradient	Main line 4 %	Depots 3 % and Stations 1 %
Radius of curves	Main line 300 m	Depots 225 m
Spacing of two tracks	4.1 m	For underground two tunnels 5.4 dia at 13 m c/c

Note: Delhi Metro from Connaught place to Palam airport is likely to have standard gauge of 1435 mm (4 feet 8.5 inches)

29.7 KOLKATA METRO

Much of the suburban traffic in Kolkata is borne by Eastern Railway for distances extending over 100 km. On the South-Eastern Railway, the suburban sections extend from Howrah to Balichak, which lie at a distance of 92 km from each other. The railway system in Kolkata terminates at Howrah on one side and Sealdah on the other, both of which lie on the periphery of the central business district of Kolkata. In view of this, commuters have to cover the distances within the city by other modes of transport or on foot.

City buses carry a major portion of commuters from one place to another. Besides the suburban rail and bus transport services, the tramways are an important mode of mass transport system in Kolkata. The Kolkata Tramway Company has about 400 trams. The underground metro railway in Kolkata started functioning from 24 October 1984 between Esplanade and Bhawanipore. It stretches from Dum Dum to Tollygunge and covers a route length of 16.45 km. The construction work of this underground railway was done by the cut and cover method in the open area. However, the *driven shield tunnelling* method was adopted wherever the metro alignment passed under residential buildings or a canal.

29.7.1 New Technologies used in Kolkata Metro

The construction of a metro rail system is a very complex process requiring the application of several new technologies in the fields of civil, electrical, signalling, and telecommunication engineering. For the first time in India, engineers, backed by their own experience and supplemented by the knowledge gained through their studies abroad, have incorporated the following advanced technologies with regard to the construction of the Kolkata Metro.

(a) Cut and cover method of construction using diaphragm walls and sheet piles
(b) Use of extensive decking to keep the traffic flowing over the cut while construction was in progress underneath
(c) Shield tunnelling using compressed air and airlocks
(d) Integration of a ballastless track using elastic fastenings, rubber pads, epoxy mortar, and nylon inserts
(e) Air conditioning and ventilation system for controlling the atmosphere of stations and tunnels
(f) Third rail current collection system for traction
(g) Underground substations with dry-type transformers and SF-6 circuit breakers
(h) Continuous automatic train control system

(i) Tunnel-train VHF-radio communication system
(j) Train control and supervisory remote control systems fitted with microprocessors for substations
(k) Automatic ticket vending and checking system

29.7.2 Benefits of Kolkata Metro

The Kolkata Metro, like all other mass rapid transit systems in the world, cannot be financially remunerative, as subsidies are inevitable in the operation of such an expansive system. However, introducing a system that covers the entire city has resulted in the following benefits to the public.

(a) The social benefits of the Kolkata Metro far outweigh the burden of the subsidies.
(b) It carries more than 0.5 million commuters per day in trains that follow each other at intervals of about 5 minutes during the peak hours.
(c) The commuters are assured of a pollution-free, safe, punctual, and comfortable journey.
(d) Commuting time is drastically reduced.

The salient features of the Kolkata Metro are listed in Table 29.5.

Table 29.5 Salient features of Kolkata Metro System

Features	Details
Total route length	16.45 km
Number of stations	17
Coaches per train	8
Maximum permissible speed	55 kmph
Average speed	30 kmph
Voltage	750-V dc
Total power requirement	94 million kwh
Method of current collection	Third rail
Travel time from	
Dum Dum to Tollygunge	33 min
Dum Dum to Girish Park	11 min
Tollygunge to Central	18 min
Capacity of each coach	278 standing, 48 sitting
Capacity of each train	2500 passengers (approx.)
Interval between trains	5.5 minutes in peak hours
Estimated cost	₹ 15.4 billion (at 1993–94 rates)
Environment control	Forced ventilation with washed and cooled air

The expansion of the Kolkata Metro is also being planned. A detailed project report for the East–West corridor from New Dass Nagar to Salt Lake Sector 5 has been submitted to the state government by the DMRC. The corridor will be 16 km long with 9 km of underground section. An interchange has been proposed for the crossing of the N-S line at Central Station. The corridor will pass below the river Hooghly at a depth of about 30 m from the ground level.

29.8 MUMBAI SUBURBAN SYSTEM

Mumbai, which lies on the western coast of India, is one of the world's largest and most crowded cities. It is also considered the commercial capital of India. Every day about 5.5 million commuters travel on the Mumbai suburban system, which is almost 50 per cent of the trips made by passengers on the entire railway system in India. During the peak hours almost 4500 passengers travel in one train as against the design capacity of 2700 passengers thereby creating an alarming situation.

The salient features of the Mumbai suburban railway are presented in Table 29.6.

Table 29.6 Salient features of the Mumbai suburban railway system

Feature	Central Railway	Western Railway
Corridors	*Main line* Chhatrapati Shivaji Terminus to Kasara, 121 km (slow corridor)	Churchgate to Borivli, 34 km (slow corridor)
	Chhatrapati Shivaji, 54 km (fast corridor) and up to Karjat, 46 km (slow corridor) *Harbour line (46 km)*	Churchgate to Virar, 60 km (fast corridor)
	Chhatrapati Shivaji Terminus to Wadala (9 km), Wadala to Andheri (12.3 km) Wadala to Khandeshwar (37 km)	
Stations	61	28
Signalling	Automatic block signalling up to Titwala/ Badlapur Belapur and absolute block signalling system beyond Titwala	Automatic block signalling
Power supply	1500 V dc	1500 V dc
No. of trains running	1090 trains	901 trains
Headway	5 min on slow corridor and harbour line, 4 min on fast corridor	3 min between Churchgate and Andheri and 4 min beyond Andheri
No. of rakes	100	74
No. of commuters travelling daily	2.9 million (approx.)	2.6 million (approx.)

Today, commuter traffic in Mumbai and its suburbs is served by the following five suburban railway corridors of Western and Central Railway.

Western railway
   (a) A pair of lines from Borivli to Churchgate, designated local lines
   (b) A pair of lines between Virar and Churchgate, designated through lines

Central railway
   (a) A pair of lines between Mumbai VT and Kalyan, designated local lines
   (b) A pair of lines between Mumbai VT and Kalyan, designated through lines (suburban sections extend up to Kasara on the North–East ghat section and up to Karjat on the South–East ghat section).

(c) A pair of lines, known as the harbour branch, between Mumbai VT and Chembur, with a chord line extending from Raoli Junction to Bandra and a single line extending from Chembur to Mankhurd.

A metro rail system is being planned for Mumbai as well. The DMRC has submitted detailed project reports for three corridors to the Mumbai Metropolitan Regional Development Authority. The first corridor stretches from Versova to Ghatkopar (12 km) and has been approved by the Government of Maharashtra. The second corridor stretches from Colaba to Charkop via Bandra, is 36 km long, and 10 km of the route from Colaba to Mahalaxmi will be underground. The third corridor extends from Bandra to Mankhurd via Kurla and is 13 km long. Both the first and third corridors will be constructed on elevated ground.

29.9 CHENNAI SUBURBAN SYSTEM

Commuter traffic on the suburban railway system of Chennai is carried on the following sections.

(a)	Chennai Central to Trivellore	42 km (BG)
(b)	Chennai Beach to Tambaram and Chengalpattu	60 km (MG)
(c)	Chennai Beach to Royapuram and Gummidipoondi	47 km (BG)

The Chennai Beach to Tambaram section carries the bulk of commuter traffic. This section is electrified and the trains comprise an EMU stock. The Government of Tamil Nadu has recently approved the construction of Chennai metro corridors from Tiruvottiyur to the airport via Chennai Central (31 km) and Ponnamallee to Chennai beach (13 km).

29.10 METRO RAILWAY IN VARIOUS CITIES OF INDIA

With the success of Delhi Metro Rail System, about 23 cities in India are either operating or constructing or planning a metro system. This includes Kolkata Metro and Delhi Metro, as mentioned in the previous sections.

Details of these metro railways are given in Table 29.7.

Table 29.7 Metro railways in operation/construction/planning in Indian cities

City	Population	Status
Delhi	16.7 million (Delhi NCR)	**Phase I and II** (189 km) - (in operation) Huda City–Jahangirpuri Noida/Vaishali–Dwarka Dilshad Garden–Rithala Mundka-Inderlok–Kirti Nagar Central Secretariat–Badarpur New Delhi–Dwarka Airport line **Phase III** (120 km)–(in progress) Mukundpur–Yamuna Vihar (55.69 km) Janakpuri (W)–Kalindi Kunj (33.49 km) Central Secretariat–Kashmere Gate (9.37 km) Jahangirpuri–Badli (4.48 km) Badarpur–Faridabad (13.5 km) Kalindi Kunj–Botanical Garden (3.5 km) **Phase IV** in planning phase

Contd.

Table 29.5 (*Contd.*)

City	Population	Status
Kolkata	14 million (Kolkata Metropolitan).	1. Tollygung–Dumdum, (17 km) — (in operation) 2. Tollygung–Garia, (9 km) (in operation) 3. Dumdum–Dakshineshwar, (5 km) — (in progress) 4. Howrah–Salt Lake sec V (14 km) — (in progress) 5. Joka–BBD Bagh, (17 km) 6. Noapara–Barasat, (19 km) 7. Baranagar–Barrackpur, (13 km)
Mumbai & Navi Mumbai	20.5 million (Mumbai Metropolitan)	**Phase I** 1. Vasova–Andheri–Ghatkopar, (11 km) — (in progress) 2. Dahisar–Charkop–Bandra–Mankhund, (32 kms) — (in progress) **Phase II** 1. Colaba–Bandra–Airport, (30 km) 2. Carmac Bunder–Ghatkopar–Teen Haath Naka (40 km) **Phase III** 1. Airport–Kanjur Marg (9.5 km) 2. Andheri (E)–Dahisar (E) (18 km)
Bengaluru	About 10 million (Bengaluru District)	1. Baiyappanahalli–Mysore Road Terminal (18 kms) — (6.7 km opened in Oct 2011 and rest in progress), 2. Nagadandra–Puttenahalli, (24 km) — (in progress)
Chennai	7.5 million (Chennai Metropolitan)	1. Washermenpet–Chennai Airport, (23 km) — (in progress) 2. Chennai Airport–St Thomas Mount, (22 km) — (in progress)
Hyderabad	About 8.0 million (Hyderabad Metropolitan)	1. Miyapur–L B Nagar, (27 km) — (in progress) 2. JBS–Falaknuma, (16 km) — (in progress) 3. Nagola–Shilparaman, (23 km) — (in progress)
Jaipur	3.2 million	1. Sitapur–Ambabari (23 km) — (in progress) 2. Mansarovar–Badi Chaupar (12 km) — (in progress)
Gurgaon	About 1.6 million (Gurgaon District)	1. Sikandarpur–DLF 3 (5 km) — (in progress) 2. Iffco Chowk–Dwarka Sec. 21 — (in progress)
Ahmedabad & Gandhinagar	About 6.5 million (Ahmedabad Metropolitan)	APMC to Mother Dairy (24 km) Thaltej–Ahmedabad Jn (11 km) Sarkhej–Indroda Circle (32 km) Vasna–Sarkehj (6 km) Sarkhej–Dholera (120 km)
Kozhikode	3.0 million	Karipur Airport to Medical College (33 km)
Lucknow	3.0 million (Lucknow Metropolitan)	Planning
Chandigarh	1.1 million (Chandigarh District)	Planning
Kochi	2.2 million	Planning

Contd.

Table 29.5 (*Contd.*)

City	Population	Status
Bhopal	2.4 million	Planning
Ludhiana	1.61 million	Planning
Nagpur	2.13 million	Planning
Patna	1.6 million	Planning
Pune	3.11 million	Planning
Kanpur	2.78 million	Planning
Indore	1.96 million	Planning
Surat	4.8 million	Planning
Vadodara	1.67 million	Planning
Rajkot	1.6 million	Planning

SUMMARY

With the rapid increase in the population of the metro cities in India, a rail-based mass transportation system has become the need of the hour. Its capacity is 10 times more than that of road transport systems and it also has a much better safety record. The Delhi Metro has been a successful experiment and other state governments are preparing for introducing a similar system in their metro cities.

REVIEW QUESTIONS

1. Distinguish between surface and underground railway systems. Enumerate the factors that favour the selection of one over the other.
2. In what way has the underground railway system assumed considerable importance and promise for the metropolitan transport system? Describe briefly the principal tasks involved in the construction of an underground railway system.
3. Briefly state the problems of a metro tube railway.
4. What do you understand by a metropolitan town? What are the problems of a metropolitan town with regard to its transport system?
5. Discuss briefly the various types of transport that can be provided in a metropolitan town and the relative advantages and limitations of the same.
6. What are the essential features of an underground railway? What do you understand by the cut and cover method of construction of an underground railway?
7. Briefly discuss the problem of transport in Delhi. Give the salient details of the MRTS scheme of Delhi.
8. What are the salient features of the Kolkata Metro Railway? Enumerate the technologies used in its construction.

Choose the correct answer from the choices given.

9. In the 'cut and cover' method of construction of underground railway, the cross section adopted is:
   (a) circular
   (b) rectangular
   (c) any of (a) & (b)
   (d) none of these

10. In the case of tube railway which is undeground, the depth should be more than:
    (a) 5 m   (b) 10 m
    (c) 15 m   (d) 25 m

11. Phase I of Delhi Metro Rail Section was completed in:
    (a) 1997   (b) 2000
    (c) 2005   (d) 2011

12. The double-line BG metro routes are spaced at a distance of:
    (a) 3.2 m   (b) 3.7 m
    (c) 4.10 m   (d) 4.75

13. The Kolkata Metro Railway became operational in:
    (a) 1980   (b) 1984
    (c) 1990   (d) 1995

Railway Tunnelling

INTRODUCTION

A tunnel can be defined as an underground passage for the transport of passengers, goods, water, sewage, oil, gas, etc. The construction of a tunnel is normally carried out without causing much disturbance to the ground surface.

The history of tunnels is very old. The first tunnel was constructed about 4000 years ago in Babylon to connect two buildings. The first railway tunnel in the world was constructed at the end of the nineteenth century to connect Switzerland and Italy. The cross section of the tunnel was in the shape of a horseshoe and its length was about 20 km.

Railway tunnelling in India started in the nineteenth century. In hilly terrain, in order to reach upto a destination, generally a maximum ruling gradient of 1 in 100 is provided on BG and grades steeper than this require extra efforts by a pusher engine. This gradient would mean laying of hundreds of kilometres of track to skirt hills depending on the contours. This can be avoided by tunnelling, which directly reduces the length of the alignment.

On Indian Railways, the first tunnel was constructed near Thane on Central Railway known as the *Parsik tunnel*. Till a few years back it was the longest railway tunnel in India with a length of about 1317 m.

Today, there are 348 railway tunnels in India. A large number of tunnels are also under construction on the Udhampur–Quazikund sector of the Udhampur–Srinagar–Baramula project.

30.1 NECESSITY/ADVANTAGES OF A TUNNEL

The necessity of constructing a tunnel may arise because of one or more of the following considerations.

(a) A tunnel may be required to eliminate the need for a long and circuitous route for reaching the other side of a hill, as it would considerably reduce the length of the railway line and may also prove to be economical.

(b) It may be economical to provide a tunnel instead of a cutting, particularly in a rocky terrain. Depending upon various factors, a rough calculation would indicate that for a small stretch of land the cost of constructing a tunnel is equal to the cost of a cutting in a rocky terrain.

(c) In hills with soft rocks, a tunnel is cheaper than a cutting.

(d) In metropolitan towns and other large cities, tunnels are constructed to accommodate underground railway systems in order to provide a rapid and unobstructed means of transport.

(e) A tunnel constructed under a river bed may sometimes prove to be more economical and convenient than a bridge.

(f) In the case of aerial warfare, transportation through tunnels provides better safety and security to rail users compared to a bridge or deep cutting.

(g) The maintenance cost of a tunnel is considerably lower than that of a bridge or deep cutting.

However, the construction of tunnels is also disadvantageous in certain ways, as enumerated here.

(a) The construction of a tunnel is costly as it requires special construction machinery and equipment.

(b) The construction of a tunnel involves the use of sophisticated technology and requires experienced and skilled staff.

(c) It is a time-consuming process.

30.2 TUNNEL ALIGNMENT AND GRADIENT

A precise and detailed survey is necessary before setting the alignment of a tunnel on the ground. A small error in setting the alignment would result in the two ends never meeting at all. When starting work, both the ends of the tunnel as well as the centre line are marked with precision on the ground so that the correct length of the tunnel can be determined. An accurate survey is then carried out to ensure that the centre line of the alignment and the levels are transferred properly to their underground positions.

The following points require special attention when deciding the alignment and gradient of a tunnel.

(a) The alignment should be straight as far as possible since normally such a route would be the shortest and most economical.

(b) The minimum possible gradient should be provided for a tunnel and its approaches.

(c) Proper ventilation and adequate lighting should be provided inside the tunnel.

(d) The side drains in a tunnel should be given a minimum gradient of 1 in 500 for effective drainage. In longer tunnels, the gradient should be provided from the centre towards the ends for effective and efficient drainage.

30.3 SIZE AND SHAPE OF A TUNNEL

The size and shape of a tunnel depend upon the nature and type of ground it passes through and also on whether it is designed to carry a single or a double railway line.

The shape of a tunnel should be such that the lining is able to resist the pressures exerted by the unsupported walls of the tunnel excavation.

If the ground is made up of hard rock, then the tunnel can be given any shape. Tunnels in rocky terrains are generally designed with a semicircular arch with vertical sidewalls. In the case of soft ground, such as that consisting of soft clay or sand, the pressure from the sides and the top must be resisted. A circular tunnel is generally best suited for resisting both internal and external forces regardless of the purpose for which the tunnel is used. Theoretically, a circular section provides the largest cross-sectional area for the smallest diameter, which provides greater resistance to external pressure. However, this type of cross section is more useful for drains carrying sewage and fluids and for aquaducts built for irrigation purposes. For railway track, the circular portion at the bottom of the tunnel has to be levelled in order to lay the track and facilitate the easy removal of muck and placing of concrete. The typical cross section of a tunnel is shown in Fig. 30.1. The sections commonly adopted for tunnelling and the purpose these tunnels serve are enumerated in Table 30.1.

Table 30.1 Shape and purpose of tunnels

Shape	Purpose
Circular	Water and sewage
Elliptical	Water and sewage mains
Horseshoe	Roads and railways
Arched roof with vertical walls	Roads and railways
Polycentric cross section	Roads and railways

The size of a railway tunnel depends upon the gauge of the railway track and the number of lines. The typical dimensions for railway tunnels are specified in Table 30.2.

Table 30.2 Size of the tunnel

Gauge (mm)	Single line		Double line	
	Breadth (mm)	Height (mm)	Breadth (mm)	Height (mm)
BG (1676)	4880 – 5490	6700 – 7320	8530 – 9140	6700 – 7320
MG (1000)	4270 – 4880	6100 – 6700	8530 – 9140	6100 – 6700

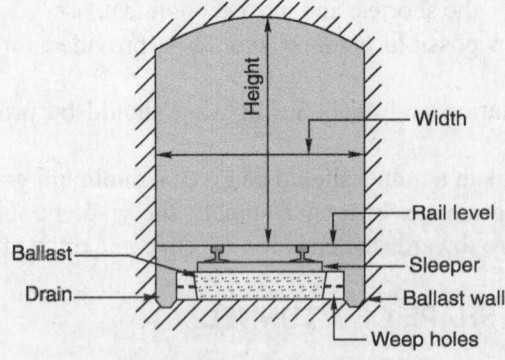

Fig. 30.1 A typical cross section of a tunnel

30.4. METHODS OF TUNNELLING

There are various methods of tunnelling. The selection of a method depends upon the size of the bore, the condition of the ground, the equipment available, and the extent to which timbering is required. Tunnelling may be basically divided into two main groups.
  (a) Tunnelling in hard rocks
  (b) Tunnelling in soft rocks

These are described in detail in the subsequent sections. Tunnelling through water-bearing strata and compressed air tunnelling are discussed subsequently.

30.4.1 Tunnelling in Hard Rocks

The following methods are generally employed for tunnelling in hard rocks.

Full face method

The full face method is normally selected for small tunnels whose dimensions do not exceed 3 m. In this method, the full face or the entire facade of the tunnel is tackled at the same time. Vertical columns are erected at the face of the tunnel and a large number of drills mounted or fixed on these columns at a suitable height as shown in Fig. 30.2. A series of holes measuring 10 mm to 40 mm in diameter with about 1200 mm centre-to-centre distance are then drilled into the rock, preferably in two rows. These holes are charged with explosives and ignited. Next the muck is removed before repeating the process of drilling holes.

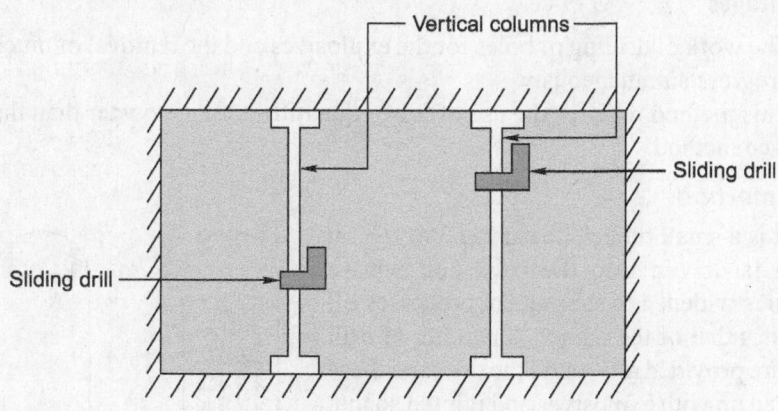

Fig. 30.2   Full face method

Advantages

  (a) Since an entire section of the tunnel is tackled at one time, the method is completed expeditiously.
  (b) Mucking tracks, which are tracks used for collecting muck, can be laid on the tunnel floor and extended as the work progresses.
  (c) With the development of the 'jumbo' or drill carriage, this method can be used for larger tunnels too.

Disadvantages
  (a) The method requires heavy mechanical equipment.
  (b) It is not very suitable for unstable rocks.
  (c) It can normally be adopted for small tunnels only.

Heading and bench method

In this method, the heading (top or upper half) of the tunnel is bored first and then the bench (bottom or lower half) follows. The heading portion lies about 3.70 m to 4.60 m ahead of the bench portion (Fig. 30.3). In a hard rock, the drill holes for the bench are driven at the same time as the removal of the muck. The hard rock permits the roof to stay in place without supports.

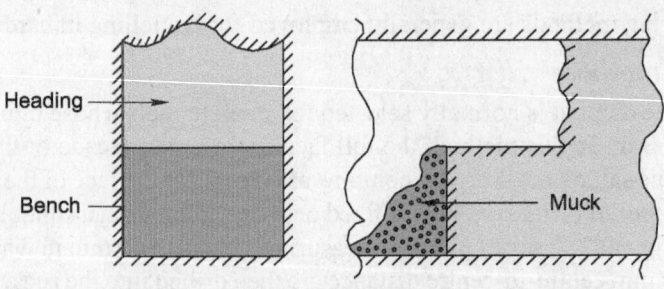

Fig. 30.3 Heading and bench method

Advantages
  (a) The work of drilling of holes for the explosives and the removal of muck can progress simultaneously.
  (b) This method requires the use of lower quantities of gunpowder than the full face method.

Drift method

A drift is a small tunnel measuring 3 m × 3 m, which is driven into the rock and whose section is widened in subsequent processes till it equates that of the tunnel. A number of drill holes are provided all around the drift and these are filled up with explosives and ignited so that the size of the drift expands to become equal to the required cross section of the tunnel.

The position of the drift depends upon local conditions; it may be in the centre, top, bottom, or side as shown in Fig. 30.4. Field experience has shown that the central drift is the best choice, as it offers better ventilation and requires lower quantities of explosives. The side drift, however, has the advantage that it permits the use of timber to support the roof.

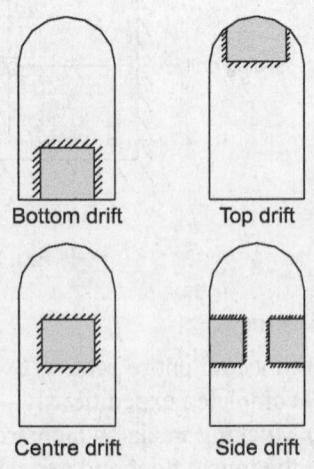

Fig. 30.4 Drift method

Advantages

   (a) If the quality of the rock is poor or if it contains excessive water, this is detected in advance and corrective measures can then be taken in time.

   (b) A drift assists in the ventilation of tunnels.

   (c) The quantity of explosives required is less.

   (d) A side drift allows the use of timber to support the roof.

Disadvantages

   (a) It is a time-consuming process, as the excavation of the main tunnel gets delayed till the drift is completed.

   (b) The cost of drilling and removing the muck from the drift is high, as the work has to be done using manually operated power-driven equipment.

Pilot tunnel method

This method normally involves the digging of two tunnels, namely a pilot tunnel and a main tunnel. The cross section of the pilot tunnel usually measures about 2.4 m × 2.4 m. The pilot tunnel is driven parallel to the main tunnel and connected to the centre line of the main tunnel with cross cuts at many points. The main tunnel is then excavated from a number of points. The pilot tunnel offers the following advantages.

   (a) It helps in removing the muck from the main tunnel quickly.

   (b) It helps in providing proper ventilation and lighting in the main tunnel.

The method, however, requires the construction of an additional tunnel and therefore the time and cost of construction are higher as compared to the methods described before.

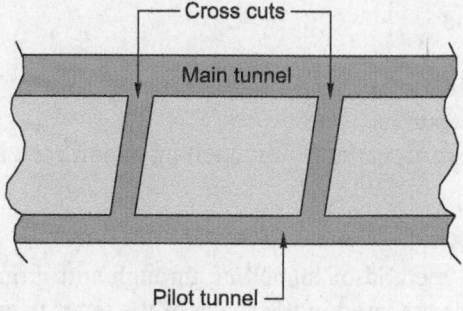

Fig. 30.5 Pilot tunnel

Perimeter method of tunnelling

In this method, the excavation is carried out along the perimeter or periphery of the section. The method is also known as the *German method*.

30.4.2 Tunnelling in Soft Ground or Soft Rock

Tunnelling in soft ground or soft rock is a specialized job. It does not involve the use of explosives and the requisite excavation work is done using hard tools, such as pickaxes and shovels. In recent times, compressed air has also been used for this purpose. During excavation, the rock requires support at the sidewalls and the roofs depending upon the type of soil. The support could be provided in the form of timber or steel plates or other similar material. The various operations involved in soft rock tunnelling are as follows.

(a) Excavation or mining
(b) Removal of excavated material
(c) Scaffolding and shuttering
(d) Lining of tunnel surface

The nature of the ground is the most important factor in deciding the method to be used for tunnelling. The types of ground which are generally encountered in the field are detailed in Table 30.3.

Table 30.3 Types of grounds

Nature of ground	*Typical quality of ground*
Running ground	Requires instant support throughout the excavation. Examples include dry sand, gravel, silt, mud, and water bearing sand.
Soft ground	Requires instant support for the roof but the walls can do without support for a few minutes. Examples include damp sand, soft earth, and certain types of gravel.
Firm ground	The sidewalls and face of the tunnel can do without support for one or two hours, but the roof can last only a few minutes. Examples include firm clay, gravel, and dry earth.
Self-supporting ground	Excavation of the tunnel section can be carried out without support for small lengths ranging from 2 to 5 m. Examples include sand stone, hard clay, etc.

In the case of soft rock, the selection of the method of tunnelling depends upon the following important factors.

(a) Nature of ground
(b) Size of tunnel
(c) Equipment available
(d) Sequence of operations

Some of the important methods of tunnelling in soft rock are described in the following sections.

Forepoling method

Forepoling is an old method of tunnelling through soft ground. In this method (Fig. 30.6), a frame is prepared in the shape of the letter A, placed near the face of the tunnel, and covered with suitable planks. Poles are then inserted at the top of the frame up to a viable depth. The excavation is carried out below these poles, which are supported by vertical posts. The excavation is carried out on the sides and the excavated portion is suitably supported by timber. The entire section of the tunnel is covered thus. The process is repeated as the work progresses.

Forepoling is a slow and tedious process and requires skilled manpower and strict supervision. The method has to be meticulously repeated in sequence and there is no short cut for the same.

Linear plate method

In the linear plate method (Fig. 30.7), timber is replaced by a standard size pressed steel plates. The use of pressed steel plates is a recent development.

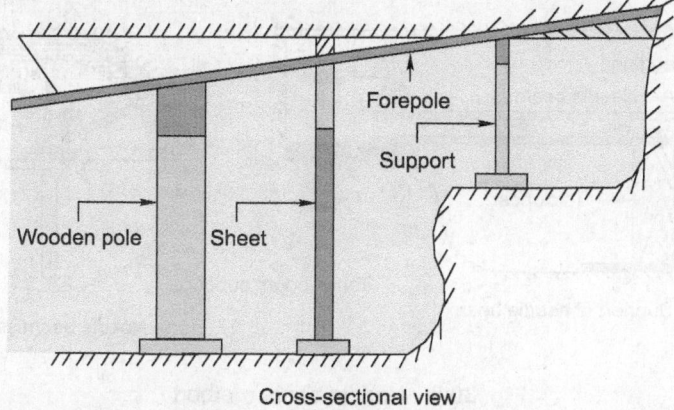

Cross-sectional view

Fig. 30.6 Forepoling method

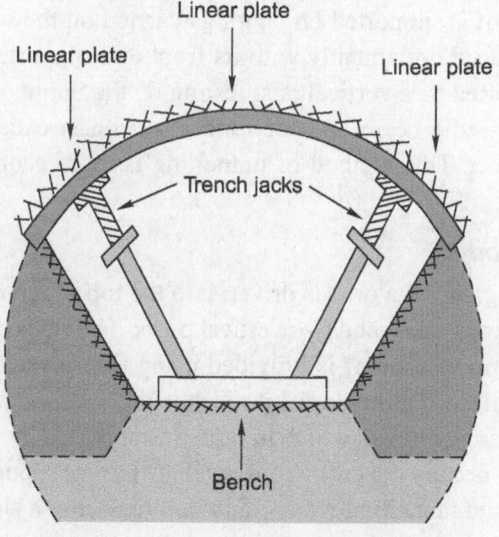

Fig. 30.7 Linear plate method

Advantages

The method has the following advantages.

(a) The linear plates are light and can be handled easily.

(b) The number of joints is less, as the linear plates are bigger in size, and as such the maintenance cost is low.

(c) The steel plates are fireproof and can be safely used while working in compressed air condition.

(d) The necessary work can be done by semi-skilled staff.

(e) There is considerable saving in terms of the excavation and concrete required.

Needle beam method

The needle beam method (Fig. 30.8) is adopted in terrains where the soil permits the roof of the tunnel section to stand without support for a few minutes. In this

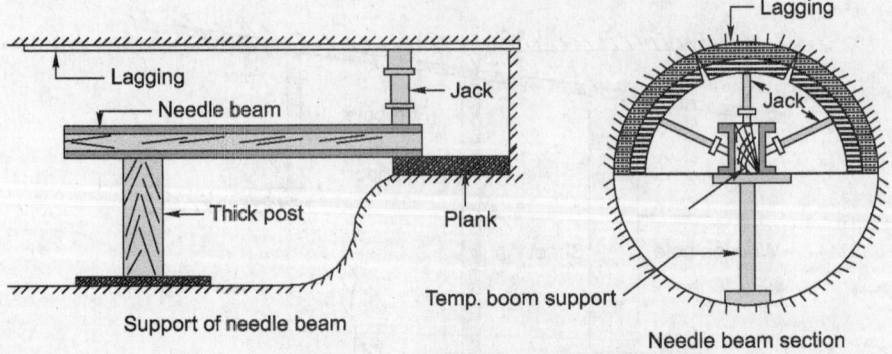

Support of needle beam

Needle beam section

Fig. 30.8 Needle beam method

method, a small drift is prepared for inserting a needle beam consisting of two rail steel (RS) joints or I sections and is bolted together with a wooden block in the centre. The roof is supported on laggings carried on the wooden beam. The needle beam is placed horizontally with its front end supported on the drift and the rear end supported on a vertical post resting on the lining of the tunnel. Jacks are fixed on the needle beam and the tunnel section is excavated by suitably incorporating timber. This method of tunnelling is more economical compared to other methods.

American method

In this method (Fig. 30.9), a drift is driven into the top of the tunnel. The drift is supported by laggings, caps, and two vertical posts. The sides of the drift are then widened and additional support is provided using timber planks and struts. The process of widening is continued till it reaches the springing level. Wall plates are fixed at the springing level, which in turn are supported by vertical posts. The vertical posts now occupy the entire roof level. The posts supporting the drift can then be removed and tunnelling work continued further in a similar manner.

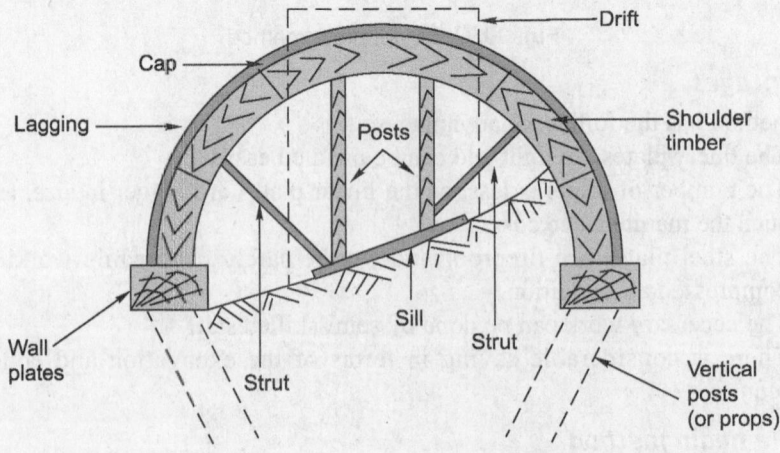

Fig. 30.9 American method

English method

This method is similar to the American method except that the roof load is supported by underpinning instead of using vertical posts. A drift is driven into the top of the tunnel about 5 m ahead of the existing arch lining. The drift is subsequently widened on both sides and supported by crown bars and posts. The work is carried on till the springing level is reached. The sill is then extended across the tunnel and the extended piece is supported by underpinning. This method requires good quality timber as well as simultaneous and frequent shifting from place to place.

Austrian method

This method is used for long tunnels, particularly those at great depths, where the walls of the excavation may yield under the weight of the cover. It involves excavating the whole section for a short length and furnishing with sidewalls and an arch.

Belgian method

This method is particularly suitable for areas where the height of the overburden is less and the surface is not to be disturbed. In this case, the heading is excavated first and supported by crown bar posts and laggings. The sides are excavated next and supported by crown bars and posts. Finally, the work of lining the arch is carried out and further excavation is done.

30.4.3 Tunnelling Through Water-bearing Strata

Tunnelling through subaqueous or water-bearing strata is quite a different job. Shield tunnelling is generally preferred in such cases. A shield is a movable frame that is used to support the face of a tunnel. The tunnel is excavated and lined under the protection of the shield.

A shield is a device meant for excavation that is to be carried out beneath the water-bearing strata. It basically consists of a cutting edge, a skin plate in the form of a shell structure, and a hood of jacks, ring girders, stiffening steel plates, ports as well as port doors, and a tail. The various methods of shield tunnelling through different types of soils are enumerated in Table 30.4.

Table 30.4 Methods of shield tunnelling in different soils

Type of soil	Method of tunnelling
Silt	One or two port doors are opened. The material is excavated and deposited at the bottom of the tunnel.
Clay	One or two ports are opened and the material flows continuously into the tunnel. Excavation is carried out and the soil is removed immediately after the excavation.
Sand	In this case, tunnelling is of the open type. The sand settles on the floor of the shield and it should be continuously removed. Proper care should be taken to ensure that the material does not block the propelling jacks and other equipment.
Running sand	The bulk head shield is used in this case. Other details regarding tunnelling in such a soil are the same as for sand.

Tunnels constructed using the shield method usually have a circular section because of the following considerations.

(a) The rotation of the shield is easy in a circular section.

(b) It grants protection to the primary lining.

(c) The circular section provides the maximum cross-sectional area with the smallest perimeter.

(d) The circular section is ideally suited to resist the semi-fluid pressure exerted by the soft ground.

30.4.4 Compressed Air Tunnelling

Compressed air tunnelling method is possibly the most modern method of tunnelling. The compressed air, which has a pressure of about 1 kg/cm^2, is forced into the enclosed space within the tunnel so that the sides and top of the tunnel do not collapse and remain in their position. The equipment for tunnelling consists of a bulk head, which is an airtight diaphragm with an airlock. The airlock is an airtight cylindrical steel chamber with a door at each end opening inwards.

Tunnelling by means of compressed air is quite a difficult process because of the following reasons.

(a) The pressure inside the earth varies from the bottom to the top of the tunnel.

(b) It is not possible to ascertain the pressure on the floor of the tunnel as it depends upon the nature of the strata.

(c) The pressure varies from strata to strata depending upon the moisture content, which is difficult to ascertain.

(d) The compressed air normally escapes through the pores and the air pressure diminishes continuously. The application of air pressure has to be varied from time to time in order to achieve a balanced value. The determination of this value depends more on experience than on technical considerations.

30.5 USE OF MODERN TECHNOLOGIES FOR TUNNELLING IN DIFFERENT SITUATIONS

These include the following.

1. Cut and cover method of tunnelling
2. Shield tunnelling
3. Tunnelling using drill and blast systems
4. Use of tunnel boring machines
5. New Austrian Tunnelling Method (NATM)

Cut and cover

This is a simple method of construction for shallow tunnels where a trench is excavated and covered with an overhead support system strong enough to carry the load of what is to be built above the tunnel. Two basic forms of cut-and-cover tunnelling are available:

Bottom-up method A trench is excavated, with ground support as necessary, and the tunnel is constructed in it. The tunnel may be of in-situ concrete, precast concrete, precast arches, or corrugated steel arches; earlier brickwork was used. The trench is then carefully back-filled and the surface is reinstated.

Top-down method   Here side support walls and capping beams are constructed from ground level by such methods as slurry walling, or contiguous bored piling. Then a shallow excavation aids in making the tunnel roof of precast beams or in-situ concrete. The surface is then reinstated except for access openings. This allows early reinstatement of roadways, services, and other surface features. Excavation then takes place under the permanent tunnel roof, and the base slab is constructed.

Shield tunnelling

In soft soil and where deep tunnels are excavated, a tunnelling shield is normally adopted. In early shield tunnelling, the shield functioned as a way to protect labourers who performed the digging, and moved the shield forward, progressively replacing it with prebuilt sections of a tunnel wall. Later shields were used for preventing slippage of earth from the sites into the excavated trench till the tunnel is constructed. The shield can then be moved forward to the next section to be taken up. The deep tunnels for the Kolkata Metro were built using this technology. The shield also can divide the workface into overlapping portions that each worker could excavate.

Tunnelling using drill and blast systems

This method is used in hard rock. It can be used for full face as well as for header excavation. Drilling jumbos go to the tunnel face and drill a set of holes in the portion to be tunnelled. These holes are then charged with controlled explosives and simultaneously blasted. There after special loaders excavate the spoils and the work proceeds cyclically ahead. This method can also be followed in soft soil tunnels, with supports and the portal being fixed at convenient points ahead of the tunnelling. This method was extensively used on Konkan Railway where drilling jumbos and special loaders, imported from Sweden, were used giving a progress of about 45 to 70 m per month. For muck removal electro-hydraulic digging arm loaders were imported from Sweden.

Use of tunnel boring machines

Tunnel boring machines (TBM) were first used for railway tunnels in the Udhampur–Srinagar–Baramulla (USB) Project, and on DMRC. Tunnel boring machines are used as an alternative to drilling and blasting methods in rock and conventional 'hand mining' in soil. TBMs have the advantages of minimizing the disturbance to the surrounding ground and producing a smooth tunnel wall. This significantly reduces the cost of lining the tunnel, and makes them suitable to use in heavily urbanized areas. The major disadvantage is the upfront cost. TBMs are expensive and can be difficult to transport. However, as modern tunnels become longer, the cost of tunnel boring machines versus drill and blast is actually less — this is because tunnelling with TMBs is much more efficient and results in a shorter project completion time.

Modern TBMs typically consist of the rotating cutting wheel, called a cutter head, followed by a main bearing, a thrust system and trailing support mechanisms. The type of machine used depends on the particular geology of the project, the amount of groundwater present and other factors.

In hard rock, either shielded or open-type TBMs can be used. All types of hard rock TBMs excavate rock using disc cutters mounted in the cutter head.

The disc cutters create compressive stress fractures in the rock, causing it to chip away from the rock in front of the machine, called the tunnel face. The excavated rock, known as muck, is transferred through openings in the cutter head to a belt conveyor, where it runs through the machine to a system of conveyors or muck cars for removal from the tunnel. In a fractured rock, shielded hard rock TBMs can be used, which erect concrete segments to support unstable tunnel walls behind the machine. In soft ground, there are two main types of TBMs: Earth Pressure Balance Machines (EPB) and Slurry Shield (SS). Both types of machines operate like Single Shield TBMs, using thrust cylinders to advance forward by pushing off against concrete segments. The cutter head does not use disc cutters only, but instead a combination of tungsten carbide cutting bits, carbide disc cutters, and/or hard rock disc cutters.

The New Austrian Tunnelling method (NATM)

It was developed between 1957 and 1965 in Austria. The main idea is to use the geological stress of the surrounding rock mass to stabilize the tunnel itself. The main features on which NATM is based are: mobilization of the strength of rock mass, and achieving *shotcrete* protection by applying a thin layer of shotcrete immediately after face advance measurements. Every deformation of the excavation must be measured, providing flexible support with a primary lining that is thin and reflects recent *strata* conditions. The tunnel is strengthened by a flexible combination of *rock bolts*, wire mesh and steel ribs, and quickly closing the invert and creating a load-bearing ring is important. This is crucial in soft ground tunnels where no section of the tunnel should be left open temporarily. This method is being used on the Delhi Metro and is also proposed to be used in the East–West Corridor of the Kolkata Metro.

30.6 VENTILATION OF TUNNELS

A tunnel should be properly ventilated during as well as after construction for the following reasons:
  (a) To provide fresh air to the workers during construction
  (b) To remove the dust created by drilling, blasting, and other tunnelling operations
  (c) To remove dynamite fumes and other objectionable gases produced by the use of dynamites and explosives

The methods listed below are normally adopted for the ventilation of a tunnel. These are illustrated in Fig. 30.10.

Natural method of ventilation   This is achieved by drilling a drift through the tunnel from portal to portal. In most cases natural ventilation is not sufficient and artificial ventilation is still required.

Mechanical ventilation by blow-in method   In the blow-in method, fresh air is forced through a pipe or fabric duct by the means of a fan and supplied near the washing face (or the drilling face; the drilling operation requires the washing of bore holes too). This method has the advantage that fresh air supply is guaranteed where it is required the most. The disadvantage is that the foul air and fumes have to travel a long distance before they can exit the tunnel and in the process it is possible that the incoming fresh air will absorb some dust and smoke particles.

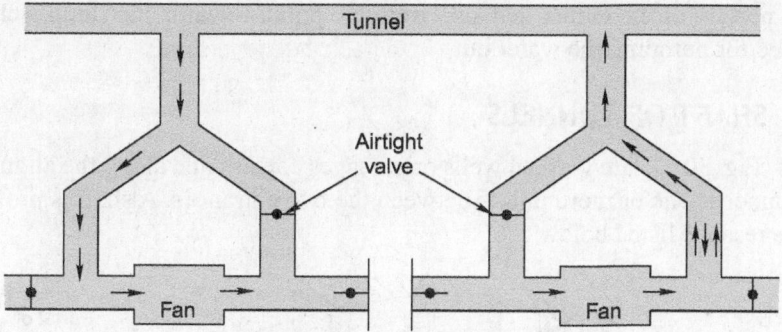

Fig. 30.10 Ventilation of tunnels

Mechanical ventilation by exhaust method In the exhaust or blow-out method, foul air and fumes are pulled out through a pipe and expelled by a fan. This sets up an air current that facilitates the entrance of fresh air into the tunnel. This method has the advantage that foul air is kept out of the washing face. The disadvantage, however, is that fresh air has to travel a long distance before it can reach the washing face during which period it may absorb some heat and moisture.

Combination of blow-in and blow-out methods By combining the blow-in and blow-out methods using a blower and an exhaust system, respectively, a tunnel can be provided with the best ventilation. After blasting the ground, the exhaust system is used to remove the smoke and dust. After some time, fresh air is blown in through the ducts and the rotation of the fans is reversed in order to reverse the flow of air.

30.7 LIGHTING OF TUNNELS

It is very important to ensure that the tunnels are well lit so that the various activities and operations involved in tunnelling can be carried out effectively and safely. The common types of lighting equipment normally used in tunnels are electric lights, coal gas or acetylene gas lights, or lanterns. Electric lights are considered the best option, as these radiate bright light of the required intensity, are free from smoke, are easily manoeuvrable from the point of view of the extension, etc.

Places where plenty of light should normally be provided are operation points, equipment stations, bottom of shafts, storage points, tempering stations, underground repair shops, etc.

30.8 DRAINAGE OF TUNNELS

Good drainage of the tunnels is very essential in order for them to operate safely and smoothly during the construction period as well as afterwards. The sources of water for this purpose include groundwater and water collected from the washing of bore holes. Water seeping in up through the ground as well as from the washing of bore holes is collected in sump wells and pumped out. If the tunnel is long, a number of sump wells are provided for the collection of water.

After the construction is over, drainage ditches are provided along the length of the portion of the tunnel that slop from the portal towards the sump well and are used for pumping the water out.

30.9 SHAFT OF TUNNELS

Shafts (Fig. 30.11) are vertical wells or passages constructed along the alignment of a tunnel at one or more points between the two entrances. A shaft is provided for the reasons listed below.

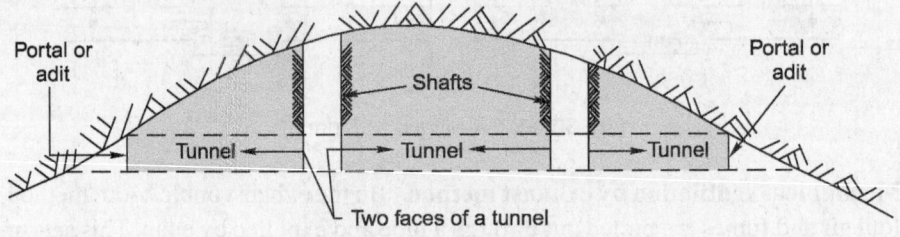

Fig. 30.11 Tunnel shafts

Working shafts These are provided for the expeditious construction of tunnels by tackling the same at a number of points. These are generally vertical and of a minimum size of 3.7 m × 3.7 m or of a diameter of 4.30 m.

Ventilation purposes In order to ensure better ventilation, these shafts are generally inclined and have a girder size of about 1.2 m diameter.

30.10 LINING OF TUNNELS

Tunnels in loose rock and soft soils are liable to disintegrate and, therefore, a lining is provided to strengthen their sides and roofs so as to prevent them from collapsing. The objectives of a lining are as follows.
  (a) Strengthening the sides and roofs to withstand pressure and prevent the tunnel from collapsing
  (b) Providing the correct shape and cross section to the tunnel
  (c) Checking the leakage of water from the sides and the top
  (d) Binding loose rock and providing stability to the tunnel
  (e) Reducing the maintenance cost of the tunnel

30.10.1 Sequence of Lining

The lining of a tunnel is carried out in the following steps.
  1. In the first stage guniting is done to seal the water in rock tunnels.
  2. Concrete lining is done either in one attempt as in the case of circular tunnels or by separately tackling the vest, the sidewall, and the arch. For small tunnels that measure 1.2 to 3.0 m in diameter, the concrete lining can be provided by the hand placing method. In the case of bigger tunnels, concrete pumps or pneumatic placers are used for placing the concrete.
  3. The concrete is cured to its maximum strength. If the humidity inside the tunnel is not sufficient, curing can be done by spraying water through perforated pipes.

The different types of lining practices adopted by Indian Railways depending upon ground conditions are depicted in Fig. 30.12.

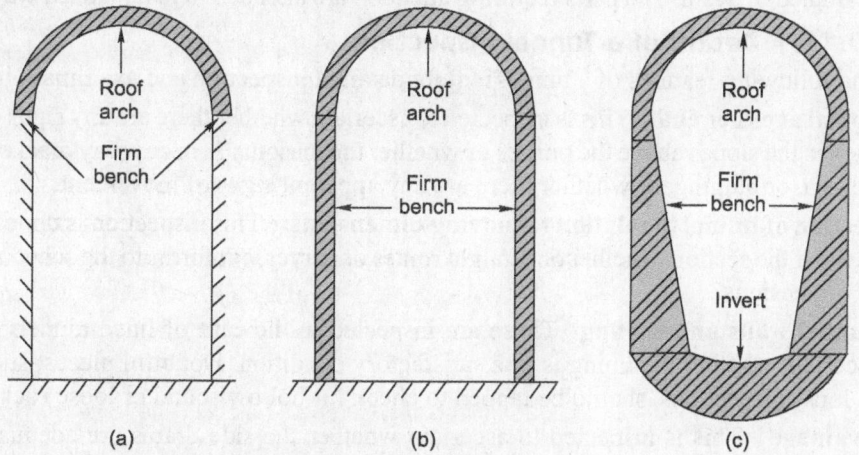

Fig. 30.12   Linings of tunnels

30.10.2 Types and Thickness of Lining

Theoretically, the lining provided inside tunnels may be of timber, iron, steel, brick, or any other construction material but in practical terms the lining provided most commonly is that of reinforced concrete or concrete surface. Concrete lining is provided in tunnels because of (a) its superiority in structural strength, (b) ease of placement, (c) its durability, and (d) lower maintenance cost.

The thickness of concrete lining depends upon various factors such as conditions of the ground, size and shape of the tunnel, soil pressure, and the method of concreting. The thickness of concrete is calculated by the following empirical formula:

$$T = 0.083D \tag{30.1}$$

where T is the thickness of the lining in centimetres and D is the diameter of the tunnel in metres.

On the basis of field experience, railway engineers have devised a thumb rule of providing 2.5 cm of lining for every 30 cm of the diameter of the tunnel. As per this thumb rule, the thickness of the lining of a tunnel with a 1 m diameter would be $(100/30) \times 2.5$ cm = 8.3 cm.

30.11 MAINTENANCE OF RAILWAY TUNNELS

The proper maintenance of railway tunnels is very important from the point of view of the safety of the trains as well as that of the travelling public. In order to ensure that the railway tunnel is in a good condition all year around, it is inspected in the following ways.

Inspection by assistant engineer   Apart from routine inspection, the tunnel is inspected in detail by the AEN once a year before the onset of the monsoons. All the substructures of the tunnel are inspected and the aspects that require attention are recorded.

Inspection by permanent way inspector   The PWI inspects the tunnel in detail once in a year after the monsoons are over, paying particular attention to the track and its components. The aspects requiring attention are taken care of immediately.

Inspection by works inspector The structural portion of the tunnel is inspected in detail once in a year by the IOW. All the sub-components of the structure are examined in detail. The parts requiring attention are attended to in a planned way.

30.11.1 Details of a Tunnel Inspection

The following features of a tunnel require detailed inspection and examination.

Portal at either end This is inspected to ascertain whether there are any signs of slips in the slopes above the portals or whether the masonry is in any way cracked, shaken, or bulging, or whether there are any apparent signs of movement.

Section of tunnel in relation to moving dimensions This inspection is done to check if the section, whether on straight routes or curves, conforms to the schedule of dimensions.

Tunnel walls and roofing These are inspected in the case of lined tunnels to ascertain whether the lining is in a satisfactory condition. Doubtful places, such as loose projections, should be tapped to check for hollow sound or loose rock.

Drainage This is inspected to ascertain whether the side drains are adequate and functional.

Ventilation shafts These are inspected to ascertain whether the ventilation shafts are adequate and free from vegetation and other growth.

Lighting equipment and special tools This inspection of these features is required to ascertain whether the lighting equipment and special tools, wherever supplied, are in a good state.

Track The track is inspected to ascertain whether its line and level are correct, including that of the approaches. Rails, sleepers, and fastenings should be particularly examined for corrosion.

Trolley refuges In order to provide safety to the engineers inspecting the tunnel on trolley as well as for the safety of technical staff carrying out work in tunnels, trolley refuges are provided so that there can be safe places, where trolley or technical staff can wait, when trains pass. The trolley refuges should be located at maximum of 100 m apart.

Trolley refuges in a tunnel or deep cut are identified by putting a rail post painted with luminous paint and a letter 'R' written on it.

For easy identification of the location of trolly refuges in tunnels and deep cuttings a distinguishing mark such as a rail post, painted with luminous paint with a mark 'R' may be erected by the side of the trolly refuge.

30.12 SAFETY IN TUNNEL CONSTRUCTION

Tunnelling is a difficult, hazardous, and time-consuming process and the whole operation has to be done systematically so that safety is ensured at all times. Normally accidents in a tunnel occur under the following circumstances.

(a) Falling rocks
(b) During the loading and hauling of muck
(c) Poor handling of explosives
(d) During shaft operations
(e) Cramped working space

The following tips are suggested for preventing accidents during tunnelling operations.

(a) Equipment and tools should be in good working condition.
(b) Regular and detailed inspections should be carried out during tunnelling operations.
(c) Visual inspections should be done to detect seams and planes of weakness so as to avoid the falling of rocks.
(d) There should be provision of sufficient support by ensuring that tunnelling is done properly in order to avoid the collapsing of the tunnel as well as falling rocks.
(e) There should be provision of good lighting and non-slippery walkways, which partially help in relieving the strain of a cramped working space.
(f) Provisions should be made for the removal of extra debris and refuge as well as for good drainage in order to avoid accidents. Efforts should also be made to provide good ventilation.
(g) Telephone facilities should be provided, particularly inside the shaft and at other places in the tunnel, to ensure smooth and accident-free operations.
(h) Firefighting equipment should be provided at all key points.
(i) Safety sign boards should be provided at all key locations.
(j) All workers should be medically fit to work inside the tunnel and they should be examined periodically. Doctors and first aid facilities should be available at the site.
(k) Wearing of helmets by all workers employed in a tunnel should be made mandatory.

SUMMARY

Tunnels are required in special circumstances such as when the track cannot be laid on natural ground or when it is not economical. The construction of a tunnel is a very tedious and specialized work and, therefore, a very judicious decision is required when it comes to tunnel selection. The method and equipment required to construct a tunnel would depend on the nature of the soil. The tunnel should be provided with adequate light, ventilation, and other safety features.

REVIEW QUESTIONS

1. Discuss the points to be considered when determining the shape and size of tunnels.
2. Why are railway tunnels necessary? Draw a sketch to illustrate a single-track railway tunnel.
3. Briefly describe the construction cycle for tunnelling. How are drainage and ventilation facilities provided in railway tunnels?
4. Why is a tunnel necessary? Describe the advantages and disadvantages of constructing a tunnel.
5. Name the various methods of tunnelling in hard and soft rocks. Describe one in each case.
6. What is shield tunnelling? Discuss the various methods of shield tunnelling adopted in different types of soils.
7. What is the importance of ventilation during tunnelling? Describe the various methods of providing ventilation in a tunnel.

8. Write notes on:
   (a) Lighting of tunnels
   (b) Drainage of tunnels
   (c) Lining of tunnels
   (d) Compressed air tunnelling
9. Why is it necessary to maintain a tunnel? Describe the various types of inspections carried out to ensure that tunnels are properly maintained.
10. Describe the various points to be kept in mind to ensure the safety of tunnels. What are the causes behind accidents in tunnels?
11. Differentiate between the following.
    (a) Full face method and drift method of tunnelling
    (b) Heading and bench method and pilot tunnel method

Choose the correct answer from the choices given.

12. The breadth of a tunnel (in mm) for a double-line BG section is:
    (a) 8200–8400   (b) 8350–8600
    (c) 8530–9140   (d) 9100–9500
13. 'Full face method' of tunnelling is normally selected for tunnels whose dimensions do not exceed:
    (a) 1.5 m   (b) 2.2 m
    (c) 3.0 m   (d) 3.5 m
14. The size of tunnel adopted by the drift method is:
    (a) 2 m × 2 m   (b) 2.5 m × 2.5 m
    (c) 3 m × 3 m   (d) 45 m × 4 m
15. In compressed air tunnelling, the pressure of compressed air is normally:
    (a) 1 kg/cm^2 (b) 1.5 kg/cm^2
    (c) 2.0 kg/cm^2 (d) none of these
16. For expeditions construction of tunnels, 'working shafts' are made at a number of points. The minimum size of these working shafts is:
    (a) 2.5 m × 2.5 m   (b) 3.2 m × 3.2 m
    (c) 3.7 m × 3.7 m   (d) 4.0 m × 4.0 m
17. The tunnels should be inspected by an assistant engineer:
    (a) once in 3 months
    (b) once in 6 months
    (c) once in a year
    (d) whenever the AEN feels the necessity
18. The trolley refuges in a tunnel should be spaced at a distance of:
    (a) 50 m   (b) 100 m
    (c) 200 m   (d) 500 m

Signalling and Interlocking

INTRODUCTION

The purpose of signalling and interlocking is primarily to control and regulate the movement of trains safely and efficiently. Signalling includes operations and interlocking of signals, points, block instruments, and other allied equipment in a predetermined manner for the safe and efficient running of trains. Signalling enables the movement of trains to be controlled in such a way that the existing tracks are utilized to the maximum.

In fact, in railway terminology signalling is a medium of communication between the station master or the controller sitting in a remote place in the office and the loco pilot (As per latest instructions of the Railway Board, 'drivers' are now called 'Loco Pilots') of the train.

The history of signalling goes back to the olden days when two policemen on horseback were sent ahead of the train to ensure that the tracks were clear and to regulate the movement of the trains. In later years, policemen in uniform were placed at regular intervals to regulate the movement of trains. Railway signalling in its current form was introduced for the first time in England in 1842, whereas interlocking was developed subsequently in 1867.

31.1 OBJECTIVES OF SIGNALLING

The objectives of signalling are as follows:
  (a) To regulate the movement of trains so that they run safely at maximum permissible speeds
  (b) To maintain a safe distance between trains that are running on the same line in the same direction
  (c) To ensure the safety of two or more trains that have to cross or approach each other
  (d) To provide facilities for safe and efficient shunting

(e) To regulate the arrival and departure of trains from the station yard
(f) To ensure the safety of the train at level crossings when the train is required to cross the path of road vehicles

31.2 CLASSIFICATION OF SIGNALS

Railway signals can be classified based on different characteristics as presented in Table 31.1.

Table 31.1 Classification of signals based on different characteristics

Characteristics	Basis of classification	Examples
Operational	Communication of message in visual form	Fixed signals
Functional	Signalling the loco pilot to stop, move cautiously, proceed, or carry out shunting operations	Stop signals, permissive Signals, shunt signals
Locational	Reception or departure signals	Reception: Outer, home, Departure: Starter, and advanced starter signals
Constructional	Semaphore or colour light signals	Semaphore: Lower quadrant or upper quadrant. Colour light: Two aspects or multiple aspects.
Special characteristics	Meant for special purposes	Calling-on signals, repeater signals, coaching signals, etc.

Figure. 31.1 shows further classification of signals and Table 31.2 lists the signalling requirements of various classes of stations.

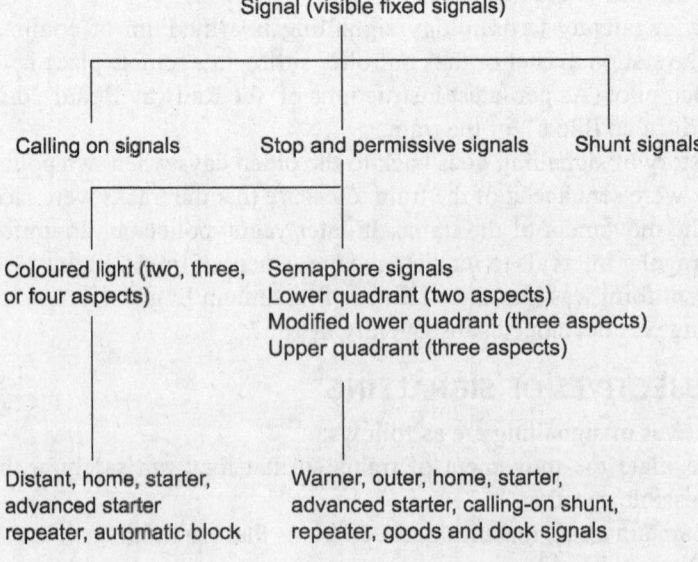

Fig. 31.1   Classification of signals

Table 31.2 Signals required at stations

Classification of station	Minimum requirement of signals	Remarks
A class	Warner, home, and starter	An outer signal can be provided after obtaining special permission
B class	Outer and home	In multiple-aspect upper quadrant (MAUQ) areas, distant home and outer signals are provided
C class	Warner and home	In MAUQ areas, the warner signal is replaced by a distant signal

Further details regarding the different types of stations are given in Chapter 26.

31.2.1 Audible Signals

Audible signals, such as detonators and fog signals are used in cloudy and foggy weather when hand or fixed signals are not visible. Their sound can immediately attract the attention of drivers. Detonators contain explosive material and are fixed to the rail by means of clips. In thick foggy weather, detonators are kept about 90 m ahead of a signal to indicate the presence of the signal to the drivers. Once the train passes over the detonators thereby causing them to explode, the driver becomes alert and keeps a lookout for the signal so that he/she can take the requisite action.

Note: Audible signals are basically used for execution of engineering works.

31.2.2 Fixed or Visible Signals

These signals are visible and draw the attention of the drivers because of their strategic positions.

Hand signals   These signals are in the form of flags (red or green) fixed to wooden handles that are held by railway personnel assigned this particular duty. If the flags are not available, signalling may be done using bare arms during the day. In the night, hand lamps with movable green and red slides are used for signalling purposes.

Caution indicators   These are fixed signals provided for communicating to the driver that the track ahead is not fit for the running the train at normal speed. These signals are used when engineering works are underway and are shifted from one place to another depending upon requirement.

Note: Hand signals and caution indicators are basically used for execution of engineering works and do not strictly come under the category of 'Signalling system'. Full details of these items can be seen in Chapter 18.

Fixed signal   These are firmly fixed on the ground by the side of the track and can be further subdivided into stop signals.

Stop signals   These are fixed signals that do not change their position. They inform the drivers about the condition of the railway line lying ahead.

In fact, stop signals permit loco pilot (driver) to cross the signal when taken off and none till next signal ahead.

The stop signals normally used on railways are semaphore signals, coloured light signals, and other such signals as explained in subsequent sections.

31.3 FIXED SIGNALS

The various types of fixed signals used on railways are as follows.

Semaphore signals

The word 'semaphore' was first used by a Greek historian. 'Sema' means sign and 'phor' means to bear. A semaphore signal consists of a movable arm pivoted on a vertical post through a horizontal pin as shown in Fig. 31.2.

The arm of the semaphore signal on the side facing the driver is painted red with a vertical white stripe. The other side of the signal is painted white with a black vertical stripe. The complete mechanical assembly of the signal consists of a spectacle arm, a pivot, a counterweight spring stop, etc., and is housed on top of a tubular or lattice post. In order for the signal to also be visible at night, a kerosene oil or electric lamp, operated through a twilight switch, is fixed to the post. The arm moves the spectacle, which contains green and red coloured glasses. The red glass is positioned at the upper end and the green glass is positioned at the lower end of the spectacle so that the red light is visible to the driver when the arm is horizontal and the green light is visible when the arm is lowered. The semaphore signal is used as a stop signal with a red square ended arm as well as a warner with a red fish tailed arm signal.

Lower quadrant semaphore signals move only in the fourth quadrant of a circle and have only two colour aspects. In order to provide the drivers with further information, multi-aspect upper quadrant signalling (MAUQ) is sometimes used on busy routes. In this system, the arms of the semaphore signals rest in three positions and the signals have three colour aspects, namely red, yellow, and green associated with the horizontal, 45° above horizontal, and vertical directions, respectively. Details of MAUQ are given in subsequent paras.

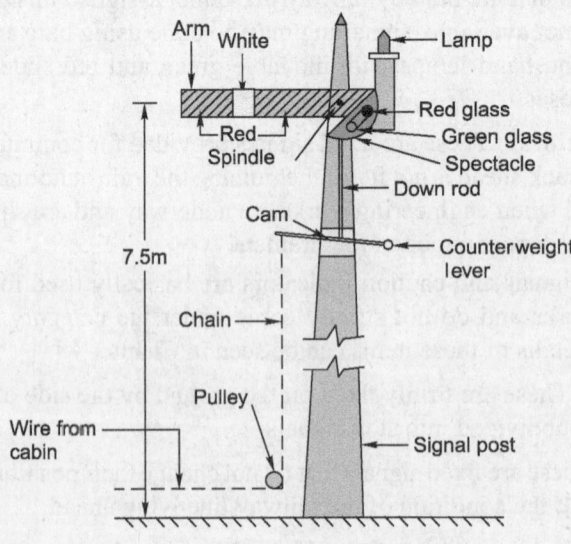

Fig. 31.2 Semaphore signal

Stop signal in MAUQ Signalling

In case of multi-aspect upper quadrant (MAUQ) signalling of semaphore stop signal with a square ended arm, there may be three situations (Figure 31.2a) as indicated below in Table 31.3:

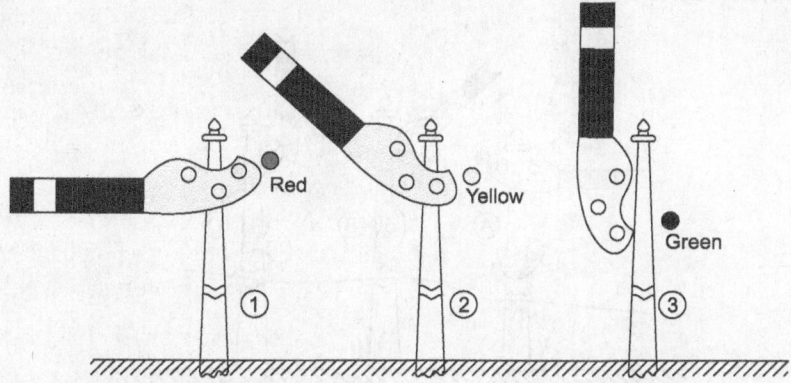

Fig. 31.2 (a)   Semaphore stop signal in MAUQ signalling

Table 31.3   Aspects and indications of stop signal in MAUQ signalling

Position	Aspect of signal	Position of arm	Colour during night	Indication
1.	ON-Stop	Horizontal	Red	Stop dead
2.	OFF-Caution	45° above Horizontal	Yellow	Proceed with caution and be prepared to stop at next signal
3.	OFF-Proceed	90° above Horizontal	Green	Proceed at maximum permitted speed

The signals are designed to be fail-safe so that if there is any failure in the working of the equipment, they will always be in the stop position. These signals are operated by hand levers or buttons located in a central cabin, which is normally provided near the station master's office. Semaphore stop signals are normally provided as outer signals, home signals, starter signals, advanced starter signals, and warner signals.

Permissive signal—warner or distant signal

In order to ensure that trains speed up safely, it is considered necessary that warning be given to drivers before they approach a stop signal. This advance warning is considered necessary, otherwise the drivers may confront a 'stop signal' when they least expect it and take abrupt action, which can lead to perilous situations. A warner or distant signal has, therefore, been developed, which is to be used ahead of a stop signal and is in the form of a permissive signal that can be passed even in most restricted conditions. In the case of a stop signal, the driver has to stop the train when it is in the 'on' position, but in the case of a permissive signal, the driver can pass through even when it is in the 'on' position.

The warner signal is similar to a stop signal except that the movable arm is given the shape of fish tail by providing a V-shaped notch at the free end; the white strip is also V shaped.

Distant signal in MAUQ signalling

In case of multi-aspect upper quadrant (MAUQ) signalling for semaphore distant signal, there is a fish tailed arm, painted yellow with a black band near the edge of the arm (Fig. 31.2b). There are three aspects and indications of the same are presented in Table 31.4.

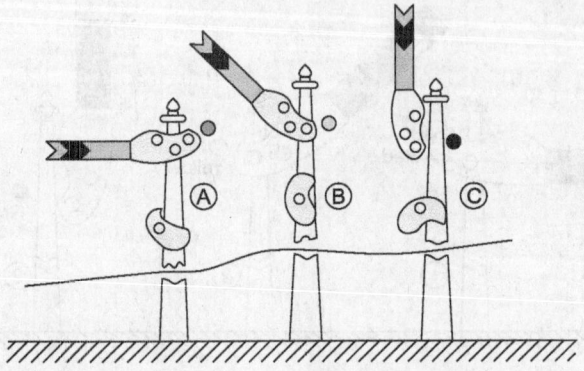

Fig. 31.2 (b) Semaphore distant signal in MAUQ signalling

Table 31.4 Aspects and indications of distant semaphore signal in MAUQ signalling

Case	Aspect of signal	Position of arms	Colour during night	Indications
A	ON—Caution	Horizontal	Yellow light	Proceed and be prepared to stop at next stop signal.
B	OFF	45° above horizontal	Two yellow lights in vertical alignment	Proceed and be prepared to pass the next signal at caution.
C	OFF—Proceed	90° above horizontal	Green light	Proceed at maximum permitted speed.

In the case of signalling using coloured light, the permissive signal is distinguished from the stop signal by the provision of a P marker disc on the signal post.

The warner signal is intended to warn the driver of a train regarding the following aspects as explained in Table 31.5.

(a) To inform that the driver is approaching a stop signal

(b) To inform the driver as to whether the approach signal is in 'on' or 'off' position

The warner signal can be placed at either one of the following locations.

(a) Independently on a post with a fixed green light 1.5 m to 2 m above it for night indication

(b) On the same post below the outer signal or the home signal

In case a warner is fixed below an outer signal the various positions of the outer and warner signals and their corresponding indications are given in Fig. 31.3.

Table 31.5 Position of warner arm or distant signal

Position	Day indication for semaphore signal	Night indication for semaphore signal*	Aspect
Caution	Arm horizontal	Red light Yellow light	Proceed with caution and be prepared to stop at the next stop signal
Attention	Arm inclined 45° in the upward direction	Two yellow lights	Proceed cautiously so as to pass the next stop signal at a restricted speed
Proceed	Arm inclined 90° in the upward or 45° in the downward direction	Two green lights	Proceed at full permissible speed

* Also day and night indication for a coloured light signal.

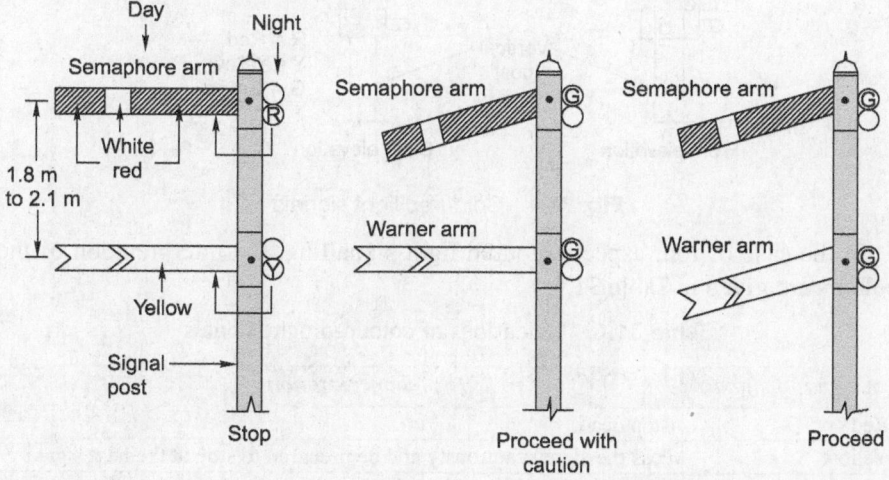

Fig. 31.3 Warner below an outer signal

Coloured light signals

These signals use coloured lights to indicate track conditions to the driver both during the day and the night. In order to ensure good visibility of these light signals, particularly during daytime, the light emission of an electric 12 V, 33 W lamp is passed through a combination of lenses in such a way that a parallel beam of focused light is emitted out. This light is protected by special lenses and hoods and can be distinctly seen even in the brightest sunlight. The lights are fixed on a vertical post in such a way that they are in line with the driver's eye level. The system of interlocking is so arranged that only one aspect is displayed at a time. Coloured light signals are normally used in automatic signalling sections, suburban sections, and sections with a high traffic density.

Coloured light signals can be of the following types.
  (a) Two-aspect, namely green and red
  (b) Three-aspect, namely green, yellow, and red
  (c) Four-aspect, namely green, double yellow, and red

In India, mostly three-aspect or four-aspect coloured light signalling is used. In the case of three-aspect signalling, green, yellow, and red lights are used. Green indicates 'proceed', yellow indicates 'proceed with caution', and red indicates 'stop' (Fig. 31.4).

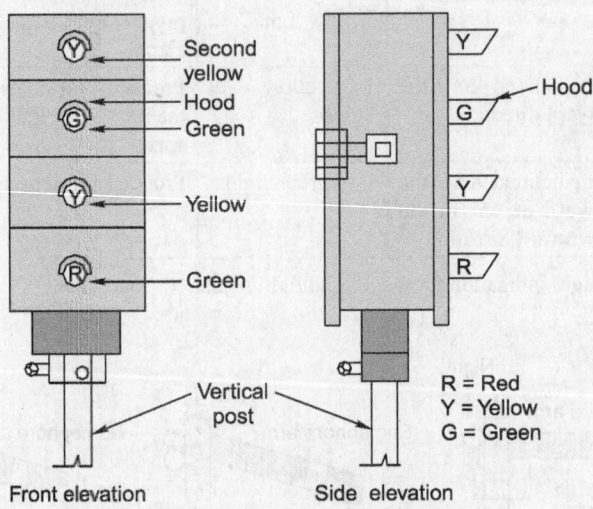

Fig. 31.4 Coloured light signals

In the case of four-aspect coloured light signalling, the interpretation of the colours are given in Table 31.6.

Table 31.6 Indications of coloured light signals

Colour of signal	Interpretation
Red	Stop dead
Yellow	Pass the signal cautiously and be prepared to stop at the next signal
Two yellow lights displayed together	Pass the signal at full speed but be prepared to pass the next signal, which is likely to be yellow, at a cautious speed
Green	Pass the signal at full permissive speed

In conventional semaphore signals, the 'on' position is the normal position of the signal and the signals are lowered to the 'off' position only when a train is due. In the case of coloured light signals placed in territories with automatic signalling, the signal is always green or in the 'proceed' position. As soon as a train enters a section, the signal changes to 'red' or the 'stop' position, which is controlled automatically by the passage of the train itself. As the train passes through the block section, the signal turns yellow to indicate the driver to 'proceed with caution' and, finally, when the train moves onto the next block section, the signal turns green indicating to the driver to 'proceed at full permissible speed'.

Thus, it can be seen that each aspect of the signal gives two pieces of information to the driver. The first is about the signal itself and the second is about the condition of the track ahead or of the next signal. This helps the driver to manoeuvre the train safely and with confidence even at the maximum permissible speed.

Calling-on signal

This consists of a small arm fixed on a home signal post below the main semaphore arm (Fig. 31.5). When the main home signal is in the horizontal (on) position and the calling-on signal is in an inclined (off) position, it indicates that the train is permitted to proceed cautiously on the line till it comes across the next stop signal. Thus, the calling-on signal is meant to 'call' the train, which is waiting beyond the home signal.

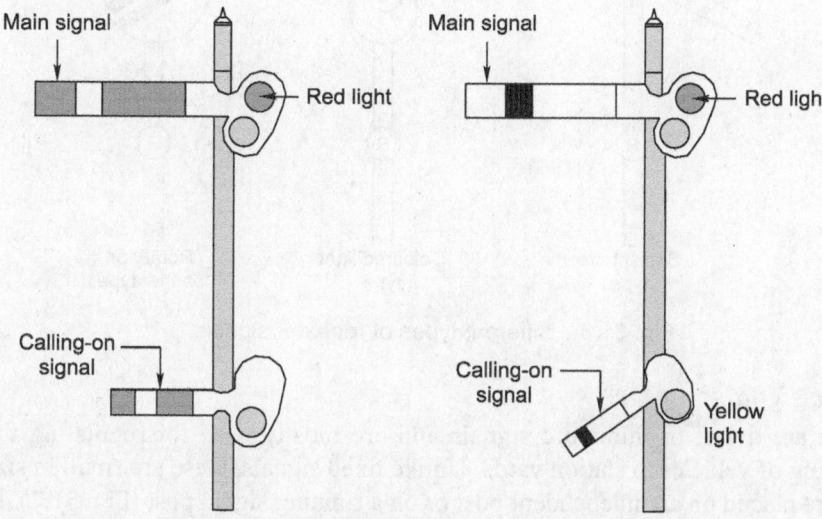

Fig. 31.5 Calling-on signal

The calling-on signal is useful when the main signal fails, and in order to move a train, an authority letter has to be sent to the driver of the waiting train to instruct him/her to proceed to the station against what is indicated by the signal. In big stations and yards, the stop signals may be situated far off from the cabin and the calling-on signal expedites the quick reception of the train even when the signal is defective.

Co-acting signal

In case a signal is not visible to the driver due to the presence of some obstruction, such as an overbridge or a high structure, another signal is used to move along the main signal on the same post. This signal, known as the co-acting signal, is an exact replica of the original signal and works in unison with it.

Repeater signal

In case where a signal is not visible to the driver from an adequate distance due to sharp curvature or any other reason or where the signal is not visible to the guard of the train from his position at the rear end of a platform, a repeater signal is provided at a suitable position at the rear of the main signal. A repeater signal is provided with an R marker and can be of the following types.

(a) A square-ended semaphore arm with a yellow background and a black vertical band
(b) A coloured light repeater signal
(c) A rotary or disc banner type signal

The 'off' positions of these three types of repeater signals are depicted in Fig. 31.6.

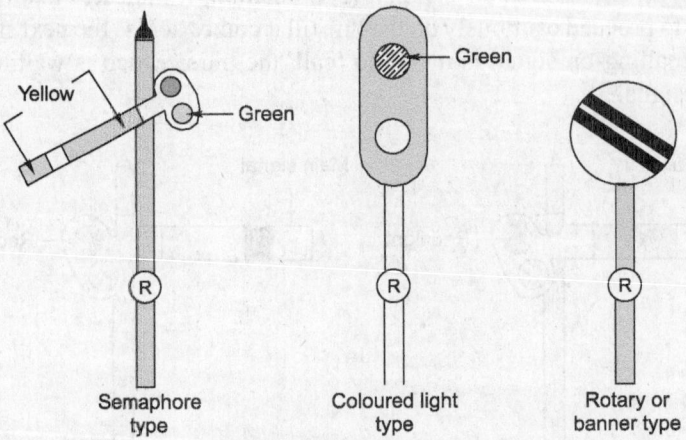

Semaphore type | Coloured light type | Rotary or banner type

Fig. 31.6 Different types of repeater signals

Shunt signals

These are dwarf or miniature signals and are mostly used for regulating the shunting of vehicles in station yards. Unlike fixed signals, these are small in size and are placed on an independent post or on a running signal post (Fig. 31.7). In semaphore signalling areas, the shunt signals are of the disc type.

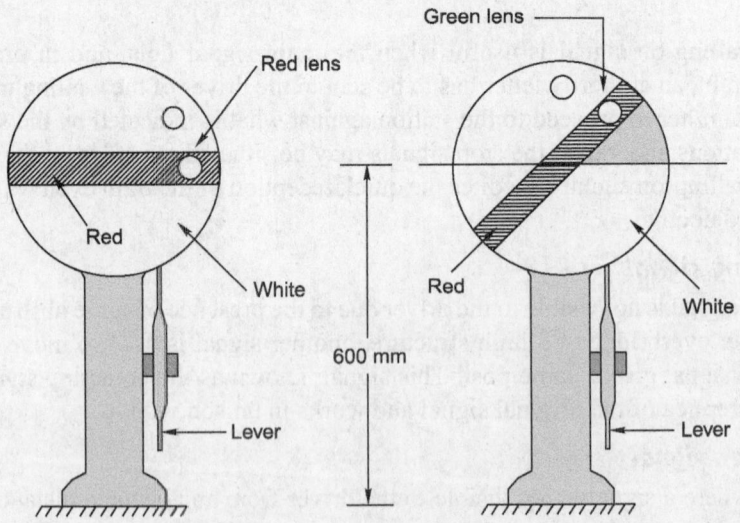

Fig. 31.7 Disc type shunt signals

The disc type of shunt signal consists of a circular disc with a red band on a white background. The disc revolves around a pivot and is provided with two holes, one for the red lamp and the other for the green lamp, for the purpose of night indication. At night, the 'on' position of the signal is indicated by the horizontal red band and the red light, indicating danger. During the day the red band is inclined to the horizontal plane and during the night the green light indicates that the signal is 'off' (Fig. 31.7).

In colour light signalling areas, the shunt signal on an independent post consists of two white lights forming a line parallel to the horizontal plane. This indicates that the signal is 'on' or that there is danger ahead, whereas two white lights forming a line inclined to the horizontal plane indicate 'off' or that the train can proceed (Fig. 31.8).

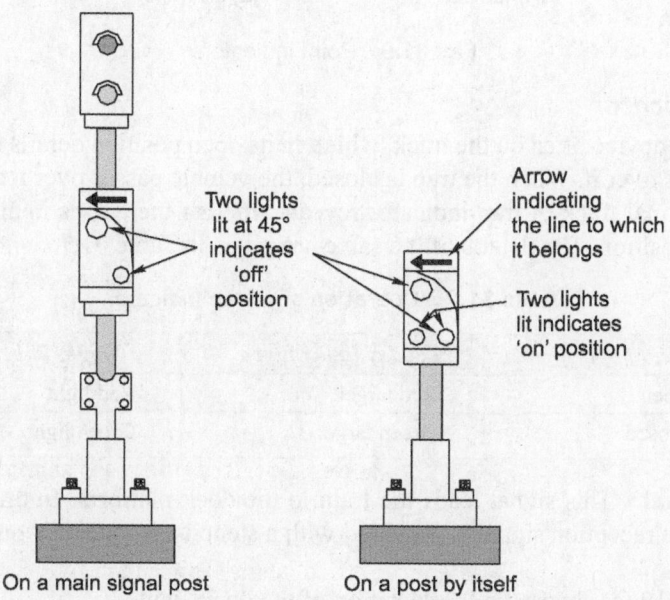

Two lights lit at 45° indicates 'off' position

Arrow indicating the line to which it belongs

Two lights lit indicates 'on' position

On a main signal post   On a post by itself

Fig. 31.8 Shunt signals in coloured light signalling area

Point indicators

These are used to indicate whether points have been set for the main line or turnout side (Fig. 31.9). It essentially consists of an open box with two white circular discs forming two opposite sides of the box and green bands on the other two remaining sides. The box rotates automatically about a vertical axis with the movement of the points. The white disc indicates that the points are set for the main line. When the points are set for the turnout side, the green bands are visible to the loco pilot (driver). At night white light indicates a main line setting and green light signifies a turnout side setting.

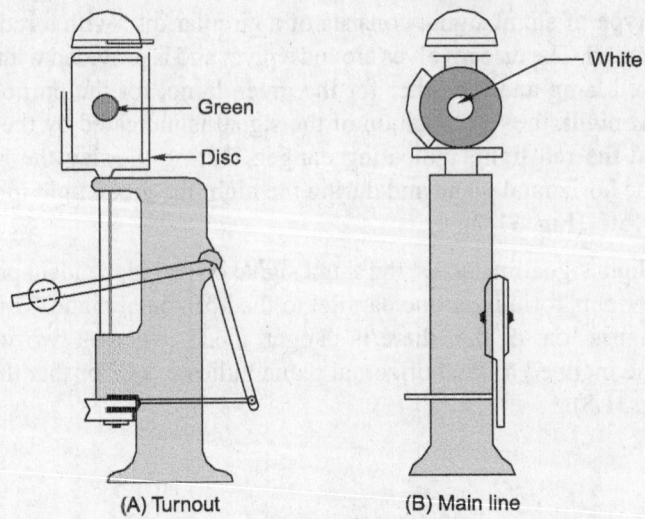

(A) Turnout   (B) Main line

Fig. 31.9 Point indicators

Trap indicator

A trap is a device fitted on the track, which in its open position derails the vehicle that passes over it. When the trap is closed, the vehicle passes over it as it would over a normal track. A trap indicator reveals whether the trap is in an 'open' or 'closed' position. The details of the same are given in Table 31.7.

Table 31.7 Operation of a trap indicator

Position of trap	Day indication	Night indication
Trap open	Red target	Red light
Trap closed	Green target	Green light

Dock signal This signal leads the train to the dock platform. In this case, the semaphore reception signal is provided with a stencil-cut letter 'D' on the signal for use.

Figure 31.10 (a) shows the 'on' position of the dock signal.

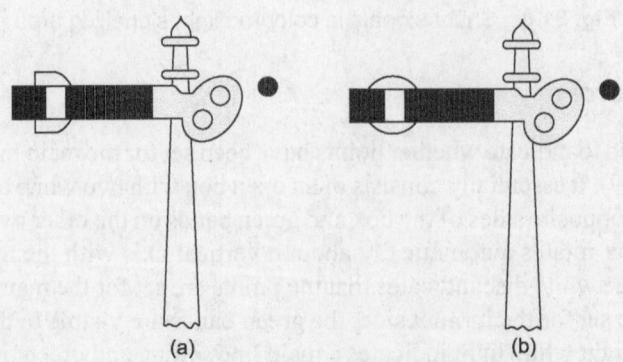

(a)   (b)

Fig. 31.10 (a) Dock signal (on position) and (b) Goods signal (on position)

Goods signal   This signal leads the train to the goods running line. In this case a stencil-cut letter 'O' is provided on the signal arm of a semaphore signal as shown in Fig. 31.10 (b).

Engineering indicators

When the track is under repair, trains are required to proceed with caution at restricted speeds and may even have to stop. Caution indicators help the driver of a train to reduce the speed of (or even stop) the train at the affected portion of the track and then return it to the normal speed once that portion has been covered.

Full details of these engineering indicators are given in Chapter 18 'Track Maintenance'.

Sighting board

A sighting board (Fig. 31.11) is an indication to the Loco pilot (driver) that he or she is approaching the first stop signal of a railway station. The function of a sighting board is to allow the driver to estimate the location of the next stop signal from the current location so that he/she starts applying brakes in case the first stop signal is in an 'on' position.

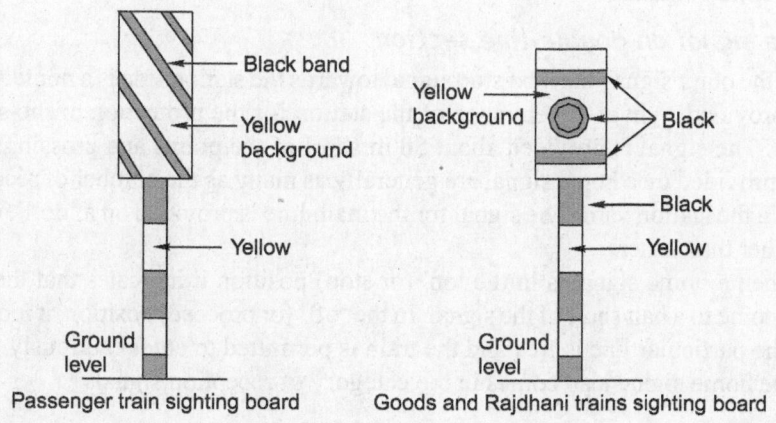

Fig. 31.11   Sighting board

As the requisite braking distance of goods trains and Rajdhani trains is greater than that of the passenger trains, the sighting boards for goods trains and Rajdhani trains are located farther and their design is different from that of sighting boards meant for passenger trains. The distances of sighting boards are listed in Table 31.8.

Table 31.8   Positions of sighting boards

Type of sighting board	Position
Passenger train sighting board	1000 m for speeds over 72 kmph for BG tracks and 48 kmph for MG tracks
Goods trains and Rajdhani sighting board	1400 m for speeds over 72 kmph for BG tracks and 48 kmph for MG tracks

31.4 STOP SIGNALS

The various types of stop signals with reference to their location on a station are discussed here in detail.

Outer signal on double-line section

This is the first semaphore stop signal at a station that indicates the entry of a train from a block section into the station limits. This signal is provided at an adequate distance beyond the station limits so that the line is not obstructed once the permission to approach has been given. It is provided at a distance of about 400 m from the home signal. The signal has one arm but has a warner signal nearly 2 m below on the same post.

When the outer signal is in the 'on' (or stop) position, it indicates that the driver must bring the train to a stop at a distance of about 9 m from the signal and then proceed with caution towards the home signal. If the outer signal is in the 'off' (or proceed) position, it indicates that the driver does not need to reduce the speed of the train if the home signal is also in the 'off' (or proceed) position, which is indicated by the 'off' position of the warner.

As the outer signal controls the reception of trains, it comes under the category of reception signals.

Home signal on double-line section

After the outer signal, the next stop signal towards the station side is a home signal. It is provided right at the entrance of the station for the protection of the station limits. The signal is provided about 50 m short of the points and crossings. The arms provided on a home signal are generally as many as the number of reception lines in the station yard. The signal for the main line is provided on a 'doll', which is higher than others.

When a home signal is in the 'on' (or stop) position it indicates that the train must come to a halt short of the signal. In the 'off' (or proceed) position, it indicates that the particular line is free and the train is permitted to enter cautiously.

The home signal also comes in the category of reception signals.

Routing signal

The various signals fixed on the same vertical post for both main and branch lines are known as routing signals. These signals indicate the route that has been earmarked for the reception of the train. Generally, the signal for the main line is kept at a higher level than that for the loop line. It is necessary for the driver of a train approaching a reception signal to know the line on which his or her train is likely to be received so that he or she can regulate the speed of the train accordingly. In case the train is being received on the loop line, the speed has to be restricted to about 15 km per hour, whereas if the reception is on the main line, a higher speed is permissible.

Signalling arrangement under Modified Lower Quadrant Signalling System
Since lower quadrant semaphore two aspect signalling system is capable of conveying limited information to the loco pilot, the arrangement was modified to be known as Modified Lower Quadrant Signalling. In this arrangement a warner is provided on the same doll as that of mainline home signal, see Fig. 31.12.

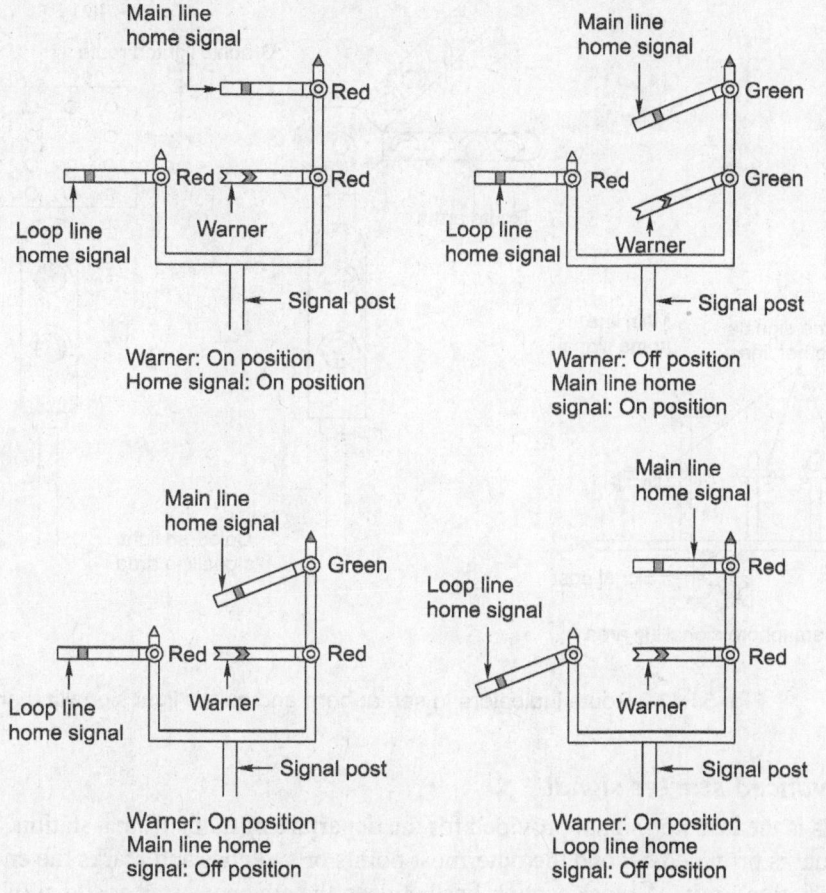

Fig. 31.12 Modified lower quadrant signalling arrangement

Route indicators can also be provided by including separate home signals for each line, with the main line home signal being placed the highest while all the other signals are placed at the same level.

In the case of coloured light signals, the home signal is provided with either a graphic lighted route indicator displaying the line number on which the train is to be received or different arms lighted by five lamps. These lamps form the arm, which is used for indicating a line, while there is no arm in the case of a main line as depicted in Fig. 31.13 (p.572).

Starter signal

The starter signal is a stop signal and marks the limit upto which a particular line can be occupied without infringing on other lines. A separate starter signal is provided for each line. The starter signal controls the movement of the train when it departs from the station. The train leaves the station only when the starter signal is in the 'off' position. As this signal controls the departure of a train, it comes under the category of departure signals.

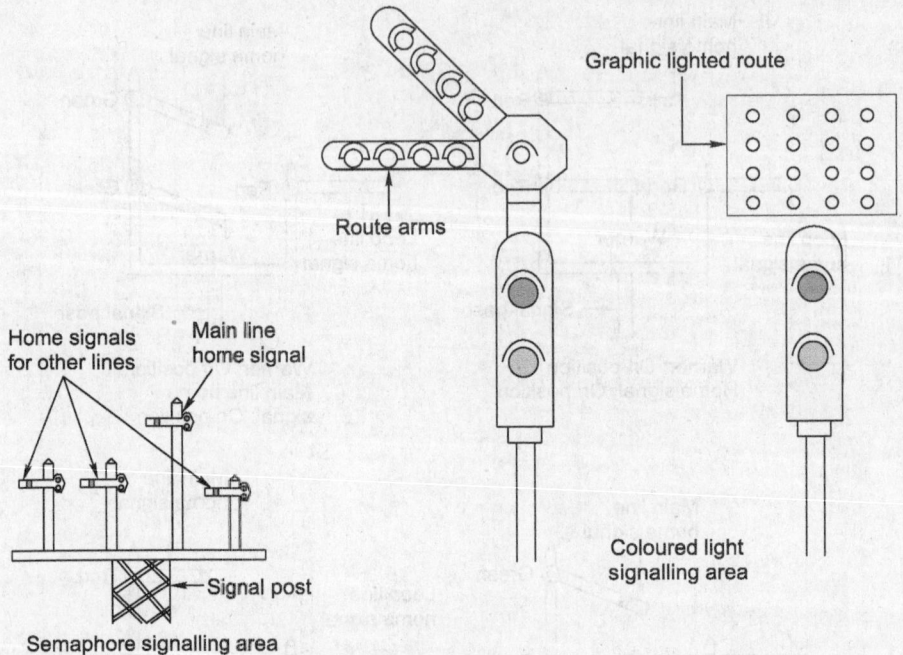

Fig. 31.13 Route indicators in semaphore and colour light signalling areas

Advanced starter signal

This is the last stop signal provided for the departure of trains from a station. The signal is provided beyond the outermost points or switches and marks the end of the station limits. A block section lies between the advanced starter signal of one station and the outer signal of the next station. No train can leave the station limits until and unless the advance starter is taken off.

31.5 SIGNALLING SYSTEMS

The signalling system can be broadly classified into two main categories.
  (a) Mechanical signalling system
  (b) Electrical signalling system
  In addition to these two main categories of signalling systems, electronic or solid-state signalling system is also in use. Each system of signalling comprises five main components.
  (a) Operated units such as signals and points
  (b) Interlocking system
  (c) A transmission system such as single- or double-wire transmission or electrical transmission through cables
  (d) Operating units such as levers and press buttons
  (e) Monitoring units such as detectors, treadle bars, and track circuits
  The comparison between mechanical and electrical signalling based on these five broad components is given in Table 31.9.

Table 31.9 Comparison of signalling systems

Component	Mechanical	Electrical
Operated units signals	Mechanically operated signals as per lower quadrant or upper quadrant and modified lower quadrant signalling	Coloured light signals with two- aspect, three-aspect or four-aspect signalling
Points	Mechanically operated points; locking with the help of point locks, stretcher bars, and detectors	Electrically operated points (by converting the rotary movement of electric motors into linear push or pull); locking with the help of slides and solid rods
Level crossing gates	Interlocking of manually operated swing leaf gate or operation and Interlocking of mechanically operated lifting barriers	Operation and Interlocking of electrically operated lifting barriers
Transmission systems	Single or double wire transmission to the requisite points by means of rods or double wires	Electrical transmission through overhead wires or underground cables
Operating units	Hand levers with a range of 500 to 2000 m used in collaboration with single wire or double wire lever frames	Push buttons, rotary switches, or electrical signalling equipment
Interlocking units	Mechanical interlocking with plungers attached with levers and tappets moving across in a locking trough	Interlocking through electromagnetic switches known as relays or solid-state switching devices
Monitoring units	Monitoring of points with the help of mechanical detectors; monitoring of the passage of trains using a treadle, which is an electro-mechanical device	Monitoring with the help of direct current track circuits, alternating current track circuits, electronic track circuits, axle counters, etc.

31.6 MECHANICAL SIGNALLING SYSTEM

The mechanical signalling system mostly involves signals and points as explained in this section.

In this system, both the signalling system and the interlocking are managed mechanically.

31.6.1 Signals

The signals used in a mechanical signalling system are semaphore signals. These signals are operated by means of either a lower quadrant or an upper quadrant signalling system.

Lower quadrant signalling system

This system of signalling was designed so that the semaphore arm of the signal could be kept either horizontal or lowered. The lower left-hand quadrant of a circle is used for displaying a semaphore indication to the driver of a train. This concept was possibly developed based on the left-hand driving rules applicable on roads in the UK and in India.

Upper quadrant signalling system

In lower quadrant signalling, the semaphore arm of the signal can only take two positions, namely horizontal or lower; it is not possible to include a third position for the semaphore arm, such as a vertically downward position due to design as well as visibility problems, since as the semaphore arm would, in that case, be superimposed on the signal post. Due to this limitation, the upper quadrant system (see Table 31.10) was developed, which can display more than two aspects. In this system, it is possible to incorporate three positions of the semaphore arm, namely (a) horizontal, (b) inclined at an angle of about 45° above the horizontal level, and (c) vertical, i.e., inclined at an angle of 90° above the horizontal level, see Fig. 31.14. The positions of the arm, the corresponding indications, and their meanings are listed in Table 31.10.

Table 31.10 Details of upper quadrant signals

Position of arm		Indication	Interpretation
Distant/warner signal			
Day	Horizontal	Caution	Proceed at a caution speed and be prepared to stop at the next signal
Night	One yellow light		
Day	Inclined 45° above horizontal	Attention	Proceed and be prepared to pass the next stop signal at a restricted speed
Night	Two yellow lights in a vertical line		
Day	Vertical	Clear	Proceed at full speed
Night	One green light		
Stop signals			
Day	Horizontal	ON	Stop dead
Night	One red light		
Day	Inclined 45° above horizontal	Caution	Proceed and be prepared to stop at the next signal
Night	One yellow light		
Day	Vertical	Clear	Proceed at full speed
Night	One green light		

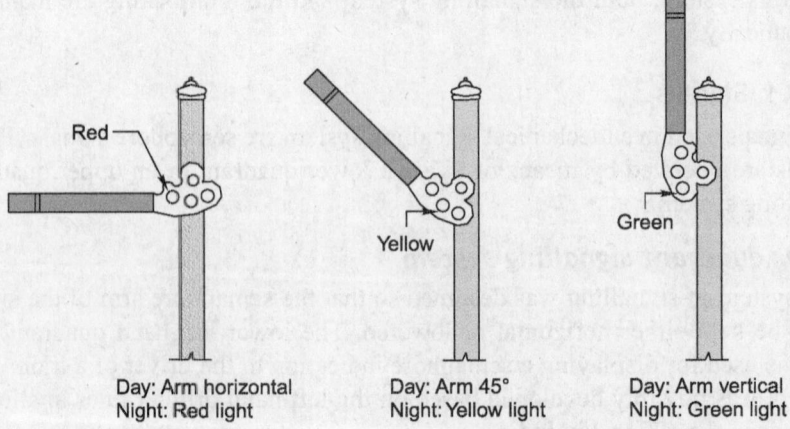

Red

Yellow

Green

Day: Arm horizontal
Night: Red light

Day: Arm 45°
Night: Yellow light

Day: Arm vertical
Night: Green light

Fig. 31.14 Upper quadrant signalling

31.6.2 Points

Points are set mechanically and are kept in locks and stretcher bars. The mechanical arrangement for operating them includes a solid rod with a diameter of 33 mm running from the lever provided in the cabin and connected to the point through cranks and compensators. Owing to transmission losses, the operating points with rods is restricted to a specified distance from the cabin.

The following devices are used to ensure that the points are held rigidly in the last operated position under a moving train and to ensure absolute integrity of the same.

(a) Point locks to hold the point in the required position and to rigidly hold the point in the position of the last operation

(b) Facing point lock with lock bars to prevent the movement of points when a train is passing over them

These devices are further discussed in the subsequent paragraphs.

Point locks

A point lock is provided to ensure that each point is set correctly. It is provided between two tongue rails and near the toe of the switch assembly. The point lock consists of a plunger, which moves in a plunger casing of facing point lock. The plunger is worked by means of a plunger rod, which is connected to the signal cabin through a lock bar. Additionally, there is a set of stretcher blades and each blade is connected to one of the tongue rails. Each blade has two notches and they move inside the facing point lock plunger casing along with the tongue rails. When the points are set correctly for a particular route, the notch in the stretcher blade rests in its proper position and the plunger rod enters the notch, locking the switch in the last operated position.

Detectors

Detectors are normally provided for all the points for the following reasons.

(a) To detect any defect or failure in the connection between the points and the lever as well as any obstruction between the stock and the tongue rail.

(b) To ensure that the correct signal, which corresponds to the point set, is lowered.

(c) A detector can be mechanical or electrical. In the case of a mechanical detector, the point is held in the position of the last operation, which is achieved de facto by virtue of its design. However, it cannot be considered as a device to keep the point locked.

A detector normally consists of a detector box, which is provided with one slide for points and another set of slides for signals. The signal slides are perpendicular to the point slides. The slides are held suitably and no vertical movement of the same is possible. The signal slide has only one notch whereas the point slide has a number of notches depending upon the number of signals relevant to the points. The detector works on the principle that a particular signal can be lowered when the notch in that particular signal slide coincides with the notch in the point slide. For example, if the points are correctly set for the main line, the point slide moves and its notch comes to rest opposite the notch of the main line signal slide. The main line signal slide can then be pulled and the main line signal lowered. It may be noted here that the point slide will move and its notch will rest in its correct position only if the points are properly set and there is no obstruction in between.

The linear type (or slide type) of mechanical detector is used for single-wire signalling (Fig. 31.15), whereas the rotary type of detector (Fig. 31.16) is used for the double-wire signalling. A double-wheel detector is a rotary-type detector that rotates in a vertical plane. It detects the correct setting of points and, in addition, locks the points in the last operated position in the case of wire breakage.

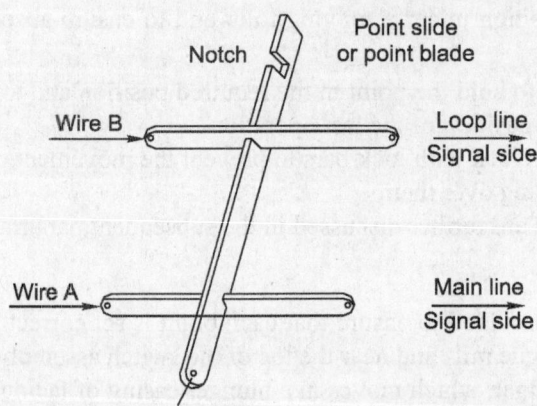

Fig. 31.15 Mechanical detector for single-wire signalling

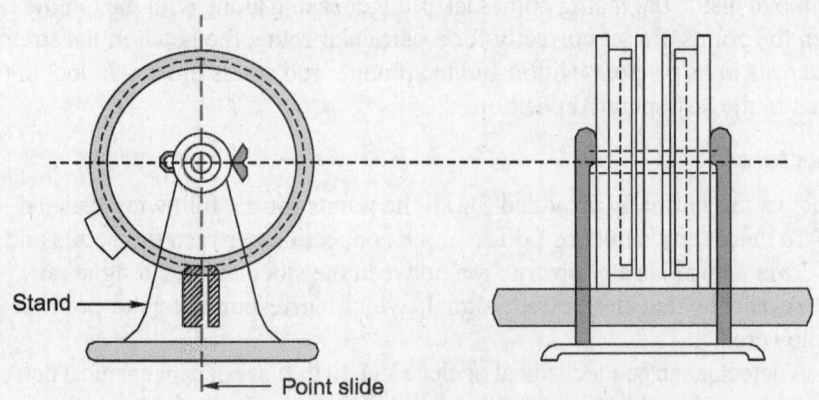

Fig. 31.16 Double-wheel rotary detector

Lock bar

A lock bar is provided to make it impossible to change the point when a train is passing over it. The lock bar is made of an angled section and its length is greater than that of the longest wheel base of a vehicle. Short revolving clips are provided to hold the lock bar in place on the inside face of one of the rails. The length of a lock bar is normally 12.8 m for BG and 12.2 m for MG sections.

The system is so designed that when the lever in the cabin is pulled to operate the locking device, the lock bar rises slightly above the rail level and then comes down. In the occurrence that a vehicle is positioned on the same location, the lock bar cannot rise above the rail level due to the flanges of the wheel and as such the point cannot be operated.

31.6.3 Types of Transmission Systems

A signal is operated by pulling the associated lever and this action is transmitted through a single-wire or double-wire system. Initially, the single-wire system was the most popular way of operating signals and, in fact, some stations on Indian railways still use this system.

In the single-wire system, only one wire is stretched between the operating lever and the signal, whereas in the double-wire system a loop of two wires that run parallel to each other is wrapped over a drum lever and this system works on the principle of the pull and push arrangement.

Single-wire transmission

In the case of single-wire signalling, transmission is done with the help of the following equipment.

Lever frame   A lever frame carries out the dual function of operating the single-wire system and actuating the interlocking in order to ensure safety. There are two types of lever frames, namely, a direct locking lever frame and a catch handle type lever frame, which are used for this purpose on Indian Railways. The levers are pulled in order to operate the signals and points, locks and lock bars.

Signal transmission wire   The entire transmission is done through an 8 or 10 SWG (standard wire gauge) galvanized steel wire. In places where the transmitting wire has to manoeuvre a turn, a multistandard galvanized steel wire is used, which is hauled over a horizontal or vertical wheel by pulling the lever. The arm of the signal remains as such while the lever is being pulled. When the lever returns to its normal position, the counterweight provided with the signal arm restores the wire back to its original position.

Cabin wire adjuster   In order to accommodate the increase or decrease in the length of the cabin wire due to its expansion as a result of temperature variation, cabin wire adjusters are provided. If this is not done, it may result in inadequate lowering or drooping of the signal, thus giving rise to unsafe conditions.

Signal parts and fitting   Signal parts and fittings are provided as per the details given in Fig. 31.2. The signal arm is brought to the 'off' position by a down rod connected to counterweight, which is pulled by a lever through a single wire. In the case of a breakdown of the signal transmission system, the signal arm is brought back to the horizontal position by means of a counterweight.

Limitations of single-wire signalling

The single-wire signalling system does not give satisfactory performance in the operation of signals located at long distances. The signals do not operate correctly as they droop and do not get properly lowered, thereby conveying misleading information to the driver. This defect develops due to the following reasons:

(a) Sagging of the wire due to its own weight
(b) Expansion or contraction of the wire due to temperature variations
(c) The elastic stretching of wires due to tension arising as a result of operating the lever
(d) Entangling of the wire due to its slipping into the pulley stake and the horizontal and vertical wheels
(e) Frequent breakage of the wire

It has been noticed that the slightest slackness in attending to any one of these problems can cause the lowering of the signal, known as 'drooping' or improperly put back to 'on' even when the lever is not pulled. On account of these defects, situations may arise in which a signal may not be lowered at all or a signal may not return to the 'on' position even after the lever has been restored to its normal position, if the wire gets entangled. As a result of these limitations, the single-wire signalling system can be used for operating signals from a range of only about 950 m.

Rod transmission   In the single-wire transmission system, the signal is lowered or set in the 'off' position by pulling the lever. The signal returns to the 'on' position due to the effect of gravity as soon as the lever is restored to its normal position and the tension in the wire is released.

Where the operation of points is concerned, the points have to be set in either the normal or the reverse position; one of these positions can be attained through pulling and the other by pushing. Solid rods of 30 mm (1 1/4″) diameter are used to connect the levers to the points. The rods or pipes move on standard roller guides fixed at about 2 m (6 ft) intervals. A suitable crank is also used at every change of direction. The rods are subjected to expansion and contraction due to temperature variations and as such are provided with rod compensators at designed intervals.

Rod temperature compensator

A rod compensator, also known as a temperature compensator, is provided to neutralize the effect of thermal variations. It consists of a pair of cranks—one acute and one obtuse—connected by a link and is so designed that it absorbs the expansion or contraction due to temperature variations. The compensator is normally placed at the centre of the rod upto a length of 36.5 m. If more than one compensator is required, these are placed at quarter points.

As can be seen in Fig. 31.17, the points, A or B may move left or right, but the total distance between them remains the same.

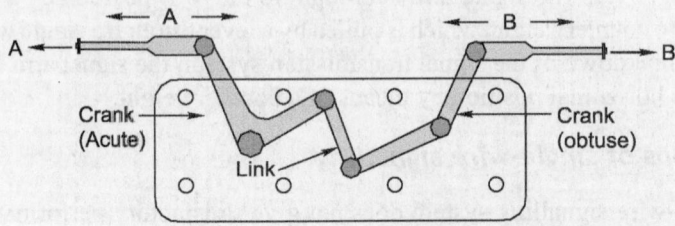

Fig. 31.17   Temperature compensator

Double-wire transmission system

In this system, power is transmitted with the help of two wires from the lever to operated units, such as signals, points, locks, detectors, and so on. Each wire consists of eight to ten SWG solid galvanized steel wires attached to pulley stakes, which are driven firmly into the ground. The two wires are connected between the lever and the signal to form a continuous loop. When the lever is operated, it leads to the wire getting pulled and when the lever is brought back to its normal position, it results in a push to the wire. This pull and push mechanism (Fig. 31.18) causes the drum to rotate in one direction when the lever is pulled and in the other direction when it is restored to its normal position. The rotary motion of the drum is then converted into linear movement by the use of cams and cranks and this finally actuates the signal. Figure 31.19 shows the complete double-wire transmission system.

It may be brought out that double-wire compensators are provided in the wire run to always keep the same tension in the wire.

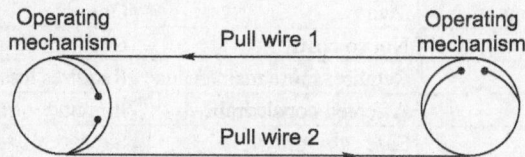

Fig. 31.18   Double-wire mechanism

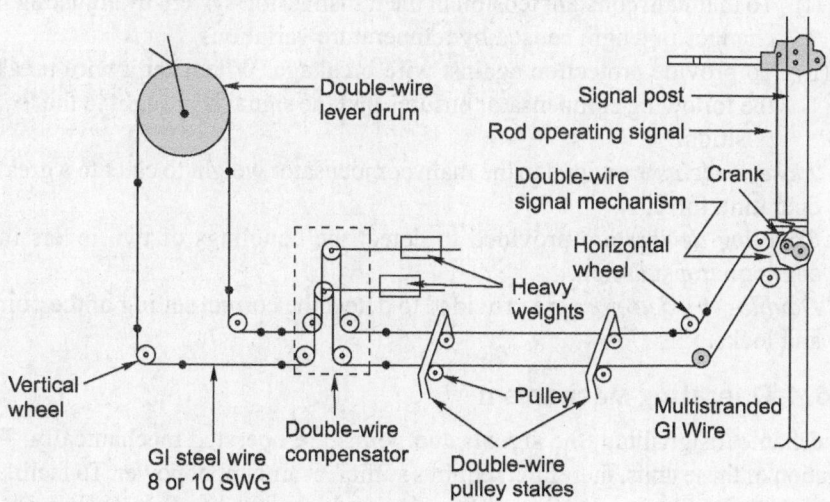

Fig. 31.19   Double-wire transmission system

Advantages of double-wire transmission

The main advantages of double-wire transmission over single-wire transmission are enumerated here.

(a) Operating a double-wire transmission system is easier.

(b) A double-wire transmission system permits multi-aspect signalling.

(c) The range of operations of the double-wire transmission system is greater.

(d) There is no drooping of signals in double-wire transmission. This results in better safety.

(e) The operating cost of a double-wire transmission system is less.

Table 31.11 compares a signal-wire and a double-wire signalling system. The special features of a double-wire signalling system are as follows.

Table 31.11 Comparison of transmission systems

Feature	Single-wire signalling	Double-wire signalling
Drooping of signals	Can occur	Does not occur
Outside interference	Higher possibility	Lower possibility
Method of signal return to the 'on' position	By the action of gravity	By pulling back the lever
Range of operation for signals	950 m	1200 m for a 500-mm drum
Damage to down rod	More frequent	Less frequent
Adjustment of length due to temperature variation	By manually operated cabin wire adjuster	Automatically done by double-wire compensator
Requirement of cabins	Two	One
Cost aspect	Not so costly	Costlier
Maintenance	Requires more maintenance	Requires less maintenance
Effect of weather	Affected considerably	Immune
Safety aspect	Safe	Safer

(a) *Double-wire compensators* are provided to perform the following functions.
   (i) To maintain constant tension in the transmission system by adjusting the changes in length caused by temperature variations.
   (ii) To provide protection against wire breakage. Whenever a wire breaks, the following compensator ensures that the signal is brought to the 'safe' position.
(b) *Jockey weights* are added to the main compensator weight to cater to a greater operating force.
(c) *Coupling devices* are provided to detect the couplings of two levers in a common transmission.
(d) *Double-wire detectors* are provided to detect the correct setting of the points and locks.

31.6.4 Operating Mechanism

In mechanical signalling, the signals and points are operated mechanically. The operation of these units, therefore, requires sufficient amount of power. To facilitate the quick and smooth operation of these units, levers are used in single-wire as well as in double-wire mechanical transmission. These levers may be installed individually near an operated unit, such as a point or a signal or may be grouped together in a lever frame depending upon the type and standard of signalling. Lever frames may be of the following types.

Single-wire lever frame

In a single-wire lever frame, the lever is a bar fulcrumed on a shaft. The longer end of the bar rests towards the lever while the shorter end rests towards the signal or the point in order to gain mechanical advantage. Pulling the lever results in a movement of 200 mm in the case of points and 300 mm in the case of signal operation. All the levers that are required to operate different units, such as points and signals are grouped together on a common fulcrum shaft in a compact lever frame, which is kept below the level of the operation platform. Only a convenient length of the lever is allowed to protrude beyond the platform to facilitate easy operation. A latch arrangement is also provided with the lever arm to maintain the lever in the position in which it was last operated.

Double-wire lever frames

Unlike single-wire systems, a double-wire transmission system is in the form of a loop, one end of which is wrapped across a drum that is provided with an arm. When this arm is rotated about the centre of the drum, it imparts a stroke of 550 mm or 600 mm depending upon the rquirement and as such the type of lever used. There are six types of double-wire levers used on Indian Railways that impart strokes of 500 mm to 600 mm.

When referring to points, normally a 500 mm stroke drum is capable of operating all the equipment satisfactorily up to a distance of 500 m. In case of any difficulty, 600 mm stroke levers may be used.

The interlocking of these operating mechanisms is discussed in Section 31.8.

31.6.5 Monitoring Units

Monitoring units are provided to ensure that points are set properly and locked for the safe passage of trains. These units are also sometimes used to monitor the passage of trains. Detectors and treadles are used as point setting monitoring devices on Indian Railways. Detectors have been discussed in Section 31.6.2.

Lock bar holding route during passage of train

Once the signal is taken off, the entire route ahead of the signal is kept locked and/held so till the signal lever is put back to normal.

It is not always possible for a train to pass immediately after the route has been set for its reception or departure and the signals have been placed in the 'off' position. There is generally a time gap between these two events. During this intervening period, the operator has to conduct many other operations in the yard, such as the shunting of some other train, monitoring the movement of other trains, and booking tickets, and in the process, the operator is likely to forget about the passage of the train on the route set by him. In order to eliminate human error, some devices are provided to ensure that it is not possible to move points till the train clears all of them in its route. For this, the route holding bars, which are similar to lock bars are provided at a distance of 180 m from the signal if the first point falls beyond this distance from the signal. The point is provided with the lock bar and facing point lock. This arrangement of placing holding bars at a distance of 180 m between each other starting from the signal continues to hold the route.

Note: Treadles were earlier used for the safe monitoring and passage of the train. At present, treadles have been phased out and track circuits are being used in its place.

31.7 ELECTRICAL SIGNALLING SYSTEM

The electrical signalling system is progressively replacing the mechanical signalling system on Indian Railways, especially with the coming up of railway electrification projects facilitating availability of electric power reading. The main reasons behind this are as follows.

(a) There are a number of movable parts in the mechanical signalling system, such as rods, wires, and cranks, which cause heavy wear and tear, frictional losses, and many of these parts can be sabotaged by unauthorized persons.

(b) The arms of the semaphore signals used in mechanical signalling afford poor visibility during the day. The night indications of these signals are also not satisfactory.

(c) The operational time of the mechanical signalling system is much greater than that of the electrical signalling system.

In the electrical signalling system, electrical energy is used for displaying signal aspects. The transmission of power is done electrically and the units are operated by electrical push buttons while the system is monitored by electrical systems. The interlocking in this system is also done electrically.

In the case of electronic or solid state signalling system, the signals are operated by the electrical method, but the interlocking is done electronically.

System of electronic interlocking For electronically managed interlocking system a computer and mouse is used for controlling the signalling system; however, a panel is also used with a switch to change the mode of control from a panel to a visual display unit (VDU).

As control of the signalling system is carried out by a VDU and a mouse, the portion of the signal and interlocking plan shall be selected for taking off the concerned signal by clicking on to the signal profile on the VDU. The system shall work within, following the same principles as in case of electrically interlocked system but analyse the requirements electronically and energize the relays in the form of the final command to operate points and make the concerned signal off.

31.7.1 Operated Units

The operated units consist of signals and points. The electrical signalling system consists of either coloured light signals or signals with semaphore arms operated by electric motors.

A point is operated by converting the rotary movement of the electrical point machines fastened on the sleepers near the point into a linear push or pull force. There are low-voltage point machines operated with a 24 V dc supply and high-voltage point machines operated using a 110 V dc supply. The operation of a point machine involves first unlocking the lock, bringing the point from normal to reverse or reverse to normal as the case may be, and then locking the point once again. The operating time of these point machines varies between three and five seconds.

31.7.2 Transmission Medium

The medium of transmission for operating electrical equipment is either an overhead alignment or an underground copper cable. The overhead alignment is used when the number of conductors is limited. In areas provided with 25 kV ac traction, it is not possible to use overhead alignments due to the induced electromotive force

(EMF) generated as a result of electrostatic and electromagnetic induction. In big yards, cables are used as a medium of transmission for the operation of point machines. The cables are either hung on hooks and run by the side of the track or laid underground. In areas provided with ac traction, underground screened cables are used.

31.7.3 Operating System

Normally push buttons and rotary switches are used for operating signalling equipment that work on electricity. The complete yard layout is represented on the face of a console. Signals, tracks, points, and the gates of level crossings are depicted in their geographical positions on this console and the positions of these switches are then marked at the foot of signals and on various tracks.

Complete interlocking is achieved through electromagnetic switches known as *relays*. The two methods of interlocking available are panel interlocking or route relay interlocking.

Pressing of two buttons on the operating console checks clearance of route; if found clear, sets the route by operating points automatically by the system and then locks the route, takes the concerned signal off. This is known as route relay interlocking, often called on smaller stations as 'panel interlocking' as a misnomer.

31.7.4 Monitoring System

A monitoring system mainly consists of point detectors, track circuits, and axle counters, all of which are discussed here in detail.

Electrical point detector

The electrical point detector detects and ensures that points are properly set. It also works on a 'slide system' as used in the mechanical system. These slides are so adjusted that a gap of 3 mm is left between the switch rail and the stock rail so that the two do not come in contact and, therefore, it is not possible to turn the signal off at any time.

Track circuit

The track circuit is an electric circuit formed along with the running rails and connected to the signal and cabin. Its function is to indicate the presence of a train (or vehicle) on the track. In order to set up a track circuit, the ends of the rails forming the circuit are isolated by insulating the rail joints. The rails are laid on wooden sleepers so that they are electrically insulated from each other but due to scarcity of wood, now concrete sleepers duly insulating the two rails are being used. The ends of the rail on one side of the track are connected to a battery through resistances, etc., while on the other side of the track, the ends of the rails are connected to a relay. When the track is free, energy from the battery reaches the relay and energizes it. As soon as the track is occupied, the two rails are short-circuited because of the wheels and axle of the train and the relay does not get any adequate feed from the battery. It, therefore, gets de-energized, thereby breaking the circuit connected with the signals, thus ensuring that the concerned signals are set to indicate danger. The various types of track circuits used on the Railways are as follows.

 (a) Direct current track circuit
 (b) Alternating current track circuit
 (c) Electronic track circuit, which are audio-frequency track circuits

Axle counters

As already mentioned, two consecutive rails need to be insulated from each other for setting up a track circuit. The most essential requirement for track circuiting is the use of wooden sleepers. Due to the shortage of wooden sleepers on Indian Railways, an attempt is being made to progressively use a device known as an *axle counter*, which can be used as a substitute for track circuiting to detect the presence or absence of a vehicle on a track. A pair of rail inductors are installed at either end of the track for counting the axles. As soon as a train enters the track section from one end, the number of axles entering the section are counted automatically. Similarly, when the train leaves the track section at the other end, the axles are counted once again at the other end. If the same number of axles are counted at both the ends, it indicates that the section is free or unoccupied. If the number of axles counted at the exit end are less than the axles counted at the entrance to the section, it means that the section is still occupied.

31.8 INTERLOCKING

Interlocking is a device or a system meant to ensure the safety of trains. With the increase in the number of points and the signals and introduction of high speeds, it has become necessary to eliminate human error, which would otherwise lead to massive losses of life and property. The points and signals are set in such a way that the cabin man cannot lower the signal for the reception of a train unless the corresponding points have been set and locked. The signal is thus interlocked with the points in a way that no conflicting movement is possible and the safety of trains is ensured.

Interlocking may therefore, be defined as a technique, achieved through mechanical or electrical means by which it is ensured that before a signal is taken 'off', the route which the signal controls is properly set, locked and held till such time the entire route is traversed by the train and at the same time all the signals and points, the operation of which would lead to conflicting movements, are locked against the feasibility of such conflicting movements.

The signal and interlocking system is so designed that the failure of any equipment results in the turning 'on' of the signal, thus ensuring train safety.

31.8.1 Essentials of Interlocking

Lever frames and other types of equipment provided for the operation and control of signals, points, etc., must be so interlocked and arranged as to comply with the following essential regulations:

(a) It should not be possible to turn a signal off unless all points for the line on which the train is to be recieved are correctly set, all the facing points are locked, and all interlocked level crossings are closed and inaccessible to road traffic.

(b) The line should be fully isolated before the signal is turned off, i.e., no loose wagons should be able to enter this line.

(c) After the signal has been taken off, it should not be possible to make adjustments in the points or locks on the route, including those in the isolated line. Also, no interlocked gates should be released until the signal is replaced to the 'on' position.

(d) It should not be possible to turn any two signals off at the same time, which can lead to conflicting movements of the trains.

(e) Wherever feasible, the points should be so interlocked as to avoid any conflicting movement.

31.8.2 Standards of Interlocking

The speed of a train depends on a number of factors such as the haulage capacity of the locomotive, the fitness of the track, the fitness of the rolling stock, the load of the train, the ruling gradient and standard to which the signalling system is provided.

The speed for a particular section is determined based on all these factors. Depending upon the maximum speeds permitted on a section, the stations are interlocked in keeping with the prevalent standards, and signalling equipment and other facilities are provided accordingly. There are four standards of interlocking based on the maximum permissible speeds prevailing on Indian Railways. These refer to the speeds over the main line with respect to the facing points and the yard.

On Indian Railways, standards of signalling and interlocking have been classified into four categories, depending upon the speed of through running trains.

  (i) Upto 50 kmph: Standard-I
 (ii) Upto 100 kmph: Standard-II
(iii) Upto 130 kmph: Standard-III
 (iv) Upto 160 kmph: Standard-IV

Signalling arrangements

The signalling equipment, manner of locking points, and operation of signals and points differ in the different standards of interlocking. The types of signalling equipment to be provided at different interlocked stations and other requirements to be met for each of the four categories do differ.

In Standards II, III and IV, the signalling now has to be of a multiple-aspect type. In Standard-I, however, two-aspect signalling is still permitted and the starter signals are not compulsorily required. Starter signals are essential for other standards of interlocking, keeping in view the high speed of operation.

Locking of points

The method of locking of points is the key locking in Standard-I. It is an indirect method of interlocking between signals and points. In Standard-II, a plunger type facing point lock is used. The plunger lock can be operated from the cabin or from the site itself. In standard-III and IV, the points are to be centrally operated and the locking between points and signals is required to be direct.

Isolation of lines

In Standard-I, isolation is optional. In Standards II, III, and IV, the main line must be isolated from all the adjoining lines.

Details of signalling as well as of interlocking for different interlocked stations, viz., Standard I, Standard II, Standard III, and Standard IV are briefly summarized in Table 31.12.

Table 31.12 Details of signalling and interlocking arrangement for different interlocking standards

Type of interlocking	Details of signals	Details of interlocking
1. Standard I Interlocking (Speed upto 50 kmph)	(i) In two-aspect signalling territories: Outer and bracketed home signals in each direction. (ii) On multi-aspect territories: distant, home, and starter signals in each direction.	(i) Interlocking between facing points and signals by means of a key lock. Signal cannot be turned off till facing points are correctly set and locked. (ii) Signals cannot be lowered till trap switch, if any, is closed. (iii) The interlocking between points and signals must ensure that no signal can be taken 'off' to admit a train on the line that has been fitted with a trap switch, unless the trap switch has been closed.
2. Standard II Interlocking (Speed upto 100 kmph)	Multiple-aspect signals with distant, home, starter, and advanced starter signals in each direction. At stations in automatic signalling territories, a semi-automatic/ manual home signal and a semi-automatic/manual starter signal in each direction. At installations on two-aspect, signalling territories, outer warner, bracketed home, starter, and advanced starter, signal in each direction.	The interlocking between points and signals may be by mechanical, electrical or electronic means. All new installations shall have electrical or electronic interlocking.
3. Standard III Interlocking (Speed upto 130 kmph)	Multi-aspect colour light signals with distant, home starter, and advanced starter signals in each direction. At stations in automatic signalling territories, a semi-automatic/ manual home signal and a semi-automatic/manual starter signal in each direction. Wherever required, more than one distant signal may be provided, i.e., distant and inner distant signals.	The interlocking between points and signals shall be by electrical or electronic means.
4. Standard IV Interlocking (Speed upto 160 kmph)	Multiple-aspect colour light signals with distant, inner distant, home, starter and advanced starter signals in each direction. At stations, in automatic signalling territories a semi-automatic/ manual home signal and a semi-automatic/manual starter signal in each direction.	The interlocking between points and signals shall be by electrical or electronic means.

31.8.3 Methods of Interlocking

There are basically two methods of interlocking as explained below.

Key interlocking

Key interlocking is the simplest method of interlocking and still exists on branch lines of small stations on Indian Railways. The method involves the manipulation of keys in one form or the other. This type of interlocking is normally provided with standard interlocking with a speed limit below 50 km per hour. The simplest arrangement of key interlocking is accomplished in the following manner.

(a) Take the example of a station with a main line and a loop line, the point can be set either for the main line or branch line.

(b) The point has two keys, the first is key A, which can be taken out when the point is set and locked for the main line. Similarly, key B can be taken out when the point is set and locked for the loop line. At any given time either key A or key B can be taken out, depending upon whether the route is set for the main line or the loop line.

(c) The lever frame operating the signals is provided with two levers. The lever concerning the main line signal can be operated only by key A and similarly the loop line signal lever can be operated only by key B.

(d) If the train is to be received on the main line, the points are set and locked for the main line and key A is released. This key is used for interlocking the main line signal lever, thus lowering the signal for the main line. Since key A cannot be used for interlocking and lowering the loop line signal, only the appropriate signal can be taken off. This type of interlocking is called indirect locking.

In case more than one point is to be operated, the key released at the first point is used to unlock and operate the second point and so on. The key released at the last point can then be used for unlocking the lever operating the appropriate signal. This type of interlocking is also known as succession locking and is also used for checking conflicting movements in shunting operations. There are other methods of interlocking with the help of keys, but all of them involve considerably lengthy trips from the point to the signal levers and from point to point, thereby leading to delays. Such arrangements are, therefore, satisfactory only for stations that handle very light traffic.

Mechanical system of interlocking

Almost 70 per cent of railway stations in the country work with the mechanical system of signalling. The interlocking arrangements for mechanical signalling system have to be mechanically oriented. There are two systems of mechanically designed signals: (i) single-wire system and the (ii) double-wire system.

A mechanically structured signal has (i) spectacle with an arm; (ii) signal post, which may be tubular or lattice. Longer posts are chosen to be lattice; and (iii) a counterweight to help pull the wire back to allow the signal to go back to its on/normal position. Such mechanical structures of signals are: (i) two-aspect semaphore signal and (ii) multiple-aspect semaphore signal.

Mechanical interlocking or interlocking on lever frames is an improved form of interlocking compared to key locking. It provides greater safety and requires less manpower for its operation. This method of interlocking is done using plungers and tie bars. The plungers are made of case hardened steel sections measuring 30 cm × 1.6 cm and have notches in them. The tappets tie bars are placed at right angles to the plungers and are provided with suitably shaped pieces of cast iron or steel that fit exactly in the notches of the tappets.

The entire arrangement of plungers and tappets is provided in a locking trough. Each lever is attached to the plunger, which has suitably shaped notches to accommodate the locking tappets.

When a lever is pulled, it moves the plunger to which it is connected. Due to *wedge action*, the tappet accommodated in the notch of the plunger is pushed out at right angles to the movement of the plunger. The motion is transmitted to all other tappets that are connected to this tappet through a tie bar. As a result of this motion, the other tappets either get pushed into or out of the respective notches of the other plunger depending upon the interlocking provided. In case the other tappet is free but slips inside the notch of the other plunger, it locks the lever connected to this plunger. In consequence, the other lever gets locked in that position, and cannot be operated. However, if the tappet was earlier positioned in the notch of the plunger, thereby locking the lever, and is now out of the notch, the other lever becomes free to be operated.

This arrangement, thus sets the predetermined sequence for pulling the levers and hence actuates the interlocking.

Electrical system of interlocking

As the signal displays fixed light illuminated by incandescent lamp or a light emitting diode (LED) signal, the operation of such system may be through mechanically operated levers or by push buttons provided on the yard layout depicted on the top of panel box to be termed as the control cum indication panel.

Under the electrical signalling system the colour light signals are used in any case operated by lever, points (i) operated by wrought iron solid 33 mm rod; (ii) operated by electric point machines, or operated by Control cum Indication Panel operating points by an electric point machine with signals being coloured light.

The system of operation of electrically operated signals by levers is hybrid and is invariable an interim measure to suit 25 kV ac traction, to be subsequently converted to oepration by control-cum-indication panel.

Electronic system of interlocking

In case of electrical interlocking the relays used are wired using a cable of copper conductors to translate the interlocking relations given in the selection table, into logic circuits. In case of electronic interlocking, these logic circuits are converted into Boolean equations and finally converted into a program loaded on the processor of the computer.

The yard layout may be loaded on the computer and the train operation may be carried out using a mouse and the VDU or the control-cum-indication panel or either of the two with both options being available, transferring the control by a switch provided on the control-cum-indication panel.

Ultimately relays are energized by the system having evaluated the interlocking relations, to transmit commands to point and the signal.

31.8.4 Typical Cases of Interlocking

The following typical cases of interlocking are usually encountered.

Normal locking In this case, pulling one lever locks the other lever in its normal position. Such locking shall be required in situation like the signal lever locking a point lever, when the signal requires the point to be moved for train movement.

Back locking or release locking In this case, when the lever is in its normal position, it also blocks the other lever in its normal position, but when this lever is pulled it releases the other lever, which can then be pulled. Furthermore, once the second lever is also pulled, the first lever gets locked in the 'pulled' position and cannot be returned to its normal position unless the second lever is restored to its normal position.

Such locking is required in situations like the signal lever requiring a point lever to be revised for a given train movement. Here the point lever shall be pulled first and then the signal lever shall be pulled to keep the point lever locked in pulled condition till signal lever is in pulled condition.

Both-way locking In this case, once a lever is pulled, it locks the other lever in its current position that is, in the normal or pulled position.

Such type of locking is normally required in situations when the lock on point is to lock the point in either position. Here, if the point is to be locked in normal condition, the point lever shall get locked as it is by pulling back the lever.

Special or conditional locking In this case, the pulling of one lever locks the other lever only when certain conditions are fulfilled, say the third lever being in a normal or pulled position as the case may be.

Such a locking is normally required when a signal leads to more than one route.

31.8.5 Mechanical Interlocking of Points and Signals of a Two-line Railway Station

To understand the mechanical interlocking of the points and signals of a two-line railway station, let us take the case of a typical railway station that has both a main line and a loop line. It is provided with a home signal 1 operated by lever 1 for the main line and a home signal 2 operated by lever 2. Point 3 is set for the main line when lever 3 is in its normal position and for the loop line when lever 3 is pulled. In its pulled position lever 4 locks point 3 in either position (normal or reverse) by pushing the plunger of the facing point into the 'lock in' position. The essentials of this interlocking system are as follows (Fig. 31.20).

(a) It should not be possible to take off both the signals, that is, 1 and 2 at the same time, that is, the train should be received either on the main line or the loop line at any given time. To achieve this, tappet A is forced out of the plunger of lever 1 when lever 1 is pulled. The tappet enters the notch of the plunger of lever 2 and cannot move out until lever 1 is pulled. This prevents lever 2 from getting pulled when lever 1 is pulled. The reverse of this situation is also true.

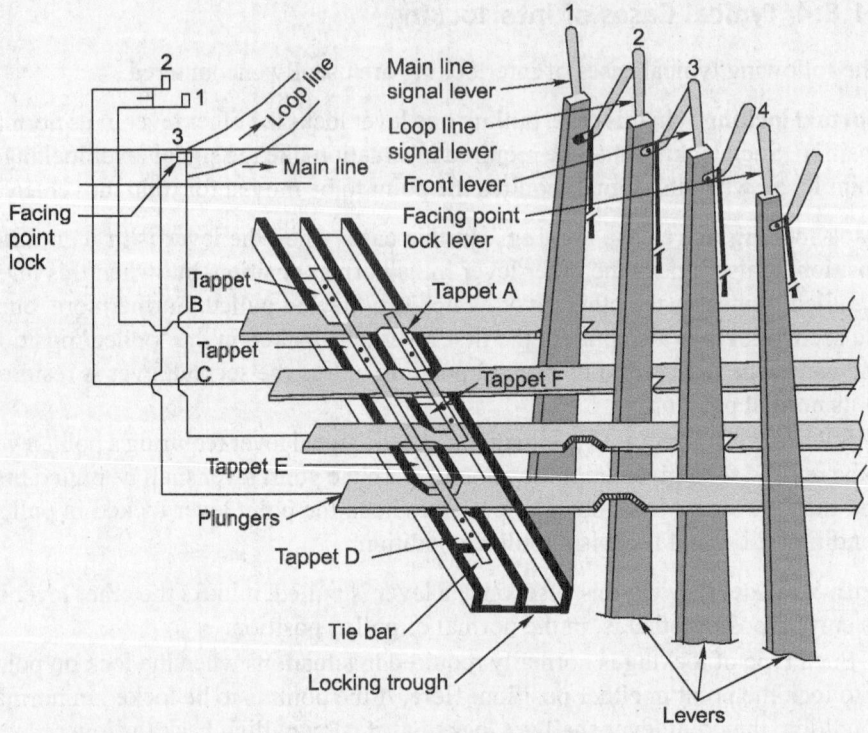

Fig. 31.20   Interlocking of points and signals of a two-line railway station

(b) It should not be possible to turn a signal off until and unless the point is set and locked. If an effort is made to pull either lever 1 or 2, the same cannot be pulled. This is because of the following reasons. Tappets B and C are rivetted with a tie bar that is in turn connected with tappet D. Since tappet D is butting against the face of the plunger of lever 4, tappet B or C cannot be moved. This plunger can move only when lever 4 is pulled, thereby bringing the notch of the plunger opposite tappet D. Once this happens, it is not possible to restore the position of lever 4 till lever 1 or 2 has been brought back to its normal position. This is a case of *release locking*. Levers 1 and 2 are released by pulling lever 4, which in turn locks the point.

(c) Lever 4 is a lock lever and locks point 3 in either position. The same can be visualized for tappet E.

(d) Similarly, the main line home signal lever 1 cannot be pulled if point lever 3 is pulled, that is., if it is set for the loop line, as the tappet F cannot be pushed, which allows lever 1 to be pulled when the notch in the plunger of lever 3 that lies opposite tappet F has been shifted from its position due to the pulling of lever 3. With lever 3 normal, lever 1 can be pulled conveniently, simultaneously resulting in the locking of lever 2 in it normal position by tappet A.

This example indicates that, with proper planning, it is possible to mechanically interlock the movement of points and signals and thus ensure complete safety.

31.8.6 Electrical Interlocking of Points and Signals of a Two-line Railway Station

Electrical interlocking is achieved through electric switches known as relays. The manipulation of relays achieves interlocking, whereas lever locks that are attached with the levers in place of plungers or in addition to plungers prevent a lever from getting pulled, or allow it to get pulled or normalized if the interlocking so permits.

Relays use the simple principle of electromagnetism, whereby a soft iron core wrapped inside a wire coil turns into electromagnet when current is passed through the wire. An armature is attached to this electromagnet, which has a number of finger contacts that come into contact with each other when the armature is attracted to the magnet and break the contact when the armature is not attracted to it. The whole system is housed in a glass or metal box and is known as *relay*.

31.9 SYSTEMS FOR CONTROLLING TRAIN MOVEMENT

The system adopted for controlling the movement of trains should be such that it allows the trains to run in either direction as well as facilitates faster trains to overtake slower trains, thus ensuring the complete safety of trains. The following systems are chiefly used for controlling the movement of trains on Indian Railways.

Time interval system

In this system, there is a time interval between two successive trains. A train is dispatched only after sufficient time has elapsed since the departure of the previous train. This system works fine just as long as everything goes well with the previous train, but if there is a mishap and the previous train is held up, the system fails, jeopardizing the safety of the trains.

Space interval system

In this system, there is a space interval between two consecutive trains. Only one train is permitted to occupy a particular length of the track. A succeeding train is permitted to occupy the same track length from either side only after the first train has cleared it. This system guarantees safety as only one train is in motion at one time.

31.9.1 Methods of Controlling Train Movement

Based on these systems, the following methods are adopted for controlling the movement of trains on Indian Railways.

One-engine-only system

This system permits only one train to remain in a section at one time. The movement of trains is controlled with the help of a *wooden staff* or a *token* with suitable identification marks, which are in the possession of the driver (loco pilot) of the train. As the same object cannot be at two places at the same time, the safety of trains is fully ensured. This system is possible only on short branch lines that have limited traffic. Normally there is only one train, which works to and fro on the same section. The system fails if it becomes necessary to dispatch more than one train in the same direction. This system does not require a 'line clear' directive.

Following train system

In this system, trains follow each other after a time interval that is generally less than 15 minutes. Trains scheduled after the first train can run at a maximum speed of 25 km per hour. As an adequate time interval is kept between two successive trains, safety is ensured to a limited extent. The system is used under the following circumstances.

(a) In the case of emergencies, such as the failure of block instruments and the telephone system
(b) In short double-line stretches

Pilot guard system

In such a system, one person, known as the *pilot guard*, accompanies the train by riding on the foot plate of the engine (or gives a ticket personally to the guard of the train, which is authority to proceed) and returns to the same station with another train. The pilot guard is normally identified by his or her prescribed uniform, which is red in colour, or the badge that he or she wears and is an authority for the train to proceed. Even in this system trains can follow each other after a fixed time interval of not less than 15 minutes. The system is applicable in short single-line sections or in the case of failure of communication between two stations.

Train staff and ticket system

This system is similar to the pilot guard system. The authority to proceed in this case is either a wooden staff or a ticket. There is only one wooden staff for a section and the same is kept at one of the two stations on that section. Each station has a ticket box, which contains printed tickets and is kept locked. The wooden staff is interlocked with the box in a way that it cannot be taken out so long as the box is locked. A train can only be dispatched from the station that has the staff. In case only one train is to leave the station, then the staff is handed over to the driver of the train. If more than one train is to be dispatched from the same station, the preceding trains are dispatched on the authority of the ticket while the last one is dispatched along with the staff. The time gap between two successive trains is not less than 15 minutes and the speed of the trains is restricted to 25 km per hour. A similar system is followed for dispatching trains from the other station. In this system, the safety of the trains is ensured on account of the fact that only one ticket can be issued at one time and the driver insists on seeing the staff before accepting the ticket as his authority to proceed.

Absolute block system

This system involves dividing the entire length of the track into sections called *block sections*. A block section lies between two stations that are provided with block instruments (explained later). The block instruments of adjoining stations are connected electrically and a token can be taken from the block instrument of a particular station with the consent of both the station masters on a single line section. On a double-line section taking the departure signal 'OFF' is the authority to proceed for the loco pilot while there is no token system.

In the absolute block system, the departure of a train from one station to another is not permitted until and unless the previous train has completely arrived at the next station, that is., trains are not permitted to enter the section between two stations at the same time. The procedure by which this system is maintained is known as the *lock and block* procedure. The instruments used for this purpose are known as block instruments.

Block instruments   Each station has two block instruments; one for the station ahead and the other for the previous station. The block instruments of two adjacent stations are electrically interconnected. These block instruments are operated with the consent of the station masters of the stations on either end of the block section, who are also responsible for giving the line clear indication. Normally, a round metal ball called a 'token' is taken as the authority to proceed in a block section. This token is contained inside the block instrument on a single-line section.

The following different types of block instruments are used on Indian Railways depending upon various requirements.

Single-line token instruments   These are meant for sections with single lines. No train is authorized to enter the block section without a token. The token can be taken out of the block instrument of the departure station only when the station master turns the handle of the block instrument towards the end labelled 'Train going to side'. This can be done only with the cooperation of the station master of the station on the other side of the block section, who turns the handle of his or her block instrument towards the end labelled 'Train coming from side'. It is not possible to turn the signal permitting the entry of the train into the block section off until the handle of the block instrument has been turned towards the 'Train going to side' label. In this situation, the handles of both these instruments get locked in the last operated position and it is not possible to normalize both the block instruments until the train arrives at the next station and the token has been inserted into the block instrument of that station. This phenomenon of keeping the block instruments locked and releasing them only during the passage of a train is the lock and block procedure discussed earlier.

Single-line tokenless block instruments   There have been occasions when a train has had to be brought to a halt because of the driver misplacing the token, causing the trains to get detained for long periods. In order to avoid such occurrences, tokenless block instruments have been developed. The same principle as that of the block system is followed here but without the use of a token. The last stop signal permitting the entry of the train into the block section, which is normally the advanced starter signal, is interlocked with the block instrument in such a way that it is not possible to turn this signal off unless the block instrument has obtained the line clear command.

Double-line block instruments   In a double-line section, traffic is unidirectional. The block instrument comprises a commutator handle and two indicator needles placed in vertical alignment. In order for the block instrument to work on a double line, the station master turns his block instrument commutator to the 'Line clear side'. This causes the electrical circuit to make contact in such a way that the advanced starter of the dispatching station can be turned off.

Working details

Take an example of a block section AB situated between two stations A and B on a single-line section (Fig. 31.21). A train is waiting at A to enter the section AB. The procedure is as follows.

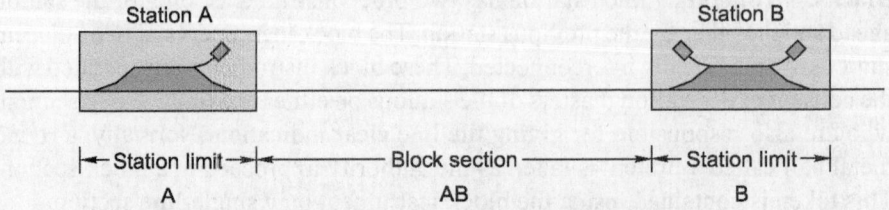

Station A   Station B

|← Station limit →|←————— Block section ————→|← Station limit →|

A   AB   B

Fig. 31.21   Block section AB between stations A and B

1. The station master of station A establishes contact over the telephone with the station master of station B with the help of the block instrument telephone and requests the station master of station B to grant a line clear, i.e., permission so that he can dispatch train A.
2. Once the station master at station B has ensured that the line is clear according to the prescribed norms, he agrees to receive the train and grants a line clear. For this, he gives a private number and operates the block instrument of his station in a prescribed manner. The station master at station A notes this private number and simultaneously operates his block instrument so that a 'ball token' is extracted from the block instrument.
3. The station master at station A then allows the point to be set, lowers the signal and hands over the 'ball token' to the driver of the train waiting at station A.
4. The station master at station B also gets the points set and lowers the signal for the line on which the train is to be received.
5. The train then starts from station A and enters the block section AB.
6. The train reaches station B. The driver of the train hands over the ball token to the station master of station B. After ensuring that the entire length of the train has been received, the station master at B pockets the ball token in the block instrument. He then informs the station master at station A of the arrival of the train on a private number as proof of the same. The points at station B are then set as they were before and the reception signals restored to their normal positions.
7. With the insertion of token into the block instrument at station 'B' the start station 'A' and at station 'B' form their commutations to normal line closed position thereby making both block instruments ready for next operation.
8. The same procedure is repeated when the train has to enter a block section BC.

The system is absolutely safe and works on the principle of providing space intervals. Most stations on Indian Railways work on this principle. The following are the essential features of the absolute block system.

(a) No train is allowed to leave a station unless 'permission to approach' has been received in advance from the block station.

(b) On double lines, permission to approach is not given until the line is clear, not only up to the first stop signal of the next station, but also upto an adequate distance beyond it.

(c) On a single line, 'permission to approach' is not given until the following conditions are satisfied.

   (i) The line is clear of trains running in the same direction, not only upto the first stop signal of the next station, but also for an adequate distance beyond it.

   (ii) The line is clear of trains running in the opposite direction.

(d) When two trains are running in the same direction on the same track, permission to approach should not be given to the second train till the entire length of the first train is within the limits of the home signal, the 'on' status of all the signals behind the first train has been restored, and the line is clear, not only up to the first stop signal of the station, but also upto an adequate distance beyond it.

Automatic block system

In the space interval system, clearing a long block section is a protracted event and the subsequent train has to wait till the preceding train clears the entire block section. This impairs the capacity of the section with regard to the number of trains it can clear at a time. In order to accommodate more trains in the same section, the block section is divided into smaller automatic block sections. This is particularly done for sections that are long and have turned into bottlenecks. The essentials of an automatic block system on a double line section are as follows.

(a) The line should be provided with continuous track circuiting.

(b) The line between two adjacent block stations may, when required, be divided into a series of automatic block signalling sections, entry into each of which will be governed by a stop signal.

(c) The track circuits should control the stop signal governing the entry into an automatic block signalling section in the following manner.

   (i) The signal should not assume the 'off' position unless the line is clear in advance, not only upto the next stop signal, but also for an adequate distance beyond it.

   (ii) The signal should automatically turn on as soon as the train passes it.

31.10 MODERN SIGNALLING INSTALLATIONS

Advancements in electrical engineering and electronics have greatly contributed to the modernization of signalling installations, leading to better safety, increased speeds, and quicker movement of trains. Some of these modern signalling installations are described below.

Panel interlocking

In panel interlocking, all points and signals are operated electrically from a central location and the switches for operating these points and signals are mounted on a panel, which also bears the diagram of the yard layout. Electrical interlocking of these points and signals is achieved by means of relays. The main advantage of panel interlocking is that the various functions of all the points and signals, even though the same cover great distances, can be centrally controlled, thereby eliminating the need for multi-cabin operations of the same. With the elimination of intercabin control and slotting, the time that is normally lost in coordination is saved and the line capacity increases so that a greater number of trains can be run by a smaller operating staff.

Route relay interlocking

Route relay interlocking is an improvement over panel interlocking. Unlike panel interlocking, where each point in the route has to be individually set with a respective switch and where the clearance of the signal is obtained by operating the signal switch, route relay interlocking involves the use of only a pair of switches to perform all these operations automatically. Using this pair of switches, the desired route for the train is set automatically by putting all the points along the route in their desired positions. The required signal is then cleared automatically too.

During this operation, it is also ascertained that there is no conflicting movement of the trains in progress or in the offing and also that the route is clear for the movement of a train, including at the overlap. One of the essential requirements for this is that the entire yard has got to be track circuited. The condition of the track circuits and the various indications of all the signals on the route are mirrored on the panel that carries the diagram of the yard. By looking at these indications, the panel operator can easily discern which portion of the track is clear or occupied and which signal has been cleared for the movement of the train. Once the route is set to allow the passage of a train, the relevant portion of the diagram on the panel gets illuminated with white lights. The lights turn red when the track is occupied by the train. When the train has cleared the track, the lights automatically go off.

Route relay interlocking is very useful in busy yards such as the Mumbai suburban section, where traffic density is very heavy. As route relay interlocked yards are fully track circuited, they ensure complete safety with regard to the movement of trains.

Solid state interlocking

(i) Historically, interlocking started with mechanical lever frames. As the size of yards and train movements increased, size of lever frames also increased. With the advent of electro-mechanical relays, lever frames gave way to relay interlocking based installations. This development resulted in faster operation with embedded housing of the interlocking installations. Today, Indian Railways has around 250 route relay interlocking (RRI) and 2800 panel interlocking (PI) installations. These installations use thousands of electro-mechanical relays requiring complex wiring and inter-connections.

(ii) With the development of fault tolerant and fail safe techniques, electronics and particularly microprocessors have found acceptance in the area of railway signalling the worldover. Railways in advanced countries have gone in for large-scale introduction of microprocessor-based solid state interlocking (SSI) system. SSI system occupies considerably less space, consumes less power, is reliable, and is easy to install and maintain.

(iii) Seeing the advantages of SSI system over conventional relay-based interlocking system, Indian Railways has taken up an ambitious programme to introduce SSI systems progressively through replacement works.

(iv) The first solid state interlocking on Indian Railways was commisioned in parallel mode on 25 July 1987 at Srirangam, a temple city in south India. After gaining confidence in the functioning of the system, SSI was finally put to use in 1989.

SSI system is now being progressionaly used on Indian Railway as a part of new technological development.

Centralized train control

The operation of all the points and signals of the various stations of a section is centralized at one place in such a system. Thus all the points and signals are controlled by a single official called the centralized train control (CTC) operator. A CTC operator virtually takes over the work of the station masters of several individual stations and operates all the points and signals at a station through remote control.

The CTC panel is normally provided at a central location and controls various stations upto a distance of about 120 km on either side. There is a separate panel provided for the operator, which depicts the entire section, including the points, crossings, and signals. The signals, routes, points, etc., are operated from the panel by means of separate knobs. This panel also depicts whether the various tracks are occupied or otherwise.

In a CTC system, panel interlocking is provided at all stations, which ensures complete safety. The CTC operator sends commands to the station equipment in the form of coded electric pulses by pressing the relevant buttons. The station equipment receives these commands, and sets the points to the desired position, and clears the appropriate signals. After the task is completed, indication signals are automatically sent back to the CTC panel in the form of coded electric pulses and the positions of the points and signals are indicated on the panel. In the route interlocking system, instructions regarding the running of trains, arrangement of crossings, etc., are issued by the control office through phone calls placed to various station masters and the actual control of the movement of the trains between the stations is exercised by the station masters. In the CTC system, all the functions of the controller and the station masters are carried out by the CTC operator, who is always aware of the position of all the trains in the section through the illuminated panel and who can remotely operate the various signals and points at all the stations. He or she can, therefore, make judicious plans regarding how the trains will move forward and how they will cross each other. Moreover, the automatic block system is always adopted in conjunction with CTC with the result that the number of trains in a block section can also be increased. The major advantages of the CTC system are enumerated below.

(a) There is considerable saving in the amount of time taken by trains to complete a run and as such the line capacity of the section is increased. In fact, with the introduction of CTC the necessity of doubling the track can be overlooked.

(b) No trained station masters, points men, etc., are required at the various stations. The CTC operator does all the work from the central panel. Thus, there is considerable reduction in the number of skilled staff members required.

(c) The system has the potential to detect any defects in the track.

Train protection and warning system

Train protection and warning system (TPWS) was mandated in the UK under the Railway Safety Regulations after 'Ladbroke Grove accident' in 1999 where 31 people died. As per this regulation all passenger train lines must be equipped with a train protection system.

The system automatically applies a train's brakes if it approaches a signal fitted with TPWS at danger too fast, or if it fails to stop at a signal at danger, or if it is travelling too fast on the approach to certain speed restrictions and buffer stops that are fitted with TPWS. It therefore reduces the risk of derailment and collisions between trains and of derailment through overspeeding.

The Railway Safety Review Committee headed by Justice H.R. Khanna has also recommended the use of TPWS for improving safety on Railways.

System description of TWPS

- TPWS is a spot transmission-based train control system to be used as an overlay on the existing signalling system. (Fig. 31.22)

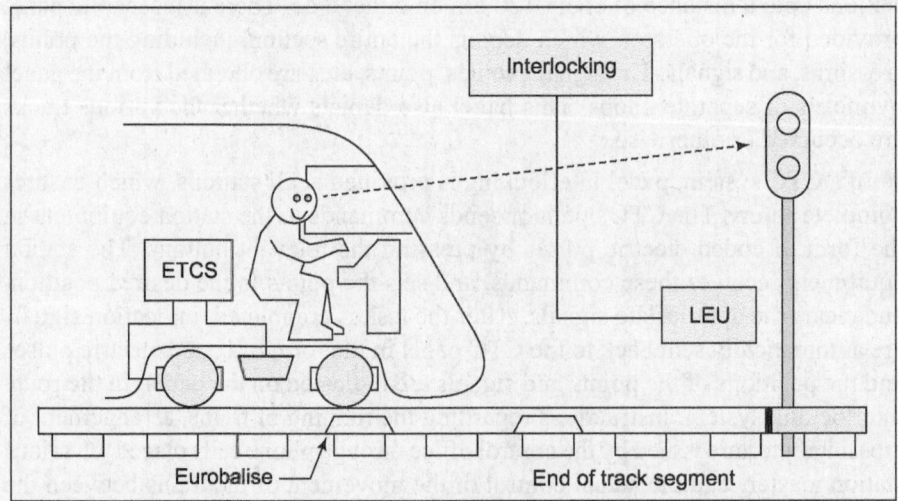

Fig. 31.22 A schematic diagram of train protection and warning system (TPWS)

- Movement authorities are generated trackside and are transmitted to the train via Eurobalises.
- TPWS provides a continuous speed supervision system, which also protects against overrun of the authority.

System of working of TPWS

The system comprising (i) track side subsystem and (ii) computer-based system evaluator with an indication panel and a tacometer, installed in loco. The passive track device, electromagnet of automatic warning system (AWS) is replaced by the Blias, a sealed microprocessor chip mounted on the sleepers. This track device is an interface between the interlocking system and the loco computer.

The computer works on two out of three selections. It evaluates the information gathered from the track device and odometer. A warning is given to the loco pilot to enable him to react within five seconds of the warning. If the loco pilot does not react in time, and the train exceeds more than 10 kilometres of permissible speed the service brakes are applied till the train is brought down to the permitted speed and if the train is not likely to stop at the given point emergency brakes are applied.

The system is for complete control of the train by monitoring (i) speed of the train, (ii) signal indications in the loco cab, (iii) automatic application of brakes when required and application of temporary speed restrictions en-route.

Last vehicle check device

One of the important features of the absolute block system is that a train should not be allowed to enter the block section unless the last train has arrived complete with the last vehicle at the station from either end of the block section. This must be verified by the operator responsible for operating the block instrument by observing the last vehicle board or last vehicle light. It has been observed that in quite a few cases the operator has ignored in certifying the last vehicle. To eliminate human error, a device known as the *last vehicle check device* has been developed. This consists of a passive equipment that the guard hangs on the last vehicle of the train. Each station is provided with a corresponding active equipment that emits waves of a predetermined frequency. This equipment is fitted at the entrance of each station. The coupling of the instruments takes place whenever the last vehicle passes by the active equipment, which is sensed by the last vehicle check device (LVCD). The operator receives an indication of the same and can now close the block instrument and issue a fresh line clear. The last vehicle check device eliminates human error in ensuring the safe arrival of the complete train.

Block proving by axle counters

- This is a device to ensure that full train has passed and the block section is clear.
- Block proving through axle counter is a system (Fig. 31.23) which provides automatic clearance of block section, by replacing conventional block instrument by a block panel. The in-count of axles at advance starter signal of the sending end station and out-count at home signal of receiving end station provides a check of complete arrival of a train. A zero resultant count indicates by inference that the block section is clear automatically.

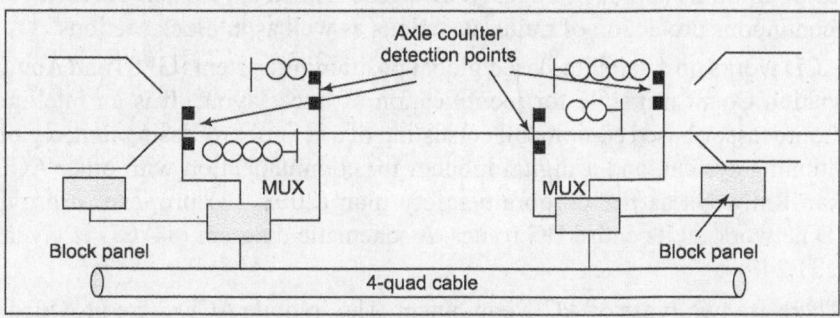

Fig. 31.23 Axle counter (schematic plan)

- The device helps in preventing collisions in mid section as the information whether the section is clear or occupied or part-train is known in advance.

- It has been experienced that provision of block proving by axle counter not only enhances safety, but also increases line capacity.

Warning system

Automatic warning system (AWS) is a technical aid which provides audio–visual warning to the driver and prevents him from passing a signal at danger. This system is sometimes also called as 'auxiliary warning system.

In the automatic warning system (AWS), the signal aspect is conveyed to the driver's cabin in the locomotive through electro magnets in the track. The electronic message is received by the antenna at the loco bottom. The signal aspect is thus available to the driver in his cabin. The train brakes are applied atuomatically if the driver fails to control the train based on the signal aspect.

Currently, AWS of 'Siemen's design' is installed in Mumbai suburban area of Western and Central Railway. A trial is being done on Delhi–Mathura section where the signal aspect will be conveyed to the driver's cabin through radio waves. Depending upon the way the information is transferred between the cab and the track, AWS can be classified into the following types:

Intermittent Here the information is provided only at designated places (signals). Information update is also done at these places only. Thus train has to travel on the basis of previous data until the next input.

Continuous In this type, the system updates information to the train on a continuous basis. Any change/deviation from the schedule can be updated immediately. This also includes any change in the running speed and alteration in the route ahead.

Anti-collision device (ACD)

To achieve the target of enhanced safety, anti-collision device (ACD) has been developed indigenously to prevent various types of collisions, (e.g., head-on collisions, side and rear-end collisions) including accidents caused due to infringement by derailed vehicles on adjoining tracks. This device also helps in detecting train parting, and provides audible and visual warning at level crossing gates when trains approach them. ACD is also named as 'Raksha Kavach'. It is for continuous protection of trains at stations as well as in block sections.

ACD works on a satellite-based global positioning system (GPS) and Angular Deviation Count principle for identification of track layout. It is an intelligent microprocessor-based equipment; consisting of a central processing unit, a global positioning system, and a digital modem for communication with other ACDs. Indian Railways as per corporate safety plan (2003–13) propose to provide ACD network on its entire BG routes. A schematic diagram of ACD is given in Fig. 31.24.

There are two types of ACD equipment. The 'mobile ACDs' are provided on locomotives and brake-vans and 'stationary ACDs' for stations and level crossing gates. All the ACDs interact with each other and exchange information within their radio zones upto 3 km. ACD interactions lead to a decision whether the loco ACD should apply brakes or not.

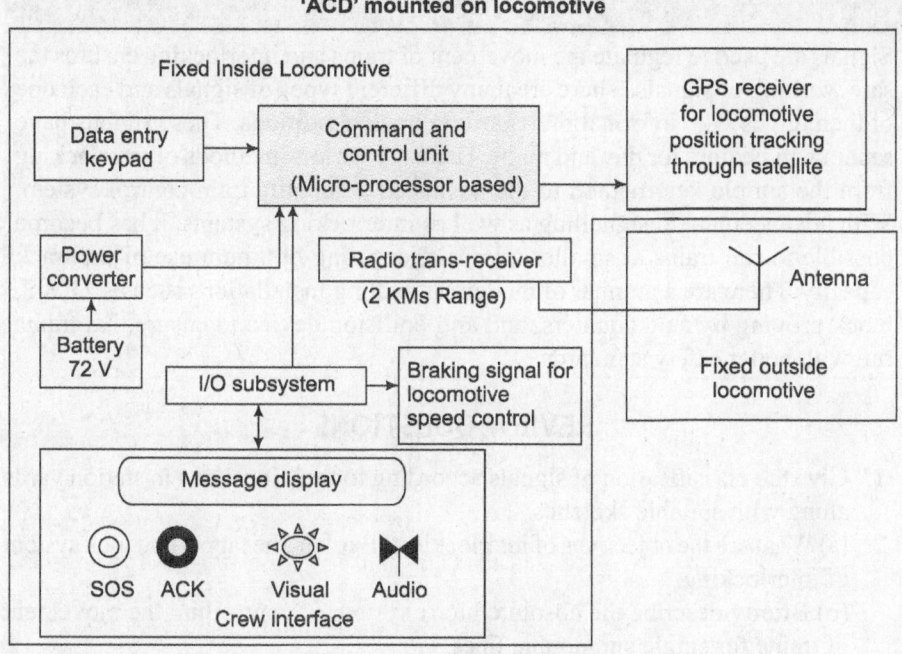

Fig. 31.24 Anti-collision device-block schematic

Functioning of ACD It functions by (i) detecting situations when two trains are not maintaining a given minimum distance creating situation for collision. Under such a situation (ii) it applies brakes, (iii) Door Dristi (Distant vision) of 3 kilometres under all weather conditions (iv) Covering loco pilots timely inaction and (v) Extending to level crossing gates. The system is provided in Locos, SLRs for the guard, level crossing gates, and stations.

Use of ACD The anti-collision device will be useful in preventing the following types of accidents:

- Two trains colliding with each other—head-on-collision or side collision and rear-end collision
- Parting of trains and rolling back
- Collision between trains and road vehicles at manned (non-interlocked) as well as unmanned level crossing gates
- Mishaps occurring due to train crew becoming inactive for some time due to any reason

Current use of ACD The anti-collision device (ACD) is still under trial. The device is not fool-proof and some improvement is being made. A pilot project of ACD to prevent cases of collisions has been used and is being tried with modified ACD for about 10,000 km on different railway systems. Once the fool-proof system is developed, ACD is likely to be extended in a big way over Indian Railways.

SUMMARY

Signals are used to regulate the movement of trains and interlocking ensures the safe working of signals. There are many different types of signals and each one of them gives vital information regarding track conditions. These signals have separate indicators for day and night. There are various methods of interlocking from the simple key method to the advanced automatic train control system. With advancement in signalling as well as interlocking systems, it has become possible to run trains at smaller intervals, ensuring optimum use of the track capacity. There are a number of modern signalling installations such as TPWS, block proving by axle counters, and anti-collision device to ensure that trains run with better safety standards.

REVIEW QUESTIONS

1. Give the classification of signals according to their locations in station yards along with suitable sketches.
2. (a) What are the objectives of interlocking? Explain the tappet and lock system of interlocking.
   (b) Briefly describe the absolute block system of controlling the movement of trains for single and double lines.
3. What essential purposes are served by signalling and interlocking? What is meant by route relay interlocking?
4. (a) Briefly describe the locations and purposes of the following signals:
   (a) warner, (b) outer, (c) home, (d) starter, (e) advance starter
   (b) What are the minimum signal requirements in each direction for a two-aspect signalling system in class A, B, and C stations?
   (c) What are the essentials of interlocking? Distinguish between direct and indirect interlocking.
5. (a) Explain the technique of interlocking in a railway system.
   (b) What purposes does the lock bar serve?
   (c) Distinguish between the following pairs of terms.
       (i) Calling on-signal and co-acting signals
       (ii) Warner and distant signals
6. Write a short note on:
   (a) TPWS   (b) AWS
   (c) ACD

Choose the correct answer from the choices given.
7. In standard II interlocking, the maximum speed of the train is:
   (a) 50 kmph   (b) 80 kmph
   (c) 100 kmph   (d) 120 kmph
8. In double-wire signalling the range of operation for signals is:
   (a) 200 m   (b) 350 m
   (c) 800 m   (d) 1200 m
9. In upper-quadrant signalling, the signals can be:
   (a) two position   (b) three position
   (c) four position   (d) none of these

10. In standard III interlocking, the interlocking between points and signals shall be by:
    - (a) mechanical measures
    - (b) electrical measures only
    - (c) electronic measures only
    - (d) electrical or electronic measures
11. Permissive signal is:
    - (a) stop signal
    - (b) warner signal
    - (c) distant signal
    - (d) warner and distant signal
12. The end of warner signal is:
    - (a) square
    - (b) angular
    - (c) V shaped
    - (d) circular

CHAPTER
32

Modernization of Railways and High-Speed Trains

INTRODUCTION

Indian Railways, in keeping pace with the advanced railways of the world, has been modernizing its railway system for quite some time. The maximum permissible speed on the BG sections of Indian Railways was increased to 120 km per hour by first introducing it on the Delhi–Howrah route in 1969. This increase in speed was possible after carrying out extensive investigations and trials on the Rajdhani route using a WDM-4 locomotive and all-coiled coaches. The study was based on the fundamental concept that safety and comfort at high speeds depend upon the interaction of the track with the vehicle. If the suspension system of the rolling stock is very good, then even though track maintenance may be of a comparatively average quality, a reasonable level of comfort will still be achieved. To a limited extent, track stability can be economically introduced on Indian Railways without carrying out major changes in the track structure, by simply selecting better locomotives and rolling stocks and adhering to a slightly higher standard of track maintenance. It was for this reason that the Rajdhani Express was hauled by a WDM-4 locomotive and included all-coiled coaches. The speed of the train, which was originally 120 km per hour, has been increased to 140 km per hour and Indian Railways is now planning to increase the speed to 160 km per hour.

Similarly, in the case of MG, the long existent limit of 75 km per hour has finally been overcome and trains consisting of a YDM-4 locomotive and all-coiled Integral Coach Factory (ICF) coaches now run at speeds of up to 100 km per hour since December 1997.

32.1 MODERNIZATION OF RAILWAYS

Railways are modernized with the objective of allowing heavier trains to run safely and economically at faster speeds, of improving productivity, and of providing better customer service to rail users. This consists of upgrading the track, use of better designed rolling stock, adopting a superior form of traction, better signalling and telecommunication arrangements, and using other modern techniques in the various operations of a railways system. A railway track is modernized by incorporating the following features in the track.

(a) Use of heavier rail sections, such as 52 kg/m and 60 kg/m and the use of wear-resistant rails for heavily used sections so as to increase the life of the rails

(b) Use of curved switches of 1-in-16 and 1-in-20 type for smoother arrival at yards

(c) Use of prestressed concrete sleepers and elastic fastenings, such as Pandrol clips to provide resilience to the track and ensure the smooth movement of trains at high speeds

(d) Use of long-welded rails and switch expansion joints to ensure a smooth and fast rail journey

(e) Modernization of track maintenance methods to include mechanized maintenance, measured shovel packings, etc., in order to ensure better track geometry, facilitate high speeds and smooth travel

(f) Track monitoring using the Amsler car, portable accelerometer, Hallade track recorder, etc., to assess the standards of track maintenance and plan for better maintenance, if required

Other aspects of modernization of the railways generally include making the following provisions.

(a) Use of better-designed all-coiled, anti-telescope ICF coaches with better spring arrangements and better braking systems for a safe and smoother rail travel

(b) Provisions of universal couples to ensure uniformity in the coupling of the coaches

(c) Introduction of diesel and electric traction in order to haul heavier loads at faster speeds

(d) Introduction of modern signalling techniques to enable trains to move at high speeds without any risks

(e) Setting up of a management information system for monitoring and moving freight traffic in order to avoid idle time and increase productivity

(f) Computerization of the train reservation system to avoid human error and provide better customer service for reservation of berths

(g) Use of computers and other modern management techniques to design and maintain railway assets more efficiently and economically, to ensure efficient human resource development (HRD) to increase productivity, and to provide better customer service

32.2 EFFECT OF HIGH-SPEED TRACK

Investigations carried out in connection with high-speed trains have revealed that an increase in speed does not necessarily result in a corresponding increase in the deformation and stresses in track components, which necessitates the use of a heavier track structure. The loads, deformations, and stresses in the track components were found to be augmented as a result of the incongruous movement of vehicles on the track including pitching, rolling, bouncing, etc., which occur when the track is poorly maintained. Therefore, it is possible to operate the same vehicles at a higher speed on a given track structure without imposing any additional loads and stresses, provided that the standards of maintenance of the track and the vehicles are sufficiently improved so as to control these inhibiting movements of the vehicle when it runs at higher speeds. The existing track structure on the Rajdhani route is considered to be of adequate standard for speeds reaching as high as 120 to 140 km per hour.

To achieve still higher speeds of the order of 160 to 200 km per hour, the standard of maintenance needs to be very high, as very close track tolerances will have to be maintained. Maintaining the existing tracks at such tolerance limits may be uneconomical and may necessitate the adoption of an improved track structure, which can be maintained at closer tolerance limits at a comparatively low cost. The modern track structure, consisting of long-welded rails with concrete sleepers, elastic fastenings, and ballastless tracks may well fulfil this requirement. The cost of this modern track may be comparatively high, but its maintenance will involve limited expenditure.

32.3 VEHICLE PERFORMANCE ON TRACK

When judging the performance of a high-speed vehicle on the track, it is ascertained that the following requirements are fulfilled.

(a) At the defective locations in the track, the variations in the vertical and lateral wheel loads should not reach a level where the wheels can get derailed because of the incidence of mounting.

(b) At defective portions of the track, the variations in the vertical and lateral wheel loads should not reach a level where they can lead to derailment as a result of the distortion to the track.

(c) It should be ensured that the passengers are generally comfortable while the vehicle plies on the track and that goods are carried without any damage.

(d) In the case of diesel and electric locomotives, it is specified that a lateral force lasting more than 2 m should not normally exceed 40 per cent of the axle load plus 2 tonnes. The values of acceleration recorded in the cab at locations should, as far as possible, be limited to $0.3g$, both in the vertical and the lateral direction. A peak value of $0.35g$ may be permitted in case the records do not indicate a significant tendency of this value to be repeated. The ride index is calculated for the purpose of comfort analysis, based on the lateral and vertical acceleration values obtained from acceleration output graphs. The ride index should normally not be greater than 4; a value of 3.75 is preferred.

In the case of carriages, the values of horizontal and lateral accelerations should generally not increase beyond $0.3g$, and an occasional value of $0.35g$ is permissible, as stipulated in the case of locomotives. The value of the ride index should not increase to 3.5, though a value of 3.25 is preferred.

32.4 HIGH-SPEED GROUND TRANSPORTATION SYSTEM

The high-speed rail ground transportation (HSGT) system is a concept developed to meet the challenges of the increasing demands of passenger transportation and recover the share of traffic from road and other modes of transport.

32.4.1 What are High-speed (HS) Trains?

The International Union of Railways (UIC) defines a high-speed train as one that runs at over 250 km per hour on dedicated tracks or at over 200 km per hour on upgraded conventional tracks. A 'high-speed line' is thus a new line designed to permit trains to operate at speeds above 250 km per hour throughout the whole journey, or at least over a significant part of the journey. Alternatively, it could also be an upgraded conventional line, suitable for carrying traffic at speeds above 200 km per hour.

32.4.2 Need for High-speed on Indian Railways

Development of high-speed tracks in India has become inevitable due to ever-increasing demand of traffic and faster services. However, proper planning should be done and routes identified so that concerted effort is made on these routes for improvement of track and track geometry by proper allocation of resources. Work to be undertaken should be properly planned and executed to the required tolerances, so that there are minimum constraints after the introduction of high-speed trains. For speeds in the range of 200 km per hour and higher, it is necessary that alternative routes are decided and new construction taken up conforming to the highest standards.

32.4.3 Introduction of Shatabdi Express

The fastest train on Indian Railways with a maximum speed of 140 km per hour, was introduced in July 1988 on the New Delhi–Jhansi–Bhopal section. The train can run on the following maximum speeds.

Section	Max. permissible speed
New Delhi–Faridabad ...	130 kmph
Faridabad–Agra Cantt ...	150 kmph
Agra Cantt–Jhansi/Bhopal ...	130 kmph

RDSO had carried out a detailed study of the track conditions on New Delhi–Agra–Jhansi section and also of the existing locomotives and coaches on use on Indian Railways. Oscillation trials were also carried out at a maximum speed of upto 160 km per hour and the following standards have been laid:

Type of loco and coaches This includes single WAP QC electric locomotive fixed with modified MK II bogies hauling ICF all coiled air braked EOC type coaches. LHB coaches have been introduced in Rajdhani and Shatabdi trains.

Track standards

> *Rails and sleepers* Minimum 52 kg rails, PRC sleepers, M + 7 sleeper density, 250 mm ballast cushion (minimum 10 mm clean ballast), compacted and stable formation.
>
> *Fastenings* This includes key fastening and CST-9 sleepers permitted provisionally for speeds upto 130 km per hour and temporarily permitted on Delhi–Agra section for speeds upto 140 km per hour.
>
> *Cant* The maximum cant deficiency 100 mm and the rate of change of cant or cant deficiency is 5 mm/sec.

Bridges These should be to broad gauge main line loadings and standards.

Traction installations   The OHE should be provided with swivelling type of cantilever assembly, having tension in the conductors regulated automatically with a pre-sag of 100 mm and a design speed potential of 160 km per hour. The movement of trains should be so adjusted that the catenary's current in any feeding section does not exceed 600 amperes for the reason of safety.

Signalling   The distant signals should be visible from a distance of at least 800 m under the worst visibility condition. The run through lines should be track circuited within a block section limit.

32.4.4 Introduction of High-speed Trains in India

Indian Railways has been planning to introduce high-speed systems and the Ministry of Railways has decided to develop high-speed corridors for 250 to 350 km per hour speeds. Six corridors have been identified. The six corridors earmarked for high-speed routes, where pre-feasibility studies are being conducted, are presented in Table 32.1.

Table 32.1   Six corridors marked for high-speed routes

Name of corridor	Approx length (km)
Delhi–Chandigarh–Amritsar	450
Pune–Mumbai–Ahmedabad	650
Hyderabad–Dornakal–Vijaywada–Chennai	664
Chennai–Bangalore–Coimbatore–Ernakulam	649
Howrah–Haldia	135
Delhi–Agra–Lucknow–Varanasi–Patna	991
	3539

Technologies for high-speed trains

There are a number of options for introducing high-speed trains.

Upgradation of existing railway lines   The existing infrastructure could be upgraded so that the railway line is made suitable for speeds of 200 km per hour and above. In this case, well-designed track, proper rolling stock, and appropriate signalling equipment have to be provided. The existing bottlenecks have to be removed, particularly for track geometry.

Construction of dedicated high-speed corridors   For this separate high-speed corridors will need to be designed and constructed. For this purpose, high-tech infrastructure, after laying down the desired standards, will have to be adopted for smooth, efficient, and safe operation at designed speeds of 250 km per hour and above.

32.4.5 High-speed Trains as per Railway Vision 2020

Railway Vision 2020 envisages the development of Indian Railways on the lines of the railways of developed countries in the world. The current effort to provide fast non-stop train services under the new brand of 'Duronto' will continue. In addition, the Vision aims at increasing the speed of regular passenger trains to 160–200 km per hour on segregated routes, which will bring about a major transformation in train travel. Vision 2020 also envisages the implementation of at least four high -speed rail projects to provide bullet train services at 250–350 km per hour, one in each of the regions of the nation and planning eight additional corridors connecting commercial, tourist, and pilgrimage hubs. In short four high-speed corridors of about 2000 km are planned to be completed by 2020 and it is proposed to plan the development of eight such other corridors.

32.5 HIGH-SPEED RAILWAYS IN WORLD SCENARIO

The construction of the first high-speed railway (HSR) was taken up in Japan in 1959 and 'Bullet trains' (Tokaido Shinkansen section) introduced in October 1964 at speeds of 210 km per hour. This revolutionary concept of high-speed railway was introduced in Europe in the 1980s and today high-speed trains operate on more than 18000 km of tracks in most of the advanced railways of the world, at speeds varying between 250 to 350 km per hour. A high-speed railway requires appropriate infrastructure, having specially-designed track, new rolling stock, and advanced signalling concepts along with other ancillary facilities.

The length of high-speed railway lines in various countries in the world operating at a speed exceeding 250 km per hour is given in Table 32.2.

Table 32.2 Important high-speed lines in the world (speeds exceeding 250 kmph)

Country	Length of high-speed lines (km)	Details of important fast trains
Japan	2678	'NOZOMI': Max. speed limit 300 kmph
France	1893	'TGV': Max. speed 320 kmph
Germany	1300	'ICE': Max. speed 300 kmph
Spain	1687	'AVE': Max. speed 300 kmph
Portugal	3320	
China	3120	High-speed C Class: Max. speed 250 kmph
Belgium	830	Thayls: Max. speed 300 kmph
Italy	890	Eurostar: Max. speed 250 kmph
USA	360	Acela Express: Max. speed 250 kmph
South Korea	412	KTX: Max. speed 300 kmph
Taiwan	245	Chiayi: Max. speed 300 kmph
World Rlys	17892	Max. speed varying 250–350 kmph

The French Railways holds the world train speed record of 574.8 km per hour, which was set in April 2007. Recently, a Japanese magnetic leviation (maglev) train achieved a speed of 581 km per hour in trials.

32.5.1 Infrastructure Requirement of High-speed Railway

A number of complex issues have to be addressed for providing sound infrastructure for high-speed trains:

Track This includes specially-designed track with properly compacted formation and appropriate track geometry.

Bridges and tunnels These include advanced designed bridges and tunnels with special emphasis on approaches.

Rolling stock This includes dedicated coaching stock and high-speed locomotives (diesel and electric).

Signalling This includes appropriate signalling technology and concepts.

Miscellaneous issues These include grade separation, fencing, and environmental protection issues.

32.5.2 Track Structure for High-speed Routes

Rails For high-speed routes, 60 kg rails are adopted by the railways world over. Standard length of 25 m in Japan, 54 m/62 m in Germany, and 108 m in France has been utilized. Continuous-welded rail (CWR) is used to improve the ride quality and to reduce noise and vibrations.

Sleepers Pre-stressed concrete sleepers have been a better choice as they have a long life of 50 to 60 years. Sleeper density of 1660 is being used on Indian Railways and could also be adequate for high-speed route as this is the maximum density to carryout machine maintenance.

Fastenings Double elastic rail fastenings are necessary for the concrete sleeper track. Rubber pads are used as cushioning material between the rail and sleepers fastened by leaf spring/wire spring/TGV Nabla/ICE Vossloh fittings for distribution of vertical load and for dampening the vibrations.

Track structure used on World railways for high-speed routes is given in Table 32.3.

Table 32.3 Track structure for high-speed railways in the world

Component	SNCF-French Railways	German Railways	Japanese Railways
Gauge	1435 mm	1435 mm	1435 mm
Rails	UIC 54 and 60 kg	UIC 60 kg	UIC 60 kg
Rail cant	1:40	1:40	1:40
Sleeper	Concrete/wooden	PSC/polyurethane foam/glass fibre	PSC/polyurethane foam/glass fibre
Sleeper density	1666	1724	1724
Fastenings	TGV Nabla/ ICE Vossloh	Leaf spring/ wire spring	Leaf springs/ ICE Vossloh

Curve for high-speed trains Flat curves are generally adopted on a high-speed track. Flat curves become necessary in view of the restriction on maximum values of cant deficiency and cant excess along with the maximum speed of operation. The minimum radius of curvature for high-speed lines generally varies from 4000 m to 7000 m for a standard gauge. Geometric parameters of the track for various high-speed railway on world railways are presented in Table 32.4.

Table 32.4 Geometric parameters of the track for various high-speed railway on world railways

Country	France		Germany		Spain		Belgium
Design speed	300	350	300	350	300	350	300
Min R of curvature (m)	4000	6250	3350	5120	4000	6500	4800
Max. cant (mm)	180	180	170	170	150	150	150

(Contd.)

Table 32.4 *(Contd.)*

Country	France		Germany		Spain		Belgium
Design speed	300	350	300	350	300	350	300
Cant deficiency (mm)	85	85	130	112	100	65	100
Max. cant gradient	35	35	40	40	12.5	25	15–21
Min. vertical radius (m)	16000	21000	14000	20000	24000	25000	20000
Transition curve length (m)	300	350	408	476	360	460	420

Level crossing/grade separation   Normally, level crossing is not suitable for high-speed train operation and therefore, for road transport, either road overbridges or road underbridges need to be planned. However, in unavoidable circumstances, level crossings may be required. Then it must be interlocked with the signals. Sophisticated arrangement of interlocking the signals of the train with that of road transport with the help of video camera is used.

Fencing   On high-speed lines, trespassing is very risky and thus not at all permitted. Therefore, the entire high-speed track is to be provided with fencing. It is noticed from the experience of high-speed corridors world around that at very high speeds, track ballast stones sometimes fly off and hit the surroundings. To avoid such incidences also track fencing is required.

32.6 TILTING TRAINS FOR HIGH-SPEED ROUTES

To overcome the limitation of speed on account of tight curves on mixed traffic routes, where it is not possible to cant the track, vehicles with a tilting suspension system having tilting mechanisms can be used. Trains that tilt, can go up to 25 per cent to 40 per cent faster around curves than conventional trains without upsetting the passengers and this can significantly increase the speed on existing lines. (Fig 32.1) Depending on the curvature and other parameters, the train tilts on the curve and gives additional superelevation to the passengers and thus they experience less cant deficiency and more comfort. With tilting trains, cant deficiency of up to 275 mm is permitted on a standard gauge. Considering a 275 mm cant deficiency and a 180 mm Ca for Indian Railways, the radius will workout to be less than the radius requirements with conventional trains. Tilting technology is used normally for speed exceeding 350 km per hour.

Fig. 32.1   Tilting trains for high speeds

32.7 MAGLEV TRAINS

High-speed trains have speed limitations, as rail–wheel contact becomes difficult to maintain with increase in speed. Developments using the concepts of linear motor and magnetic levitation, have made commercial maglev trains a practical proposition. The concept plan of Maglev guideway trains and its energization system is given in Fig. 32.2.

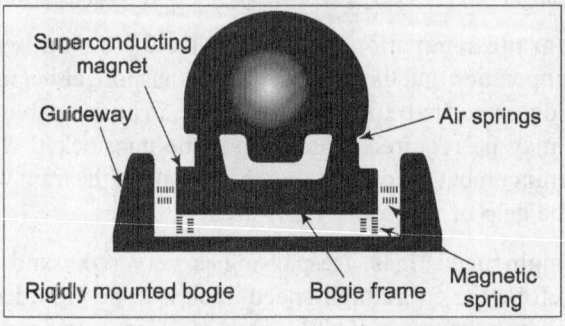

Fig. 32.2   Rigidly mounted bogie of a Maglev train

Brief details of Maglev trains

Maglev trains are more akin in operation to aeroplanes, than to conventional railways. These trains are provided with retractable landing wheels and horizontal guide wheels, which come into operation whenever the speed comes down below 200 km per hour, the minimum to achieve levitation.

The trains are propelled using linear motor power; the three phase coils forming the starter of the linear motor are installed in the sidewalls of the guideway.

The speed of the train would be in the range of 450 to 550 km per hour and this is controlled by varying the frequency of the power passing through the coils.

The entry to the carriages is from the top; thus overhead platforms will be built for passengers to board the train.

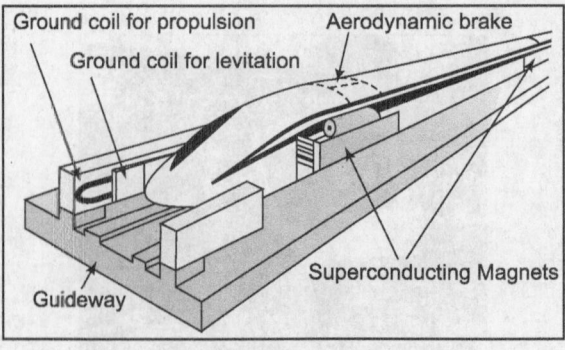

Fig. 32.3   Maglev train

Technology of Maglev trains

When electric current is passed through propulsion coils on the ground, a magnetic field (north and south poles) is produced. The train is propelled forward by the attractive forces of the opposite poles and the repulsive forces of the similar poles, acting between the ground coils and the superconducting magnets built into the vehicles, based on the linear motor concept.

Advantages of Maglev trains   These are as follows:
- Floats 10 mm above the guideway
- Propelled by guideway
- Low maintenance cost
- Low noise or high speeds
- High speeds upto 500 km per hour

One disadvantage of Maglev trains is the high corridor cost.

World record of 502 km per hour   The Yamanashi Maglev test line is 18.4 km long (16.0 km tunnel + 2.4 km open section) having double line with maximum grade of 40 per thousand and minimum curve radius of 8000 m. In such a short length, it achieved the world speed record of 502 km per hour on 24 December 1997.

Super Maglev trains at 3700 km per hour speed

 (i) With the Maglev technology being on the anvil, new ideas and concepts are coming forward and by 2050 supersonic trains appear probable.
 (ii) The concept envisages enclosing the Maglev train in a tube from which air will be evacuated, thus creating a system without aerodynamic resistance. The power requirements to both levitate and propel the train are minimized using superconducting magnets.
(iii) Using these two engineering concepts of vacuum tube and superconductivity. dedicated tubes for each direction could be assembled about 50 m (below the sea bed or to surface floats. Cost of laying such a tube will be about 4 million UK pounds per km.
(iv) The proposed planning envisages unified acceleration for the first six minutes of the route and about 15 minutes for deceleration to the terminus, and the remainder of the journey could be at a speed of 3700 km per hour making a journey, say from the UK to the west coast of the US in less than two hours.

REVIEW QUESTIONS

1. Write short notes on the track requirements for high-speed trains.
2. Enumerate all the important works which may have to be undertaken for strengthening and improving an existing track so that higher speeds are permissible on it.
3. What are the various measures normally taken to improve the track to accommodate high speeds?
4. Write short notes on:
   (a) Tilting train   (b) Maglev trains
   (c) Super Maglev trains

Choose the correct answer from the choices given.

5. The maximum speed on Indian Railways was achieved at 120 kmph in the Delhi–Howrah route in:
   - (a) 1951
   - (b) 1960
   - (c) 1969
   - (d) 1970

6. The maximum permissible speed of the Shatabdi train is:
   - (a) 120 kmph
   - (b) 130 kmph
   - (c) 140 kmph
   - (d) 150 kmph

7. The world record of a high-speed train is about:
   - (a) 300 kmph
   - (b) 400 kmph
   - (c) 480 kmph
   - (d) 575 kmph

8. Most of the high-speed trains in the world run on:
   - (a) metre gauge (1000 mm)
   - (b) standard gauge (1435 mm)
   - (c) broad gauge (1676 mm)
   - (d) none of these

9. For a high-speed train the sleepers used are:
   - (a) wooden sleepers
   - (b) cast iron sleepers
   - (c) steel sleeper
   - (d) concrete sleepers

Dedicated Freight Corridor and Other New Developments

INTRODUCTION

Indian Railways has recently planned a new corridor called 'Dedicated Freight Corridor (DFC), which is meant for carrying only freight traffic at higher speeds and increased axle loads. There have been many other new developments on Indian Railways such as the following:

  (i) Dedicated Freight Corridor
 (ii) Konkan Railway Project
(iii) Kashmir Railway Valley Project
 (iv) Track Management System
  (v) Increase of axle loads, grinding of rails, ballastless track, and so on.
 (vi) High-speed railway and Metro Railways in India.

All these are discussed in detail in this chapter.

33.1 DEDICATED FREIGHT CORRIDOR

In order to meet the increasing demand of freight traffic, Indian Railways is planning to have specially dedicated new double lines to carry the increased traffic with higher axle load.

Dedicated freight corridors are planned for the entire Golden Quadrilateral and its two diagonals by laying two new parallel double lines exclusively for freight traffic, thereby making the existing system a passenger corridor. The approximate length of the six DFCs (four sides of the quadrilateral plus two diagonals) will be 11,500 km of double line (23,000 km of rail track).

The existing axle loads on the Indian Railways system are 20.3 tonnes and have recently been enhanced on some selected routes to 22.9 tonnes. DFCs are being designed to take axle loads of 30 tonnes and for the current 25 tonnes axle load wagons are proposed to be run for feeder routes.

33.1.1 Details of DFC Project

Indian Railways, in the first stage, has taken up only two projects. These are Western corridor of 1490 km and Eastern corridor of length 1805 km. Details of these corridors including funding arrangement and proposed year of completion are given in Table 33.1.

Table 33.1 Details of Western and Eastern corridors of DFC

Items of corridor	Sections covered	Year of completion	Finding
Phase I of Western Corridor	Rewari–Vadodara (920 km) D/L	2009–2016	JICA (Phase-I)
Phase II of Western Corridor	Vadodara–JNPT (430 km) D/L	2010–2017	JICA (Phase-II)
Phase III of Western Corridor	Rewari–Dadri (140 km) D/L	2010–2017	JICA (Phase-III)
Total of Western corridor	Total length 1490 km	Work to be completed by 2017	
Phase I-APL1 of Eastern Corridor	Khurja–Kanpur (343 km) D/L	2009–2016	World Bank
Phase II-APL 2 of Eastern Corridor	Kanpur–Mughalsarai (390 km) D/L	2010–2016	World Bank
Phase-III-APL 3 of Eastern Corridor	Khurja–Ludhiana (397 km) S/L	2011–2016	World Bank
Phase IV of Eastern Corridor	Dankunj–Sonnagar (550 km) D/L	2011–2016	On PPP mode
Phase I(a) of Eastern Corridor	Sonnagar–Mughalsarai (125 km) D/L	2010–2016	Ministry of Railways
Total of Eastern corridor	Total length 1805 km	Work to be completed by 2016	

Note: D/L = double line, S/L single line.

33.1.2 Main Advantages of Dedicated Freight Corridor

The construction of DFCs will greatly enhance capacity and mobility mainly due to the following:

(a) Slow moving goods trains (75 kmph/100 kmph) will not come in the way of passenger trains (100 kmph/150 kmph) nor the slow moving freight trains will have to wait for giving precedence to faster moving passenger trains. This will improve average speeds of both the passenger and the freight trains and the sectional capacity will markedly improve.

(b) Higher axle load wagons plying on DFCs will carry more load per train.

(c) Existing corridors (passenger corridors) will be relieved of heavier freight trains thus giving relief in maintenance especially to old existing bridges.

(d) Since there will be no level crossings, (complete grade separation) the passenger corridors could also run tilt body passenger trains upto speeds of

200 kmph, by suitably fencing the existing tracks and providing cab signalling. The Rajdhani which currently run, at a maximum speed of 140 kmph and takes about 16 hours for a journey between New Delhi and Mumbai and about 28 hours between New Delhi and Chennai may be able to cover these distances in about 11 and 18 hours respectively.

These main advantages can be summarized broadly as follows:

(i) To create rail infrastructure to carry high levels of freight traffic and thus, increase the share in freight market

(ii) To increase throughput by higher axle loads, better track loading density, and improved pay load/tare ratio

(iii) To speed up freight train operations, achieve higher productivity through better utilization of railway assets

(iv) To reduce the unit cost of transportation, inventory costs and greater customer satisfaction

(v) To relieve existing rail corridor for additional passenger traffic

33.1.3 Standard of Construction of DFC

The standard of construction of DFC is presented in Table 33.2

Table 33.2 Standard of construction of DFC

Gauge	1676 mm
Rails	60 kg 110 UTS, 20 rail panel (260 m) to be handled by mechanical track laying equipment.
Sleeper	PSC 1660 sleepers per km. for main line & 1540 per km. for loop line.
Points & crossings	60 kg rail with 1-in-12 curved switches and CMS crossings on PSC sleepers and thick web switches.
Ballast	300 mm cushion (Machine crushed) with current RDSO specification.
LWR/CWR/ Welding	• 20 rail panels are to be converted into LWR/CWRs with mobile flashbutt welding/gas pressure welding. • SKV welding should be avoided strictly. All in-situ welds to be joggle fishplated.
Gradient	• Flat territory: Mid section–1 in 400 or flatter (compensated) Yards–1 in 1200 • Semi Ghat territory: Mid section–1 in 200 compensated or Yards–1 in 1200 • Block section: at a convenient locaton gradient of 1 in 1200 to be provided for future crossing station
Curvature	• Flat territory: Maximum curvature–1°. • Semi Ghat territory: Maximum curvature–2°.
Formation	• Top width of embankement–7.5 m with 2:1 side slopes • Track centre–5.3 m • Complete embankment should invariably be provided with turfing
Cutting	• Cutting width including drains–11.0 m. Side slops to be designed depending on earth material. • Erosion, boulder fall, earth slips blocking the drain, etc. to be totally avoided

(Contd.)

Table 33.2   (*Contd.*)

Bridges	• Ballasted deck bridges with RCC slab/RCC box/PSC slab/PSC box girder. • To ensure high-quality concrete, use only ready mix concrete. Mobile ready mix plants can be planned which can be shifted at suitable interval. • Use only high-grade concrete with suitably designed admixtures to create economical structures.
Road crossings/level crossings	• As far as possible, there shall be no level crossing. • Complete length to be fenced on both sides.
Maximum speeds	100 kmph (freight train)
Type of traffic & axle load	25 tonne-double stack container movement with 15,000-tonne trailing loads: 30 tonne for bridges.

33.1.4 Track Maintenance

The standards of track maintenance on DFC are summarized below.

 (i) Fully mechanized maintenance of the track and structure is suggested to keep human resources at a minimum level.

 (ii) Dedicating a maintenance block of four hours that will be given daily in conversation with engineering department.

(iii) Through tamping of track and tamping of points and crossings will be done with CSM (Continuierliche Stopf Maschine, English version—continuous action tamping machine) and UNIMAT machine.UNIMAT stands for all Plasser tamping, levelling, living machines for points and crossings and plain track.

(iv) Ballast cleaning machine (BCM) will be deployed for deep screening once in ten years. There will be shoulder cleaning of ballast once in five years by the shoulder ballast cleaning machine (SBCM).

 (v) Isolated track defects will be attended by using lightweight offtrack tie tampers moved by the Rail Mounted Vehicle (RMV) similar to the tower wagon.

(vi) Use of RMV and lightweight motor trolley is recommended for quick and efficient attention of isolated track defects.

(vii) The track maintenance system can be divided into three tiers.

- **The top tier** will be the backbone of the maintenace system comprising CSM machine for tamping of plain track and UNIMAT machine for tamping points and crossings.
- **The middle tier** consists of mobile maintenance gang (MMG) units, which are responsible for tamping of isolated spots conventionally known as slack picking.
- **The bottom tier** comprises track maintenance and monitoring gangs under the sectional SE/JE.

33.1.5 Container Traffic

Container traffic will be growing faster than normal traffic. Accordingly, running of double stack container trains is contemplated on the DFCs and overhead clearances of structures such as road overbridges (ROB), foot overbridges (FOB), and tunnels, are being planned accordingly, also keeping in mind the aspects of electrification of the lines now or in future. For heavier traffic, electrification of routes prove to be a more economical proposition in addition to an overall reduction in the consumption of energy, vis-a-vis use of fossil fuels.

New wagons for better productivity

Most of the existing wagons on the Indian Railways have a tare-to-payload ratio of 1 : 2.7 and efforts are on to ensure an improved ratio of a tare-to-payload ratio of about 1 : 4 by having lighter steel alloys to enhance the carrying capacity of wagons for a given axle load.

The Basic Design Features of Proposed Dedicated Freight Corridor (DFC) vis-a-vis Existing Standards of Indian Railways are given in Table 33.3.

Table 33.3 Basic design features of proposed DFC versus the existing standards on Indian Railways

Basic features	Details of Indian Railways	Details of DFC routes
Moving dimensions (height × width)	4.265 m × 3.20 m	7.10 m × 3.66 m (For western corridor) 5.10 m × 3.66 m (for Eastern corridor)
Train length	700 m	700 m/1500 m
Train load	4000 tonnes	150000 tonnes
Axle load	22.9 tonnes/25 tonnes	32.5 tonnes /25 tonnes for track superstructure
Track loading density	8.67 tonnes/m	12 tonnes/m
Maximum speed	76 kmph	100 kmph
Grade	Up to 1 in 100	1 in 200
Curvature	Up to 10°	Up to 2.5°
Traction	Electrical (25 KV)	Electrical (2 × 25 KV)
Station spacing	7 – 10 km	40 km
Signalling	Absolute automatic with 1 km spacing	Automatic with 2 km spacing
Communication	Emergency sockets/ mobile train radio	Mobile train radio

33.1.6 Completion of Project

The completion of the project in a time schedule of ten years is likely to be a difficult and challenging task involving construction of about 2300 km of rail line every year. (Currently Indian Railways executes about 200 km of double-line projects in a year, which is the average of tenth year plan). It will be desirable to execute the DFC project through a separate organization in view of its importance and effect on the national economy.

It is proposed to complete the first stage of DFC project in a time span of about 10 years.

33.2 KONKAN RAILWAY LINE

Konkan Railway line runs along the Konkan coast of India. It was constructed and is operated by the 'Konkan Railway Corporation'. It runs from Mangalore in Karnataka to Roha in Maharashtra through Goa, along the west coast of India and the Western Ghats.

The Konkan is a coastal strip of land bouded by the Sahyadri hills on the east, and Arabian Sea on the west. It is a land with rich mineral resources, dense forest cover, and a landscape fringed with paddy, coconut, and mango trees. Konkan Railway as such has a difficult terrain and construction of a rail link in this difficult terrain was quite a challenging task.

The formidable terrain to be conquered and the short construction period meant that the project could only be completed with the help of several technological innovations. Some of these innovations/special technical features are given in the subsequent pages.

The Konkan Railway is the missing link between India's commercial capitals, Mumbai and Mangalore. The 760 km line connects Maharashtra, Goa, and Karnataka— a region of criss-crossing rivers, plunging valleys, and mountains that soar into the clouds. Konkan Railway is the biggest railway project undertaken in the Indian subcontinent in the twentieth century. In terms of sheer magnitude, technical challenges, and innovative financing, it has no parallel.

The Konkan Railway line is a single-line track, and is not electrified. Although it has been designed for high-speed traffic of 160 km per hour, the fastest train on the route, the Trivandrum Rajdhani Express, currently runs at a maximum speed of 130 km per hour. The route is open to both freight and passenger traffic. The line, which runs parallel to the Arabian Sea coastline, offers some of the most spectacular views of any Indian Rail journey. The Konkan railway route intersects the national highway NH 66 at many places.

There are fifty-six stations on the entire line. Although the route is currently a single line, Konkan Railway and South Western Railway lines run parallel from Majorda to Madgaon in Goa, making that section a double line.

Konkan Railway project was started immediately after the formation of Konkan Railway Corporation on 19 July, 1992. The project took about seven years to complete, as originally planned.

The railway line from Mumbai to Madgaon became operational on 26 Janauary 1988. An alignment plan of Konkan Railway is given in Fig. 33.1. Some the salient features of the Konkan Railway are given in Table 33.4.

33.2.1 Technical Challenges of Konkan Railway Line

Major challenge of Konkan railway was to limit the ruling gradient at 1:150 and to avoid sharp curves to allow higher speeds and heavy train loads. Trains running up to 160 km per hour could not meander up and down mountains; they had to cut right through the terrian. Hundreds of tunnels and bridges had, therefore, to be constructed.

Bridges, were constructed with meticulous planning and design inputs coupled with quick solutions to technical problems arising at the site. Similarly, tunnels were planned after extensive soil surveys. Engineers found themselves confronting with nine soft soil tunnels, where water gushed in and the mud caved in. The digging of the soft soil laterite tunnels was the project's most serious problem, and foreign experts who were brought could not help much. In fact, similar problems were recently experienced during tunnel construction on the west coast of Sweden.

Fig. 33.1 Map showing places traversed by Konkan rail (Source: Konkan Railway Technical Bulletin)

Table 33.4 Salient features of Konkan Railway

Item	*Detail*
Gauge	Broad gauge (1676 mm)
Route length	760 km
Ruling gradient	1:150 (Compensated) 0.67 %
Rails	52 kg 90 UTS (welded rails)
Sleepers	PSC mono-block sleepers
Number of curves	320
Number of major bridges	179 (linear waterway 21:50 km)
Number of minor bridges	1819 (linear waterway 5.73 km)
Number of road crossings	300 (Road overbridges/road underbridges)
Longest bridge	Across the river Sharavati in Honnavar (2065.8 m)
Tallest viaduct	Panval Nadi (64 m high)
Total number of stations	59
Signalling	Panel interlocking with colour light signal
Telecommunication	State-of-the-art optic fibre with digital communication

Other important features of technical innovations in the Konkan Railway line are given below:

Speed   This line is designed for an unrestricted speed of 160 kmph.

Track structure   This includes the following:

(i) Use of concrete sleepers on the entire route including points and crossings

(ii) Use of thick web tongue rails for switches of points and crossings to ensure smooth running of trains and minimize maintenance costs

(iii) Long welded rails on the entire route to ensure noiseless travel and minimum maintenance

Bridges and viaducts   This includes the following.

(i) Pre-casting of pre-stressed concrete bridge girders followed by launching

(ii) Adoption of hollow reinforced concrete piers for tall viaducts constructed by slip forming

(iii) Reinforced concrete framed configurations for viaducts

(iv) All bridges (except three spans across the navigational channels of the rivers Zuari and Mandovi) with concrete decks and ballasted track in order to maintain uniformity of track structure and reduce maintenance costs

(v) India's first incrementally launched continuous pre-stressed concrete box girder bridge at the tallest viaduct at Panval Nadi

(vi) First 124.2 m-span steel through girder with welded fabrication launched by a unique technique of lifting the 680 tonne girder by a floating crane and placing on piers to span across the navigational channel at Mandovi and Zuari bridges in Goa

Tunnels

(i) Ballastless track in long tunnels to ensure practically maintenance-free track and safe operation of trains

(ii) Forced ventilation of long tunnels to ensure desirable environment inside.

Signalling and telecommunications

(i) Panel interlocking and four-aspect colour light signalling to enhance operational efficiency

(ii) Optic fibre communication links along the alignment to facilitate communication at electronic speeds and in turn improve the working efficiency of the system during the operational phase

(iii) Computer network along the route to facilitate efficient train movements

33.2.2 Bridges

There are a total of 1998 bridges on Konkan Railway line consisting of 179 major bridges and 1819 minor bridges. All these bridges are in difficult terrain. The details of longest span of the bridges are given below.

(i) For concrete bridges : 53.5 m (PSC Box Girder)

(ii) For steel bridges : 124.2 m (Open web steel through girder)

(iii) Longest bridges : Across river Sharavati in Honnavar (2065.8 m)

(iv) Tallest viaduct : Panval Nadi (64 m high)

33.2.3 Tunnels

There are 92 tunnels aggregating to a total length of 83.6 km with nine tunnels being longer than 2 km. It was the first time that such massive tunnelling work was attempted for vehicular tunnels in India. Of these, 74 km was through hard rock, 8.4 km through soft soil, and the balance 1.2 km through cut and cover construction. Details of important tunnels are given in Table 33.5.

Table 33.5 Important tunnels of Konkan Railway project

Name of tunnel	Length of tunnel (m)	Cost of tunnel in milion ₹
Natuwadi	4389	71.4
Chiplun	2033	38.4
Sawarde	3404	62.7
Parchuri	2628	71.0
Karbude	6506	126.6
Tike	4077	60.2
Berdewadi	4000	57.0
Barcem	3343	55.3
Karwar	2950	47.9

33.2.4 Earthwork

The earthwork for Konkan Railway was a huge task. Some important details are as follows.

(i) Maximum height of an embankment : 25 m

(ii) Deepest cutting : 28 m

(iii) Total earthwork : 88.77 million m³

33.2.5 Benefit of Konkan Railway Line

1. The following are the benefits of Konkan Railway. There is considerable saving of distance and time as per details given below.

Table 33.6 Saving in distance and time on account of Konkan Railway Line

Item	Saving in distance (km)			Saving in travel time (Hrs)		
	Other routes	Konkan Railway	Savings	Other routes	Konkan Railway	Savings
Mangalore–Mumbai	2041	914	1127	41	15	26
Mangalore–Ahmedabad	2653	1358	1295	--	--	--
Mumbai–Cochin	--	--	--	36	24	12
Mumbai–Goa	--	--	--	20	10	10
Mangalore–Delhi	3033	2249	784	--	--	--
Cochin–Mumbai	1849	1336	513	--	--	--

2. Linkages of the Konkan ports with the rest of the country
3. Making possible the extraction of the region's rich, largely untapped mineral, agricultural, and educated manpower resources
4. Making possible the industrial investments that depend on an efficient transport facility.
5. Freeing of rail transport capacity on the north/west to south routes by diverting mail/express trains
6. The enhanced economic opportunities including increased employment resulting from better transport facilities

33.3 KASHMIR RAIL LINK PROJECT

With a view to provide an alternative and a reliable transportation system to Jammu and Kashmir, the government of India planned a 326 km-long railway line joining the Kashmir Valley with the Indian Railways network. This project has been declared as a national project.

The Jammu–Udhampur–Katra–Qazigund–Baramulla Railway line (JUSBRL) is the biggest project in the construction of a mountain railway since Independence. The length of this new rail line from Jammu to Baramela is 326 km.

An alignment plan of the Kashmir Rail link (JUSBRL) is given in Fig. 33.2.

The route crosses major earthquake zones, and is subjected to extreme temperatures of cold and hot, as well as inhospitable terrain, making it an extremely challenging engineering project. The railway line has been under construction since 1983 by various railway companies in India. The project has had a long and chequered history but good progress was made only after it was declared a national project in 2002. The scheduled date of completion was 15 August 2007. However, several unforeseen complications have pushed back the final completion deadline to 2017.

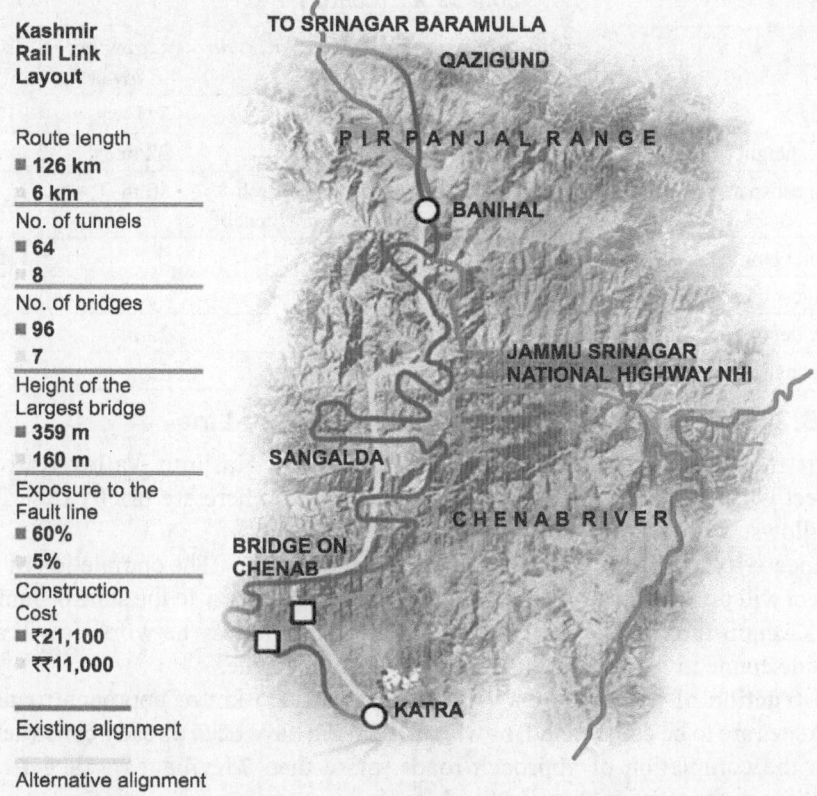

Kashmir
Rail Link
Layout

Route length
- **126 km**
- **6 km**

No. of tunnels
- **64**
- **8**

No. of bridges
- **96**
- **7**

Height of the
Largest bridge
- **359 m**
- **160 m**

Exposure to the
Fault line
- **60%**
- **5%**

Construction
Cost
- **₹21,100**
- **₹₹11,000**

Existing alignment

Alternative alignment

TO SRINAGAR BARAMULLA
QAZIGUND

PIR PANJAL RANGE

BANIHAL

JAMMU SRINAGAR
NATIONAL HIGHWAY NHI

SANGALDA

CHENAB RIVER

BRIDGE ON
CHENAB

KATRA

Fig. 33.2 Alignment plan of Jammu–Udhampur–Srinagar–Baramulla rail link
(*Source:* Northern Railway Technical Bulletin about Kashmir Railway)

The USBRL project is divided into four sections as per details given in Table 33.7:

Table 33.7 Details of the Kashmir Rail Link Project

Details of section	Length of section km	Progress and completion time
Leg 1, Jammu – Udhampur	53	Completed in April 2005
Leg 2, Udhampur – Katra	25	Under construction, to be completed by 2013
Leg 3, Katra – Qazigund	129	Under construction, to be completed by 2017
Leg 4, Qazigund – Baramulla	119	Completed in October 2009

33.3.1 Salient Features of Udhampur–Baramulla Section

The stretch from Udhampur to Baramulla is 273 km and has been divided into three sections. The salient features of these sections are listed in Table 33.8.

Table 33.8 Salient features of Udhampur–Baramulla section

Item	Udhampur–Katra	Katra–Qazikund	Qazikund–Baramulla	Total
Route length (km)	25	129	119	273
Max curvature (degree)	2.750	2.750	2.750	

Contd.

Table 33.8 *(Contd.)*

Item	Udhampur–Katra	Katra–Qazikund	Qazikund–Baramulla	Total
Bridges	38	72	811	912
Max. height of bridge	85 m	359 m	22 m	
Longest span	154 m steel girder over river Jhajjar	467 m steel arch over river Chenab	45 m	
Tunnel length (km)	10.2	105.6	0	115.8
Longest tunnel	3.15 km	10.96 km		
Max depth of cutting	20 m	40 m	12 m	
Stations	3	11	15	29

33.3.2 Advantages of Kashmir Valley Railway Line

Apart from strategic and political considerations, Kashmir Valley Railway Project will have great socio-economic advantages. There are briefly described as follows:

Connectivity of Kashmir valley to rest of the country The completion of this project will provide an all-weather and reliable connectivity to the state of Jammu and Kashmir through the rest of the country by the railway network; it will also provide connectivity by rail to far flung areas of the state.

Construction of access roads A total of about 235 km of approach roads to worksites are to be constructed, of which 188.6 km have been already constructed. With the completion of approach roads, more than 73 villages will get road connectivity benefitting thousands of people.

Employment generation It also provides direct and indirect employment to thousands of local people for the day-to-day requirement of the project. Also, there is provision for permanent job in railways to one of the family members, if more than 75 per cent of their land has been acquired.

33.3.3 Challenges of the Project

The challenges of building a railway, where none has gone before, are naturally unprecedented. With surprises and constraints around every corner, innovation is a key element in progress, requiring not only keen engineering skills, but also ingenuity and an ability to utilize what is available at hand to overcome everything from tricky geology to extreme climate to sheer logistical constraints. From the laying of the tracks to the transport of the rolling stock, the valley posed unique problems that needed on-the-spot solutions.

Kashmir Railway is perhaps the most difficult new railway line project undertaken on the Indian subcontinent by the Government of India. The terrain passes through the young Himalayas, which are marred with geological surprises and numerous problems. The alignment of the line presents one of the greatest railway engineering challenges ever faced, duetoits requirement to pass through the Himalayan foothills and the mighty Pir Panjal range, with most peaks exceeding 15,000 feet (4,600m) in height.

The route includes many bridges, viaducts, and tunnels. The Railway is expected to cross a total of 900 overbridges and pass through over 100 km of tunnels, the longest of being about 11 km in length.

Bridges

The greatest engineering challenges involve the crossing of the river Chenab, which involves building a 1,315 m-long (4,314 ft) bridge, 359 m (1,178 ft.) above the river bed, and the crossing of the Anji Khad, which involves building a 657 m-long (2,156 ft) bridge, 186 m (610 ft) above the river bed. The Chenab bridge will be the highest railway structure of its kind in the world, 35 m higher than the Eiffel Tower in Paris. Both bridges are to be simple span bridges. Cor-Ten Steel is planned to be used to provide a environment -friendly appearance and eliminate the need to paint the bridge. The design and structure is very similar to the New River Gorge bridge. The project is being managed by the Konkan Railway Corporation. The completion is scheduled for 2012, four years after the first isolated section of the route was opened for local passenger services, and it requires the use of 26,000 tonnes of steel.

The sector Qazigund–Barmulla has maximum number of bridges. An incredible 811 bridges features in this section, comprising 64 major ones including the 430 m-long veith bridge over the river Jhelum with 25 m-high piers and 737 minor ones for the train, as well as 10 road overbridges. Due to these bridges, the number of level crossings has been kept at six. The design of the bridges includes ballasted composite decks on curves and an open web girder (OWG) and plate girders for other major bridges and modified broad gauge (MBG) loading.

For major bridges however a well foundation was deemed necessary with multiple span arrangements over rivers and gorges. Well foundations between 25 m to 30 m deep (going below the liquefiable to soil depth of 15 m to 20 m) were decided as they would be able to resist the heavy vertical and lateral forces caused by the movement of trains.

Raft foundations are used for higher structures, which would settle in case of an earthquake, and will not break. The box bridges on raft foundations were thus made of 2 m-long precast segmental elements, with a design chamber of 300 mm to 450 mm at the centre so that levels would be maintained after the embankment was constructed.

Foot overbridges (FOB) at all stations are also provided on raft foundations with design mix cement concrete, with steel members and railings. The staircases and floors of the walkway have RCC slabs covered with pre-cast cement concrete anti-skid tiling. The roofing is done with steel trusses, covered with ridged sheeting. The steel handrails are covered with a specific PVC mould.

Tunnels

Construction of tunnels has been one of the most challenging jobs, particularly as the terrain passes though young Himalayas, which are full of varied, geological surprises, slips and many other difficult problems.

The most difficult section for tunnels was Katra–Quzigund section, which has around 82 per cent of its length in tunnels.

All tunnels including the New Banihal Tunnel will be constructed using the New Austrian Tunnelling method. Numerous challenges have been encountered while tunnelling through the geologically young and unstable Shivalik mountains. This has required some innovative solutions using steel arches and several feet of shotcrete.

Curves and ruling gradient

In order to ensure high speed, the maximum curatives used is 2.75° and a ruling gradient of 1 in 100. Even though the line is being built through a mountainous region, a ruling gradient of 1 per cent has been set to provide a safe, smooth, and reliable journey. More importantly, bankers will not be required, making the journey quicker and smoother. It will be built to the Indian Standard broad gauge of 1676 mm, laid on concrete sleepers with continuous welded rail and with a minimum curve radius of 676 m. The maximum line speed will be 100 km per hour (62 mph). Further, provision for future doubling of the track will be made on the major bridges. Additionally, provisions for future electrification will be made, though the line will be operated with diesel locomotives initially, as Kashmir is an electricity scarce region at present. There will be 30 stations on the full route, served by 10 to 12 trains per day initially.

Weather conditions

The weather too, poses a big problem. Work on the track between Mazhom (Rajwansher) and Baramulla came to a halt in the winter of 2008–09 as heavy snowfall made it impossible to recoup missing fittings on sleepers. Tamping of tracks by machine and inspection was also hindered as the rails and sleepers lay obscured by the snow, and labour did not turn up for work. In such situations the Ballast Regulatory Machine, which broomed extra ballast from the sides of the rail fittings was used. The same principle was used to remove snow also.

Transport of coaches and locomotives

The section from Qazigund to Baramulla was completed in Octobar 2009 and had to be opened for traffic. The intervening sections Udhampur – Katra (25 km) and Katra–Qazigund (148 km) was however, still under construction.

In order to open the Qazigund–Baramulla section, it was a big job to carry by alternate routes the locomotive and rolling stock. The carriage of rails, sleepers and other materials even though difficult, was manageable.

One of the most massive tasks involved the transport of the DEMUs and passenger coaches from the plains, upto Kashmir, crossing the mighty Pir Panepal range. The 3.2 wheel trailer with independent axle movement and hydraulic arrangement for lateral shifting, and its 10-wheel Volvo truck are marvels of innovation.

Signalling and communications

Three-aspect colour light signalling is being installed on the route to maintain train safety. GSM-R equipment may be installed in the future to improve the quality of the system.

Security

Security for the line has been a major concern, with the regions the line passes through continuing to face terrorist challenges. The presence of the international border with Pakistan close by aggravates these challenges. Plans have been made. for closed-circuit cameras at all major bridges, tunnels, and railway stations. Lighting is provided on all major bridges and inside tunnels. Additionally, a special security detail to protect the infrastructure has been contemplated.

33.3.4 Cost and Expenditure

The total cost of the project as per the latest estimate is 1,95,000 million. Table 33.9 lists the sections-wise costs.

Table 33.9   Cost and expenditure of the Kashmir Valley Railway line

Section	Cost (million)
Udhampur–Katra	11130
Katra–Qazigund	148930
Qazigund–Baramulla	34940
Total	195000

33.3.5 Time Schedule

The latest target dates for different sections are listed in Table 33.7:

33.4 TRACK MANAGEMENT SYSTEM

Track management system (TMS) is a computerised tool for planning, implementing and monitoring the track maintenance works. It prioritizes maintenance inputs based on track condition thereby ensuring need-based maintenace.

Track management system is a central server-based web-enabled software, which intergrates various track structure data, inspection data, work data, etc., to assist railway engineers in ascertaining the correct level of maintenance and renewal inputs to be made at requisite location with the objective to maximize benefits of input given to track. It addresses the need for the integration of all required information with the help of computerized programs and presents it to engineers to help them in planning track maintenance activities. TMS provides a means to rapidly access inspection data, track diagram, reports and multi-media information commonly used to plan inspections and maintenance inputs and generate reports that can be taken into the field.

Information being provided by TMS also plays vital roles in track maintenance. It specifies where renewals are required and it initiates and controls the activities required for maintaining track in satisfactory condition. It is felt that TMS also will have a profound impact on optimum utilization of scarce track maintenance resources and improving the general efficiency of engineering department. The system will also help in planning of deployment of costly track machines for maintenance and renewals.

33.4.1 Advantages of Track Management System

The following are the various advantages of TMS:

Optimum utilization of track maintenance efforts TMS will be able to ensure that track maintenance inputs are need based and demand responsive based on condition of track geometry and track structure degradation.

Track maintenance efforts for increased volume of traffic with heavier axle load With the introduction of more trains and heavy axle loads, the requirement of track maintenance is becoming more stringent. Unless automation of track maintenance efforts is introduced, it may be almost impossible to maintain track to the required standards. The need of TMS is considered essential to meet the increased demand of track maintenance efforts.

Ensuring track geometry in safe condition TMS will ensure that track maintenance is planned based on track geometry and other technical inputs and the condition of track does not deteriorate due to unacceptable conditions, which sometimes may be even unsafe.

Powerful database management tool TMS will maintain a complete database of the track structure and its condition. This database will be updated periodically as track condition deteriorates due to usage of the track or improves whenever a maintenance operations is carried out. All this data shall be used to present the user with changes in track quality, both history and extrapolated. The goal is to generate the extensive data and decide about the best possible remedial actions.

Saving in manpower and machine hours The Three-tier system will work on need-based maintenance principle, thus bringing about considerable savings in manpower and machines. The tamping requirement and, therefore, tamping cost is expected to come down. This will also reduce pulverization of stone giving increased life to ballast. Savings due to elimination of manual compilation of data and information and communication will also be substantial.

Permanent way materials management The system will help in management of permanent way materials, that is rails, sleepers and fittings required for track that is renewals and maintenance. Better planning for procurement and movement of materrials system will reduce debit and credit of permanent way materials within the subdivisions and outsite subdivisions.

Analysis of rail/weld failures The system will help in analysing the rail/weld failure defect rate, which in turn will help in better planning of renewals or corrective actions to be taken to correct the incidence of failures and reducing the defect rate.

Advance warning system TMS will provide advance alerts for bad location needing attention, overdue renewals, overdue inspections, attention to inspection notes of various officials. This will help the engineers and field staff to pay attention on all such activities and flickering on alerts.

33.4.2 Details of Track Management System

Track management system is a computer-based dynamic system to take care of all aspects of track matters. With this system in place, information on track structure, welding, renewal planning, track location needing attention, inspections due, compliance of inspection, store data, etc., can be seen at any given points. The system is web-based and all track officials viz junior engineers (permanent way) AENs, DENs, Senior DENs, officers, i.e., AENs/ XENs/Dy CE/CEs/Pr.CE and all engineering officers of the Railway Board will be connected. All officers/office of Engineering Department will have access to this system with a proper password. Further all the inspection registers and ledgers maintained by JEs/SE/SSEs will be withdrawn and all information have to be uploaded on to the system on day-to-day basis. This will give a great relief to the field staff in keeping records of track maintenance, track maintenance planning, track renewal planning, material management, etc.

The system has the following modules

Assets section This icon of assets register will have information, of all relevant issues about track assets such as SEJ, LWR, P&C, curves buffer rails, weld details, level crossing, glued joint, sand hump, rail sleeper, ballast fastenings, fish plate, track type, rail pad, liner, land boundary, formation, protection work, drainage works, erosion control measures, weak formation and formation treatment, and so on.

Inspection section The module gives full details of various issues where inspection is done. These are as follows.
(i) LWR inspection (ii) Points and crossing inspection (iii) Level crossing inspection (iv) Push trolley inspection (v) Foot plate inspection (vi) Track diagram (vii) Track planning (viii) Miscellaneous section (ix) Reports section (x) Innovations section (xi) PCDO section (xiii) Purchase order section.

33.4.3 TMS Software Design and Methodology

TMS software is a web-enabled central server-based software developed in J2EE platform. The software is developed in a modular fashion in such a way that any module developed in future can be easily integrated into it. Master data of section details, and various components of track structures are stored into the system. Data on the condition of various components is also stored and updated by various manual and mechanized inspection data, work data, and material change data.

TRC data is electronically transferred in the system while other inspection data is fed manually into the TMS. This updates the condition data of various components. To avoid duplicate entry of inspection data in registers and TMS, forms have been developed so that officials can record their inspection directly into netbooks (small laptops) and then upload them into TMS. Works carried out in field by sectional gangs, machines and through special works are also fed into TMS.

33.4.4 Implementation of TMS on Pilot Divisions

TMS has been planned to be deployed in six pilot divisions of Indian Railways in different geographical locations, mainly Agra, Waltair, Bilaspur, Secunderabad, Bengaluru, and Salem at a total cost of ₹ 100 million.

33.4.5 Future of TMS

TMS has been implemented on the Agra division as a pilot project. Based on the encouraging results of TMS on the project, Indian Railways has decided to implement TMS on all divisions at a cost of about ₹ 400 million.

TMS has been very successful and results have not only shown better quality of track, but also reduction in the cost of track maintenance efforts.

The future of TMS on Indian Railways is very bright and it is likely to be a very progressive step in streamlining of work of the TMS.

33.5 INCREASE OF AXLE LOAD*

Indian Railways has decided to increase the axle loads on its various routes. This had become necessary for better utilization of its assets and thereby improving the finances of Indian Railways.

It may be brought out that the main freight and passenger carrying routes on Indian Railways are clogged with traffic levels much beyond their carrying capacities. Decongesting the staturated routes with new capacity augmentation works is capital intensive and time-consuming. Indian Railways has followed a conscious strategy to augment capacity by allowing the rolling stock of the railways to be loaded to their optimum load ability.

In 2005**, Indian Railways permitted running of BOXN wagons loaded up to CC+2 and subsequently to CC+8+2 on certain identified iron ore routes (Axle load 22.82 tonnes) as a pilot project, where CC is the designed carrying capacity of BOXN wagons (axle load 20.32 tonnes). The current position is that CC+6+2 (22.3 tonnes axle load) have been universalized, trains with CC+8+2 (22.9 tonnes axle loads) have been permitted on more than 21,000 route km and trains with 25 tonnes axle loads have been introduced on five routes covering more than 1000 km.

The increase of axle load from 20.3 tonnes to 22.9 tonnes and thereby increasing the carrying capacity of wagons to CC+8 had augmented the carrying capacity of wagons from 58.8 tonnes to 69.8 tonnes, i.e. an increase of about 17 per cent. This has been a major cause for increasing the freight traffic and turning around of finances of Indian Railways.

Note: As per rough estimates, one tonne of extra loading per wagon implies an additional revenue of ₹ 5000 milliom per annum. The logic was that each wagon could be loaded about 60 times a year. Average wagon turnaround being a little over six days, it means that over 160,000 wagons could potentially benefit from this; additional loading could reach upto 10 million tonnes per annum. At an overage freight earning of ₹ 500 per tonne, Indian Railways could earn an additional ₹ 500 million.

The implications of increasing the axle load to 22.9 tonnes on the existing tracks, bridges, and wagons is an issue which remains to be examinted in detail particularly to check if it results in over-stressing of the track and bridges.

---

* Study report of IIM Ahmedabad and Railway Staff College Vadodara on the turnaround of Indian Railways – July 2006

** Railway Board letter no. 2003/CE-II/TS/S Vol-I dated 2.5.2005 and 4.5.2005, Also Railway Board letter no. TCR/1394/2004/2 dated 10.5.2005.

33.5.1 Initiatives taken by Civil Engineering Department

Various initiatives, that have been taken by Civil Engineering Department to allow running of higher axle loads on sustainable basis are briefly summarized below:

(i) A policy has been made to provide track, bridge and formation for most of new track renewals, doublings, gauge conversions and new lines fit for carrying 25 tonnes axle load wagons.

(ii) Thick web-switched and weldable CMS crossings are planned to be used.

(iii) Rail grinding will be introduced progressively to get better life out of 90 UTS rails.

(iv) Lubrication of rails will be done with mechanized means on selected locations especially in the Ghat sections and on sharp curves.

(v) System for mechanized ultrasonic flaw detector examination with digital recording will be introduced progressively.

33.6 RAIL GRINDING

To maintain a good track geometry particularly on heavy axle load routes, it is not sufficient to carry out only regular track maintenance, such as tamping and lining. It is equally important to address the problem of irregularities of rail surface due to spalling, scabbing and corrugation, etc., by grinding on rail.

Since the oscillation caused by corrugations and waves exerts a considerable influence on the track maintenance costs and on the riding quality, it is absolutely necessary to remove these faults occuring on the rail surfaces and the running edges. In view of this, rail grinding is done by modern rail grinding machine to treat corrugation and waves and to provide a good logitudinal profile of rails. Rail grinding is considered the single most effective maintenance practice to control the effects of rolling fatigue, remove rail undulations, restore rail profile and maximize rail life.

33.6.1 Rail Grinding Trains

To remove corrugations and other surface defects like pitting and, scabbing, grinding of rails is being done by rail grinding trains on many railway systems. These rail grinding trains are provided with a number of rotating grinding wheels, which are able to remove corrugations upto a depth of about 0.2 to 0.5 mm in a single pass at a speed of 0 to 6 m/s. Usually, five to six passes are made to completely grind off the corrugation. These grinding trains are also equipped for complete re-profiling of the rail table, often needed to treat the worn-out rails at curves.

However, grinding of rails does not offer any permanent solution to the rail corrugations as the corrugations reappear and need grinding again.

During the grinding operation, vertical, horizontal, and angular movement of grinding unit, are computer controlled using a set of rail grinding patterns, which are programmed by the operator to accomplish the grinding method required.

Indian Railways has recently procured a rail grinding machine from M/s Loram, (US). It is being used for rectification of rail surface defects and for rail profiting heavy haul routes of South Eastern Railway. The broad features of the rail grinder are as follows:

(i) Make	:	LORAMS X - 16-switch and crossing grinder
(ii) Size	:	12142 mm × 3351 mm × 3810 mm
(iii) Travelling speed	:	64.4 kmph in both directions
(iv) Speed while grinding	:	6.4 kmph
(v) Water system	:	Spray bars at each end. Storage capacity 800 gallons
(vi) Output per working hour	:	Removal of defects of the order of 0.4 mm in one pass on the track.
(vii) Usage	:	Can effectively work on plain track as well as on points and crossings

Indian Railways is planning to purchase more number of grinding trains as it is felt that grinding of rails has become a technical necessity particularly in view of heavier axle load trains running on Indian Railways.

33.6.2 Types of Grinding

Rail grinding can be generally classified into three main types as indicated below:

Initial grinding   This is done on rails that are freshly laid during new construction or after rail renewal. Initial grinding corrects construction damage and removes the decarbonized surface area of the rail head where the mechanical properties are poorer than the rails' deeper layers. The removal of 0.30 mm from the surface layer of the rail, guards against the damage to rail, which is likely to take place in the formation of squats.

Preventive grinding   This is done at a stage when the idea is to treat the rail when damage is at the initial stage. This approach is based upon cyclical timing so that grinding is done from time to time. The grinding campaigners are steered in accordance with the cumulative loading.

Corrective grinding   This is based on symptom-related interventions. Campaigners are directed by monitoring damage against present levels as removing short pitch corrugation once it reaches a 0.05 mm depth.

33.6.3 Planning for Future

It may be highlighted that world over, periodic grinding of rail head profile has been successfully adopted for controlling rail corner fatigue, corrugation and shelling as well as for otpimizing rail wear. This maintenance practice has helped in controlling rail fractures and prolonging the life of rails. Rail grinding also becomes essential with the adoption of heavy axle loading on Indian Railways and increase in traffic density. Indian Railways has already procured one rail grinding machine as already brought out earlier.

Indian Railways has procured two more rail grinding machines (RGMs) and two more are programmed for procurement. To sustain the traffic growth forecast, it would be necessary to bring the entire track on Indian Railways under rail grinding regime by 2020.

By rail grinding, the railway track will get desired rail profile and which in turn reduces stress concentration at same spots, thereby extending the life. It is assumed that from same rail, railway may get double the life. Rail grinding will help in reduction of fractures and defect rate.

33.7 BALLASTLESS TRACK

The technical concept of a railway track consisting of ballast, sleepers, and rails is very old and has stood the test of time. Such a system is simple and can be rapidly extended, renewed, or dismantled. The general problem that occurs with ballasted tracks is that the ballast under the pressure exerted by the load caused geometrical unevenness and clogging of the ballast bed by fine particles. Therefore, regular maintenance is needed to restore track alignment. The experience in Germany and other countries has been that the conventional track may be used for speeds of up to 250 km per hour but not beyond that. For higher speeds, the construction of a ballastless track is required. In a ballastless track, the rails are directly fastened to the concrete slab using elastic fastenings. A ballastless track is expensive but is likely to require little or no maintenance during its lifetime.

Ballastless track is being adopted for the metro rail system as well as for some of the high-speed tracks on various developed railway systems of the world. It is preferred because once properly constructed, ballastless track requires very little day-to-day maintenance. In India too, Delhi Metro as well the Calcutta Metro Rail System have used ballastless tracks for special locations. Important characteristics of ballasted and ballantless track are given in Table 33.10.

Table 33.10 Comparision of ballasted track and ballastless track

Characteristics	Ballasted track	Ballastless track
Reliability of method	Known and proven method; easily available Technology for maintenance	The technology of construction is still in infant stage: New research is however being done.
Absorption of impact forces	Absorbed by ballast and formation	Elastomeric pads absorb the impact
Maintenance due to correction of track geometry	Packing and leveling the track while using ballast as medium helps in correction of surface geometry	Maintenance of correction of surface geometry not easily possible.
Maintenance due to settlement and loss of elastic property of track	The pulverization of ballast and its sinking in the formation makes maintnance difficult	Normally no maintenance required except periodical replacement of elastic components.
Quality of construction	Quality is average but construction defects can be attended by mechanized maintenance	Highly sensitive to construction defects; if constructed properly, the quality is very high.
Resistance to lateral and logitudinal forces	Limited resistance but can be slightly improved by mechanized maintenance.	High resistance to lateral and longitudinal forces.
Construction cost	Average	Very high (sometimes even 3 to 4 times)

33.7.1 Types of Ballastless Track Assemblies

These are several type of ballastless track assemblies as discussed below.

Plinth type ballastless track assembly

The assembly has the following components (Fig 33.3).

(i) 6 mm-thick elastic rail pad

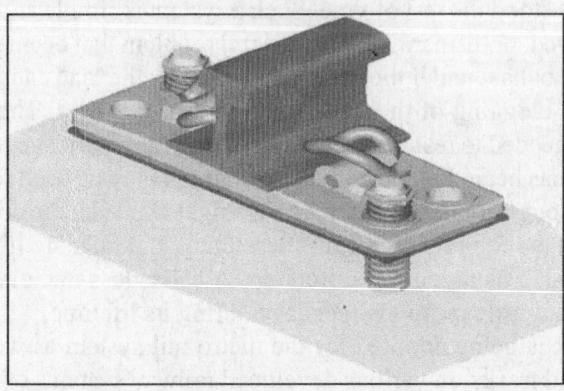

Fig. 33.3 Plinth type ballastless track assembly (Used in Calcutta Metro)

(ii) Cast iron base plate for load distribution

(iii) An elastomeric pad of 12 mm thick to function as ballast

(iv) Steel plates for vertical adjustment

(v) Fitting and fastening consisting of elastic rail clips, high tensile steel bolts screwed into high-density polythene inserts, triple coil spring washer, eccentric insulating washers for insulation and lateral adjustment.

This assembly has all indigenously developed components and construction can be done by manual labour without the use of major machinery. The assembly has given satisfactory service and there is no problem of excessive vibrations and noise.The assembly has, however, a limitation that on sharp curves bolt fixing arrangements create problems.

Vossloh 336 ballastless track assembly

The assembly has components as given below: (Fig 33.4)

(i) 6 mm-thick elastic rail pad

(ii) Malleable cast iron plate for load distribution

(iii) 10 mm-thick elastic base plate pad and 5 mm thick plastic pad

(iv) Fitting and fastening, viz. Vossoloh elastic rail clip, high tensile steel anchors cast in concrete; triple coil spring washers and eccentric insulating washers for insulation and lateral adjustment.

The assembly has given excellent service in Delhi Metro, with well-designed pads and good-quality track construction. The thickness of high-density polythene pad can be changed for vertical adjustment. The construction is also quite simple.

Logwell ballastless track assembly (LM-1)

This new type of assembly is developed by M/S Logwell Forge Ltd. incorporating the best features of both Delhi Metro and Calcutta Metro ballastless track. The assembly has the following main parts: (Fig. 33.5)

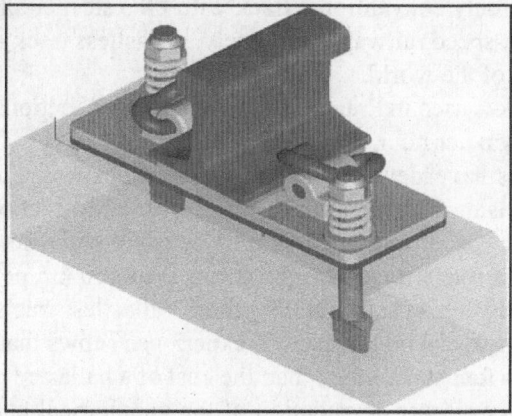

Fig. 33.4   Ballastless track assembly (used in Delhi Metro)

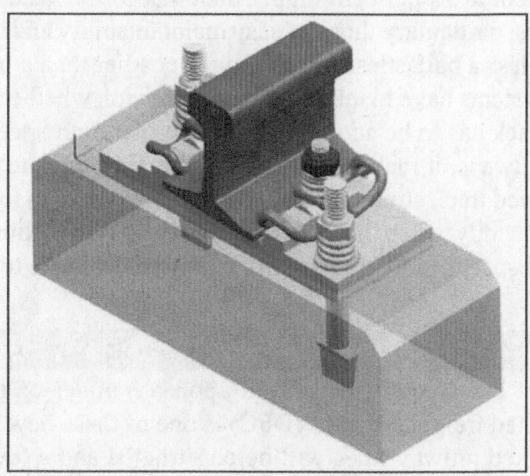

Fig 33.5   Logwell ballastlesstrack assembly

(i) Base plate is fastened by six clips having a toe-load of 900 to 1200 kg.

(ii) Base plate pad 10 mm thick and plastic pad of 5 mm thick

(iii) The base plate is anchored to concrete by having stronger fixing arrangement.

The manufacturers claim that this assembly is more economical as it uses an indigenous design. Also, the assembly has relatively lesser number of components for economy and better maintenance.

33.8 BALLAST TRACK FOR HIGH-SPEED ROUTES

The leading railways operating at high speeds use conventional track consisting rails fastened on pre-stressed concrete (PSC) sleepers with elastic fastening which are supported on ballast. About 90 per cent of the high-speed track in the world is on conventional ballasted structure. French TGV marked a record of 525 km per hour on conventional ballasted track; and conventional track is strong enough to bear the stresses of a speed of up to 300 km per hour. It is not only a heavy structure which is required for high-speed rial (HSR), but also the high standards of maintenance.

Though earlier, only conventional track with PSC sleeper and ballasted decks were used for high-speed railway lines, lately, ballastless track is also being used by many railways of the world.

Ballastless track is used in France in the underground sections where trains run at a speed of 220 km per hour.

Many Railways have developed high-speed ballastless track. In particular, in Germany a decision has been taken recently to build sections of high-speed lines (or lines with speeds above 200 km/h) by using ballastless track, except at locations where the trains travel at speeds less than 200 km per hour, such as at stations, etc. Initially, the cost of building these ballastless tracks greatly exceeds the cost of building tracks on ballast, but experience shows that the maintenance costs, especially in tunnels, are less than the cost of a ballasted track (by an order of 1/5th), due to the slower degradation of the geometrical parameters of these tracks. The German experience shows that the cost of building ballastless track is between 50 per cent to 75 per cent higher than that for ballasted track. There are certain advantages, particulary those of easy maintainability and increased service life, but nevertheless a ballastless track is much costlier than a ballasted deck.

The railway systems have to take a strategic decision whether a ballasted deck or a ballastless track has to be adopted for a particular high-speed railway line.

For Indian Railways, it may be desirable in the first stage to have a ballasted track with improved track structure of upgrading of speeds up to 200–250 kmph. However, subsequently when Indian Railways considers dedicated high-speed corridors for high speeds (250–350 kmph), it may be desirable to use a ballastless track for the same.

SUMMARY

Indian Railways has recently embarked upon a number of new innovative projects. Dedicated freight corridor (DFC) is one of these new projects, where separate earmarked railway lines will be constructed and a 'dedicated freight corridor' is being developed, where goods trains will run at higher speeds and heavier axle load. Similarly, other such challenging projects are Konkan Railway Project and the Kashmir Rail Link. There are a number of other new developments such as track management system which will help Indian Railways to manage the infrastructure more efficiently and safely.

REVIEW QUESTIONS

1. What is the purpose of dedicated freight corridor? Give brief details of the DFC project.
2. (i) What are the technical challenges of Konkan Railway? Give the benefits of Konkan Railway.
3. Describe the main challenges and economic advantages of the Kashmir Rail link.
4. Give brief details of the Track Management System (TMS). What are the main advantages of TMS?
5. Why is rail grinding done? Give brief details of rail grinding trains.
6. Which type of ballastless train is being used by the Delhi Metro. Compare the merits/demerits of a ballastless track, vis-a-vis a ballasted track.

Choose the correct answer from the choices given.

7. The maximum gradient being adopted in a flat territory on DFC is:
   (a) 1 in 150   (b) 1 in 250
   (c) 1 in 300   (d) 1 in 400
8. The maximum speed contemplated of a goods train on DFC is:
   (a) 60 kmph   (b) 75 kmph
   (c) 100 kmph   (d) 120 kmph
9. The maximum axle load permitted on DFC would be:
   (a) 20 tonnes   (b) 22.5 tonnes
   (c) 25.0 tonnes   (d) 30.0 tonnes
10. The approximate length of Konkan Railway line is?
    (a) 500 km   (b) 650 km
    (c) 760 km   (d) 900 km
11. The first train from Mumbai to Madgaon on Konkan Railway began operation in:
    (a) 1995   (b) 1998
    (c) 2000   (d) 2003
12. The tallest viaduct on the Panval Nadi is about:
    (a) 40 m   (b) 48 m
    (c) 64 m   (d) 100 m
13. The Jammu–Udhampur railway line was completed in:
    (a) 2000   (b) 2003
    (c) 2005   (d) 2008
14. The maximum curvature on Kashmir Railway lines is:
    (a) 4.0°   (b) 3.5°
    (c) 2.75°   (d) 2.00°
15. The ruling gradient of the Kashmir Railway line is:
    (a) 1 in 250   (b) 1 in 200
    (c) 1 in 150   (c) 1 in 100
16. TMS stands for:
    (a) Track Modernization System   (b) Track Maintenance System
    (c) Track Management System   (d) Mean Speed Track
17. In initial grinding of rails, rails are grinded to an extent of:
    (a) 10 mm   (b) 20 mm
    (c) 30 mm   (d) 40 mm
18. Cost of construction of a ballastless train compared to a ballasted track is:
    (a) more   (b) equal
    (c) less   (d) insufficient information to predict
19. The first metro railway constructed in India was in:
    (a) Delhi   (b) Kolkata
    (c) Chennai   (d) Bengaluru

Appendix A

ANSWERS TO MULTIPLE CHOICE QUESTIONS

Chapter 1
10. (c); 11. (b); 12. (c); 13. (c); 14 (b); 15. (b);
16 (a); 17. (d); 18. (b); 19. (b); 20. (a)

Chapter 2
8. (b); 9. (c); 10. (d)

Chapter 3
5. (c); 6. (d); 7. (c)

Chapter 4
7. (a); 8. (b); 9. (b); 10. (d); 11. (a); 12. (d);
13. (c); 14. (c); 15. (b)

Chapter 5
9. (b); 10. (a); 11. (b); 12. (b); 13. (c)

Chapter 6
7. (c); 8. (d); 9. (a); 10. (c); 11. (a); 12. (c);
13. (a); 14. (c)

Chapter 7
14. (a); 15. (c); 16. (c); 17. (c); 18. (b);
19. (a); 20. (a); 21 (b); 22 (c); 23(c); 24
(c); 25 (d)

Chapter 8
8. (b); 9. (b); 10. (a); 11. (b); 12. (a); 13. (a);
14. (a); 15. (a)

Chapter 9
8. (c); 9. (d); 10. (d); 11. (b); 12. (a); 13. (c)

Chapter 10
13. (b); 14. (a); 15. (b); 16. (b); 17. (b);
18. (a); 19. (a); 20. (a); 21. (a); 22. (c);
23. (c); 24. (b); 25. (c); 26. (a); 27. (a);
28. (b); 29. (c); 30. (a); 31. (b); 32. (c);
33. (a); 34. (b); 35. (a)

Chapter 11
7. (a); 8. (d); 9. (a); 10. (b); 11. (c); 12. (a)

Chapter 12
5. (c); 6. (c); 7. (b)

Chapter 13
24. (c); 25. (d); 26. (b); 27. (d); 28. (c); 29. (d);
30. (d); 31. (c); 32. (b); 33. (c); 34. (c)

Chapter 14
19. (b); 20. (c); 21. (b); 22. (b); 23. (a);
24. (d); 25. (d); 26. (d); 27. (a); 28. (c);
29. (b); 30. (b); 31. (c).

Chapter 15
13. (c); 14. (d); 15. (b); 16. (a); 17. (c);
18. (b); 19. (a)

Chapter 16
11. (a); 12. (b); 13. (b); 14. (c); 15. (b);
16. (c); 17. (b); 18. (c); 19. (a)

Chapter 17
12. (c); 13. (d); 14. (c); 15. (d); 16. (d);
17. (d); 18. (c); 19. (c); 20 (c); 21 (c)

Chapter 18
11. (c); 12. (a); 13. (c); 14. (c); 15. (a);
16. (c); 17. (b); 18. (c); 19. (b)

Chapter 19
6. (d); 7. (d); 8. (d)

Chapter 20
9. (c); 10. (d); 11. (d); 12. (b); 13. (b);
14. (c); 15. (c); 16. (d); 17. (b); 18. (b);
19. (b); 20. (a); 21. (b)

Chapter 21
8. (b); 9. (b); 10. (a); 11. (d); 12. (a); 13. (d);
14. (e); 15. (a); 16. (a); 17. (c)

Chapter 22
8. (c); 9. (d); 10. (c); 11. (c); 12. (c);
13. (a); 14. (b); 15. (a)

Chapter 23
5. (c); 6. (b); 7. (b); 8. (b); 9. (c); 10. (d);
11. (c); 12. (a); 13 (a)

Chapter 24
9. (c); 10. (c); 11. (b); 12. (c);

Chapter 25
11. (b); 12. (c); 13. (a); 14. (c)

Chapter 26
6. (d); 7. (b); 8. (d); 9. (d); 10 (a); 11 (d);
12 (a); 13. (b)

Chapter 27
6. (b); 7. (c); 8. (a); 9. (c)

Chapter 28
10. (c); 11. (c); 12. (d); 13. (d); 14. (c);
15. (b); 16. (b); 17. (d); 18. (b); 19. (c)

Chapter 29
9. (c); 10 (d); 11 (c); 12 (c); 13 (b)

Chapter 30
12. (c); 13. (c); 14. (c); 15 (a); 16. (c);
17. (c); 18. (b)

Chapter 31
7. (c); 8. (c); 9. (b); 10. (d); 11. (d); 12. (c)

Chapter 32
5. (c); 6. (d); 7. (d); 8. (b); 9. (d)

Chapter 33
7. (d); 8. (c); 9. (d); 10. (c); 11. (b); 12. (d);
13. (c); 14. (c); 15. (d); 16. (a); 17. (c);
18. (a); 19. (b)

Index

About the Authors

Satish Chandra is currently Professor, Department of Civil Engineering at the Indian Institute of Technology Roorkee. A PhD from IIT Roorkee, Dr Chandra is an expert in the area of highway materials and traffic-flow modelling and has published over 100 research papers.

M.M. Agarwal retired as Chief Engineer, Northern Railway. Besides a career with Indian Railways spanning 33 years, he has also served German Railways and Zambia Railways. He is an expert on track technology and has published about 50 technical papers on track modernization and other related track matters, many of which have been discussed in international and national conferences/seminars. He has also authored *Indian Railway Track: Design, Construction, Maintenance and Modernisation*. He is the recipient of the national award for special work done on track maintenance and the President's award for the best original book on railway tracks in Hindi.

Plate 1

Abt rack railway system (Chapter 3, page 50)
(*Courtesy:* A.M. Hurrell This file is licensed under the Creative Commons Attribution-Share Alike 3.0 Unported license; http://creativecommons.org/licenses/by-sa/3.0/deed.en)

The rack railway system consists of three rails, i.e., one extra toothed rail in the middle in addition to the two normal rails. The locomotive also has a toothed pinion wheel whose teeth fit into the grooves of the central toothed rail. (Chapter 3, page 49)

Plate 2

Welding machine APT 1500 R (Chapter 16, page 311)
(*Source:* www.plassertheurer.com)

Dynamic tamping express—Plasser machine 09-4X (Chapter 20, page 381)
(*Source:* http://www.123people.com/s/plasser+theurer)

Plate 3

Four-sleeper mode of a dynamic tamping express (Chapter 20, page 381)
(*Source:* Plasser & Theurer Technical Literature)

Dynamic track stabilizer (Plasser-DGS 62 N) (Chapter 20, page 381)
(*Source:* Plasser & Theurer Technical Literature)

Plate 4

Plasser American BDS 100/200 ballast distribution system sitting on 'no hump' track at BNSF Northtown Yard (Chapter 20, page 381)
(*Source:* Plasser & Theurer Technical Literature)

WG steam locomotive (Chapter 24, page 435)
(*Source:* http://sambrandist.blogspot.in/2011/01/dining-elizabethan-on-gcr.html)